Wissenschaftsethik und Technikfolgenbeurteilung
Band 23

Schriftenreihe der Europäischen Akademie zur Erforschung
von Folgen wissenschaftlich-technischer Entwicklungen
Bad Neuenahr-Ahrweiler GmbH
herausgegeben von Carl Friedrich Gethmann

Springer

Berlin
Heidelberg
New York
Hong Kong
London
Milan
Paris
Tokyo

C. Streffer · H. Bolt · D. Føllesdal · P. Hall
J. G. Hengstler · P. Jakob · D. Oughton
K. Prieß · E. Rehbinder · E. Swaton

Low Dose Exposures in the Environment

Dose-Effect Relations and Risk Evaluation

With 68 Figures and 28 Tables

 Springer

Editor of the series

Professor Dr. Dr. h.c. Carl Friedrich Gethmann
Europäische Akademie GmbH
Wilhelmstraße 56, 53474 Bad Neuenahr-Ahrweiler, Germany

For the authors

Professor Dr. Dr. h.c. Christian Streffer
Auf dem Sutan 12
45239 Essen, Germany

Editing

Friederike Wütscher
Europäische Akademie GmbH
Wilhelmstraße 56, 53474 Bad Neuenahr-Ahrweiler, Germany

ISBN 978-3-642-05923-0

Bibliographic information published by Die Deutsche Bibliothek
Die Deutsche Bibliohek lists this publication in the Deutsche Nationalbibliografie; detailed bibliographic data is available in the Internet at <http://dnb.ddb.de>.

Springer-Verlag is a part of Springer Science+Business Media
springeronline.com

© Springer-Verlag Berlin Heidelberg 2010
Printed in Germany

The use of general descriptive names, registered names, trademarks, etc. in this publication does not imply, even in the absence of a specific statement, that such names are exempt from the relevant protective laws and regulations and therefore free for general use.

Product liability: The publisher cannot guarantee the accuracy of any information about dosage and application contained in this book. In every individual case the user must check such information by consulting the relevant literature.

Coverdesign: deblik, Berlin

Printed on acid-free paper   62/3020hu – 5 4 3 2 1 0 –

Europäische Akademie

zur Erforschung von Folgen wissenschaftlich-technischer Entwicklungen
Bad Neuenahr-Ahrweiler GmbH

The Europäische Akademie

The *Europäische Akademie zur Erforschung von Folgen wissenschaftlich-technischer Entwicklungen GmbH* is concerned with the scientific study of consequences of scientific and technological advance for the individual and social life and for the natural environment. The Europäische Akademie intends to contribute to a rational way of society of dealing with the consequences of scientific and technological developments. This aim is mainly realised in the development of recommendations for options to act, from the point of view of long-term societal acceptance. The work of the Europäische Akademie mostly takes place in temporary interdisciplinary project groups, whose members are recognised scientists from European universities. Overarching issues, e.g. from the fields of Technology Assessment or Ethic of Science, are dealt with by the staff of the Europäische Akademie.

The Series

The series "Wissenschaftsethik und Technikfolgenbeurteilung" (Ethics of Science and Technology Assessment) serves to publish the results of the work of the Europäische Akademie. It is published by the Academy's director. Besides the final results of the project groups the series includes volumes on general questions of ethics of science and technology assessment at well as other monographic studies.

Acknowledgement

The project *Environmental Standards. Low Dose Effect Relations and their Risk Evaluation* has been partially supported by the Bundesministerium für Bildung und Forschung.

Preface

The Europäische Akademie Bad Neuenahr-Ahrweiler GmbH is concerned with the scientific study of the consequences of scientific and technological advance for the individual, society and the natural environment and, therefore, not least with the study of consequences of recent developments in life-sciences and medical disciplines. The *Europäische Akademie* intends to contribute to find a rational way for society to deal with the consequences of scientific progress. This aim is mainly realised by proposing recommendations for options of action with long-term social acceptance. The work of the *Europäische Akademie* mostly takes place in temporary interdisciplinary project groups, whose members are recognised scientists from European universities and other independent institutes.

In the light of recent discussions on the validity of scientific assumptions concerning the extrapolation of harmful effects into the low-dose range the Europäische Akademie set up a project on "Low-Dose Exposures in the Environment their Risk Assessments and Regulatory Processes" in January 2001. Experts of radiobiology, toxicology, medical epidemiology, modelling, jurisprudence, psychology and philosophy from different European countries were brought together to discuss current scientific developments in environmental standard setting and risk evaluation of substances and ionising radiation at very low doses in order to ensure rational, efficient, and fair decisions.

I am glad that by means of this memorandum, that continues our work on environmental standards and after our recommendations with respect to the regulation of combined agents we can now focus on the effects and the regulation of harmful agents in the very low-dose range.

I hope this memorandum will get the credit that it deserves. It should be able to enhance the awareness that scientists and regulators are faced with various uncertainties when harmful and especially genotoxic agents have to be regulated. In this regard, the project group developed many valuable recommendations.

For more than two years, the members of the project-group had made a very challenging and productive effort on an interdisciplinary project. I would like to thank the members of the project-group and their chairman, Professor Dr. Dr. h. c. Christian Streffer, for this commitment. I am also grateful to Dr. Kathrin Prieß who co-ordinated the project on behalf of the Europäische Akademie.

Bad Neuenahr-Ahrweiler, December 2003 Carl Friedrich Gethmann

Foreword

Exposures to physical and chemical toxic agents have increased with the rise of the industrialization and of the density of human populations in wide regions of the world. During the last decades concerns of the public have been voiced that health effects may be caused by these exposures. In the range of low exposures the induction of cancer is the most serious concern. In order to limit or optimally to avoid possible health effects environmental standards and regulations are necessary. A complete avoidance of health effects is not possible if a dose response without a threshold dose does exist. The procedure of developing environmental standards must be based on solid scientific data. However, exposure limits will be usually set in dose ranges where no significant health effects can be measured. Under these circumstances extrapolations are performed from higher dose ranges with measurable data to the lower dose ranges without measurable effects for risk estimates. Such processes bear uncertainties which lead to questions not only within the scientific community but also within the public. For these and other reasons it is generally necessary also to consider ethical, sociological, economic as well as legal aspects in these processes. An interdisciplinary approach and a broad discussion is wanted. This approach has been pursued in a project group with scientists from various disciplines and various European countries. The regulations for environmental standards usually differ for physical agents like ionising radiation and carcinogenic chemicals. It has been tried during the discussions of the group therefore whether and under which conditions a certain harmonization is possible for these agents with carcinogenic activities.

In January 2001, Professor Per Hall (Department of Medical Epidemiology, Karolinska Institute, Stockholm), Professor Eckard Rehbinder, (School of Law, University of Frankfurt/M), Professor Christian Streffer (Institute for Science and Ethics, University of Essen) and Dr. Elisabeth Swaton (IAEA, Vienna) met to discuss a project of the Europäische Akademie GmbH. Further experts, Professor Hermann Bolt (Institute of Occupartional Health, University of Dortmund), Professor Dagfinn Føllesdal (Department of Philosophy, Oslo University/Stanford University), Professor Jan-Georg Hengstler (Center for Toxicology, University of Leipzig), Dr. Peter Jacob (GSF National Research Center for Environment and Health, Neuherberg), and Professor Deborah Oughton (Department of Chemistry and Biotechnology, Agricultural University of Norway, Aas) subsequently broadened the group's intellectual spectrum. The group agreed to carry out an exemplary study on the effects on human health after exposures to ionising radiation and chemical carcinogens in the low dose range. Finally the impact of the individual expertise for the setting of environmental standards were to be developed in an interdisciplinary manner.

The members of the project group prepared draft texts from the perspective of their disciplines. The drafts, together with additional expertise given by invited scientists served as the basis for the interdisciplinary discussions of the group in order to work out a compilation of general agreement. In this context, the working group would like to thank Professor Dietrich Henschler (Institute of Toxicology, University Würzburg) and Professor Wolfgang Lutz (Institute of Toxicology, University Würzburg) for their valuable contributions in the field of toxicology and modelling of dose responses. The project was co-ordinated by Dr. Kathrin Prieß (Europäische Akademie GmbH) who also contributed actively to the project group's work as co-author and as collaborator in the final shaping of the project. Finally, as a result of 15 meetings and several years work, the project group approved the texts in the present form.

In summer 2002 the work of the project group was advanced so that the results were presented to a review panel of experienced colleagues for a critical review. The project group's midterm meeting took place in September 2002 in Bad Neuenahr-Ahrweiler. The meeting allowed discussions of the results of the project group with a broader interdisciplinary expert panel. Reviews were given by Professor Maria Blettner (University of Bielefeld, now University of Mainz), Professor Monika Böhm (University of Marburg), Professor J. Philip Day (University of Manchester), Professor Anne Fagot-Largeault (Collège de France, Paris), Professor Ludwig Feinendegen (Lindau), Professor Bernd Hansjuergens (UFZ Centre for Environmental Research Leipzig), Professor Bernhard Irrgang (Technical University of Dresden), Professor Georg F. Kahl (University of Goettingen), Professor Jos Kleinjans (University of Maastricht), Professor Wolfgang Köck (UFZ Centre for Environmental Research Leipzig), Professor Suresh H. Moolgavkar (Fred Hutchinson Cancer Research Centre, Seattle), Dipl. Päd. Holger Schütz (Research Centre Juelich). We are indebted to all the participants for the helpful discussions and advice that improved the manuscript to its present form.

We also would like to thank Dr. Vesna Prokić (GSF Neuherberg) for collaboration and Mrs Christa Schotola (GSF Neuherberg) for producing graphs for the chapter on mathematical models. We also appreciated the discussions with Dr. Wolfgang Heidenreich (GSF, Neuherberg) on the TSCE model.

A big thank-you is addressed to Mrs Birgitta Jerresten from the Karolinska Institute for assistance at the meeting in Stockholm as well as staff members of the Norwegian Academy of Sciences in Oslo at the meeting in Oslo. Our acknowledgements are also gratefully addressed to Mrs Friederike Wütscher of the Europäische Akademie who was responsible for the proof reading of the final text, as well as to Mrs. Margret Pauels of the Europäische Akademie who cared for our meetings in Bad Neuenahr-Ahrweiler.

The collaboration with all mentioned colleagues and with the Europäische Akademie was an enjoyable undertaking and pleasure.

Essen, December 2003   Christian Streffer

Table of Contents

1 Introduction

1.1
Background

Since the beginning of life on earth, all living beings have been exposed to potentially harmful agents like ionising radiation or chemical substances from various natural sources. In fact the evolution of life arose, in part, as a consequence of these, at first sight, hostile conditions and most organisms, including humans, have developed effective defence mechanisms against many harmful agents. At present, however, the types of changes brought about by exposure to radiation or chemical substances, be they man-made or natural, are generally considered to be detrimental to both environment and human health.

The tremendous increase of the human world population during the last 100 to 150 years – which augmented about fivefold during the last century – together with the increasing technological progress with its unavoidable release of toxic substances led to densely populated areas and high pollution loads in many regions of our planet. Nevertheless, the technological progress brought forth many conveniences, which the overwhelming majority of the population in industrialised countries would not want to miss. Ever since, society showed an enthusiastic belief in the benefits of progress. However, with increasing industrialisation, the damage to the environment caused by these industrial as well as by some naturally occurring contaminants could not be overlooked. Finally, even human health effects were noticed, yet under certain extreme conditions, which raised considerable concerns. Over the past 50 years there has been an increasing awareness of the need to regulate potentially harmful agents in order to achieve the qualitative objectives of protecting the environment with regard to clean air, clean drinking water, etc. The uses of radiation and chemical agents in industry and medicine have always been subject to particularly stringent regulations.

The first occupational effects of pollutants were observed centuries ago, for example, in Saxon silver mine workers that had been exposed to radon – unknown at that time- or in chimney sweepers that had been exposed to smoke. However, the awareness and understanding of such effects certainly only arose during the second half of the last century. Scientific investigations about the toxicological effects of physical and chemical agents have been carried out for many years, and mechanisms and dose-responses of toxic chemicals and ionising radiation have been studied. Regulatory standards for toxic agents were derived for conditions at working places. Later, standards to protect the public and the environment were also introduced. The increased scientific understanding of the toxicology of various substances enabled scientists to voice warnings of possible detriments before damages were noticed in the environment or in humans. For instance, in the 1960's, after scientists had measured the radioactive fallout from atmospheric atomic bomb testing, many demanded the immediate cessation of those tests. This initiative finally led to several international agreements (i. e. the "Treaty Banning Nuclear Weapon Tests in the Atmosphere, in Outer Space and Under Water" 1963).

1.2
The problem

Initially, regulatory limits on exposures for both workers and the general public were set according to Paracelsus' statement "the dose makes the poison"[1], which is frequently interpreted as meaning that there was always a threshold below which no effect would occur. To introduce a margin of safety, regulatory limits were set at least one to two orders of magnitude below the levels at which health effects had been observed. However, it appears that in many cases, such a threshold level cannot be determined. Some effect is assumed at any dose and thus complete protection is not possible, even by setting environmental standards at a very low level. Hence, standards are set in general by applying safety factors. These factors are set according to risk estimates that can be obtained by extrapolations from studies that have been carried out in dose ranges far above the regulatory limit. For the protection of the general public standards are usually set at a lower dose range than is the case for those at work places.

Nowadays, the public is confronted almost every day with alarming announcements of newly discovered health risks of unforeseen extent, be it dioxin in milk, irradiated fruits, bovine spongiform encephalopathy (BSE), or acrylamide in food. These announcements are often contradictory. This can, at least, partly be attributed to the lack or the uncertainty of knowledge about certain risks. For, inevitably, the process of standard setting includes a major degree of uncertainty. Some risk might persist in the range of very low exposures, though not detectable, and the intermediate and long-term effects of exposure to potentially harmful agents are often poorly understood. To some extent, public concern is also provoked by contradictory or unreliable statements prompted by conflicting interests of the different stakeholders.

The increasing knowledge of harmful effects of exposure, the enhanced sensitivity of detection methods, the deeper insights into the mechanisms of damage, but also the increasing awareness that there is still a lot of research to do on how to reduce uncertainty, generate scientific debates that are often controversially portrayed in the media. In turn, the situation leads to considerable uncertainty, worries and doubts in the different parts of the society. It has been linked to society's shift from an uncritical acceptance and support of progress to an ever increasing scepticism towards new technologies, often implying a claim for "zero release" or "zero burden" that determines essential parts of the risk debate in the field of exposures caused by technology and human lifestyle.

---

[1]   Philippus Aureolus Theophrastus v. Hohenheim named Paracelsus (*1493/94–†1541): *„Alle ding sind gifft und nichts ist ohn gifft. Allein die Dosis macht das ein ding kein gifft ist. Als ein Exempel: ein jetliche speiss und ein jetlich getranck so es über sein dosis eingenommen wirdt, so ist es gifft. "*

1.3
Example – the acrylamide case

The recent problem of the discovery of acrylamide in baked or fried starch-containing food provides an example of the manifold problems that arise when a new issue needs to be addressed by policy. With enhanced analytical skills, acrylamide was detected in food by Swedish scientists in April 2002 (Swedish National Food Administration 2002). This substance is formed when starch-containing food is fried or baked at high temperatures, and exposure had certainly occurred for many generations without being perceived as harmful. However, acrylamide is a known carcinogen in animals and it is regulated on this basis. The maximum dose that is allowed to pass from packing material to food devoted to human consumption is 10 $\mu g \, kg^{-1}$. On November 3, 1998, the European Union (1998) set a limit of 0.1 $\mu g \, l^{-1}$ for drinking water in accordance with the WHO recommendations. The recent measurements in fried and baked food suggested that these concentrations were often exceeded up to 10^5-fold (according to the *Bundesamt für Verbraucherschutz und Lebensmittelsicherheit*, German Federal Agency for Consumer Protection and Food Safety). To reduce the risk to "as low as reasonably achievable" (i. e. by applying the ALARA principle) the German authorities for consumer health tried to appeal for a self-commitment:

– of the food industry, to reduce these values for instance by changing the temperature of production and
– of the public, through advice to follow general dietary recommendations and to survey the temperature of baking and frying.

Still, a lot of open questions persist, many of which go beyond purely scientific arguments, and appear, in one or another form, in the debates of risks of exposures caused by technology (Granath and Törnquist 2003). Some of them are: Is acrylamide really dangerous to human health and does it cause cancer in humans, as animal experiments would suggest? At what dose is it dangerous in food? How does acrylamide act in the human body? Is there a possibility of predicting effects according to general models? How many people could have been affected by cancer due to acrylamide? These questions concern the basic scientific assessment of mechanisms, effects and dose-effect models as well as the justification of transferring animal data to humans. Is there a need for further action? Is there a moral obligation to continue the risk evaluation process, knowing that acrylamide is carcinogenic to animals? What are the consequences of setting standards? At first sight, this seems to be a technical and economic problem, as some food simply would need to be forbidden or new techniques of food treatment would need to be developed. What would happen if nothing were done? Humankind has lived for generations with this risk, without experiencing high mortality, so why should we care? This addresses normative aspects of risk evaluation.

Is there enough evidence to set standards as for drinking water or is the ALARA approach sufficient? Are the standards plausible? How are they justified? This involves the problem of scientific evidence or plausibility that often gives rise to very controversial discussions.

Is there a legal obligation to set standards for acrylamide or does existing legislation already protect the public in an appropriate way? How should the public be effectively informed without causing undue insecurity? Risk communication is a very sensitive issue as it is quite easy to obtain the counter-effect of what was intended with the communication process.

All the above are examples of the types of fundamental questions about environmental carcinogens that this report will address, hopefully illustrating that the setting of environmental standards and regulations is something that cannot be performed by scientists alone. The questions concern problems that can only be solved in a rational process involving natural sciences, ethics, social sciences, economics and law to obtain broadest possible consensus between science, policy and society. Hence, an interdisciplinary discussion and participation is appropriate and necessary. It appears to be of high relevance to seek uniform principles for the risk evaluation of different toxic agents such as chemicals and ionising radiation. This should be the foundation for rational standard setting in order to protect the environment and human health adequately. In our democratic systems, acceptance of these procedures and results by a wide public is needed and has a great value. This goal is supported and can probably only be achieved by a high degree of transparency and acceptability during the various steps of standard setting.

1.4
Preliminaries

Chemical and physical agents present in the environment can influence human health in a multitude of known and unknown positive and negative ways. They might cause acute effects such as respiratory problems, skin irritation, numbness or other neuronal disorders, that usually disappear after exposure and do not cause permanent harm. These effects are certainly undesirable, but since they are reversible they might be acceptable. Of much more concern are of course health effects that will not vanish after exposure like allergies, cancer, and endocrine disruption, neuronal defects (e. g. disturbed mental and intellectual development, paralysis...), hereditary effects, sterility or developmental disorders (e. g. teratogenesis). These effects mostly develop long time after the chronic exposure to agents that are diluted in the human environment. Moreover, they are in general not perceptible without instrumental analysis. Hence, people often feel an insidious threat.

The presence or absence of a threshold dose leads to the distinction of two fundamentally different types of effects with respect to dose-responses and different approaches in regulation. When there is a threshold, risks through the corresponding toxic agent can be completely avoided by setting the dose limit below the observed threshold dose. In these cases toxicologists base their recommendations on the so-called "no observed adverse effect levels" (NOAEL) usually derived from experimental data. Because the individual susceptibility of test animals or humans varies across a range of threshold doses, the threshold is not an absolute quantity that can be measured directly. Instead, it must be estimated indirectly from experimental dose-response studies that determine the NOAEL in several dose groups of test animals, from clinical observations or from epidemiological studies. By applying a certain number of extrapolations and safety factors, standards to protect human health may be derived (examples: Kalberlah and Schneider 1998).

Where there is no threshold this extrapolation is not possible. These so-called stochastic effects occur by chance, often after long-term low level (i. e. chronic) exposure to agents and generally without a threshold level of dose. In the low-dose range, their probability is usually proportional to the dose. In that case, one attempt to quantify risks is the quantitative risk analysis (for details see Anderson et al. 1983; EPA 1986) with the aim to give differentiated information about the magnitude of risk. To do so, the concept of "unit risk" (UR) is introduced, which represents the slope of the linear part of the dose-effect curve (Mosbach-Schulz 1989) or in other words the factor of proportionality between exposure and the risk of an adverse health effect (i. e. tumour). For genotoxic effects or human carcinogenicity, standards are derived based on this quantitative risk analysis. However, not all advisory bodies accept the concept of "unit risk", as the low quantitative reliability of some estimates is a major point for criticism. Hence, some advisory bodies just distinguish between carcinogens and other substances without deriving any standard for the first group. In that case, the ALARA principle ("as low as reasonably achievable") is applied to reduce the risk.

Agents that are considered to act without a threshold are genotoxic carcinogens, mutagens, and probably sensitising (allergy creating) agents. In certain cases, some embryotoxic effects can also occur with dose-responses without a threshold, especially when exposures take place in very early developmental stages. However, non-carcinogenic agents with long term effects are often not considered in regulations (Neus et al. 1998a). Bearing this in mind, the report will focus on genotoxic agents with hereditary and carcinogenic effects in humans. For these effects, often databases are available that allow to draw rational conclusions. For other agents, sound information is still scarce. The focus of this report is mainly on health effects in humans as humans are the main object of most standard setting procedures. Health effects on animals are considered in connection with experimental data for the establishment of dose-effect relationships and the study of mechanisms.

Ionising radiation is certainly one of the most intensively studied carcinogenic agent in humans. The comparison with chemicals will help to find common mechanisms and hopefully to close some gaps in the understanding of mechanisms in all carcinogenic agents. With selected examples of high practical relevance, reasonable criteria for a common classification and evaluation of ionising radiation and chemical agents without threshold will be elaborated and evaluated on the basis of general mechanisms.

Quality of extrapolation, evaluation of residual risks and limit value setting strongly depends on the applied dose-effect models. Especially the shape of dose-effect curves in the very low-dose range drawn from dose-effect models may lead to considerably different risk estimates at low doses. (Fig. 1.1 A and B curves, see Streffer et al. 2000 for a more extensive discussion). In radiation protection, major advisory bodies use the widely adopted conventional linear-no-threshold (LNT) model. Notwithstanding the LNT model, several groups (i. e. BEIR 1990; ICRP 1991) have included a dose and dose rate effectiveness factor (DDREF) of about two for chronic exposures to γ- or X-rays (ICRP 1991). This divides the risk per unit dose at low doses or low dose rates by two, compared to the risk obtained from the linear extrapolation. However, recently, a lot of criticism arose in the radioprotection community against this widely used LNT model. Diverging opinions on overestimation and even underestimation of residual risks by using the LNT model have been put forth.

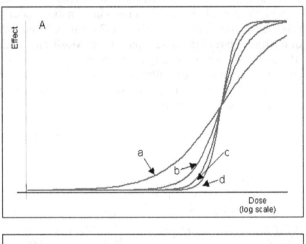

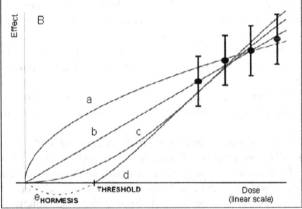

Fig. 1.1 Presentation of different dose-effect curves. A: semi-logarithmic scale. B: Double-linear scale. a = supra-linear; b = linear; c = sublinear (linear-quadratic) d = effect with threshold; e = hormesis

In this report, criteria for exposures to low doses are established based on a mechanistically sound classification. Despite the uncertainty of risk estimates in the low-dose range, these criteria will improve the risk evaluation of toxic agents. Models based on experimental as well as on epidemiological data today are powerful tools to develop an improved hypothesis on mechanisms. Some relevant examples will be presented to demonstrate the strengths and weaknesses of frequently applied models that may help to compare risk evaluations for genotoxicity due to ionising radiation and particular chemicals. It is crucial to understand the mechanisms of the different damage patterns of agents instead of relying solely on a simple extrapolation from effects at high doses to those at low doses and from effects in animals to those in humans. Experimental animal data, as well as cellular and molecular studies with mammalian cells (including *in vitro* studies with human cells), will assist to understand these mechanisms. The influence of various biological processes like DNA repair can be analysed under these circumstances. Taking all these data together, attention must also be paid to the uncertainties in risk factor evaluation in the low-dose range. The influence of exposure conditions including exposure rates, DNA repair, adaptive response, low-dose hypersensitivity, apoptosis, induction of genomic instability and further biological processes need to be evaluated for the establishment of risk factors.

Our efforts to protect the environment should be governed by different ethical principles. The crucial problem, the normative ethical aspects of different options for action under ignorance or high uncertainty, will be examined. Alternatives need to be compared and actions, values and norms have to be tested for consistency. At the very basis of all environmental policy stands the discussion of the normative notion of acceptable risk and the society's use of laws, taxation and economic incentives to implement these normative considerations in a feasible way. As a democratic tool, policy needs to address risk and its acceptability so as to achieve a lasting regulation beyond the coincidental come and go of changing fashions of acceptance. Solid scientific foundations of standard setting are of high significance also from the point of view of the responsibility of sciences for society.

In addition, philosophy addresses theoretical questions concerning methodology and especially extrapolation and hypothesis testing. In some cases sciences, by carrying out a research project or by refusing to do so, can create facts that generate, enhance or inhibit social processes. Scientists need to recognise that, already by the mere choice of the object of research they may generate social effects that they are responsible for. Today this is the case for all topics dealing with the impairment or the protection of the environment.

Public perception of risks is a very important issue especially in the field of low-dose exposure. Here, the uncertainty of the effect and/or the severity of harm are dominant, the public therefore often feels insecure which can lead to a situation where the public perception of threat might be biased. People would feel more threatened due to uncertainty than in the case where the certainty of harm and effect are established and clear counter measures can be taken. In addition, those risks are influential instruments for the mobilisation for or against certain interests since the likelihood of negative effects of low-dose exposures is barely refutable. Beyond this, statements about possible negative consequences of exposures in the low-dose range meet the intuitive principle of "better safe than sorry" and raise calls for a stop of this activity in case of uncertainty. The public generally asks for clear answers from the scientists. It often perceives setting standards as an entirely scientific venture ("hard science") derived from precise scientific knowledge. In their understanding, science can and must determine at what dose symptoms arise; engineers know the actual state of technique to limit ambient concentrations. Hence, this limit will be set as standard (see Winter 1986 for further discussion). However, standard setting in general implies scientific suppositions or suspicions, socio-economic statements are used, and value judgements are made. Scientific judgements are burdened by uncertainty. Aside of the uncertainty of the applied extrapolation model, statistical uncertainties (significance level, statistical power, sample size) and the uncertainties inherent to the methods (epidemiology vs. animal experiments), the multiplicity of effects (combination) is a serious problem (see Streffer et al. 2000 for further discussion).

Not only incorporation of an agent, but also its fate in the human body, especially the extent of accumulation and the effect on the metabolism, are relevant for risk evaluation. However, the information on these biokinetic and biochemical parameters is often sparse or even lacking, as only few human data, mostly in the high-dose range, are available, and as an extrapolation from animal experiments (mostly carried out in high-dose ranges) is often difficult and in a number of cases not feasible, since metabolic and other processes are too different.

In the context of risk evaluation and its judgement for standard setting, risk communication and risk perception plays an ever-increasing role. The communication effort must not be restricted to the mere description of scientific facts but needs to take into account the psychological and social mechanisms of risk perception which are often based on subjective judgements which have an impact on societal acceptance. Therefore, the pressure on politicians increases to take countermeasures and to aim for restrictive regulatory standards, independently of scientific plausibility. This process might lead to undesired effects as high subsequent costs may arise from a supposed "over-regulation" or divert form the real problems. On the other hand, this can also lead to a sort of paralysis in the face of uncertainty, hence inhibiting politicians to act. This can undermine trust in policy in the long term as is demonstrated by a lot recent examples in Europe.

Law – German as well as European – is instrumental in setting the framework for tackling risk problems related to exposure to low doses of hazardous agents effectively, efficiently and equitably. Since the scientific ideal of providing full proof of dose-effect relationships cannot always be fulfilled, the general recognition of the precautionary principle in health-related environmental law is crucial. It enables the competent authorities to act already in situations of uncertainty about the adverse effects of exposure to low doses, which includes uncertainty about the existence of a threshold. However, the precautionary principle does not offer clear guidance as to the contents and limits of risk reduction policy. Several questions need to be clarified: the minimum degree of scientific evidence that may trigger precautionary action; the extent as to which variation among affected persons (susceptibility) must be considered in risk assessment and standard setting; the legitimacy and extent of considering economic, social and ethical factors, including distribution of risks and benefits as well as risk perception, in taking decisions on tolerability of risk and risk reduction measures; and finally the crucial choice of strategies to be used to reduce risk.

It is the aim of this report to study all these problems in a rational interdisciplinary discussion and to evaluate the potential for enforcement and the consequences of environmental standards protecting human health in the low-dose range. The report is divided into two main sections. Following an introduction on ethical issues of risk management, the first part addresses the largely scientific issues concerning studies and investigations about the mechanisms and magnitude of risks associated with exposure to carcinogens. The second part of the report focuses on philosophical and legal aspects of risk assessment and management, including risk perception and communication. The report helps in identifying and in closing recognised gaps in knowledge. On that basis, procedures, which may be considered as rational for the processes of setting environmental standards more efficiently, will be recommended.

1.5
References

Anderson EL, Carcinogenic Assessment Group of the U.S. Environmental Protection Agency (1983) Quantitative approaches in use to assess cancer risk. Risk Anal 3: 277–295

Committee on the Biological Effects of Ionizing Radiations [BEIR] (1990) Health effects of exposure to low levels of ionizing radiations. Board on Radiation Effects Research (BEIR V). Natl Acad Sci USA, Natl Res Council. National Academy Press, Washington, DC

European Union (1998) Council Directive 98/83/EC of 3 November 1998 on the quality of water intended for human consumption. OJ L 330, 05/12/1998: 32–54

Granath F, Törnquist M (2003) Who knows whether acrylamide in food is hazardous to humans? J Natl Cancer Inst USA 95: 842–843

Kalberlah F, Schneider K (1998) Quantifizierung von Extrapolationsfaktoren. Schriftenreihe der Bundesanstalt für Arbeitsschutz und Arbeitsmedizin. Wirtschaftsverlag NW Verlag für Neue Wissenschaft, Bremerhaven

International Commission on Radiological Protection [ICRP] (1990) ICRP Publication 60. Recommendations of the International Commission on Radiological Protection. Annals of the ICRP 21(1–3), Pergamon Press, Oxford, 1991

Mosbach-Schulz O (1998) Probabilistische Modellierung in der Prioritäten- und Standardsetzung. In: Umweltbundesamt (ed) Aktionsprogramm Umwelt und Gesundheit. UBA Berichte 1/98. Erich Schmidt, Berlin, pp 571-594

Neus H, Ollroge I, Schmid-Höpfner S (1998) Synopsis der in Deutschland bestehenden Verfahren zur Erarbeitung gesundheitsbezogener Umweltstandards. In: Umweltbundesamt (ed) Aktionsprogramm Umwelt und Gesundheit. UBA Berichte 1/98. Erich Schmidt, Berlin, pp 201–446

Streffer C, Bücker J, Cansier A, Cansier D, Gethmann CF, Guderian R, Hanekamp G, Henschler D, Pöch G, Rehbinder E, Renn O, Slesina M, Wuttke K (2000) Umweltstandards. Kombinierte Expositionen und ihre Auswirkungen auf die Umwelt. Wissenschaftsethik und Technikfolgenbeurteilung, Bd 5. Springer, Berlin

Swedish National Food Administration (2002) Acrylamide is formed during the preparation of food and occurs in many foodstuffs. Press release. April 24, 2002

Winter G (1986) Einführung. In: Grenzwerte. Interdisziplinäre Untersuchungen zu einer Rechtsfigur des Umwelt-, Arbeits- und Lebensmittelschutzes. Winter G (ed) Umweltrechtliche Studien 1, Werner-Verlag, Düsseldorf, pp 1–26

US Environmental Protection Agency [EPA] (1986) Guideline for carcinogenic risk assessment. Environmental Protection Agency, Bureau of National Affairs, September 1986. Fed. Reg. 51 FR 33993 (1986) 39: 3101–3111

2 Ethical Aspects of Risk

2.1
Introduction

The establishment of standards for tolerable, or acceptable risks in connection with exposure to noxious agents might seem to be exclusively a matter of sound scientific methodology in estimating the probabilities for damages and reducing the uncertainties involved in extrapolating from the known into the unknown. However, the setting of standards for acceptable risk is not just a matter of probabilities and degrees of uncertainty. Acceptability is a normative notion and involves a number of factors having to do with values and human autonomy that will be discussed in the following. Whoever makes the decision, whether an individual, a group, an industrial firm or a public institution or agency, there is a number of norms and values that come into the picture, such as consent, distributional justice, and many others that will be discussed in this chapter. Some of these normative factors have been included in laws and regulations in various ways that often differ from country to country, while others are not laid down in law, but remain as ethical considerations that any decision maker should take into account.

2.1.1
Risk

The notion of risk reflects this tension between mere probability and a notion that also involves values. The pure probability version of risk is often used when an event with a clear negative value, for example, getting cancer, is discussed. Usually the number of cancer cases in two situations is compared, one with the normal background exposure and one with exposure to a carcinogen. Several good examples are given in chapter 5, where this notion of risk is explained as follows: Risk estimates are often presented as excess relative risk (ERR) or excess absolute risk (EAR). These terms represent the increased rates relative to an unexposed population, i. e. background rates, measured on proportional and absolute scales. For example, an ERR of 1 corresponds to a doubling of the cancer rate, while an EAR may be expressed as the extra-annual number of cancers per 10,000 persons. Under the assumption of a linear dose-response relationship the risks may additionally be expressed as amounts per unit dose, e. g. ERR Sv^{-1}. The risks could also be given for a specific dose, e. g. ERR at 1 Sv.

This notion of risk is useful when one is studying the causal relevance of various factors for the development of an effect, in this case cancer. One main challenge in this kind of studies is to sort out the different factors by finding suitable groups, in epidemiological studies, or constructing suitable experimental settings, in experimental work.

If the aim is to make policy decisions, normally values have to be included and often the notion of risk is defined as the product of probability and the (negative) value, expressed in numerical terms (see e. g. Streffer et al. 2000, p 13; for more on 'product' and numerical values, see next section). Usually, more than two alternatives have to be considered, in fact one of the main challenges is to create alternatives, which are as good as possible, given the probabilities and the values that are at stake. Each alternative will normally have numerous consequences, some positive, some negative, whose probability will vary depending on which alternative we choose. In order to arrive at a good decision one has to compare these alternatives, taking the probabilities and values of the different consequences into account. How this shall be done, will be discussed in the next section.

2.2
Rational decision making

To survey the different normative factors we will start from a model for rational decision-making[1] (Fig. 2.1).

Briefly, the model for rational decision tells us that the right thing to do is to choose the alternative which has the highest expected value, where the "expected value" of an alternative A_i is, very roughly, indicated by $\Sigma_j (p_{ij}v_j)$. In words: the expected value of an alternative is determined, very roughly, through the following procedure: (1) for each consequence, "multiply" the probability of that consequence, given that we choose that alternative, with the value of that consequence, (2) add all these products.

---

[1]   The following model is sometimes called cost-benefit analysis. It is widely used by economists. Unfortunately, many of them take only money into account in their analysis. Of course, there are many other values than money, and a main point in what follows is that neither the notion of benefit nor the notion of cost should be taken as including just money, but the whole spectrum of values. The misuse of the model by those who regard money as the only value has been justly criticised. However, some critics have believed that the model can only deal with money as a value. Opposition to the model has led Kristin Shrader-Frechette and Lars Persson to argue that the proposed new ICRP proposals which move from radiation doses that are as low as reasonably achievable (ALARA) to doses that are as low as reasonably practicable (ALARP) are to be rejected because "what is *achievable* is a function of current *science*, while what is *practicable* is a function of *economics* and benefit-cost analysis" (Shrader-Frechette and Persson 2002, 155). Shrader-Frechette and Persson assume that cost-benefit analysis takes only money into account, and they also seem to think that consequentialism and utilitarianism is the same. There are many other kinds of costs and benefits than money and utility. The point of cost-benefit analysis is that one must use one's reason and find the best alternative, taking into account *all* the probabilities and values involved. The word 'reasonable' in both formulations (ALARA and ALARP) is in our opinion crucial. Whether one uses 'achievable' or 'practicable' is unimportant, as long as one takes all values fully into account. See also 10.4.7 below.

   What worries Shrader-Frechette and Persson is probably that the old word 'achievable' more than the new word 'practicable' stresses our obligation to do what we can to find the best possible alternative. We agree with the aim of their article, to warn against overlooking important values. But we think that the decision theoretic model – when properly used, and not based on taking money as the only value – is one of the best tools we have to make sure that we take all probabilities and values properly into account and intensify our search for ever better alternatives.

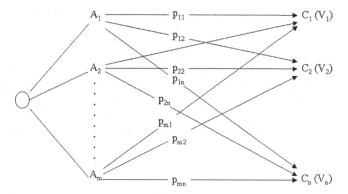

Fig. 2.1 Model for rational decision making. Here, "A" stands for alternative, "C" for consequence, "p" for probability and "v" for value.

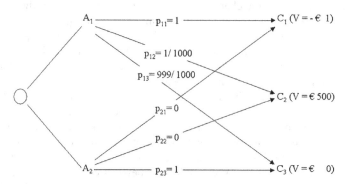

Fig. 2.2 Simple example of a model for rational decision making. Consider the case where you are offered to buy a ticket for 1 E in a lottery. One thousand tickets are sold and one prize of 500 E is offered.

As a simple example, consider the case where you are offered to buy a ticket for 1€ in a lottery. One thousand tickets are sold and one prize of 500 € is offered (Fig. 2.2).

Consider the two alternatives A_1, to buy a ticket, and A_2, to buy no ticket (see Fig. 2.2). For each alternative, there are three consequences, C_1: you pay 1 € for the ticket, C_2: you win 500 and C_3: you win nothing. If you choose A_1, the probability p_{11} of C_1 is 1, and – assuming that the lottery is not rigged – the probability p_{12} of C_2 is 0.001 and the probability p_{13} of C_3 is 0.999. If you choose A_2, the probability p_{21} of C_1 is zero, the probability p_{22} of C_2 is zero and the probability p_{23} of C_3 is one. The expected value of alternative A_1 is: $(1 - 1 €) + (0.001 \times 500 €) + (0.999 \times 0) = -0.50 €$, while the expected value of alternative A_2 is: $(0 - 1 €) + (0 \times 500 €) + (1 \times 0) = 0 €$. Hence, the rational alternative to choose is alternative A_2. This is also what we know about lotteries; more is paid in for tickets than what is paid out for prizes.

Let us not jump to the conclusion that people who buy tickets in lotteries are always irrational. There may be more values coming into the picture than those that have been mentioned. Here are some values that might make it rational to participate in the above lottery: (1) The proceeds of the lottery go to a good cause (and one might not so efficiently support this cause in another way, for example by giving one's money directly). (2) One very badly needs something that costs more than what one has (or can easily loan), but less than 500 €, and one or a few Euro make very little difference. (This illustrates that money is not an ultimate value and that the value of money is not always proportional to the amount of money). (3) One enjoys a little bit of suspense and gladly spends one Euro to get it. In all of these situations, an additional positive value has to be added to alternative A_1.

The phrase 'very roughly' in the account of 'expected value' in the penultimate paragraph above is important. The use of the arithmetical notions of product and sum there was only meant to indicate a rather complicated relationship between the values and the probabilities on the one hand, and the expected value of the alternative, on the other. One simple characteristic of the relationship is that the expected value goes up (or at least not down) when the value of a consequence goes up, and also goes up (or at least not down) when the probability of a good consequence goes up, and inversely for declining values and probabilities. The use of the model does not presuppose that values can be measured on a common scale, for example, money. Usually, however, we can at least speak of more or less of a value. However, not even this is required for the use of the model, although there would then be many situations where we cannot make a rational decision. Note, however, that even in such cases, if we nevertheless make a decision, for example, because we intuitively regard one alternative as better than the others, such decisions implicitly assign a value to the outcomes, although the value assigned is not uniquely determined by our decisions. Thus, for example, two persons can agree in their judgements about probabilities, also agree on which is the best alternative, although they have different ethical views, and therefore assign different values to the various consequences. Very often, however, there is disagreement about the probabilities. One reason for using the above diagram is that it can be of help in practical cases of disagreement, by enabling us better to locate the sources of the disagreement. Is there a disagreement about what the alternatives are, about the probabilities of the various consequences given the various alternatives, or is it a disagreement about norms and values, and if so which ones?

Note that then the model does not tell us what the values are. We cannot just put in money and neglect other values. This is often done in applications of this model, and it may give an appearance of rationality, but this is pseudo-rationality. One has to take into account the whole range of values (see sections 2.2–2.2.11 below). Moreover, these values have to be justified (section 2.2.12). The misuse of the model has led some discussant to reject the model and rely upon intuitions without proposing any systematic model to guide them or checking whether their intuitions are inconsistent. For legislative purposes such a procedure is particularly problematic as it leaves free scope for arbitrariness and subjective biases and as it lacks the transparency that is crucial for a social decision procedure (see sections 2.3, 2.5 and also 10.5.3).

2.2.1
Decision makers

Who the decision-makers are varies from situation to situation. Some decisions are made by an individual, some by a group of individuals. Some groups have become established by accident, for example a group travelling together, others are more permanently organised, some of these may have been given powers to act on behalf of others, for example political bodies or shareholders' representatives in a company. Decisions concerning acceptable risk are usually made by political bodies, often in a process where the preparatory work is done by one group or some individuals and the final ratification is made by another group. It is not always clear which group or body makes the decision. Furthermore there is the question which is central in studies of power: how are various political bodies, groups and individuals able to influence the decisions made by others?

When it comes to evaluating the decisions from an ethical point of view, it is important that consideration is not only given to how the decisions are made, but also to the effects of the various alternatives on all the different stakeholders, especially the silent ones, for example those who are handicapped or have little access to media. One very important group of stakeholders whose interests are not always given full consideration is future generations. They will be more fully discussed in section 2.2.13.

2.2.2
Alternatives

In a given choice situation, where a decision-maker is to decide between various alternatives, the first question to ask is "Which are the alternatives open to us"? The answer to this question requires scientific insight: one must know what alternatives are possible, given what we now know about the situation at hand. However, the answer is not given by a simple application of scientific theory. It also requires creative fantasy and is in that way more like the construction of a new scientific theory. When we are exploring what the alternatives are, we should not be confined to previously used solutions, but must always try to find better ones. In practice, one should start by exploring some alternatives which initially seem better than the others, to see what consequences they seem likely to have. Seeing that some consequences are bad, others good, one should then try to get ideas concerning possible improvements that will reduce the probability of bad consequences and increase the probability of good consequences. We must apply the model for rational decision we just sketched in a dynamic way; we should not just go through the model once from left to right, but iterate the procedure. Each time we go through it, we let the analysis from the previous analyses guide us in searching for ways in which our alternatives can be improved.

An example that illustrates the importance of using creative imagination to find better alternatives is the development of the waterfalls in Aurlandsdalen in Norway to get hydroelectric power. The city of Oslo had acquired the water rights, and in 1970 proposed a development plan, according to which most of this beautiful valley would be destroyed by dams, roads and power lines. The political constellations were such that one could expect the plan to be given the approval by the Norwegian parliament. Then, however, an engineer working for the Norwegian Nature Conservancy League studied the proposal carefully and found that by placing the dams in side valleys, putting most of the roads in tunnels, and keeping the power lines away from the most beautiful parts of the valley, one could get as much power as with the original plan, at about the same cost and with great gains for natural beauty and recreation. Since this new plan was better than or at least as good as the original plan with regard to all values involved, nobody voted for the original plan and the new plan went through. This example illustrates not only the importance of creative imagination, but also how it is sometimes possible to compare alternatives even when several mutually incomparable values are involved, such as amount of power produced, monetary costs, natural beauty and recreational possibilities.

Another whole group of examples is provided by the substitution of less noxious chemicals for noxious ones, as for example has happened in the case of unleaded gasoline. In the case of radioactivity, obviously the development of better methods for waste treatment would contribute greatly to making nuclear power more acceptable. However, one also has to be aware of the potential new adverse effects of the alternatives that are proposed to avoid the old problems.

Decision analysis can hence be an inducement to further technical innovation. Indeed, science does not only help us decide what the consequences of the various alternatives are, science can also give us new insights, which can enable us to develop new and better alternatives.

2.2.3
Probabilities of consequences. The precautionary principle

Another main area where science is important for risk evaluation is the estimation of the probabilities of the various consequences. A major challenge in the study of low-dose effects is the difficulty one encounters in extrapolating from the study of higher doses, where the effects have been measured, to lower doses, where at present one has no measurable effects. We can know that there are changes, but not whether these have effects on the health of those who are affected. Within this realm, issues of sound scientific methodology become a major concern, and one often is left with ignorance about the likelihood of the various consequences rather than with well-estimated probability. In such cases, one has to go into the theory of decision under ignorance.

One main point here is that when great negative values are at stake and we know little about the probabilities, we should act as if the probability of bad consequences is at the high end of the spectrum. That is, if all we know about the probability that something seriously bad is going to happen is that it lies between 0 and 0.9, we should act as if it were 0.9. This pessimistic decision principle is called the 'precautionary principle'. Later research may tell us that the probability was in fact very low, and that our precaution was therefore not necessary. However, the precaution was rational; it was the best thing to do in view of what we knew at the time the decision was made.

An example of the use of the precautionary principle was the moratorium on recombinant DNA experiments that the world's biologists agreed upon in 1974.

They were uncertain whether very harmful organisms could come out of such experiments, for example, bacteria that could spread dangerous, perhaps unknown illnesses very rapidly. During the moratorium period the biologists studied the risk in order to determine how large it was and how it could be reduced. After about a year, they decided that the danger of creating highly hazardous organisms was very small, and that one could further reduce the risk by using biological and physical barriers: one experimented on bacteria that could only survive under very special conditions, very different from those outside the laboratory (biological barrier), and one required the laboratories to be so constructed that nothing should escape from them, for example the experiments took place in a low-pressure area, so that if an organism should escape, it would be sucked in rather than get out of the laboratory (physical barrier). Assuming that the three factors are independent, the probability that a dangerous organism should escape and live on outside the laboratory would then be $p_1 p_2 p_3$, where p_1 stands for the probability that a very dangerous organism shall emerge from the recombination, p_2 for the probability that it shall escape from the laboratory and p_3 for the probability that it will continue to survive outside the laboratory. Here, p_2 is well known and very small, p_3 is not so well known, but likely to be very low, and p_1 is still unknown, but likely to be lower than one feared initially. Since the total probability is the product of the three probabilities, and one of them, p_2, is well-known and very small, and another, p_3, is also low, the product will definitely be very low. Then this kind of experiments was started. Gradually one understood better what happened and became better able to determine p_1, which also turned out to be fairly low. Some biologists, who had been critical of the moratorium, accused the proponents of the moratorium for having prevented important research. However, from the point of view of decision theory, the moratorium was the rational choice. Rationality is based on what one knows, or can find out without too large costs at the time the decision is made, and not on what one finds out later.

Another example of the use of the precautionary principle is found in section 3.6, where it is stated:

> For practical radioprotection where prospective estimates of risk have to be made and for the regulations which are based on such estimates it seems to be appropriate further to use a linear dose-response without threshold doses for stochastic radiation effects although the uncertainties are high and there is no solid scientific proof for such a shape of the dose-effect curve. This may be an overestimate of risk under certain conditions which is also justified by taking into account precautionary principles.

Sections 10.1.2 and 10.3.1 below survey applications of the precautionary principle in European and in German law. Thus, for example, section 10.1.2 quotes article 95 in the EC Treaty:

> Where there is uncertainty as to the existence or extent of risks to human health, the Commission may take protective measures without having to wait until the reality and seriousness of those risks become apparent.

The shape of the dose-effect curve is a good example of a scientific hypothesis about which there is disagreement between competent and well-informed scientists. See the discussion in section 4.5.2 of the differences between the studies by Peto et al. (1991) and Williams et al. (1993, 1996, 1999, 2000). Given the many factors that seem to affect the existence of thresholds, such as species, sex and tissue (sections 4.5.5 and 4.5.7), one should clearly be ready to admit that still further factors may be discovered in the future. As is mentioned in chapter 8, the confidence we can have in a scientific theory depends on many features: simplicity, thoroughness of testing, existence of plausible mechanisms, etc. (see also section 4.6.1).

It is important, when the precautionary principle is used, that not every imaginable scenario justifies precaution. In estimating the probability that something very bad may happen, only theories that have some scientific plausibility should be taken into account. Often even the best theories tell us that there is a certain risk, and sometimes a field is so new that reputable scientists differ in their estimates of probability. Sometimes one even has different, competing scientific theories that have all survived the scientific tests so far. What the precautionary principle says is that in such cases one should be cautious, one should use the most pessimistic scientific estimate of the probability. Then, if the value at stake is high, one will abstain until the estimates are more reliable.

The precautionary principle has been very much debated in connection with the ethics of decision making. One source of this debate is lack of clarity as to what is meant by the precautionary principle. It is necessary to stress that when we say that it is possible that an activity will have disastrous consequences, 'possible' does not mean 'imaginable in the worst science-fiction scenario'. The principle does *not* say:

> (A) If it is possible that an activity will have catastrophic consequences, we should refrain from this activity although the probability of catastrophic consequences is very small.

If this were the principle, we would have to live a very passive life, indeed, and even the passivity might have catastrophic consequences, for example that a meteor might hit just where we are standing. Instead, the principle should be stated thus:

> (B) If we do not *have very good reason* to believe that the probability is very small that an activity will have catastrophic consequences, we should abstain from that activity (for a discussion of this principle, see Føllesdal 1979).

As noted in section 10.1.1 below this is reflected by a recent decision of the German Federal Constitutional Court on protection from electromagnetic waves. While apparently extending the duty to protect to the outer limits of precaution, the Court denied a duty to protect against "purely hypothetical hazards to human health".

We shall now turn to the question of values, which is the central ethical issue concerning the setting of standards for acceptable risk.[2]

2.2.4
Utilitarianism and consequentialism

The model for rational decision that we have just sketched is, unfortunately, sometimes called utilitarianism. However, it is common and useful to distinguish this

---

[2]   There is a rich literature on the precautionary principle. An article where it is discussed in the European context in connection with scientific uncertainty and environment-related technologies may be found on the following web page: http://www.jrc.es/pages/iptsreport/vol68/english/GOV1E686.html.

model, which could be called consequentialism, from utilitarianism. Consequentialism is defined by saying that the rightness or wrongness of an act depends exclusively on the values of the consequences of the act, as in the above model. The doctrine called "utilitarianism" is a special kind of consequentialism, which has three further defining features (Sen 1979; see also Sen 1982, 1983):

1. Everybody counts equally.
2. Utility is the only value.
3. Distribution does not matter.

The first feature means that it does not matter who is affected by the action. Prince and pauper count equally.

The second feature, utility, is vague. Some of the early utilitarians identified utility with pleasure. John Stuart Mill (1863), in *Utilitarianism*, objected against that some kinds of pleasure are more valuable than others:

> It is better to be a human being dissatisfied than a pig satisfied; better to be Socrates dissatisfied than a fool satisfied. Moreover, if the fool, or the pig, is of a different opinion, it is because they only know their own side of the question. The other party to the comparison knows both sides.

Some utilitarians, such as the Berkeley economist Harsanyi, have defined utility so widely as to comprise anything that anybody prefers.

The third feature of utilitarianism, that distribution does not matter, follows from the former: utilitarians hold that a fair distribution is not a value by itself. All that matters is the utility. (Note, however, that according to feature 1 above utilitarianism is fair in the following sense: everybody counts equally: the amount of utility is all that counts, it does not matter whoever benefits or suffers.) Many consequentialists would hold that the distribution of the costs and the benefits between the people involved is itself a value. A just distribution is a positive value that one should strive for and that should be weighted into the values of the various alternatives. According to utilitarianism, if one can produce more utility by taking something from somebody who has little in order to give it to somebody who already has much, then this would be ethically right. One could, for example, imagine a case where a wine-lover, who gets very much enjoyment out of expensive wines, should get money to buy such wines from poor people because he can get more utility out of the money. Utilitarians defend themselves by saying that such cases would be unlikely. Normally, in view of the principle of marginal utility, a poor person would get more utility out of a certain amount of money or other goods than one who already has a lot of it. Nevertheless, the claim that distribution does not matter is one of the weak spots of utilitarianism and has contributed to give it a bad name. Another reason that the word 'utilitarianism' is often used as a derogatory term, is that 'utility' is such a vague term, that easily can be shaped according to what one wants to achieve in a debate, often it is simply used to stand for money.

Independently of what might be meant by 'utility', there are, however, two varieties of utilitarianism: One holds that it is the *total* utility that matters; the other regards the *average* utility as decisive. An average-utilitarian would prefer a society with relatively few people who can all have a high standard of living. The total-utilitarian would gladly have a larger population where the average was lower, if the higher number of people brought the total higher. The difference comes out in population policy. The principle of marginal utility tells us that one who is poor will get more utility out of a resource than one who is well off. A total-utilitarian would therefore hold that a fairly large world population would be good, as long as each individual got more utility out of the resources than one would get out of them if there were fewer individuals. An average-utilitarian would have a rather small total world population. If anybody who is already well off can get even a little better off by receiving still a little more of the world's resources, then this would be preferable to having some more people, even if they would get a lot of enjoyment out of the resources that give the well-to-do a slight lift.

Leaving this special ethical view behind, let us now return to the more general model of consequentialism. Note that in this model the notion of consequence should be taken in a broad sense, so that it includes everything that happens as a consequence of our deciding to act in a certain way. Notably the consequences include the action itself and all the later events that are caused by that action. Both are important, the nature of the actions, as deontologists claim, and the nature of the consequences. Focusing exclusively on the actions, as some deontologists do, is too confined, and focusing only on the later causal effects of the action, as utilitarians and some narrow consequentialists do, is also too limited. In comparing alternatives, we have to take into account both the characters of the actions and their causal effects. Some actions are so awful that they can hardly be justified by any good effects. On the other hand, deontologists like Kant had great problems explaining why a small lie could not be justified when it could save lives. To evaluate and compare the alternatives we therefore have to bring in values of all kinds, including the values of actions.

2.2.5
Values

The consequentialist's model for rational decision just sketched is often used in business decisions and also in political decisions. Unfortunately, in most such applications money is the only value that is taken into account. However, there is a wide range of other values, and it is highly important that they all be brought into the picture if sound decisions are to be made. It is sometimes said about economists that they know no other value than money. This is unfair; it is to the economists that we owe some of the best analyses of other values than money.

Let us note that money by itself does not even qualify as a value. It is what we can do with money or get for money that makes money something worth striving for. Thus, as noted, the utilitarians claimed that utility is the only value. In their view, the value of money is derived from the degree to which it can be useful. Let us now leave utilitarianism, turn to the more general consequentialism, and look at the various values that should be taken into account in the weighing of the alternatives.

2.2.6
Time

One value that is seldom discussed, is the value of time. Time comes in already in the flow of the decision process itself. It takes time to make a careful decision, and the time it takes must be weighed when we are considering the different alternatives. Often it is crucial to act quickly, without going through all alternatives and consequences. Thus, for example, if somebody is about to drown, it would normally be best to act as soon as we find a way of getting the person up, rather than considering at length various alternative ways of doing it. Very quickly, all alternatives will get an immensely high negative value: the person drowns.

In less drastic cases it may be worth thinking carefully through the alternatives, and in the case of important decisions, we may obtain more information before we act. However, the time and other resources it takes to get this information has to be included among the costs, and in many cases it may be more rational to act on the basis of the limited information we already have. How much time and effort we ought to spend on getting more information is itself a decision theoretic question, where we have to weigh the various alternatives against one another?

A related issue, that tends to be overlooked by many regulators, is the time it takes to go through the procedures for getting activities approved. Cumbersome procedures that require much paperwork and time-consuming efforts by many people, whether they be applicants or regulators, are immoral if just as good decisions – or almost as good decisions – could be reached in alternative, less resource-consuming ways. Here, too, decision theory has to be applied to the decision procedure itself. One has to find alternatives that make the required procedures as simple and resource saving as possible.

Time is also central in our valuation of many of the other values that we are going to consider as factors, thus, for example, both life and health are valued in proportion to how long they last. Finally, time itself may be regarded as a value, one of our most important resources, that of which life is made.

2.2.7
Life

We would all agree that life, especially human life, is a most important value that has to be taken into account when we make our decisions, both because lives can be saved through the activities we propose, and because we are evaluating actions that may put lives at risk. In order to facilitate decisions one often has tried to put a value on life. Several different approaches have been proposed, but all are unsatisfactory, among them are the following three:

1. Society's investment in a person. (Among its many drawbacks: this gives higher value to older and well-educated people than to children).
2. Market value: the value people put on their own lives when they are willing to take jobs where they risk their lives for a higher salary, etc. (Drawback: persons in severe need may agree to do things that they would not have done if they had had other alternatives; they may consent to something in a market which they would not have chosen if the market had been different: consent *in* a market must be distinguished from consent *to* a market. Also, people do many risky things, for example, hang gliding, not for money, but because of other values that they cherish, for example, because they like the thrill.)
3. Utility to society. (Drawback: same as for utilitarianism in general, depends on a notion of utility that tends to be too narrow.) A non-utilitarian variant of this would be to equate the value of a person's life with the sum of all the value that this per-

son can be expected to bring about in his/her life, taking into account the whole range of values, not just utility. This would give a higher value to young, healthy and well-educated people than to people who are old and sick or unemployable. It would give an especially high value to creative artists and scientists and also to people who act ethically and do a lot of good to others, both to those who are close to them and to society as a whole. This proposal fits in better with many people's intuitions than the previous two. However, it clashes with two other intuitions: that all human lives have the same value, and that this value is not dependent upon anything that we produce or bring about. This will be discussed further in section 2.2.11 below. Moreover, the question with which we started remains: what is the value of a life compared to other values, in particular the value of money?

So far, there has been no satisfactory proposal for putting a monetary value on life. Nor has one found any satisfactory way to set a negative monetary value on the shortening of a life, for example due to cancer or other kinds of damage to the DNA, which is one of the risks one has to consider in connection with the setting of low-dose standards. In many decision situations, one has to weigh life against life. What should then be taken into consideration? Age? Ability? Number of dependents? Merit, for example: if several patients need blood transfusions and there is shortage of supply, should then one who frequently has donated blood be given preference? Alternatively, are all lives equal? We shall come back to these difficult questions in sections 2.2.12 and 10.4.6.

2.2.8
Health

Health is a value that can be at risk in connection with low doses, but also a value that can be furthered through research and technical developments. Here, the first question is: What is health? A good definition of health has turned out to be difficult. An attempt is the World Health Organisation's definition: "a state of complete physical, mental, and social well-being". It has its obvious problems. There is now a great literature on the subject. However, although the notions of health and health improvement are difficult, it is somewhat easier to agree on what constitutes deterioration of health. For our purposes, we need both. We are willing to take risks in order to improve our health, for example when we bicycle to and from work. However, when we discuss risk we are even more often concerned with activities that pose a threat to health.

A second question, here as in connection with the other values we are considering, is how much weight should be given to health when it is weighed against other values. Again, we have no good way of converting health into money, but health is nevertheless a most important value that has to be taken into account in all deliberations about acceptable risk (see also section 2.2.12).

2.2.9
Environmental values

An issue that is getting ever more attention, also in the case of noxious agents, is damage to the environment, to animals or plants, or to the usefulness for recreation or sheer beauty. Does the environment, its animals, plants and its natural beauties have a value in themselves? Or do they have a value only in virtue of their being means for enhancing the quality of life for humans? Whatever justification is given for the value of environment, there will be agreement that it has some value and that this value must be taken into account when decisions about acceptable risk of radiation and toxic substances are made.

It is often difficult to estimate probabilities of damage in this context. Ecosystems are complicated, and small changes can bring about large and unexpected consequences, such loss of a species . In this area, the precautionary principle (section 2.2.3 above) is therefore particularly relevant (Shrader-Frechette K 1981).

A special aspect of the environment is how well one feels in a given environment. Living in an environment where one is exposed to risks may, for example make one feel uncomfortable. This feeling may often be out of proportion to the actual risk to which one is exposed. Thus, for example, minute cancer risks, or risks of flying, may often affect a person strongly. One question to ask is then: to what extent should these kind of "irrational" risks be taken into account in ethical decision-making? The answer, here as in the case of the many other factors that we discuss, is that what matters is not only how great a risk people are exposed to, but also how they *experience* the risk. It may sound paradoxical, but rational decision making has to take into account factors that we regard as irrational. An important aim of risk communication therefore is to try to make people aware that some of their fears seem irrational, given the probabilities and the values of the people exposed to the risk. Irrational fears often hamper social projects and put burdens on others. Yet, to the extent that they are not evidently unreasonable, they have to be included in the weighing when decisions are made[3].

2.2.10
Known individuals or anonymous ones

A further problem that arises in connection with both loss of life and damage to health, is that our emotional reactions and thereby our sense of urgency is strongly affected by whether we know the victims or not, and also by how close our ties to the victims are. For example, one may be reluctant to invest large amounts of money to make mines safer for the workers. On the other hand, if there is an accident, where survivors are trapped, the victims are no longer anonymous, and one is much more willing to spend large amounts of money on rescue actions to get the trapped miners out. Here, as in many other areas, the strength of our feelings is no good measure of the goodness or badness of the consequences. Given these biases in our feelings, it may be advisable to ask what one would do if one self, one's spouse, parents, children, or friends were among the victims. What dose limits will one permit if one's children are to be exposed?

---

[3] This is a much-debated question. Many hold that the fears and aversions that are experienced about a risky action have no place in the rational conception of risk. See, for example, Gethmann 1994, 24.
  However, Streffer et al. 2000 argue that people's risk perception should be taken into account (p. 460, recommendations 16, 17 and 18). See also section 10.4.3.

2.2.11
Free choice

One important consideration, when one is going to decide what is acceptable risk, is whether those who are exposed to the risk are free to choose whether they should be exposed to the risk or not. Smoking and hang gliding are typical examples of risks that may be ethically acceptable in a society to the extent that they do not harm others, while passive smoking may not be acceptable. Pollution and many kinds of low-dose risks fall in the second category, and the fact that the exposure to such risks is not freely chosen gives them a highly negative value.

In many countries there is discussion concerning the acceptability of high risks that some people are willing to take, such as base jumping, riding motor cycle without a helmet, driving without a safety belt, etc. The reason that such activities are outlawed in some countries is that they have consequences for others, for example for relatives, for rescue workers who sometimes get exposed to high risks in order to help risk-takers who have got into trouble, for hospitals, health workers, insurance companies, etc. There is also the factor of society's investment in the person and the positive contributions the person could make to society and to other people.

2.2.12
Consent

For a choice to qualify as free in the proper ethical sense, it has to be informed. This involves having a full grasp not only of the consequences of the action that is being considered but also knowledge of what the alternatives and their consequences are. Such information is required out of respect for the autonomy of human beings. It is also a way of making us aware of the preferences of those people who are affected by a decision. In addition, good procedures for getting informed consent are crucial for the trust people have in science and the applications of science.

The question of consent is a complicated one, not only because of the problem of information, but also because life would be unbearably cumbersome if we always should get informed consent every time we carry out an activity that exposes others to risk. Thus, for example, in industrialised societies car accidents is one of the major causes of death among otherwise healthy people. Therefore, when one is driving a car means exposing others to considerable risk. Yet, before one starts driving one's car, how should one go about getting informed consent from everybody who is exposed to this risk?

We have here a typical case of something that often happens in a society and especially in modern, industrialised societies: There are activities that are carried out by everybody, or almost everybody, which bring many benefits to the individuals and to society, and where getting everybody's consent would be unfeasible, or at least a waste of resources. In such cases one often leaves it to political bodies to work out an arrangement and gives implicit consent by electing representatives whom are then entrusted to make such decisions on one's own behalf. It is then up to the representatives and the political bodies to which they belong to go through the decision procedure that we have outlined above: survey the different alternatives, estimate probabilities and take a decision in view of the values of the people they represent. In the case of car driving, they may introduce, for example, a system of driver education, driver tests, checks for car safety, speed limits, and speed controls, etc.

Decisions concerning the acceptability of radiation levels and levels of exposure to toxic substances are usually left to political bodies. These bodies have a responsibility for going through the decision procedure that was outlined above and find the best alternative. Often such an alternative will be a system of general principles and safeguards which include rules, that require persons, firms or agencies, who want to initiate processes that expose people to risk, to get their individual informed consent before any such process may start.

Note, by the way, that getting informed consent does not free the decision-maker from responsibility. If something goes wrong, the duty to compensate the victims and care for them is still there.[4]

2.2.13
Distribution

The distribution of burdens and benefits is highly important for the acceptability of risks, and it is one of the factors that are most often neglected. When those who are exposed to the risk are the same as those who benefit from the risky activity, levels of risk may be acceptable that would be ethically illegitimate if those who are exposed to the risk are others than those who benefit (see also section 10.4.3). Exposing somebody to a risk in order that somebody else shall benefit is almost never acceptable. Yet, this is what often happens. Those who are exposed to pollution are, for example, usually people who live in dilapidated areas and are already badly off, while those who benefit from the polluting activity often can afford to live far from the pollution and are among the better off already before the questionable activity starts.[5] A good question to ask before starting a risky project is "How do the worst off in society fare from this project?"[6] If they are affected from it in a negative way, the project will be ethically highly problematic, although from purely utilitarian considerations, without regard for distribution, it might seem to be highly desirable. Particularly if those who benefit from the project are relatively few and already well off, will it be very hard to justify such a project.

One sometimes finds the view expressed that whatever measure one introduces, it should be acceptable to all stakeholders. Thus, for example, as is mentioned in

---

[4] Informed consent is a major topic, especially in medical ethics. From the wealth of literature, at least two fairly recent books should be mentioned: Beauchamp and Childress (1997) and Berg et al. (2001).

[5] Not everybody agrees with this. Thus, for example, Herman B. Leonard and Richard J. Zeckhauser of the John F. Kennedy School of Government at Harvard claim that this is not so, but they give no evidence to support their claim, they only set it forth as their opinion. The main reason they see for including distributional considerations in project analysis is that "it may improve public legitimacy, which is ultimately a crucial component of successful application of this approach", Leonard and Zeckhauser (1983, 12).

[6] This question is inspired by Rawls' "second principle of justice", Rawls (1971, 65):
The intuitive idea is that the social order is not to establish and secure the more attractive prospects of those better off unless doing so is to the advantage of the less fortunate.
A good discussion of this issue may be found in Scanlan (1982, 1998).

section 10.4.2, the European Commission communication on the precautionary principle (2000) postulates that its application

> ... must include a cost/benefit assessment advantages/disadvantages with an eye to reducing the risks to a level that is acceptable to all the stakeholders. ... If the elimination of a risk ... involves ... *shifting the risk to another population group*, the decision-making agency, according to the Commission paper, must consider to take no measures at all (our italics).

The view that social measures are acceptable only if they make nobody worse off is sometimes called Pareto optimality.[7] A Pareto optimum is an allocation of resources such that no reallocation can make one consumer better off without making at least one consumer worse off. From an ethical point of view this is, however, a very dubious criterion for acceptability. Many societies have a very unjust distribution of risks and benefits. Thus, for example, there are often population groups that are exposed to lots of pollution and risks because of activities that benefit others who are very well off. Any transfer of resources from those who are best off to those who are worst off would violate Pareto optimality.

One final point: one important group of people, who are often affected by a project, is the future generations. One should always ask how they will be affected by the present decision. Decision makers have to make sure that they are not made worse off for the benefit of some who are now well off.

2.2.14
Compensation

Many projects that can be highly beneficial to a society, such as building of roads, airports, power plants, dams, etc., do impose new burdens on people who are badly off, for example indigenous populations who live in the area that will be affected. In such cases, compensation may be a way to go. That means to compensate those whose situation is worsened by the project. In order to be ethically acceptable, the compensation must not just be enough to keep them quiet, but must take into account their values, for example their emotional ties to what they have to give up, the effect it will have on their culture, etc. In addition, it must give them a fair share in the benefits of the project, compared to what the ones who benefit the most from it stand to gain.

In many cases, flaws in the distribution of costs and benefits can be compensated for. One question here is whether compensation should be given to everybody exposed to the risk or only to those who actually suffer if something goes wrong. Various ethical considerations support the view that one should be compensated for the unease one feels when one is exposed to a risk, and for the fact that many of one's plans have to be tentative when one is exposed to risk.

If one actually becomes a victim, it is important that the compensation does not require to prove that the accident was the cause of one's suffering, but rather that the burden of proof lies on those responsible for the risk: in cases of damage, compensation must be paid unless the agent responsible for the risky activity can show that activity did not contribute to the damage. In cases where the risk is due to several

---

[7] Introduced by the Italian economist Vilfredo Pareto (*1848–†1923) in Pareto (1896–97).

different activities and perhaps also partly to background factors, it seems reasonable that the ones who are responsible for the risky activities pay a compensation that is proportional to their share of the total risk (see section 8.4.3). This is particularly important concerning low-dose risks, where cause-effect relationships are hard to ascertain and where victims in most cases will not have the resources to prove that the damage was caused by the risky activity.

A further consideration in connection with legislation concerning risky activities is that there have been cases where firms have taken a calculated risk, which involved their going bankrupt if things went wrong, but where they could reap good profits as long as no accident happened. By limiting their maximal loss they could thereby take chances where the losses would have to be carried by those who were exposed to the risk.

2.2.15
Background exposure

One factor that is often brought into discussions of acceptability is the background exposure to radiation or to toxic substances that we are already experiencing. The extra risks created by a human activity are often compared to this naturally occurring risk. If the man-made risk is negligible compared to the naturally occurring risk, it is sometimes argued that the man-made risk is therefore acceptable. This is, however, a *non sequitur*. What matters ethically are the consequences of those risks that arise through human action. The naturally occurring risks may be bad for us, and we may have an obligation to work to reduce them. What is ethically relevant is always what we can do, what alternatives are open to us, and an evaluation of these alternatives along the lines we have sketched above.

In section 10.4.1, a similar problem is discussed with regard to German law and court practice:

> German courts tend to take the view that the additional risk is tolerable when it increases the overall risk to which the individual in question is already exposed only insignificantly. Such a *de minimis* approach appears practicable. However, it is subject to doubts if the extent of the pre-existing risk is not considered. This has been recognised by a recent Administrative Court of Appeal case which in a dictum questions whether in view of significant pre-existing cancer risk a relatively small additional risk could be negligible.

2.2.16
Value of actions

As mentioned at the end of section 2.2.4, the positive or negative values of our actions must be taken into account when we evaluate and compare alternatives. Of course, actions are often assigned a value in virtue of the values of their consequences, they have an extrinsic value, which we have discussed in the previous sections. We shall here concentrate on their intrinsic value, the value they have *per se*, regardless of consequences.

A whole spectrum of such values has been discussed by ethicists through the centuries. Two main groups of values are (1) those that spring from the dignity of human beings and (2) those that depend on the mind-set or motivation we have in doing the action. Kant has been one of the main contributors to both topics, but he has been preceded and followed by numerous other philosophers.

The view that every person has a human dignity that should not be violated runs deep in moral philosophy and is reflected in moral thinking at all times. Democritus (*~460–†~370 B.C.) claimed that respect for our fellowmen is a moral duty. Several of the stoics, especially the Roman stoics (first century A.D.) had similar views, and in the Middle Ages and in the Renaissance there are interesting discussions of what we now would call human rights. However, the central development is due to Kant (*1724–†1804), who formulated one version of his famous categorical imperative as follows (Kant 1785): "Act in such a way that you always treat humanity, whether in your own person or in the person of any other, never simply as a means, but always at the same time as an end".

As a consequence of this, we should not take anybody's life, regardless of what benefits might ensue from it. We should not humiliate anybody; we should never lie. Everybody would agree that it is wrong to take lives and to lie, at least in most cases. While Kant justified these precepts by pointing to human dignity, other philosophers have tried to justify them from other approaches to ethics. Thus, for example, so-called rule-utilitarians have pointed out that in the end society will benefit by people not lying. Even though there might be some situations where lying seems to have better consequences than not lying, we should abstain from lying because lying will weaken people's trust in one another. When we consider the larger picture, then the more remote negative consequences of lying would therefore outweigh the short-range positive consequences. Rule-utilitarians regard lying as unethical because it tends to undermine a highly valuable institution in our society. Lying should therefore be a last way out, to be used only in situations where telling the truth would have large negative consequences.

Kant's uncompromising rejection of lying creates problems in situations where other important values are at stake. Thus, for example, if during the Second World War someone were hiding some Jews who were pursued by the Gestapo, it seems that one is under no obligation to tell the truth. Kant might accept that one is silent. However, in such an extreme case, lying might be the right thing to do.

On the other hand, Kant's insistence that human beings shall never be used as mere ends accords well with the feeling that it is inappropriate to assign a value to human life. Many decisions involve an implicit comparison of the value of a human life to other values, for example decisions about allocations of money to hospitals and to traffic safety. Such decisions are very difficult, and usually one does not like to make it explicit that one is thereby assigning implicitly a value to a human life. It is sometimes maintained that it is immoral to assign a value to human lives, to health, etc. (see e. g. Streffer et al. 2000, 37). However, do we not implicitly do so when we make such decisions? Are we just behaving like the ostrich and closing our eyes to what is going on? On the other hand, can one give philosophical arguments for there being an ethical difference between explicitly and implicitly assigning a value to life? This question is closely connected with other issues of so-called "sacred values" which create difficulties for cost-benefit theory (see Føllesdal 1986, 45 ff. for more on this and for references to further discussion).

Kant had also much to say on the second topic, the relevance of our mind-set, or our motivation, for the evaluation of our actions. According to Kant, the maxim upon which we act is decisive for the ethical value of the act. Two actions can lead to the same consequences. However, one may spring from our sense of duty, while the other springs from our desire for fun or other self-interest. In such a case, according to Kant, the former has the greater value, and – on some interpretations of Kant – the second has no value at all.

From the point of view of setting standards for acceptable risk the considerations in group (1), concerning human dignity, are of great importance, and are already reflected in some of the other values listed above, for example the requirement of informed consent. The considerations in group (2) may be central in judging the actions of each individual, but they are probably less significant for the task of developing good legislation.

2.2.17
Comparison of values. Ultimate justification

Decision analysts often treat money as the only measure of value. Money, like probability, can straightforwardly be expressed in numbers, and one can easily carry out calculations and compare alternatives. When it comes to values that cannot easily be measured in money, many decisionmakers do not know how to deal with them, and gradually they get into the habit of not taking them into account.

How then can one bring values of other kinds into the decisions? In some cases, like the hydroelectric power example in section 2.2.2, one can arrive at a decision without having to weigh values of different kinds against one another. In most cases, however, one has to carry out such weighting, and then one ultimately has to base one's decisions on one's intuitions. When one is faced with a decision and compares alternatives, one often intuitively feels that one alternative is better than the others. Why then not just follow one's intuitions? Why does one need decision analysis?

Mainly for two reasons, one interpersonal, the other personal: First, people do not always agree in their intuitions. Whom should one then listen to? Should one immediately take a vote? Alternatively, should one sit down and discuss the matter? If one discusses the matter, what is there then to discuss? There can be one thing or another. However, what one *ought* to discuss, are the factors that matter to making the right decision. The model is intended to bring to our attention all the factors that enter into our decision theoretic diagram: the alternatives, the probabilities and the values. If there should be more factors, the model would have to be revised, so as to take account of them. As was pointed out in section 2.2, the diagram can be of help in practical cases of disagreement, by enabling us better to locate the sources of the disagreement. Is there a disagreement about what the alternatives are, about the probabilities of the various consequences given the various alternatives, or is there a disagreement about norms and values, and if so which ones?

In addition, on the individual level, the diagram may help us. Our intuitions as to what is right and what is wrong are often determined by just a few features of the situation and may change when we learn more about the consequences, who is affected and how they experience what happens. In addition, more thinking about possible alternatives may influence our intuitions about what to do. Both on the interpersonal and on the individual level it is important not to overlook aspects of the situation that are ethically relevant. Systematic exploration of the various factors is therefore crucial for reaching the right decision.

Our reflection has to take into account not just the present decision, but also our previous decisions. We may discover that we have been quite inconsistent in our assignment of value to an event from one occasion to the next. Thus, for example, the negative value we attach to getting cancer may seemingly vary a lot from occasion to occasion. On some occasions we are willing to forego smoking, on others not. Is our inconsistency just due to weakness of the will, are the situations different, so that other values come into play, or are we actually changing the negative value we assign to cancer, and if so, why? Inconsistencies tend to be experienced as disturbing[8].

By analysing different situations and reflecting on the positive or negative values we assign to different outcomes, we may adjust these values, trying to reach consistency. This reflective process has two aspects: it emphasises consistency, and it emphasises comprehensiveness: we want our values to be consistent over the whole range of situations we might possibly encounter. The process should not only be intra-subjective; values are supposed to be inter-subjective, applying equally to all, not just to a single person. We therefore try to arrive at consistency not just between the values we ourselves assign to various outcomes, but also between the values different persons assign to these outcomes. Such comprehensive consistency makes us more confident that we have reached a satisfactory ranking of the various values.

This process, where we go back and forth between our reactions to particular situations and the general valuations and normative principles we arrive at through the analysis of the various situations, is called the method of "reflective equilibrium". It is very similar to the procedure used in natural science where we go back and forth between empirical observations and tentative general hypotheses. This approach to justification appears in glimpses in Aristotle and other classical philosophers. It was formulated especially clearly in Goodman (1955, 65–68), who applies it to the justification of deductive logic. It is applied to ethics in Rawls (1971), especially in sections 4, 9 and 87. The label "reflective equilibrium" is due to Rawls.

Justification of values and weighting of values is therefore a complicated and cumbersome process. Decision analysis is just one among many tools one needs for this process. One also has to know which factors are ethically relevant in each situation. Some of these, such as consent, distribution, etc., have been briefly discussed above. Careful analysis of concrete examples often reveals more factors and distinctions that have to be taken into account, such as, perhaps, the difference between actively bringing something about and just allowing it to happen and many other features of various situations.

To make recommendations about policy without knowing the systematic work that has been going on in ethics for years is from a methodological point on a par with making recommendations about cancer treatment without knowing the systematic work that researchers have been carrying out in that area.

---

[8]   Gethmann (1991) introduces the label "Principle of pragmatic consistency" for the principle that one should avoid this kind of inconsistencies. See also Føllesdal et al. (1986) esp. § 59: Entscheidungen unter Unsicherheit und bei schlecht meßbaren Werten.

2.3
Implementation: ethics, law and economics

In the previous sections, we have repeatedly talked about feasible alternatives. In section 2.2.1, we also mentioned how decision time counts among the costs of the various alternatives. Generally, when we are comparing alternatives in order to decide what to do, we must, as mentioned in section 2.2.2, ask how realistic the various alternatives are. Not only must we ask which alternatives are technically feasible and which alternatives give the best use of the resources, we also have to ask about the social and legal chances of having them implemented. Some alternatives depend only on our own private decisions, but others require the consent of others (section 2.2.7) and often their collaboration and economic support. Sometimes everybody can be made to see that a given action will benefit everybody. Other times, if one leaves everybody to their individual choices one may see that the result will be suboptimal: there will be an alternative that is better for everybody than the one each tends to choose. The situation called "prisoners' dilemma" is an example of such a situation.

In order to avoid such outcomes, there are several different ways to go: One can change the expected value of the different alternatives by introducing taxes or incentives in order to get everybody to choose the better alternative. Alternatively, one can use the legal system and introduce laws and regulations that forbid certain alternatives. One may also try to instill in the decision makers concern for others, so that they think through the consequences of the various alternatives and take into account how these consequences are experienced by the various people affected. In a society with well-established norms and strong feelings of community and solidarity such a solution may work and will often be the best one. Normally, however, one has to try to frame laws, regulations and economic incentives that bring us as near to the ideal solution as possible. Such solutions always have their costs, in money and in time, and these costs have to be taken into account when we compare the various alternatives. Complicated solutions also tend to be less transparent and this may lead to distrust in the legal system (see section 2.5). The existing laws tell us what is legal at present. However, they reflect the time in which they were made and the attitudes and sensitivities of the people who made them, and often they need to be supplemented or modified in order to reflect the ethical concerns of everybody involved.

In section 2.2.2 it was emphasised how important it is to find good alternatives. It is a great challenge for politicians and other framers of laws and regulations to be creative and resourceful in developing good instruments for implementing our intuitions about what a good society should be like. A good example of this kind of work is Sandmo (2000), who discusses and develops various kinds of taxation and other economic incentives to protect the environment.

2.4
Summing up

2.4.1
Questions to ask

All of the above are important values that should be taken into account when we deliberate about the rightness or wrongness of our projects. We will end with some questions that one should always ask in such deliberations:

- What are the *alternatives?*
- *Who* is affected? Do we know *who* they are, or are they *anonymous, statistical* victims?
- How are they *experiencing* what happens? Can they *control* what happens to them? Have they given *informed consent?*
- How are the benefits and burdens *distributed?* How are the *worst-off* affected? How are the *silent victims* heard? Are *future generations* taken into account?
- Are there *better alternatives?*
- *Back to the first question again*[9].

2.5
Conclusion

Acceptability depends on all the factors that have been reviewed in the previous sections. Acceptability is a normative notion, and it is never a matter of probabilities alone. What is acceptable in one situation, where only those who benefit are exposed to the consequences, and where there is free choice and informed consent, may not be acceptable in another situation, where, for example, those who are exposed to the risk are not receiving their fair share of the benefits.

These ethical considerations have consequences for law and legal systems. What risks are legally acceptable shall have to vary from case to case, depending on the various circumstances that have been mentioned above (see also chapter 10). One will therefore not normally end up simply with dose limits. Activities that impose risks on others will have to be justified by showing that those who are exposed to risk themselves benefit from the project, directly or indirectly. Their benefit must be fair, especially compared with the benefits to those who are not exposed to risk. Further, the other ethical considerations, such as consent, etc. must be satisfied where applicable. What risks are acceptable will depend strongly on these other ethical factors[10].

One should make sure that when these decisions are made, the bodies making them should contain representatives who can make sure that full attention is paid to the following features of the situation:

---

[9]   Remember the dynamic nature of rational decision making, which was mentioned in Section 2.2.2 above.

Scientists, who should be familiar with:

- the feasible alternatives,
- the probabilities of the various consequences. Where there is disagreement between competent and well-informed scientists, all main views should be represented.

Ethicist(s), who should be familiar with:

- the latest research on what different factors are ethically relevant in connection with the issue at hand, and
- how they should be weighted against one another.

Legal specialists, who have knowledge of:

- the likely social and economic impact of different kinds of laws and regulation, and
- are familiar with legislation in this area.

The various individuals and groups that are *affected*, in order that one can get a first-hand understanding of:

- how they experience the risk they may be exposed to.
- One should see to it that "silent victims" are represented, such as under-privileged and handicapped, who often are forgotten when policies are made.
- It is also highly important that *future generations*, that cannot themselves be present, be adequately represented. See also sections 10.5.2 and 10.5.3.

One might also often need other experts, for example *economists* and *ecologists*, who:

- could help clarify the often complicated economic and ecological effects of the various alternatives, and
- who could also suggest and evaluate various economic incentives and compensation schemes that might be considered.

It is highly important that both the composition of these bodies and the procedures they follow in their deliberations and decisions are "transparent": it must be as clear as possible how they reach their decision, so that the general public can be confident that their interests are well taken care of.[11] Such transparency is particu-

---

[10] The fact that legislation so often opts for a simple dose limit, without concern for how the risks and benefits are distributed and the various other ethical factors that have been mentioned, may be due to unawareness of the importance of these other factors, or lack of competence to deal with them. However, lack of competence, dangers of corruption, etc., may also be a reason for preferring a straight dose limit rather than a decision that requires judgment. Shrader-Frechette and Persson (2001, 15), refer to Lochard and Grenery-Boehler (1993, 16):

> As French experts note, the vagueness in the optimization principle is one reason that most countries prefer to base their measures of radiation control on compliance with dose limits rather than on some sort of optimization of protection that takes social and economic factors into account.

[11] The procedures must be organised in such a way "daß ihre Rationalität von den betroffenen Bürgern nachvollzogen werden kann". (Gethmann and Mittelstraß 1992, p 24). See also section 10.5.2.

larly crucial in the approach to ethics called "contractualism", which bases ethics on one's "ability to justify one's actions to others on grounds they could not reasonably reject". As John Stuart Mill (1861a) put it in *On representative government* (chapter 5), a representative assembly should be such that "those whose opinion is overruled, feel satisfied that it is heard, and set aside not by a mere act of will, but for what are thought superior reasons"[12].

2.6
References

Beauchamp TL, Childress JF (1997) Principles of biomedical ethics, 3rd ed. Oxford University Press, Oxford

Berg JB, Appelbaum PC, Parker LS, Lidz CW (eds) (2001) Informed consent: legal theory and clinical practice, 2nd ed. Oxford University Press, Oxford

European Commission (2000) Communication from the Commission on the precautionary principle. COM (2000) 1. Brussels

Føllesdal D (1979) Some ethical aspects of recombinant DNA research. Social Science Information 18: 401–19

Føllesdal D (1986) Risk: philosophical and ethical aspects. In: Brøgger A, Oftedal P (eds) Risk and reason: risk assessment in relation to environmental mutagens and carcinogens. Progr Clin Biol Res 208: 41–52

Føllesdal D, Walløe L, Elster J (1986) Rationale Argumentation. Ein Grundkurs in Argumentations- und Wissenschaftstheorie. De Gruyter, Berlin, New York

Gethmann CF (1991) Ethische Aspekte des Handelns unter Risiko. In: Lutz-Bachmann M (ed) Freiheit und Verantwortung. Berlin, pp 152–169

Gethmann CF (1994) Handeln unter Risiko: Probleme der Verteilungsgerechtigkeit. Essener Unikate 4/5: 20–28

Gethmann CF, Mittelstraß J (1992) Maße für die Umwelt. Gaia 1(1): 16–25

Goodman N (1955) Fact, fiction, and forecast. Harvard University Press, Cambridge, MA

Kant I (1785) Grundlegung zur Metaphysik der Sitten. 2nd edition, Riga 1787. Several English translations

Leonard HB, Zeckhauser RJ (1983) Cost-benefit analysis applied to risks: its philosophy and legitimacy. Working paper RC-6, Center for Philosophy and Public Policy, University of Maryland, Adelphi, MD

Lochard J, Grenery-Boehler MC (1993) Optimizing radiation protection: the ethical and legal bases. Nucl Law Bull 52: 9–27

Mill JS (1861a) Considerations on representative government. Edition 1958, with an introduction, by Shields CV. Bobbs-Merrill, Indianapolis, IN

Mill JS (1861b) Utilitarianism. Fraser's Magazine, London

Pareto V [1896–97] Cours d'économie politique professé à l'Université de Lausanne. Rouge/Pichon, Lausanne/Paris

Peto R, Gray R, Brantom P, Grasso P (1991) Effects on 4080 rats of chronic ingestion of N-nitrosodiethylamine or N-nitrosodimethylamine: a detailed dose-response study. Cancer Res 51: 6415–6451

Rawls J (1971) A theory of justice. Revised edition, 1999. Harvard University Press, Cambridge, MA

Sandmo A (2000) The public economics of the environment. Oxford University Press, Oxford

Scanlan T (1982) Contractualism and utilitarianism. In: Sen A, Williams B (eds) Utilitarianism and beyond, pp 103–128

Scanlan T (1998) What we owe to each other. Harvard University Press, Cambridge, MA

Sen A (1979) Utilitarianism and welfarism. J Philos 76: 463–489

Sen A (1982) Rights and agency. Philos Publ Aff 11: 3–39

Sen A (1983) Evaluator relativity and consequential evaluation. Philos Publ Aff 12: 113–132

---

12 A very carefully worked out version of contractualism is Scanlan (1998).

Sen A, Williams B (eds) (1982) Utilitarianism and beyond. Cambridge University Press, Cambridge

Shrader-Frechette K (1981) Environmental ethics. Boxwood, Pacific Grove, CA

Shrader-Frechette K, Persson L (2001) Ethical problems in radiation protection. SSI rapport 11, Swedish Radiation Protection Institute, Stockholm

Shrader-Frechette K, Persson L (2002) Ethical, logical and scientific problems with the new ICRP proposals. J Radiol. Prot 22: 149–161

Streffer C, Bücker J, Cansier A, Cansier D, Gethmann CF, Guderian R, Hanekamp G, Henschler D, Pöch G, Rehbinder E, Renn O, Slesina M, Wuttke K (2000) Umweltstandards: kombinierte Expositionen und ihre Auswirkungen auf die Umwelt. Wissenschaftsethik und Technikfolgenbeurteilung, Bd. 5. Springer, Berlin

Williams GM, Gebhardt R, Sirma H, Stenback F (1993) Non-linearity of neoplastic conversion induced in rat liver by low exposures to diethylnitrosamine. Carcinogenesis 14: 2149–2156

Williams GM, Iatropoulos MJ, Jeffrey AM (2000) Mechanistic basis for nonlinearities and thresholds in rat liver carcinogenesis by the DNA-reactive carcinogens 2-acetylaminofluorene and diethylnitrosamine. Toxicol Pathol 28: 388–395

Williams GM, Iatropoulos MJ, Jeffrey AM, Luo FQ, Wang CX, Pittman B (1999) Diethylnitrosamine exposure-responses for DNA ethylation, hepatocellular proliferation, and initiation of carcinogenesis in rat liver display non-linearities and thresholds. Arch Toxicol 73: 394–402

Williams GM, Iatropoulos MJ, Wang CX, Ali N, Rivenson A, Peterson LA, Schultz C, Gebhardt R (1996) Diethylnitrosamine exposure-responses for DNA damage, centrilobular cytotoxicity, cell proliferation and carcinogenesis in rat liver exhibit some non-linearities. Carcinogenesis 17: 2253–2258

3 Effects of Ionising Radiation in the Low-Dose Range – Radiobiological Basis

3.1
Introduction

Biological risks of ionising radiation were identified only a few years after the discovery of X-rays by Roentgen (1895) and of radioactivity by Becquerel (1896). At the end of the 19th and beginning of the 20th century radiation-induced skin cancers have been described (Trott and Streffer 1991). Therefore it was very soon necessary to determine standards and dose limits for ionising radiation for persons at working places and later, also for individuals of the general population. The first standards were defined in the twenties and developed during the 20th century based on that knowledge about the action of ionising radiation, which was available at the corresponding times. For risk evaluations, the shape of the dose-effect curves is of eminent significance. In radiation research as well as in toxicology of chemicals in general two principal categories of dose-effect relationships have been described (ICRP 1977). The shape of these curves is based on manifold experimental studies of radiation effects which have been investigated with molecular structures, cells and animals after radiation exposures and obtained from clinical experiences as well as from epidemiological studies which have been observed after the exposure to ionising radiation in man (UNSCEAR 1977, 1988, 1993, 1994, 2000, BEIR 1990). The principal and fundamental difference of these dose-effect relationships is very important in the low-dose range. The most significant feature is whether a threshold dose does exist – these effects are defined as deterministic or non-stochastic effects – or whether no threshold is assumed – these effects are defined as stochastic effects (Fig. 3.1) (ICRP 1977). The so-called linear-no-threshold (LNT) concept has been proposed and used for risk evaluation in the low-dose range (ICRP 1991).

In the first case radiation effects are only observed when the radiation doses are higher than the corresponding thresholds. Below the threshold no biological radiation effects of that kind occur. Then the number of the effects rise rapidly with increasing radiation doses above the threshold dose but also the severity of the radiation effects becomes stronger with increasing dose. These radiation effects are called deterministic or non-stochastic radiation effects (Fig. 3.1) (ICRP 1977, 1991; Streffer 1991). Especially the acute radiation effects, which are mainly caused by cell killing in the corresponding organs and tissues, fall into this category. Acute radiation effects are therefore preferentially observed in such organ systems in which a high cell renewal takes place. The renewal of blood cells in the bone marrow, the cell renewal in epithelia of the skin and the intestines, for example, are inhibited or reduced by radiation doses higher than 1 Gy. The acute radiation effects occur within days or a few weeks after the exposure (Scherer et al. 1991). But there

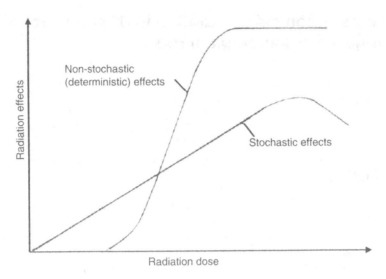

Fig. 3.1 Dose effect curves for deterministic and stochastic effects after exposure to ionising radiation (principal, schematic shapes)

are also some other somatic, deterministic effects which become manifest only after months or years, cf. the induction of cataracts in the eye lens, fibrotic changes in almost all tissues, damage of the blood vessels, necrotic processes in the skin and other epithelia fall also into the class of deterministic effects (Scherer et al. 1991).

For the development of most deterministic radiation effects, cell death and loss of cellular functions are the most significant processes. In all cases many cells have to be damaged or changed in order to cause these effects (Streffer 1991). Based on the knowledge of the mechanisms, which are involved in the development of these radiation damages, the dose ranges for the threshold can be described. They are found for almost all effects above acute radiation doses of 0.5 Gy low-LET radiation. After chronic or fractionated exposures to low-LET radiation, these threshold doses are higher (ICRP 1991; Scherer et al. 1991; UNSCEAR 1993; Hall 1994). The dose limits for workers in radiation facilities, as well as for individuals of the general population, should be fixed below the threshold doses so that these effects can be avoided.

For the second type of dose-effect relationship, it is assumed that a threshold dose does not exist, and the consequent effects are called stochastic effects. Into the category of stochastic effects fall genetic effects, the induction of cancer (ICRP 1991) as well as some developmental changes after prenatal radiation. For the development of these radiation effects, non-lethal changes of the genome in the corresponding cell nuclei are of significance (mutation, cell transformation). For the description of the mechanisms on which such dose-effect relations are based it is assumed that genetic changes in a single cell are sufficient in order to induce this type of damage (Mole 1992; Müller et al. 1994; Streffer 1997; UNSCEAR 2000). For risk evaluation, it is assumed that stochastic effects increase proportionally in the low and medium dose range with radiation dose. In the higher dose range the

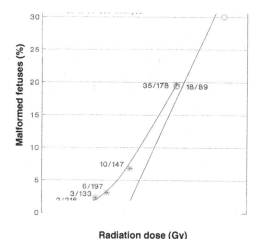

Fig. 3.2 Dose effect curves for the induction of a malformation (gastroschisis) in the HLG mouse strain after X-irradiation of zygotes (1-cell stage) or later preimplantation embryos (32- to 64-cell stage) (Müller et al. 1994)

frequency of radiation effects may decrease when cell killing occurs and dominates the dose-effect curve (ICRP 1991).

Studies on the induction of a specific malformation (gastroschisis) have shown that a dose-response without a threshold dose is only observed when the radiation effect develops from one damaged cell. In this case the mouse zygote (1-cell embryo) was irradiated 1 to 3 hours after conception and this resulted in a dose-response curve without a threshold while the irradiation of a multicellular embryo (32- to 64-cell embryos) resulted in a dose-response with a threshold (Fig. 3.2) (Müller et al. 1994).

Such a mechanism is obvious for genetic, carcinogenic and under certain conditions for developmental effects. If the genome is changed in one single germ cell and if this germ cell becomes fertilised or fertilises a female mature germ cells, organisms will develop which carry the genetic mutation. Similarly, it is assumed that cancers can develop from one damaged cell after a radiation exposure. Measurable effects have been observed, however, only after radiation doses of around 100 mSv. Only in some special cases can doses below 100 mSv induce such developmental effects.[1]

For the induction of cancer, this mechanism is not so obvious as for genetic mutations. With experimental studies on animals and epidemiological studies on irradiated humans, it has been clearly shown that ionising radiation can cause an

---

[1] Sievert (Sv) is the unit for the equivalent dose. It is obtained by multiplying the energy dose in Gray (Gy) with the radiation weighting w_0 (quality factor Q) of the corresponding radiation quality. As already mentioned low-LET radiations have a w_0 (quality factor) of 1, therefore the numerical values of the energy dose and of the equivalent dose are equal for these radiation qualities.

increase of cancer rates. The data with statistically significant effects of this kind are only found in the range of acute doses of 100 mSv and higher but not in a dose range below 100 mSv for a general population with all age groups (UNSCEAR 1988). After chronic exposures (exposures over a longer period usually weeks, months or even years) the dose ranges may be even higher before measurable effects occur. This has been observed for instance with radiation exposures to nuclear workers for solid cancers (Cardis et al. 1995; Muirhead et al. 1999) and populations with exposures to high background radiation (Wei et al. 1997). However, an increased radiosensitivity is seen in certain readiosensitive groups. Thus, the induction of thyroid cancers after exposure to iodine-131 (I^{131}) from the Chernobyl accident was higher in children compared to adults (Jacob et al. 1999; UNSCEAR 2000). On the other hand, the radiation doses to which humans are exposed in the environment from natural or man-made sources, or even at working places, are usually below 100 mSv. For the risk evaluation in these low-dose ranges it is therefore necessary to have the best possible knowledge about the shape of the dose-effect curve in order to perform extrapolations from the dose ranges with measurable effects into the lower dose ranges which are important for exposures to workers and individuals from the general population.

Table 3.1 gives an overview of radiation doses to which the population in industrialised countries like Germany is exposed. The highest exposures occur in medical treatment of cancers by ionising radiation where tumour cells have to be killed. But also in diagnostic applications of X-rays or of radioactive substances appreciable doses can be necessary in order to obtain the wanted information. Radiation doses

Tab. 3.1 Radiation exposure of humans in different areas of life.

1. Medical use of ionising radiation (dose per treatment)	
Therapy	several 10,000 mSv (mGy)
Diagnostics, local, regional	1–50 mSv

2. Working places (dose per year)	
Staff in control areas	average 4–5 mSv
Flying staff (North Atlantic)	8 mSv
Welders (electrodes with T_h)	6-20 mSv
Working places with high R_n–concentrations (e. g. water industry, Fichtelgebirge, Erzgebirge)	6-20 mSv

3. Environment (dose per year)	
Average of natural exposure in Germany	2.4 mSv
High regional natural exposure in Germany	8–10 mSv
High regional natural exposure in India	15–50 mSv
Nuclear facilities Germany (BMU)	<0.01 mSv
Exposures from the Chernobyl accident in Germany 1990 (BMU)	0.025 mSv

for single applications are given in Table 3.1. At working places with exposures to ionising radiation the average radiation doses per year are in the range of around 4 to 20 mSv. All individuals of the population are exposed to ionising radiation from natural sources by external exposures (cosmic γ-rays, γ-rays from the radioactive decay of nuclides in the soil and other materials) as well as by internal exposures (uptakes of radioactive substances with food and water as well as inhalation of radon and its radioactive decay products). These exposures sum up in average to around 2.4 mSv yr^{-1} in Germany. This average value is observed world-wide when large regions are considered. However, it can be much higher when smaller regions are considered. In Germany, it can be four times higher than the average value (Tab. 3.1). In some countries like India, regions with even much higher exposures have been found.

During recent years it has been intensively and passionately discussed whether the shape of the dose-effect relationship can be described with a linear dose-response and no threshold (LNT) or whether a threshold exists below a certain dose range in which no health effects can be measured directly. The LNT is proposed by most international bodies, which are responsible for radioprotection, like ICRP. There are numbers of biological processes, as DNA repair, adaptive response, immune response and apoptosis which may modify the primary molecular and cellular radiation damage in such a way that the health effects of ionising radiation are lowered and an apparent threshold occurs (Streffer 1997; 2001). This means that the question of a linear dose-effect relation without a threshold remains open and is a matter of debate (Workshop Report: Cellular Responses to Low Doses of Ionising Radiation, United States Department of Energy and the National Institutes of Health, Feinendegen and Neumann 1999). One strong view has been brought forward by the largest scientific society for radiation research, the Health Physics Society of the United States, which has made the following statement (Health Physics Society 1995):

> The Health Physics Society recommends against quantitative estimation of health risk below an individual dose of 5 rem (50 mSv) in one year or a life time dose of 10 rem (100 mSv) in addition to background radiation. Risk estimation in this dose range should be strictly qualitative accentuating a range of hypothetical health outcomes with an emphasis on the likely possibility of zero adverse health effects.

There have been a number of arguments presented which have been raised against the linearity of a dose-effect relationship without a threshold for radiation induced carcinogenesis. On the other hand, there are also a number of arguments in favour of a dose-response curve without a threshold (Streffer 1997). These points will be discussed in the following sections.

The uncertainties for the radiation effects in the low-dose range are high especially for carcinogenesis. In order to improve these extrapolations, knowledge about the development of stochastic radiation effects and their mechanisms is very important and must be improved. One of the most crucial questions is whether a radiation-induced cancer develops from a single cell damaged by ionising radiation. For general practices in radioprotection, a linear dose-effect curve without a threshold is assumed for the stochastic effects. Under these assumptions, there is no radiation dose without any risk. However, the general principle is valid that the rates of radiation effects become less with decreasing radiation doses. This has been frequently neglected in the public debate about radiation risks. A number of biological phenomena may modify the dose-response curve in the low-dose range, and it is necessary to evaluate these phenomena.

All living cells have very efficient systems in order to repair radiation damage in the DNA (UNSCEAR 2000). With radiation of low-LET (X-rays, γ- and β-rays) the ionising events in the exposed material have a comparatively low density, with relatively large distances between these events. Besides these radiation qualities with low ionising density (low-LET) other radiation with a high ionising density or a high-LET exist. Neutrons and α-rays which are released after the decay of some radioactive substances (cf. of uranium and plutonium), and also heavy ions (heavy atoms which have lost all their electrons) fall into these classes. The heavy ions are found especially in very high atmospheric altitudes and are significant for space flights, but they have also been used for cancer therapy in recent years. It has been found that radiation damage of the DNA after exposures to these high-LET radiation qualities is less repairable as the damage occurs in more complex clusters (see below). On the other hand, the ionising radiation of high-LET have a much shorter range of migration through the material than ionising radiation of low-LET (UNSCEAR 2000).

If a dose-effect curve without a threshold dose for carcinogenesis exists, cancers could develop from one damaged cell, probably with an initial mutation, which can lead to malignant cell transformation. The manifestation of a clinically diagnosed cancer is the endpoint of a sequence of mutations connected with changes of the regulation of cell proliferation and its stimulation. Under the consideration that cancer is induced through several mutation steps (Fig. 3.3; Fearon and Vogelstein 1990) it is not clear to which extent intermediate cells in the process of malignant cell

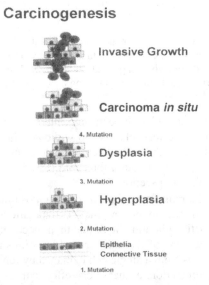

Carcinogenesis

Invasive Growth

Carcinoma *in situ*

4. Mutation

Dysplasia

3. Mutation

Hyperplasia

2. Mutation

Epithelia
Connective Tissue

1. Mutation

Fig. 3.3 Multistep mechanism of carcinogenesis with several mutation steps after exposure to ionising radiation

transformation appear, at which time the first malignant cells are really formed and clonal growth of malignant cells starts for the development of a cancer. The question is unsolved whether these processes start already during the first steps of malignant cellular transformation or whether these processes develop only in a pre-neoplastic lesion several years after radiation exposures.

While cell transformation can be investigated *in vitro*, the later processes of carcinogenesis can only be studied indirectly in animals and humans. Investigations of cancer incidence and mechanisms suggest that four to six independent steps are necessary for most cancers. For the development of leukaemias the number of mutation steps may be smaller. In many studies during recent years it has been shown that genomic instability develops many cell generations after radiation exposures (Pampfer and Streffer 1989; Little 2000) and these phenomena may accelerate the development of cancer, as the mutation rate in tissues with these genetically instable cells is increased.

An apparent threshold dose can be caused by the long latency period between a radiation exposure and the manifestation of cancer (Raabe et al. 1995). This latency period may be longer than the lifetime of an exposed individual. These considerations show clearly that it is absolutely necessary to obtain a better knowledge about the mechanisms for the development of cancer after radiation in order to evaluate the measurements and extrapolation procedures for the low-dose ranges. For a better understanding of the mechanisms, it is necessary to describe the molecular and cellular effects of ionising radiation, which are connected to the development of cancer. Further experimental data with animals will be reported which are important for the evaluation of these mechanisms. The development of genetic mutations will be considered in the same way. In this connection, it is necessary not only to discuss the effects of low-LET radiation but also to analyse the effects of high-LET radiation and to include these investigations into the risk evaluation. Furthermore, it should be mentioned that large parts of the exposures from natural sources are caused by high-LET radiation (e. g. radon and its decay products).

During recent years, it has become increasingly apparent that individual radiosensitivity can be very different. A number of genetic syndromes have been recognised with increased individual radiosensitivity (ICRP 1998). These have been especially observed with patients who received radiotherapy and showed a remarkable increase of radiosensitivity. Cellular and molecular investigations with biological material from such patients make it possible to study radiation effects in smaller dose ranges than with persons of normal radiosensitivity. Therefore, it is important to include such studies with hypersensitive patients into the discussion. Considerable progress has been achieved during recent years by such experiments in order to describe radiation effects on a molecular and cellular level in a better way.

3.2
What is a low dose?

It has already been pointed out that health effects of ionising radiation cannot be observed in the dose ranges to which our populations are exposed in the environment or usually at working places. Dose distribution is dependent on the physical processes, which occur when the energy of ionising radiation is absorbed by inter-action with atoms or molecules of the living organism. The spatial and temporal distributions of these events and the following biological processes are important for the development of radiation effects. Therefore, physical and biological consid-erations must be discussed for the decision about the question "what is a low dose?".

3.2.1
Microdosimetric considerations

With the transfer of energy of ionising radiation within materials, energy absorption takes place, covalent bonds are broken and ions or radicals are formed. These reac-tions can directly occur in biologically essential molecules (direct radiation action) like DNA or with water molecules of which the number in living cells is largest. In the latter case radicals of the water (cf. H, OH) are formed which then can react with other molecules like DNA and lead to corresponding damaging events with chemical reactions (indirect radiation action). The transfer of energy occurs in cells and tissues in discrete energy packages (Fig. 3.4). For the development of stochas-tic radiation effects the changes in the DNA by both direct and indirect radiation actions are considered important.

The principal physical unit in order to describe the energy deposition in organs and tissues is the absorbed radiation dose given in Gy. This dose unit is defined as the average energy which is absorbed in a target tissue or organ divided by the mass of the irradiated tissue or organ (see footnote 1 in this chapter, p. 39). This average energy dose, however, does not describe the large variability of energy absorption in micro-regions which results from the stochastic nature of energy deposition

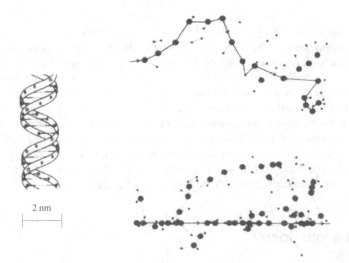

2 nm

Fig. 3.4 Schematic simulated tracks of low LET-radiation (upper panel) and high LET-radia-tion (lower panel), a section of DNA is shown on the left side in order to give a comparison for the dimensions (Goodhead et al. 1994)

events in individual cells and molecules especially when the energy dose is considered in the low-dose range.

In the medium to high-dose ranges of low-LET radiation (100 mSv and higher) a relatively homogeneous exposure of cells and tissues occurs with low-LET radiation. This changes in the low-dose range when the effects of single ionising particles have to be considered. On the cellular and subcellular level microdosimetric considerations have to be introduced under these conditions. Thus the absorbed dose in a single cell nucleus amounts in average to 1 mGy for cobalt-60 (^{60}Co) γ-radiation when in average one ionising particle of this radiation passes through the spherical cell nucleus with a diameter of 8 μm (ICRU 1993; UNSCEAR 2000). When a tissue with many cells (several hundred millions cells g[-1] of tissue) receives an averaged dose of 1 mGy, 63.2 % of the cells in the tissue will be hit, 36.8 % of the cells experience a track of at least one particle. This is equivalent to 58.2 % of the hit cells are hit by one particle and further cells will be hit by several particles (Tab. 3.2). From these considerations it follows that an average number of 18.4 % of the cells will receive two tracks. On the other hand, with an average radiation dose of 1 mGy to a tissue 36.8 % of the cells will not be irradiated at all. This percentage of unhit cells increases with decreasing dose (Tab. 3.2). Under the assumption that the sensitive target volume is to be seen in the cell nucleus, these processes of dose distribution are of high significance.

If the energy deposition in a single cell nucleus is sufficient for the induction of radiation damage and an interaction between damaged cell nuclei is not necessary for the cancer development, it is very probable that a dose-effect relationship without a threshold dose exists. However, the possibility must be considered that especially for low-LET radiation at least two independent particles have to pass the cell nucleus in order to develop radiation damage. The number of particle tracks, which pass through the cells, follows a Poisson distribution. The average number of tracks of ionising particles is proportional to the absolute dose. For an average statistically distributed tissue dose of 0.2 mGy low-LET radiation (cf. [60]Co γ-radiation) a spher-

Tab. 3.2 Proportions of a cell population traversed by tracks for various average doses[a] from γ-rays and α-particles (UNSCEAR 2000)

Mean tracks per cell	Percentage of cells in population suffering					
	0 track	1 track	2 tracks	3 tracks	4 tracks	>5 tracks
0.1	90.5	9	0.5	0.015	–	–
0.2	81.9	16.4	1.6	0.1	–	–
0.5	60.7	30.3	7.6	1.3	0.2	–
1	36.8	36.8	18.4	6.1	1.5	0.4
2.	13.5	27.1	27.1	18	9	5.3
5	0.7	3.4	8.4	14	17.5	56
10	0.005	0.05	0.2	0.8	1.0	97.1

[a] Approximately 1 mGy for γ-rays and 370 mGy for α-particles per track passing through a cell nucleus on average.

ical cell nucleus with a diameter of 8 μm would have on the average about 0.2 tracks of ionising particles. In this case only around 18 n 1 track. With an average tissue dose of 0.1 mGy about 9 % of the cells will be hit and around 0.5 % of the cells will have more than one track of ionising particles (Tab. 3.2). The microdosimetric arguments for a low dose should be evaluated with respect to linearity of the dose-effect relationship for such biological effects for which the radiation effects are induced only in those cells which have been passed by at least one ionising particle. This is apparently the case for cell killing, for the induction of chromosomal aberrations and for mutations in single cells. It is, however, unclear whether this phenomenon is also valid for the transformation of normal to malignant cells. It further has to be considered that unhit cells can show an altered gene expression when a cell was hit in the neighbourhood. Thus, an increased expression of the protein p21 was observed even in unhit cells (Little 2000). The mechanism of these phenomena is unexplained up to now. In addition, the development of other radiation effects have been observed in non-irradiated cells in the neighbourhood of irradiated cells (bystander effect) (Feinendegen and Neumann 1999).

The situation with respect to dose distribution is completely different for the exposure to densely ionising radiation with high-LET (Fig. 3.4). α-Particles have a very short range in tissue, that is dependent on the energy of the α-particles which are formed through radioactive decay of the corresponding radioactive isotopes. Thus for α-particles which are released after the radioactive decay of radium-226 (Ra^{226}) with its radioactive daughter products with energies of up to about 7.8 MeV a maximal range of around 80 μm is observed in mammalian tissues. For 5 MeV α-particles of plutonium-239 (^{239}Pu) the maximal range is around 40 μm in biological tissues. If one considers that the diameter of cell nuclei of human cells is in the range of 5 to 10 μm and the diameter of the cells in the range of 10 to 30 μm this demonstrates that α-radiation can reach on average around 1 to 2 and maximally up to 5 cell layers from their place of origin. The energy, which is deposited by one single α-particle passing through the cell nucleus, which is thought to be the radiosensitive target of the cell, is extremely variable. The energy dose can vary from very small doses (in the range of mGy) up to more than one Gy even between microregions of the same cell nucleus. These considerations demonstrate clearly that the definition of average tissue doses is an oversimplification for energy deposition of high-LET radiation especially. Individual cells in a tissue will experience very different radiation doses. It is therefore very important how the α-emitting radioactive isotopes are distributed within the tissue. Very frequently only the surfaces of an organ within a body will be reached by such radiation qualities especially when the radioactive substance is located in a neighbouring tissue or organ. This is valid for instance for radon and its radioactive decay products in the lung and for ^{239}Pu as well as ^{226}Ra, which are deposited in the skeleton.

Under these circumstances only less than 0.2 % of the cell nuclei are hit by an α-particle if the cells of a tissue receive in average a dose of 1 mGy of α-radiation, while more than 60 % of the cells are hit by ^{60}Co γ-radiation at the same average tissue dose. With such an average radiation dose of α–radiation, around 99.8 % of the cells experience no radiation event at all (Tab. 3.2). On the other hand, when an α-particle (with energy of around 5cleus a high-energy deposition will occur in the corresponding cell nucleus on average in the range of 370 mGy. In individual cell nuclei, the dose of such a radiation quality can reach values up to 1000 mGy (UNSCEAR 2000).

Thus in the low-dose ranges (average tissue doses of 1 mGy or smaller) ionisation events will occur only in part of the cells and the number of hit cells depends significantly on the radiation quality (radiation energy and type of radiation). This means that small doses can be defined based on these microdosimetric considerations and they are very heterogeneously distributed on the cellular and subcellular level.

By means of computer programs which are based on Monte-Carlo calculations it is possible today to calculate the exact position of ionisations and excitations in the track of ionising particles. If one traces the tracks of these particles in the tissue, one can observe that low-LET ionising particles, like electrons, meandrise in tissues through the processes of diffraction and they can migrate into very different·directions. In contrast to electrons, heavy charged particles (cf. helium, carbon, and argon nuclei) migrate on more straight tracks in well-defined modes through the tissue. They transfer their energy so that secondary electrons originate which then can deviate from the track of the primary particles. The ranges of these secondary electrons depend on the energy of the particles. Thus, around 99 % of the energy of 0.3 MeV protons are deposited within a range of 30 nm from the centre of the track. For a 20 MeV proton only around 2 % of the energy is deposited at more than 1 μm from the centre of the particle track. The structures of the tracks and the energy deposition in the environment of the tracks are of great significance for the development of biological effects. Clusters of energy deposition and of primary chemical changes occur which then develop into biological effects (Fig. 3.4; UNSCEAR 2000).

Each track of a radiation with low-LET like X-rays or γ-rays induces only a relatively small number of ionisations at the passage through a cell nucleus of medium size. – The ionising events are caused by secondary electrons, which are released by the interaction of the photons from X- or γ-rays with cellular molecules. – Thus, in average, around 70 ionisations occur through these electrons at a nuclear passage of a γ-quantum of ^{60}Co γ-radiation. This corresponds to an average observed energy dose of 1large variability of these processes has already been discussed. In contrast radiation exposures with high-LET, cf. 4 MeV α-radiation, leads to many thousand ionisations and therefore yields a relatively high dose in an individually hit cell nucleus. In case of such a radiation, around 25,900 ionisations occur when the particle passes through a cell nucleus. This corresponds to an observed dose of around 370 mGy (UNSCEAR 2000).

In a cell, the indirect effects, which are mainly caused by formation of water radicals and their reactions with biological macromolecules (e. g. DNA) also, occur in a range of small distances of several nm. The presence of radicals at a certain time is very limited in tissues due to the high chemical reactivity of the radicals. Although it is difficult to estimate the contribution of direct or indirect effects for the damage of DNA through low-LET radiation, studies with radical scavengers demonstrate that around 35 % of the primary DNA damage comes exclusively from direct effects and 65 % of the damage is caused by contributions through indirect effects. It is not completely clear whether the molecular nature of the damage in biological molecules, cf. in the DNA, differs from direct or indirect radiation damage (UNSCEAR 2000).

3.2.2
Biological considerations

Another possibility exists to describe the low-dose range based on biological effects. After the exposure to low-LET radiation (β-rays, γ- and X-rays) the extent of radiation effects can be described by dose-effect relations with a linear and a quadratic term of the dose cf. for chromosomal aberrations, somatic mutations and cell transformation. Further, it has to be considered that biological effects, which are observed after radiation exposures, like chromosomal aberrations, mutations or cancer, already occur without any radiation (spontaneous effects). For this reason, a constant term 'C' has to be included in possible equations. A dose-effect curve can then be written in the form:

$$E(D) = \alpha D + \beta D^2 + C. \qquad \text{(eq. 3.1)}$$

In this formula α and β are constant coefficients for the linear and quadratic term of the dose respectively. These coefficients vary for different endpoints and possibly also for various defined radiation conditions. Such dose-effect relationships have been studied after radiation exposures especially for chromosomal aberrations, mutations, and cell killing. Frequently α/β-ratios of around 200 mGy have been observed for ^{60}Co γ-radiation. This corresponds to a medium radiation dose after which the linear and the quadratic terms contribute to the radiation effects to about the same extent. From such a value calculation demonstrates that the action of radiation increases in a linear way in the low-dose range with radiation dose up to around 20 mGy, as the contribution of the quadratic term is low in this dose range (UNSCEAR 2000). Then the contribution of the quadratic term amounts to around 9 % of the whole radiation effect. Even after 40 mGy the contribution of the quadratic term is only around 17 % of the total radiation effect.

On this basis and convention, a radiation dose in the range of 20 to 40 mGy has been called a low dose (UNSCEAR 2000). In an earlier UNSCEAR report (1993) experimental data were analysed for the carcinogenesis after irradiation (especially of mice) with various dose rates of a low-LET radiation. It was concluded and proposed, based on these data, that a dose rate of around 0.06 mGy min^{-1} can be considered as a low dose rate when the exposure lasted for some days or even weeks. With such a dose rate, the induction of the tumour frequency was reduced in comparison to higher dose rates when equal total doses were compared. With smaller dose rates than 0.06 mGy min^{-1}, no further reduction of the tumour rate per dose unit was obtained. The UNSCEAR committee therefore concluded that a dose rate of 0.05 mGy min^{-1} could be considered as a low dose rate.

The evaluation of epidemiological data for carcinogenesis in humans led to the decision that a radiation dose of less than 100 mGy (mSv) of low-LET radiation was considered as a low dose by UNSCEAR (1993), as no radiation effect can be observed in this dose range for general populations (both sexes, all age groups). The international committee of the United Nations (UNSCEAR) defined doses below 200 mSv as a low dose UNSCEAR (1993). Under the assumption that also for humans a linear-quadratic dose-effect relationship may exist for carcinogenesis it can be concluded from the observed epidemiological data that within a dose range of 200 mGy the quadratic dose term is responsible for around 10 % of the effect (UNSCEAR 1993).

It has been proposed and recommended by ICRP to introduce a "dose and dose rate effectiveness factor" (DDREF) under these conditions (ICRP 1991). A DDREF of two was used for the risk factor for cancer induction by low-LET radiation. After exposures to high-LET radiation, dose-effect relations are generally observed where only the linear dose term of eq. 3.1 is relevant and the quadratic term becomes very small. Then equation 1 is modified to equation 2. The impact of dose rate is apparently not very important for high-LET radiation. Therefore, DDREF should not be used for risk factors of such radiation qualities (ICRP 1991).

$$E(D) = \alpha D + C. \qquad \qquad \text{(eq. 3.2)}$$

3.3
DNA damage and repair

Ionising radiation disrupts covalent chemical bonds in molecules like DNA. Thus, DNA strand breaks (breaks of the polynucleotide chain), as well as changes or loss of DNA bases can occur. DNA strand breaks can be formed in one DNA strand (single strand break, SSB) or in both DNA strands near to each other (double strand breaks, DSB). With low-LET radiation these events can be singular, isolated processes. Such isolated damaging events can generally be repaired very rapidly by various cellular enzymes (within minutes and few hours). This is especially the case for base damage and SSB. The repair of DSBs is more complex and takes a longer time and misrepair is more frequent for DSB than for SSB or base damage (Hall 1994; UNSCEAR 2000).

However, clusters of damaging events in the DNA can also occur which lead to more complex DNA damages. Thus, a second DSB, SSB or base damage can occur in the direct neighbourhood of a SSB or DSB in the DNA (complex SSB or complex DSB, Tab. 3.3). Such clusters are only slowly repaired or misrepaired. The repair of these clustered damages is more difficult than that of isolated sites of DNA damage. The complex radiation damaging events in the DNA occur to a much larger extent through an ionising particle with high-LET than through low-LET

Tab. 3.3 Complexity of DNA Strand Breaks after Exposure to Ionising Radiation (after D. Goodhead, MRC Harwell, UK)

Type of DNA damage	Electrons		a-Particles
	1 keV	100 keV	2MeV
Base Damage (%)	68.9	81.8	53.3
SSB (%)	25.2	16.9	23.1
Complex SSB (%)	3.250	0.71	8.70
DSB (%)	1.810	0.47	4.01
Complex DSB (%)	0.790	0.12	10.95
SSB: complex / total	0.114	0.040	0.274
DSB: complex / total	0.304	0.203	0.732
SSB / DSB	13	30	2

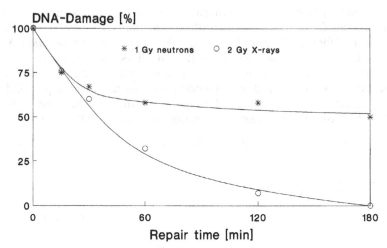

Fig. 3.5 Repair of DNA damage after irradiation of tumour cells with X-rays or fast cyclotron neutrons (~6 MeV). The DNA damage and its repair was measured 0 to 180 minutes p.r. by the comet essay (Müller et al. 2001).

radiation. Clusters are more frequent and the number of damaging events in these clusters is higher (Tab. 3.3). This is probably the reason that DNA damages which occur after exposures to high-LET radiation are repaired to a smaller extent than DNA damages caused by low-LET radiation and the repair takes longer. (Fig. 3.5; UNSCEAR 2000). The figure shows the DNA repair measured through the comet assay (Müller et al. 2001) in human cancer cells after exposure to X-rays or cyclotron neutrons (6 MeV). In principle, the same occurs in all living cells including normal cells.

These clusters of ionising radiation destroy the DNA structure largely around the damaged site. The interaction of ionising radiation with DNA leads to varying types of damage as described before. In many cases the products have been identified with respect to their chemical nature and classified. In Tab. 3.3 and 3.4, the most important categories of damage to DNA and chromosomes, which have been measured after exposures to low-LET radiation, are described and numbers are given. These quantitative data are rough estimates after an absorbed dose of 1 Gy. The data in Tab. 3.4 show that after a radiation dose of 1 Gy of low-LET radiation around 70,000 ionisations occur in the cell nucleus of which around 2,000 are directly occurring in the DNA. These primary physical events lead to the listed chemical/biochemical changes in the DNA. Mainly SSB and base damages are formed (Tab. 3.4). For the development of biological effects the DSB, however, are of greatest significance and are most relevant for the development of radiation effects. Studies show this with correlation of complex DSB, which remain unrepaired or which are misrepaired, and biological effects like chromosomal aberrations and cell death after irradiation. The numbers make clear that due to the efficient DNA repair apparently only a small part of the DNA damaging events leads to severe genetic changes like chromosomal aberrations and cell transformations which can cause health effects like cancer and mutations. It can be assumed that the

Tab. 3.4 Some types and numbers of the damage in a mammalian cell nucleus from 1 Gy of low-LET radiation (after Goodhead 1994)

Initial physical damage	Numbers of damage
Ionisations in cell nucleus	~ 70,000
Ionisations directly in DNA	~2,000
Excitations directly in DNA	~ 2,000
Selected biochemical damage (Ward 1988)	
DNA single-strand breaks (SSB's)	1,000
8-Hydroxyadenine	700
T* (thymine damage)	250
DNA double-strand breaks (DSB's)	40
DNA-protein cross links	150
Selected cellular effects	
Lethal events	~ 0.2–0.8
Chromosome aberrations	~ 1
HPRT-mutations	$\sim 10^{-5}$

number of these damaging effects decreases in a proportional manner with a decreasing radiation dose.

Thus, the severity and rate of radiation health effects does not result from the total number of primary damaged sites in the DNA but from the remaining unrepaired or misrepaired damages. The probability for repair and non-repair of the dann DNA damages is an important factor. Therefore, the complex types of radiation damages are of special significance. A good agreement has been observed for the prediction of the numbers of SSBs based on models and direct measurements; but for other types of DNA damages, this agreement is not so good. It is difficult to quantify complex forms of radiation damages. The group of Goodhead et al. (1994) has performed measurements and calculations, which have led to rough estimates. Such data are listed for electrons with energies of 1 keV, of 100 keV as well as for α-particles with energy of 2 MeV (Tab. 3.3).

Electrons with energy of 1 keV cause somewhat more than 10 % of the SSBs as complex SSBs and around 30 % of the DSBs as complex DSBs. With 100 keV electrons (somewhat lower LET than 1 keV electrons) the complex SSBs only amount to about 4 % and the complex DSBs to only 20 % after a radiation dose of 1 Gy. These differences result from the different radiation qualities with different densities of ionising events since the 100 keV electrons have a lower ionising density than the 1 keV electrons. After radiation exposures with α-particles with energy of 2 MeV around 27 % of all SSBs and about 73 % of all DSBs are complex damaging events (Tab. 3.3). These data demonstrate that complex damages especially occur with DSBs and that this is especially the case with high-LET radiation as has been pointed out before.

Some of the DNA damages, which are caused by ionising radiation, are similar or equal to those damages, which develop from endogenous processes, especially from metabolically formed oxidative radicals in the cell. This is especially valid for ionising radiation with low-LET. These "spontaneous damaging events" are caused through the thermic instability of DNA as well as through endogenous oxidative and enzymatic processes (UNSCEAR 2000). Oxidative radicals are produced within cells by a large number of metabolic processes, like amino acid oxidase and others. These radicals attack the DNA and lead to base damages as well as to nucleotide breaks. However, these events are randomly distributed over the whole genome as single events. The probability that complex damaging events, which have been observed after the exposures to ionising radiation, can develop through spontaneous metabolic processes is low. A similar situation is valid for chemical substances, especially in the low ranges of concentrations, because the reactions of single molecules (cf. alkylation of DNA bases) are again more randomly distributed in the DNA and occur as singular events. Clusters of damaged sites are less probable form such radicals. Therefore, changes of the DNA through ionising radiation are different from other damaging processes of DNA as by oxidative radicals especially in low-dose ranges.

Although the velocity and the extent of the repair depends on the structural nature of the damage and many different mechanisms occur, it is possible to describe some general principles of DNA repair. A simplified classification of these processes can be evaluated if one considers the capability of enzymes which use the base sequence of the undamaged DNA strand complementary to the damaged DNA strand enabling the repair complexes to restore the information in the damaged sites. On this basis damaged sites (cf. damaged DNA bases) in single polynucleotide strands can be removed from the DNA. A resynthesis of gaps can be performed as the information in the non-damaged polynucleotide strand can be used as a template and the open ends of the DNA will be connected by the enzyme ligase. Through such a mechanism, some SSBs can be repaired quickly by one enzymatic step.

When DNA damaged sites appear in close neighbourhood in both polynucleotide strands (DSB) a repair is much more difficult as described before and different enzymatic processes are necessary. In order to repair such damages successfully, several enzymes from different repair pathways may be needed. A large number of DNA repair enzymes have been described as cellular proteins. Special enzymatic proteins recognise the damaged positions, cf. damaged DNA bases, and these positions can be removed from the DNA polynucleotide chain by nucleases. A new DNA repair synthesis then follows and the broken polynucleotide chain is closed again. The principle enzymatic processes are the same or similar to those which perform the normal DNA metabolism (cf. DNA polymerases, DNA ligases etc.). However, very different types of DNA polymerases and ligases have been identified which have different functions concerning the repair of different damage classes.

Processes that are much more complex have to be conducted in order to repair a DSB or other damaging events, which appear in both polynucleotide strands. Theses repair processes are not verified in all steps; especially the processes of regulation of the connected enzymatic processes are only partly understood. In mammalian cells, including human cells, processes of recombination of DNA helices have apparently significant importance (UNSCEAR 2000). The numbers of enzymes, which are necessary for such events, are much larger. It is also possible that illegitimate recombination and misrepair processes occur under these condi-

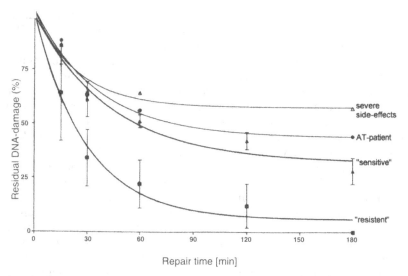

Fig. 3.6 Repair of DNA damage after irradiation of human lymphocytes *in vitro* again measured by the comet assay. The lymphocytes were taken from healthy persons ("resistant" to "sensitive"), from a patient with ataxia telangiectasia (AT) and a patient with severe side effects (hypersensitivity) (Müller et al. 2001)

tions. Under these circumstances it is possible that changes of the DNA sequence and loss of DNA sequences occur. Especially illegitimate recombination processes, which occur very rapidly after high radiation doses, can lead to mistakes (UNSCEAR 2000). The complex severe damages which apparently occur also in the low-dose range, although with less probability, lead to the situation, however, that a part of these damaging effects, after exposure to ionising radiation, cannot be repaired or misrepair occurs even after low radiation doses. In this context it is of special interest that the number of measured DSBs does not show large differences when they are measured directly after the exposures to ionising radiation with comparable energy doses of low or with high-LET. Such differences, however, are seen when the cells have had time for DNA repair (Fig. 3.5). In order to compare the radiation damage of different radiation qualities with different LETs the relative biological effectiveness (RBE) has been introduced[2]. The RBE-values for DSBs are almost one directly after radiation exposures but they increase generally into the range of 2–10 or in special cases even higher when chromosomal aberrations, mutations and other health effects are measured (UNSCEAR 2000). This finding underlines that the DNA damage observed directly after radiation exposures is not as relevant for the risk evaluation as the DNA damage after completion of DNA repair.

The damaged DNA sites are apparently recognised very quickly in living cells by specialised enzymes and can be repaired to a large extent if the complications dis-

---

[2] The *RBE* is obtained through the ratio of the radiation doses of the two radiation qualities which yield the same biological effects (radiation dose of the reference radiation – usually 200 kVp X-rays – divided through the radiation dose of the studied radiation quality).

cussed above do not occur (UNSCEAR 2000). The irradiation of human lympho-cytes *in vitro* with 2 Gy X-rays (low-LET radiation) and the observation of DNA repair have demonstrated that the repair occurs in normal persons with an efficient DNA repair capacity within about 3 hours to a large percentage of the damage. However, in these experiments only the disappearance of DNA strand breaks has been investigated investigated (Müller et al. 2001). This also includes misrepair. The repair capacity can vary strongly between individual persons. A number of syn-dromes have been described where through a genetic predisposition certain DNA repair pathways are reduced or are deficient. Frequently it has been observed that in such individuals the DNA repair of radiation damage is less and the radiosensitivity is increased (Fig. 3.6) (ICRP 1998). It has been observed by molecular genetic stud-ies that the genes which are necessary for the expression of DNA repair enzymes can be mutated or deleted in human individuals which leads to genetically based diseases (cf. xeroderma pigmentosum, ataxia telangiectasia AT, cokayne-syndrome, Fanconi-anaemia and others). Further, it has been shown that a number of enzymes, which participate in DNA repair, have an influence on the regulation of the cell pro-liferation cycle.

After a mitotic cell division, proliferating daughter cells have to go through sev-eral phases before a next cell division can take place. During the G_1-phase, the cells prepare with RNA- and protein synthesis for DNA synthesis in the S-phase. After the doubling of the DNA follows the G_2-phase again with RNA- and protein synthe-sis and the preparation for the next mitosis (Fig. 3.7). These processes are regulated by a number of proteins frequently phosphorylated. One of these key proteins is p53, which is the gene product of a tumour suppressor gene. This gene is mutated and produces therefore an inactive protein in around 50 % of human cancers (Green-blatt et al. 1994). The active *p53* also stimulates apoptosis. After radiation expo-sures, a block in the cell cycle is observed before the cell enters into the S-phase (G_1-block) or into mitosis (G_2-block). These blocks apparently give the cells addi-

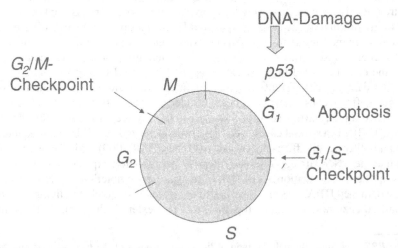

Fig. 3.7 Cell proliferation cycle and the influence of radiation-induced DNA-damage on the regulation of the cycle

tional time for DNA repair and they are therefore called checkpoints (Molls and Streffer 1984; Kastan et al. 1991; UNSCEAR 2000). This means that regulating processes of the DNA repair are frequently connected with those of cell proliferation. These processes are very efficient and they have a high significance for the development or inhibition of radiation effects especially in the low-dose range.

In conclusion, living cells are equipped with very rapid and efficient DNA repair systems. They recognise radiation-induced DNA damage very quickly. They try to eliminate the damage and to achieve a restitution of the original structure and genetic information. There exist many different repair mechanisms for the various types of DNA damage, which differ with respect to the velocity and the fidelity of the repair. Misrepair can occur which happens apparently especially with complex DSBs. Damage after exposure to high-LET is less efficiently repaired than damage after low-LET radiation. This is one decisive reason why high-LET radiation is biologically more effective than low-LET radiation when the same energy doses are compared. The regulation of DNA repair systems is very important. As many genes are involved, DNA repair deficiencies occur by mutations of these genes with a comparatively high frequency in human populations. Such genetic disorders lead to individual differences in radiosensitivity.

3.4
Cellular damages

Many investigations have been performed with mammalian including human cells after exposure to ionising radiation *in vitro*. Cell survival, chromosomal aberrations, cell transformation and gene mutation have been studied. Cell survival usually has been observed after doses above 0.5 Gy low-LET radiation whereas an appreciable number of investigations of the other endpoints have also been published with lower radiation doses. For all these radiation effects, irreversible DNA damage is decisive which has not been repaired or is misrepaired. In the case of cell death, two main mechanisms can be distinguished. On the one hand chromosomal damage develops from direct ionising events in the cell nucleus, the cells go despite this chromosomal damage through several cell cycles and the cells die in later cell generations, since they loose the ability of clonogenicity. This process is called reproductive death. On the other hand through signals at the cell membrane and signal transduction, a cascade of hydrolytic enzymatic processes is started which leads to hydrolytic degradation of cellular macromolecules. This process is called apoptosis (cell suicide). During mitotic cell divisions after radiation exposures chromosomal breaks are observed, also translocations occur which can be the basis for cell transformations. This is apparently the first step for the development of a cancer. If the radiation damage of the DNA is not so severe, the cell can further proliferate, it does not loose its clonogenicity and a clonogenic cell with a mutation develops.

These effects are important for the development of health effects after exposures to ionising radiation. Experimental data with respect to the mechanisms, which are the basis for such radiation effects, demonstrate that especially DSBs in the DNA are the decisive primary radiation events from which cell transformation and chromosomal aberrations, as well as gene mutations, develop. For these effects interactions between several damaged cells are apparently not necessary and recent investigation,s in which single cell irradiations have been performed, have yielded that the passage of one ionising particle through the cell is sufficient in order to cause such damages. This is the case even after exposures to radiation with low-LET. A threshold dose for such effects can only be expected if a complete DNA repair could be performed. As has been discussed already before, there are no indications that such a perfect DNA repair exists. However, under certain circumstances it is possible that DNA repair can apparently be stimulated and then a so-called "adaptive response" can be observed. These phenomena will be discussed later.

3.4.1
Chromosomal aberrations

For the investigation of chromosomal aberrations, peripheral human blood lymphocytes have been used in many cases. The studies of dicentric chromosomal aberrations after the exposure to ionising radiation are dominating. These types of aberrations are relatively specific for ionising radiation as they are induced predominantly by ionising radiation but only by a few chemical substances. Very extended investigations of the dose relationships after exposures to X-rays or γ-radiation have yielded that radiation doses in the range of 100 increase of dicentric chromosomal aberrations. Linear dose relationships have been found in the low-dose range for X-rays under very careful studies and the evaluation of large numbers of cells (Fig. 3.8). With high efforts, it is possible to detect also radiation doses below 100 mGy (Tab. 3.5).

In dose ranges higher than several 100 mGy, the dose relationship can often be better described by linear/quadratic equations after exposures to low-LET radiation as mentioned before. Studies in which six very experienced laboratories participated resulted in radiation induced increases of aberrations down to radiation doses (dose rate 1 Gy min^{-1}) in the range of 20 mGy (Lloyd et al. 1992). Based on these careful and extended studies it can be expected that an observation of significant effects for chromosomal aberrations is not possible

Tab. 3.5 Lowest doses at which chromosome aberrations and mutations have been detected in experimental system exposed to low-LET radiation (evaluated from UNSCEAR 2000)

System	Endpoint	Radiation	Lowest Dose[a] (mGy)
Human lymphocytes	Unstable chromosomal aberrations	X-rays	20
Human lymphocytes	Stable chromosomal aberrations	γ-rays	250
C3H10T½ cells	Cells transformation	X- and γ-rays	100
Mouse	Pink-eye mutation	X-rays	10
TK$_6$ cells	*hprt* and *tk* mutation	X-rays	250
Tradescantia	Pink mutation	X-rays	2.5

[a] acute exposures (minutes)

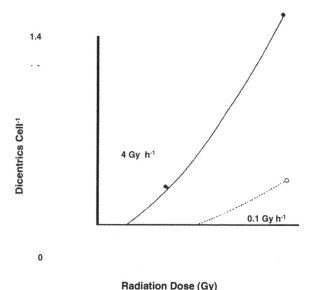

Fig. 3.8 Chromosomal aberrations (dicentrics) in human lymphocytes after exposure to low LET-radiation with high dose rate (4 Gy h^{-1}) and low dose rate (100 mGy h^{-1}) in dependence of radiation dose

below 20 mGy. Therefore, the possibility of a threshold dose in the range around below 10 mGy cannot be excluded for this effect in principle. However, extrapolations give the best fit with dose-effect relations without a threshold. Irradiation with high-LET radiation lead to linear dose-effect curves over the whole dose range of low doses up to around 1–3 Gy (Fig. 3.9). As can be seen from this figure the radiation effects per Gy increase with increasing LET. Extrapolations from the dose ranges in which a significant increase of chromosomal aberrations can be measured into the low-dose range can be described by linear dose-effect curves. A reduction of this effect with decreasing dose rates does not occur in the case of high-LET radiation. Therefore, the occurrence of threshold doses can apparently be excluded for such radiation qualities. In contrast, irradiation with low dose rates lead to reduced effects in comparison to high dose rates after exposures to low-LET radiation.

In the medium dose range (0.1 to 1.0 Gy) *RBE* values for neutrons and α-particles have been generally observed in a range of 3 to 5 for the induction of chromosomal aberrations. Such chromosomal aberrations have generally been measured in the first mitotic division after radiation exposure. It is assumed that these chromosomal aberrations are caused by DSBs in the DNA. However, it has not been exactly proven that such a developmental chain exists. Recent studies have yielded that also in the following mitotic cell divisions new chromosomal aberrations are formed. Such changes are apparently not caused by immediate DSB's but radiation damaged sites have been processed in cycling cells so that new DSB's develop during the following cell cycles after irradiation and then further chromosomal aberrations can

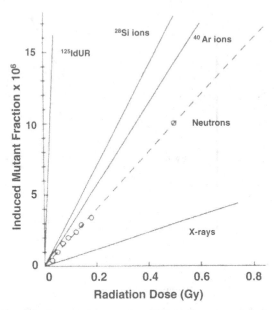

Fig. 3.9 Dose effect curves for the induction of gene mutations (HPRT) after irradiation of cells *in vitro* with various radiation qualities of low LET-radiation (X-rays) and high LET-radiation

be manifested. The resulting chromosomal aberrations show higher *RBE* values than the chromosomal aberrations which are observed in the mitotic division directly after irradiation (Weissenborn and Streffer 1988). As the dose-effect relations are linear for high-LET radiation and those for low-LET radiation have frequently a linear-quadratic form of the dose-effect curve, the *RBE* values become dose dependent. In the low-dose range, the *RBE* values increase with decreasing radiation dose. This phenomenon occurs because the slope of the dose-effect relationship becomes shallower with low-LET radiation due to DNA repair. In the low-dose range, the dose-response curve is more or less exclusively described by the linear term with the constant α (eq. 3.1). Under these circumstances "maximal" *RBE* values have been estimated (Tab. 3.6).

Tab. 3.6 Estimated RBE_m values for fission neutrons compared with γ-rays

Endpoint	RBE_m
Cytogenetic studies, human lymphocytes in culture	34-53
Cell transformation	3-80
Genetic endpoints in mammalian systems	5-70
Life shortening (mouse)	10-46
Tumour induction	16-59

3.4.2
Cell transformation

Cell transformations after the exposure to ionising radiation and the damage of the genetic material e. g. by forming DSBs have been understood as the next initiating step for the development of cancer. With the present knowledge, DSBs and possibly their misrepair are apparently participating in these processes (UNSCEAR 2000). These events have been studied in manifold experiments with cell systems *in vitro*. Normal cells, like fibroblasts, grow in *in vitro* cultures in monolayers. However, when cultured cells are transformed to malignant cells these cells grow in a multi-layer manner, they form large clones. It has also been shown that these transformed cells can form tumours when injected into mammals e. g. murine cells into mice (UNSCEAR 2000). Most studies of cell transformation have been performed with the cell lines C3H10T½ and BALB/C3T3. Both cell lines originate from fibroblasts, which have been obtained during the prenatal development of mice.

In the course of recent years it has been shown that these cell systems are not ideal for the investigation of tumour development in humans as they develop into fibrosarcoma after the transformation *in vitro* and transplantation of the trans-formed cells into mice (UNSCEAR 2000). However, in humans epithelial cancers and not fibrosarcoma are increased after irradiation. The search for an appropriate human cell system was not very successful until recently. Therefore, the data which have been obtained with murine cell lines will be reported here.

Generally, it has been shown that radiation doses of 100 mGy and higher are nec-essary in order to observe significantly increased effects. As with chromosomal aberrations linear and linear-quadratic dose-effect curves have been described for these radiobiological effects. After exposure to low-LET radiation, a dose rate effect has been found again as has been described for chromosomal aberrations. In a larger study with six laboratories, the dose dependence of cell transformation with low-LET radiation in C3H10T½ cells has been studied (Mill et al. 1998). The irra-diation and the cell cultures have been performed under identical conditions in these experiments. The irradiation with 250 kVp X-rays in a dose range of 250 mGy to 5 Gy and the culture was performed in one of the six laboratories. The evaluation of the cultures was performed in all six laboratories in a parallel procedure. With these extended studies a linear dose-effect relationship was found in the investi-gated dose range when the frequency of transformed cells was calculated based on surviving cells. In case of the calculation of transformation frequencies based on the number of irradiated cells, a bell shaped dose-effect curve with a maximum in the range of around 2 Gy was observed. In the lower dose range the rate of transfor-mations still increased linearly.

These data and the analysis of the resulting dose-effect curves do not give an indication that a threshold or a super-linear shape of the dose-effect curve occurs in the low-dose range. The dose-effect curve for the rate of transformations per surviv-ing cell shows a linear term of the dose (value for the constant α) in the range of $0.83 \pm 0.08 \times 10^{-4}$. As the same methods have been used in all six laboratories, the deviations were smaller than a factor 2 in the individual laboratories. Therefore, a very good agreement has been found for these transformation rates between the dif-ferent laboratories. In contrast, differences with a factor of 20 can be found in reports when varying experimental conditions have been used in different laboratories (Mill et al. 1998).

With higher radiation doses the transformation rate per number of irradiated cells decreased as more and more cells were killed by the ionising radiation. These data support the function of a linear dose-effect relationship without a threshold for cell transformation. However, the smallest dose with a measurable effect was only 250 mGy.

For the effect of cell transformation, an adaptive response has also been described. In a dose range of 1 to 100 mGy, given as adapting dose, it was possible to reduce the cell transformation rate with C3H10T½ cells after higher doses than it was seen with cells without an adapting dose. Higher frequencies of transformed cells were observed after high-LET radiation than after low-LET radiation comparing the same absorbed radiation dose. The phenomenon of dependence of radiation effects on radiation quality is also seen for cell transformation as it has been described with other radiation effects.

Studies of single cell irradiation again with C3H10T½ cells have also been performed with α-particles. The effects of single α-particles were compared with the effects of up to 8 α-particles per cell. If only one α-particle passes through a single cell nucleus an increase of the transformation rate per cell of about 30 % was observed. However, this effect was not statistically significant. In contrast, the transformation rate increased significantly in a cell population when the cells were hit by one α-particle on the average. However, under these circumstances, it could happen that in some single cells several α-particles per cell could pass the cell nucleus. The possibility exists that under these conditions an appreciable number of transformed cells was caused in cells which were passed by several α-particles and that these multiple passages were the decisive mechanisms of cell transformation (Miller et al. 1999).

3.4.3
Gene mutations in somatic cells

The investigation of gene mutations after radiation yielded principally very similar data as has been observed for cell transformation. This is valid for low-LET as well as for high-LET radiation (Fig. 3.9). Again, linear dose-effect relationships with the smallest dose of 250 mGy as a statistically effective dose were observed. For α-particles, an *RBE* of 4 was found (Albertini et al. 1997). Gene mutation is a relatively rare event. Experimental systems have been developed in which a gene mutation leads to the loss of an enzyme (gene product) which usually metabolises a substrate so that it becomes toxic for the cells. If the enzyme is lost, the metabolic reaction cannot take place and the corresponding cells survive (UNSCEAR 2000). With respect to gene mutations, frequently the mutation rate in the gene for the enzyme hypoxanthine-guanine-phosphoribosyltranferase (HPRT) was studied. After mutation of this gene an inactive enzyme protein is synthesised, the cells lose the ability to metabolise 6-thioguanine and thus gain resistance against this drug and the cells with the mutation survive. In a similar mode, mutations can be studied in the gene for the enzyme thymidin-kinase. These enzyme systems are found as very appropriate biological systems for such studies as the survival test is comparatively simple.

It has already been pointed out that DSBs in the DNA are the essential damaging events for the development of cellular radiation effects (chromosomal aberration, cell transformation and gene mutation). Many experimental studies, however have indicated that there is no stringent proof for such a developing chain. Comparative studies on the induction of DSBs in DNA and for various cellular effects after irradiation with different radiation qualities have demonstrated, that the *RBE* values for these various endpoints do not agree with each other if DNA repair has not yet taken place.

While *RBE* values in the range of around 5–20 have been observed for the induction of chromosomal aberrations, cell transformation and genetic mutations, the *RBE* values for primary DSBs measured directly after irradiation (before DNA repair could take place) which have been reported for α-particles and heavy ions in mammalian cells are in the range of around 0.5–2. Only in few cases, *RBE* values have been observed which exceed values of 2 for DSBs. Most of the *RBE* values for this effect are in the range of 1 (UNSCEAR 2000). It is of interest in this connection that the *RBE* values after exposures to 250 kV-X-rays as well as 6 MeV cyclotron neutrons change in dependence on the cell proliferation if chromosomal aberrations are analysed. For chromosomal aberrations, which have been determined during the first mitosis after radiation, a *RBE* value around 3 is found. In the second and third mitosis after irradiation, the *RBE* values increase and reach values of five to eight (Weissenborn and Streffer 1988). It may be possible that such latent DNA damages are of special significance for the development of health effects. This question needs further clarification.

A very radiosensitive mutation assay is the pink-eyed unstable mutation in the mouse (Schiestl et al. 1994). In this assay the reduction of the pigment in the eye is studied as a result of gene duplication and its reversion to the wild type by deletion of one copy. The reversion frequency is much higher than the rate of other recessive mutations. Female mice, homozygous for the reversion, were irradiated with X-ray doses of 10 to 1,000 mGy and the number of reversions measured. Even after a radiation exposure with 10 mGy the reversion rate was increased threefold. Another sensitive mutation system has been studied in the stamen hair of the plant *Tradescantia*. The normal dominant blue colour can be mutated to the recessive pink. Dose-response curves for pink mutations have a linear dose-response between 2.5 to 50 mGy for 250 kV X-rays and between 0.1 to 80 mGy for 0.43 MeV neutrons (Sparrow et al. 1972).

All the microdosimetric as well as the molecular and cellular studies give no indications for threshold doses in the dose-response curves of the described effects. In most cases, it is not possible to measure radiation effects significantly in dose ranges below 50 to 100 mSv and therefore extrapolations are necessary from the measurements in the medium dose range above 100 mSv. However, the best fit is usually achieved for the discussed biological endpoints by dose-response relations without a threshold.

3.4.4
Modifying phenomena

The described biological radiation effect can be modified by a number of biological processes and chemical substances. The interaction between radiation exposures and chemical substances has been discussed in an earlier project of the "Europäi-

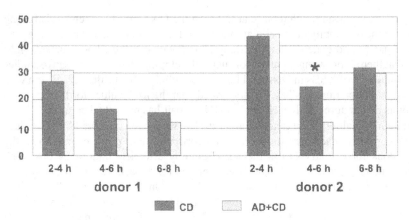

Fig. 3.10 Adaptive response measured via chromosomal aberration in human lymphocytes, two individuals (donors) with different response. Multiple fixation times AD: 0,05 Gy, 40 h, CD: 1.5 Gy, 50 h, h: time of Colcemid treatment p.r., *: difference significant with p < 0.05

sche Akademie zur Erforschung von Folgen wissenschaftlich-technischer Entwick-lungen" (Streffer et al. 2000; English version 2003). Some of the relevant modifying biological processes have already been mentioned but shall be evaluated some-what deeper. These are especially, apart from DNA repair, which has been discussed earlier: adaptive response, apoptosis, genetic instability (bystander) and genetic predisposition.

In a number of experiments it has been found that irradiation of cells with low radiation doses can lead to an enhanced resistance of these cells. This increase of radioresistance is manifested by the observation that after a small radiation dose of 10–100 mGy (adapting dose) a following radiation dose in the range of 1 to 2 Gy (challenging dose) results in smaller radiation effects than with cells which have not received the adapting dose in the small dose range (Fig. 3.10, donor 2). This phe-nomenon has been termed adaptive response (UNSCEAR 1994; Wojcik and Stref-fer 1994). The data reported in the literature can be summarised in the following manner: Adaptive response to low-LET radiation exposure has been found widely distributed with living organisms: in bacteria, yeast, plants and animals. It alsohas been observed in a number of different human as well as rodent cells. Mainly chro-mosomal aberrations have been studied, but also gene mutations, cell survival/cell death by the clonogenic assay as well as apoptosis have been investigated (Streffer 2002).

The following parameters are very important and the conditions have to be kept in certain ranges with respect to time or radiation dose in order to obtain an adaptive response which results in an increased cell survival or reduction of mutated includ-ing transformed cells. All parameters have decisive influence on the degree of adap-tive response:

– Adapting Dose (AD)
– Challenge Dose (CD)
– Interval AD-CD

- Persistence of Adaptive Response
- Cell Proliferation/Cell Cycle
- *In vitro/in vivo*-Situation
- Induction of DNA Repair
- Induction of Protein Expression
- Cell Type
- Stimulation of Immune System
- Genetic Disposition

The radiation effects were frequently reduced in the sense of adaptive response when an adapting radiation dose in the range of 5 to 200 mGy was given some hours (usually 6 to 18 h) before a challenge dose in the range of 1 to several Gy was applied. In addition, the dose rate of the adapting dose must be kept in certain ranges in order to achieve an adaptive response. There exists cross-adaptation between some toxic chemicals such as alkylating or DNA strand break producing agents and low-LET radiation (UNSCEAR 1994).

The adaptive response was especially successful in mammalian cells when the conditioning radiation exposure took place during the S/G_2-phases of the cell cycle. The cellular response was transient and lasted for about three cell cycles or two to three days. The adaptive response was mainly studied in lymphocytes of humans or rodents. It was also observed in fibroblasts and bone marrow cells.

There is no definite proof for the underlying mechanism leading to the induction of an adaptive response until now. Nevertheless, a number of experimental data support the assumption that the induction of DNA repair is very probably involved. For this mechanism an enhanced protein expression is necessary as has been shown with inhibitors of protein synthesis which also prevent adaptive response (UNSCEAR 2000; Streffer 2002).

No adaptive response has been observed during the prenatal development of rodents and this can probably be extrapolated to humans. Furthermore, there is an enormous variation in the quantitative response between individual donors (Fig. 3.10, donor 1 and donor 2). No adaptive response could be observed at all in cells from some individuals. This is especially valid for individuals with a genetically caused deficiency of DNA repair as observed with ataxia telangiectasia (AT) patients. Adaptive response and its development under low radiation exposures is very much dependent on the individual genetic disposition.

Further, there is no good evidence for an adaptive response after exposures to high-LET radiation. However, it appears if there is any adaptive response for high-LET radiation, it will be considerably smaller than with low-LET radiation. The lack of adaptive response in connection with these radiation qualities would be in agreement with the findings that little DNA repair occurs after such exposures (Streffer 2002).

In conclusion, in many biological systems the development of an adaptive response in the direction of an increased radioresistance has been observed after small doses of low-LET radiation. A dose-modifying factor in the range of 1.5 to 2.0 has usually been found. However, these processes do not occur as an universal principle in all individuals and in all developmental stages of a living organism. This is especially the case for individuals with an increased radiosensitivity. These circumstances have to be taken into account if adaptive response is discussed for practical radioprotection. Under these considerations it appears doubtful whether adaptive response should be included with respect to general regulations and risk factors in radioprotection (Streffer 2002).

Investigations on survival of mammalian cells have demonstrated a higher radiosensitivity in the dose region below 0.5 Gy low-LET radiation. This phenomenon has been termed hyper-radiosensitivity (HRS). This HRS has been observed down to radiation doses below 100 mGy. It precedes a dose region (around 0.5 to 1 Gy) in which the radioresistance of the cells increases and the cell killing is almost constant although the radiation dose increases (increased radioresistance, IRR) (Joiner et al. 2001). The phase of IRR shows some analogy to adaptive response, although the dose range is somewhat higher. The adaptation process can be suppressed by inhibitors of protein synthesis (Streffer 2002). These phenomena have been studied with a number of normal and malignant cells and it has been found that the HRS as well as the IRR differs from cell line to cell line. Generally, the phase of HRS is most expressed by the cells, which become more radioresistant at the higher dose ranges. The phenomena of HRS and IRR have also been observed with respect to radiation response with tissues *in vivo*. The induction of DNA repair, which leads to IRR, reduces to normal values within hours after radiation exposures (Joiner et al. 2001).

As has been mentioned before, studies with irradiated mammalian cells demonstrated that there exist different forms of cell death. It was observed that proliferating cells usually go through several mitotic cell divisions when irradiated with doses of 0.5 to 5 Gy, then these cells develop chromosomal aberrations and die from this damage (reproductive cell death). On the other hand, non-proliferating, radiosensitive cells can die after radiation doses of 0.5 to 5 Gy without going through a mitotic division (interphase death, apoptosis). This latter cell death is induced by the signal transduction mechanism and is also called programmed cell death. An activation of intracellular hydrolytic enzymes leads to a breakdown of biological macromolecules including DNA. Apoptosis is very important for normal developmental processes but also for the elimination of damaged and possibly malignant cells. It has therefore been proposed that apoptosis can reduce radiation risk and this is supported by the induction of the apoptotic activity after radiation exposures (Feinendegen and Neumann 1999).

This has especially been demonstrated with respect to malformations after pre-natal irradiation during the organogenesis with 2 Gy X-rays (Norimura et al. 1996; Kondo 1998). The tumour suppressor gene *p53* is involved in the triggering process of apoptosis. In *p53*-knockout mice, the rate of malformations was higher than in wild type mice. In this connection, it is of interest that unrepaired DSBs apparently stimulate apoptosis. Recent data demonstrate that different trigger mechanisms exist for the induction of apoptosis and these behave differently with respect to the induction by irradiation (UNSCEAR 2000). Further it has been observed that very little apoptosis occurs in tumour cells with *p53* mutations (Oya et al. 2003). Such *p53* mutations have been found in around 50 % of all human tumours (Greenblatt et al. 1994).

Adaptive response has also been seen for the induction of apoptosis in human lymphocytes, but again the individual variability for this effect was very large (Cregan et al. 1999). The *RBE* of fast neutrons was 1 for apoptosis in mouse thymocytes (Warenius and Down 1995), however, for intestinal crypts the *RBE* values for 14.7 MeV and 600 MeV neutrons were 4 and 2.7 respectively (Hendry et al. 1995). The influence of apoptosis on radiation risk is certainly an interesting approach, however, apparently it is not an universal mechanism and there are many open questions.

During recent years, investigations on the induction of genomic instability by ionising radiation have obtained high interest. This phenomenon has been observed in skin fibroblasts from mouse fetuses, which were irradiated 1 to 3 hours after conception during the zygote stage (1-cell stage). Around 30 to 40 cell generations after this radiation exposure the fibroblasts developed new chromosomal aberrations (Pampfer and Streffer 1989). This effect could only be explained by an increase of genomic instability, which is induced by ionising radiation. In further experiments, genomic instability after irradiation has been observed with quite a number of cell systems *in vitro* and *in vivo* for chromosomal aberrations, cell survival and gene mutations (Little 2000). Such effects have been observed with low as well as with high-LET radiation. Thus, the phenomenon is generally accepted today although the mechanism of this effect is unknown.

It has been demonstrated further that these effects do not only occur in cells which have been hit by ionising particles but also in neighbour cells of the hit cells when irradiation of single cells was performed with α-particles. This effect is called bystander effect (Azzam et al. 1998). At present, the mechanisms of induction of genomic instability and bystander effects by ionising radiation are not understood. It is also not clear whether only one mechanism exists for the various endpoints. Genomic instability is apparently not directly connected with a well-defined gene or chromosomes; it appears as a general phenomenon in which the whole genome is affected. Many possibilities are discussed. As the frequency of this effect is higher than the rate of mutations on the cellular level, an epigenetic mechanism has been discussed (Little 1998; Wright 1999). In addition, the involvement of free radicals, especially reactive oxygen species, has been proposed. Further attractive proposals are that the development of genomic instability is connected to the structure of telomeres which stabilise the structure of chromosomes or to imprinting processes in the DNA which are involved in regulating the activation and deactivation of genes (Schofield 1998). DNA repair may be impaired under these conditions. In this connection it is of interest that quite a number of syndromes with increased radiosensitivity and enhanced frequencies of cancers through genetic predisposition show also an increased genomic instability (ICRP 1998). The increased genomic instability induced by ionising radiation can be transmitted to the next generation in mice (Pils et al. 1999).

All stochastic late effects of ionising radiation are multistep processes. Best defined are these events for carcinogenesis. Many experimental and clinical studies have demonstrated that cancer develops through several sequential mutation steps. Such a model has been proposed for colorectal cancer (Fig. 3.3) (Fearon and Vogelstein 1990). For such a model, the development of genomic instability would be of very high significance, as these phenomena increase the mutation frequency and therefore increase the probability of a second or third mutation in an individual cell (Little 2000). As mutation frequencies are usually very low even after irradiation, genomic instability would facilitate the multistep processes considerably when the mutation frequency can be increased by a factor of 10 to 10,000 as it has been shown (Little 1998). The increased genomic instability in individuals with a genetically predisposed higher radiosensitivity may also explain partly the higher cancer risk in these individuals (ICRP 1998). Therefore the evaluation of the mechanisms of how genomic instability develops after irradiation would be extremely important in order to improve our understanding of the mechanisms of carcinogenesis especially and stochastic radiation late effects in general. This is essential for a better evaluation of radiation risks in the low-dose range.

It has been mentioned that individual radiosensitivity can vary very much. This has especially been observed with cancer patients in radiotherapy when a patient suffers from acute radiation effects after radiation doses which are tolerated by the vast majority of patients without any serious adverse symptoms (ICRP 1998). It may be assumed that the distribution of the radiosensitivity of our populations follows a Gaussian mode, but it also has been proven that there are some individuals in the range of several percent who are hypersensitive. These persons react with severe side effects in radiotherapy (overreactors) (Fig. 3.11). In general, children are more radiosensitive than adults. This has been found with thyroid-, breast- and other cancers (UNSCEAR 2000; see chapter 5). The effect can be explained on the basis that in these original tissues cell proliferation is higher in children than in adults and it can be stimulated by endocrine functions. As has already been mentioned, cancer develops through several steps which are mainly determined by mutation and changes in the regulation of cell proliferation. On the basis of these mechanisms it is also understandable that radiation-induced carcinogenesis can be enhanced by promoting substances like phorbol ester (TPA) and oestrogens (UNSCEAR 2000; Streffer et al. 2000).

In recent years, cellular and molecular studies *in vitro* have contributed to the understanding of those mechanisms which are responsible for the increased radiosen-

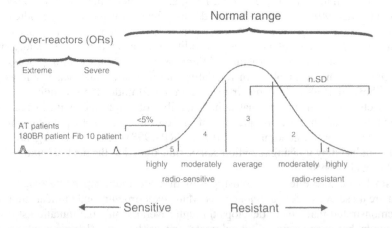

Fig. 3.11 Variability of radiosensitivity in a population measured by normal tissue response during radiation therapy. Normal range grouped in 5 classes and over-reactors (patients with a genetic predisposition for hypersensitivity: retinoblastoma (Rb) and ataxia telangiectesia (AT)

sitivity. Generally, deficiencies of specific DNA repair pathways (Fig. 3.6) and disturbances of the regulation of the cell cycle in proliferating cells have been observed
mainly in fibroblasts and stimulated lymphocytes from radiosensitive patients. After
irradiation specific blocks occur within the proliferation cell cycle in the G_1-phase
(before DNA synthesis starts) and in the G_2-phase (before mitosis starts) (Fig. 3.7).
These blocks give the damaged cells some more time in order to repair the damage.
Therefore, both processes (DNA repair and regulation of the cell cycle) are connected
with each other. From these investigations it has been possible to define and describe
the following syndromes, which are genetically inherited (ICRP 1998):

- Ataxia telangiectasia
- Fanconi anaemia
- Li Fraumeni syndrome
- Neurofibromatosis
- Nevoid basal cell carcinoma syndrome
- Nijmegen breakage syndrome
- Retinoblastoma

These syndromes are usually inherited in a recessive trait. Therefore they are
only manifested in the homozygous carriers which have frequencies in the range of
1:40,000 to 1:100,000. An increase of radiosensitivity by a factor of 5 to 10 may
occur in patients with these genetic syndromes (ICRP 1998). No data are available
for an increased cancer rate in the low-dose range but enhanced frequencies of secondary cancers have been observed in patients who have received radiotherapy for
a treatment of retinoblastoma. In these studies, it clearly was found that the radiation-induced cancer risk was increased in patients with a genetic defect in the rb
gene. In mice a dramatic increase of the tumour rate and a reduction of the latency
period for tumours have been observed when the $p53$ gene was knocked out (ICRP
1998). The cellular and molecular studies further showed that the heterozygotes are
also more radiosensitive than normal individuals, however, this is much less than
with the homozygotes. Nevertheless, the heterozygous carriers appear in the range
of some percent of the total population. Since it is assumed that a Gaussian distribution of individuals exists with respect to radiosensitivity, hypersensitive individuals
are located far outside of such a scheme. Around 5 % of cancer patients are more
sensitive in that they express severe symptoms with a general radiotherapy which is
tolerated without any symptoms by the vast majority of patients (Fig. 3.11). Such
data have been obtained with cytogenetic studies of lymphocytes from patients with
breast and colorectal cancers (Baria et al. 2002). It is also of interest that individuals with the described genetic syndromes have an increased genomic instability and
an enhanced rate of certain "spontaneous" malignancies (ICRP 1998).

In conclusion, a modification of radiation response by various mechanisms is
possible. A number of these modifying factors lead to an increase of radioresistance
like adaptive response, apoptosis and induction of immune response, but also
increased radiosensitivity takes place as by induction of genomic instability. Most
effects have been observed with cellular systems *in vitro*. These effects are usually
not studied in the low-dose range but data are only available in the medium or

higher dose range in most cases. The effects are bound to comparatively narrow conditions with respect to time windows and dose ranges. Under conditions where dose modifications are observed after exposures to low-LET radiation, these effects are not seen after exposures to high-LET radiation. The extent of modifying processes varies very much from individual to individual; it is connected to the genetic disposition. If dose-response curves are modified by adaptive response, apoptosis etc. the slope of the dose-response curve changes but this does not touch the principle problem of whether a threshold dose does exist or not.

3.5
Animal experiments

Experimental studies with animals have a great importance for the knowledge about the expected late effects of radiation in humans. Considerable information can be obtained for radiation-induced cancer and hereditary diseases from such investigations. However, there are certain limitations which have to be taken into account. It is not possible to use such experimental data for quantitative risk estimates, especially in cancer induction. However, mechanisms can be evaluated from such studies and especially the influence of various factors such as radiation quality, dose protraction or fractionation as well as dose rate, the sensitivity of individual organs and tissues as well as the age at exposure on the tumour response. The principal form of dose-response relationships can be evaluated over a wide range of radiation doses. Further, it is possible to get a better understanding of the molecular and cellular mechanisms, which are important for the development of cancers after radiation exposures (UNSCEAR 2000).

Most of these investigations on carcinogenesis have been performed with rodents (especially mice) and dogs. Hereditary effects of radiation have been extensively studied in the offspring of irradiated animals. Again most of these investigations have been performed with mice. As no clear evidence has been obtained from observations in humans on the risk of radiation-induced hereditary diseases, animal studies give the only information on dose-response relationships in this case. These studies have also been used for the evaluation of quantitative risk estimates for humans therefore (UNSCEAR 1988; UNSCEAR 2001).

3.5.1
Carcinogenesis and life shortening

The vast majority of experimental animal studies have been performed with inbred strains of rats and mice with the presence of spontaneous diseases, that can be different in some cases from those which are observed in humans. The frequency of spontaneous tumours and the radiosensitivity of different mouse strains can be very variable. In different strains of mice and rats different tumour entities can dominate. In addition, the differences between gender and age with respect to the incidence of specific tumour types can vary considerably. Nevertheless a number of cancers which also occur in humans like myeloid leukaemia and cancers of the breast, lung, ovary, pituitary and thyroid have been successfully studied mainly in mice and extrapolations to humans are possible under the above mentioned precautions.

There are also some investigations available with larger animals as dogs and other species including primates.

By analysing these studies it has to be taken into account that the life-span of most experimental animals compared with humans is rather short and that often the rates of cell turnover are quite different. Also a number of modifying factors are different and can influence the cancer rates without radiation and after radiation exposure. Laboratory animals are kept under very well defined conditions, which is not the case for a human population under study. This certainly also has a considerable influence on the experimental outcome. However, the principal mechanisms for the development of cancer spontaneously and after radiation exposures are the same or analogous in humans as well as in laboratory animals.

The possibility to detect radiation effects on cancer induction after low radiation doses depends very much on the number of animals, on the spontaneous incidence of cancers in the various tissues and organs as well as on the radiation sensitivity of these tissues and organs with respect to cancer induction. Tab. 3.6 (UNSCEAR 2000) gives some very instructive examples of this situation. This table shows especially that very large numbers of animals are needed if a chance should exist to detect radiation effects in dose ranges of 1 to 10 mGy. Even with radiation doses of 100 mGy large numbers of animals are necessary in most cases. The calculations in Tab. 3.6 also show the strong dependence on the spontaneous cancer incidence. As the incidence of myeloid leukaemia is very low in CBA-mice, only 300 animals would be needed in order to detect the effect of 100 mGy. However, in RFM-mice 120,000 animals would be necessary for the same radiation dose.

Nevertheless, there are also some advantages for studies with laboratory animals in comparison to epidemiological studies with humans. Such experiments can be planned in a much better manner. The numbers of animals for various dose groups and other experimental conditions can be exactly determined through the design of the experiment in order to get the best possible information. The animals can be exposed under very controlled conditions with well-defined dose rates or fractionation schedules and the dose estimates have a high certainty. Further the exposed animals are genetically more homogeneous and also the conditions of feeding and other parameters of lifestyle are more constant and well-defined than with human populations.

A large number of such animal studies have been performed. However, studies with good dose-response relationships, with variable dose rates and comparisons of high as well as low-LET radiation have been performed only by few studies so that the numbers, which can be used for critical evaluations is rather limited for the purposes of this analysis. These studies have been reviewed in a number of reports of UNSCEAR (1993, 2000) as well as by other scientific committees (BEIR 1990). In a number of studies it has been shown that after exposures to low or medium radiation doses (below 1 Gy) up to a range of doses of a few Gy (low-LET radiation) life shortening of experimental animals is mainly the result of an increased cancer development and mortality from these diseases. Only after higher radiation doses, radiation effects on the renewal of blood cells and on the vascularisation as well as radiation effects on other tissues become so significant that animals die from such causes.

Tab. 3.7 Lowest acute doses at which significant increases in cancer have been observed in mice (evaluated from UNSCEAR 2000)

Cancer	Mouse strain	Sex	Irradiation	Dose (Gy)
Myeloid leukaemia	RFM	Male	X-rays	0.25
		Male	γ-rays	1.00
		Female	γ-rays	1.00
		Female	γ-rays	2.00
	CBA/H	Male	γ-rays	0.50
	BC3F$_1$	Male	γ-rays	1.50
		Female	X-rays	1.00
Thymic lymphoma	RFM	Male	γ-rays	1.00
		Male	X-rays	3.00
		Female	γ-rays	1.00
		Female		2.00
Lung adenocarcinoma	BALB/c	Female	γ-rays	0.50
	SAS/4	Both	X-rays	2.50
Mammary adenocarcinoma	BALB/c	Female	γ-rays	0.20
Ovarian tumour	BC3F	Female	X-rays	0.16
	BALB/c	Female	γ-rays	0.25
	RFM	Female	γ-rays	0.50

As the death of an animal is a very precise biological endpoint and easy to measure, the investigation of life shortening after radiation exposures can be very informative with respect to cancer induction. A review of a larger number of such investigations with mice, which received single radiation doses of X- or γ-radiation demonstrated, that the life shortening ranged from 15 to 81 d Gy^{-1}. Most of the values were found between 25 and 45 d Gy^{-1} with an overall weighted average of 35 d Gy^{-1}. (Tab. 3.7; UNSCEAR 2000; Grahn and Sacher 1968). In BALB/c mice, a linear function of dose was found between 0.25 and 6 Gy for life shortening after exposures to γ-radiation (^{137}Cs). The life shortening in these studies was 46.2 ± 4.3 d Gy^{-1} (Maisin et al. 1983). Studies with other mammalian species have been summarised in an UNSCEAR report (1982) and have come to similar conclusions.

Some studies have been performed after dose fractionation. In general it seems that dose fractionation has only a small effect on life shortening. On the other hand, protracted exposures over several months to years result in a reduced effect on life shortening. Factors for the necessary dose enhancement between about 2 to 5 can be calculated in order to get the same effects for chronic exposures as for exposures with high dose rates. Studies have also been performed after exposures to high-LET radiation and have come in principle to the same conclusion (UNSCEAR 1993).

Again a linear dose-response has been observed with no effect by dose fractionation and also by dose protraction after exposures with high-LET radiation as this has already been described for cellular effects (UNSCEAR 2000).

Very extensive studies on tumour induction in mice have been published by Ullrich and Storer (1979 a, b, c) after exposures to low-LET radiation. Numbers of cancers have been investigated like myeloid leukaemia and cancers of the breast, lung, ovary, pituitary and thyroid. Female RFM mice were exposed to γ-rays (^{137}Cs, 0.45 Gy min^{-1}) with radiation doses from 0.1 to 3he animals were autopsied and diagnosed for the various types of cancers.

Significant increases of acute myeloid leukaemia (*I*) were obtained after radiation doses of 1 Gy and above and a linear dose-response curve with a very steep slope was observed:

$$I = 0.63 + 1.4\,D.$$
(eq. 3.3)

However, the dose-response curve could also be described by a linear – quadratic model for the data of these experiments (Ulrich and Storer 1979):

$$I = 0.69 + 0.86\,D + 0.00227\,D^2.$$
(eq. 3.4)

The combined analysis with a further study resulted in a significant increase of myeloid leukaemia after a radiation dose of about 0.5 Gy. In this study the data could also be fitted to a threshold – linear model. The threshold dose would be 0.22 Gy under such an evaluation. Although 18,000 mice were used in these studies the increase of myeloid leukaemias in the lower dose range was uncertain, as the number of leukaemia cases was very small. Several studies of the induction of mammary cancers after radiation exposures demonstrated that the sensitivity of various mouse strains was very different. The same was observed with rats (UNSCEAR 2000). Generally, a linear dose-response curve was found after exposures to X-rays as well as to 0.5 MeV neutrons. Ullrich et al. (1996) observed that the variability in sensitivity of the various mouse strains correlated with the sensitivity of the breast epithelial cells to radiation-induced malignant transformation.

For the induction of lung tumours it was found that the frequency of cancers was less after low dose rates in comparison to high dose rates of γ-radiation (^{60}Co). After exposures with a high dose rate (0.4 Gy min^{-1}) a linear dose-response curve was observed. After a low dose rate (0.06 mGy min^{-1}) the slope of the dose-effect curve was smaller. The dose-response could also be described by a linear response curve. In addition, a linear-quadratic dose-response was possible for both dose rates (0.4 Gy min^{-1} and 0.06 mGy min^{-1}; Ullrich and Storer 1979c). Maisin et al. (1983) found a dose-response curve with a threshold after a single or fractionated dose of γ-radiation (^{137}Cs) in the dose range from 0.25 to 6 Gy when the induction of thymic lymphomas was studied in mice. On the other hand Ullrich and Storer (1979c) found, with a dose rate of 0.45 Gy min^{-1}, a quadratic dose-effect curve with a significant increase of lymphomas after radiation doses of 0.25 Gy and higher and a threshold for a response of 0.1 Gy could not be excluded.

For the induction of ovary cancers data were also obtained in mice which could be described by a dose-effect curve with a threshold. The form of the dose-response was discussed under the aspect that a substantial killing of ovary cells occurs at higher doses and conditions, which lead to an appreciable change in the hormonal

Tab. 3.8 Statistically determined sample sizes of irradiated and control mice needed to detect a significant increase in tumour risk[a] (UNSCEAR 2000)

		Sample size for certain doses			
Mouse strain	Tumour	1,000 mGy	100 mGy	10 mGy	1 mGy
RFM	Thymic lymphoma[b]	1,300	1.2×10^5	1.2×10^7	1.2×10^9
RFM	Myeloid leukaemia[c]	1,700	1.2×10^5	1.2×10^7	1.3×10^9
CBA	Myeloid leukaemia[d]	30	300	4,000	1.3×10^5

[a] $p = 0.05$.
[b] Spontaneous incidence 1.3×10^{-1}; risk of 3×10^{-2} Gy^{-1} assumed.
[c] Spontaneous incidence 7×10^{-3}; risk of 7×10^{-3} Gy^{-1} assumed.
[d] Spontaneous incidence 7×10^{-4}; risk of 1×10^{-1} Gy^{-1} assumed.

status of the animals. With bone tumours, dose-effect curves with a threshold have also been found in a number of experimental studies (Ullrich and Storer 1979c).

However, a number of further experimental studies on cancer induction after exposures to ionising radiation have shown dose-effect curves without a threshold as mentioned before. In a review, an analysis was undertaken in order to find the lowest exposure levels after which significant increases in risk of leukaemia and solid cancers could be observed (Tab. 3.8; UNSCEAR 2000). It can be seen from this table that the lowest dose with a significant effect on cancer induction again varies considerably in the various studies. The results certainly depend on factors which influence the statistical power, such as the number of mice and the spontaneous cancer rate in these investigations.

Most of the studies showed that after high-LET exposures linear dose-response curves without a threshold are dominating. This is the case in the lower and medium dose range (0.1 to 0.5 Gy neutrons). The smallest doses of low-LET radiation after which cancer induction is significantly increased was found for mammary carcinomas and ovarian cancers (Tab. 3.8). This is interesting as these two cancer entities are very much dependent on the hormonal status of the animals which has apparently a significant impact on carcinogenesis in these cases and has been described in connection with promoting effects in a review on combined effects of chemicals (cf. hormones) and radiation (Streffer et al. 2000). The lowest radiation dose after which a significant increase of myeloid leukaemia was observed in mice was 0.25 Gy (X-rays). This was only found in one study. In further studies the doses with a significant enhancement of leukaemias were higher (Tab. 3.8).

The highest radiation exposure from natural sources comes from radon and its radioactive daughter products. In uranium miners it has been shown that radon and its decay-products induce lung tumours (Lubin et al. 1995). Therefore it was of interest to see whether radon and its decay-products will also be able to induce lung cancers in animals and whether it will be possible to study radon exposures in a low-dose range which is found in human residences. Such studies have demonstrated that animal tumours in the respiratory tract can also be observed after exposure to radon (Cross 1992). An increase of tumours in rats was observed after exposure levels below 100 WLM (0.35 Jh m^{-3}) even at exposures down to 25 WLM (0.08 Jh m^{-3}). For the exposure to radon and its radioactive decay-products the unit "working level month" (WLM) is used. Working level (WL) is defined as "any combined amount of radon and of short lived radon daughters in one liter of air that will result in the ultimate emission 1.3 × 10^5 MeV of potential α-energy" (BEIR 1988). (1 WL = 2.08 × 10^{-5} J m^{-3}). One working level month is the exposure equivalent to 170 hours at 1 WL concentration (1 WLM = 3.5 × 10^{-3} Jh m^{-3}).

It is of great interest that in these animal studies an inverse dose rate effect was observed. This means that a long duration of exposure at a lower dose rate yielded more lung cancers than exposures for a shorter duration at a higher dose rate. Rats were exposed to 50 or 500 WLM per week and in most exposure groups there was a significantly higher frequency of cancers in the groups exposed to 50 WLM than to 5000 WLM per week when the same total doses were compared. In addition, the cancer spectrum shifted somewhat to the epidermoid carcinomas when a lower dose rate was used. In a later series of experiments it was found, however, that protracted exposures over 18 months at an α-energy of 2 WL (0.0042 mJ m^{-3}) resulted in a smaller frequency of lung tumours in rats (0.6 % 95 % CI: 0,32–2,33) than an exposure with an α-energy of 100 WL protracted over four months (2.2 %) or over six months (2.4 %) (Morlier et al. 1992, 1994). These data show that the outcome of a protracted exposure can be quite variable in comparison to an acute exposure, especially for high-LET radiation.

In a recent review animal data were carefully analysed with respect to a possible appearance of a threshold dose for carcinogenesis of skin tumours (Tanooka 2001). It was demonstrated that under certain conditions the resulting dose-response curves for cancer induction can best be described with a threshold or a "practical threshold". Skin tumours in mice can be induced only with comparatively high doses of β-radiation from ^{90}Sr-^{90}Y. There is apparently a wide dose range where no tumours are induced. This effect is mouse strain dependent and cancer induction can be enhanced when the radiation is applied in a fractionated manner in one mouse strain. In addition, with high-LET radiation the time factor can be important. Besides the mentioned inverse dose rate effect a reduction of the cancer rate with a decreasing dose rate is also possible. Apparently a host tolerance to tumour induction can develop which may be caused by immunological responses. It is not surprising that such diverse situations have been observed for cancer induction, as the development of cancers is very complex and long-lasting and mechanisms are manifold in different tissues. Many factors can influence the connected biological mechanisms. A promoting effect of ionising radiation is also possible. The question whether such a mecanism can also occur at very low doses. The genetic disposition is very important although not the only determining factor. Cancer development is a multifactorial process (UNSCEAR 2000).

3.5.2
Hereditary diseases

Up to now there are no risk estimates for hereditary diseases available in humans after the exposures to ionising radiation. Apparently the size of irradiated populations and the radiation doses are too small in order to obtain significant increases in hereditary effects in the following generation of an exposed population. Therefore it is necessary to extrapolate risk estimates for humans from relevant animal experiments. For such an extrapolation of data on hereditary damage in mice and other animals to humans some preconditions have to be fulfilled:

– The amount of genetic damage induced by ionising radiation must be the same in human germ cells as for the effects, which have been observed in animal experiments.
– Considering the homologous stages of germ cells, various biological and physical factors (cf. dose, dose rate, LET) should influence the observed damage in a similar way in both, humans and animals (Ehling 1987).

From very extensive studies of hereditary effects in mice it can be concluded that a linear dose relationship is found at low doses and low dose rates of low-LET radiation for the frequency of genetic effects (Russell et al. 1958; Russell and Kelly 1982; UNSCEAR 1988, 2000). Under these conditions it is possible to extrapolate from a medium range of doses to very low radiation doses when keeping the dose rate constant. For various mutation types doubling doses have been determined. The doubling dose is reached when the radiation induced mutations result in the same frequency which is found for the spontaneous mutation frequencies in the investigated animal population. Most national and international organisations have come to a doubling dose of 1 Gy for low-LET radiation and chronic exposures after evaluation of the various animal experimental data (UNSCEAR 1988, 2000). The latter estimate (UNSCEAR 2000) is based on spontaneous mutation rates from humans and radiation-induced mutation rates from mice.

The most comprehensive information on hereditary effects after radiation comes from studies after exposures of male mice in which specific locus mutations were measured. Russell et al. (1982) studied the mutation in these mice at dose rates of 0.72–0.9 Gy min^{-1} with radiation doses between 3 Gy and 6.7 Gy and compared the data with those from dose rates of 8 mGy min^{-1} and smaller for doses between 0.38 and 8.61 Gy. For the studies with the lower dose rate the dose-response curve was less steep. The slope of the curve was reduced by a factor of about 3. This is apparently due to a more extensive repair of damage at lower dose rates. However, it is of high interest that a further reduction in the dose rate to 0.007 mGy min^{-1} did not yield a further reduction of mutations. Apparently some kind of saturation or optimisation of repair of radiation damage has been achieved at a level of dose rates of around 8 mGy min^{-1} so that a further reduction of the dose rate has no effect (UNSCEAR 2000).

Very similar results were obtained with ^{60}Co γ-rays with radiation doses in the range of 6.2 to 6.4 Gy. An acute radiation dose of 6.4 Gy at a dose rate of 0.17 Gy min^{-1} yielded a mutation rate of 13.1×10^{-5} per locus (Lyon et al. 1972). When a dose of 6.30 equal daily fractions the mutation rate decreased to 4.17×10^{-5} per locus which was very similar to the frequency of mutations which resulted from a chronic exposure at 0.08 mGy min^{-1} and a total dose of 6.2 Gy (3.15×10^{-5} per locus) (Lyon et al. 1972).

Searle (1974) summarised data from a number of publications on specific locus mutations in spermatogonia of mice after chronic exposures to γ-rays. The dose range was 0.38–8.6 Gy. The summarised data could be very well fitted to a linear dose-response curve for the mutation frequency represented by the following equation:

$$I = 8.34 \times 10^{-6} + 6.59 \times 10^{-6} D \text{ (assuming 1 Gy} \approx 100 \text{ Roentgens).} \quad \text{(eq. 3.5)}$$

This is in very good agreement with the dose-response curve, which was obtained by Russell and Kelly (1982). Acute doses between 6.7 and 10 Gy yielded a reduction of the mutation frequency at high dose rates, which may be connected to an extended killing of spermatogonia at these high radiation doses. Data on specific locus mutations in mice have also been reviewed after neutron exposures. The exposures were given in an acute and chronic form with dose rates ranging between 0.01 Gy min^{-1} and 0.79 y min^{-1}. The data points in a dose range of 0.5 to 2.1 Gy produced a data set, which fitted the equation:

$$I = 8.30 \times 10^{-6} + 1.25 \times 10^{-4} D. \quad \text{(eq. 3.6)}$$

From these data it follows that for neutrons no dose rate effect occurred which is in agreement with the general finding that DNA repair takes place only to a minor degree after neutron exposure. Comparison of the effects after exposures to low and high-LET radiation resulted in an *RBE* of 19 (Searle 1974).

However, a very pronounced dose rate effect was observed for the genetic mutations after irradiation of pregnant mice with X-rays just before birth. The pregnant mice received an X-ray dose of 3 Gy at dose rates of 0.73–0.93 Gy min^{-1} and in a second experiment of 7.9 mGy min^{-1} 18.5 days after conception (around 2 days before birth). The mutation frequency (specific locus mutations) in the offspring decreased from 8.7×10^{-6} to 6×10^{-7} mutations Gy^{-1} per locus when comparing the high dose rate with the low dose rate exposure which equals a reduction factor of almost 15. The mutation rate of the low dose rate experiments did not differ significantly even with a dose of 3 Gy from that in the controls, even after a total radiation dose. This demonstrates that there exists a very efficient DNA repair in these developmental stages.

Similar experiments, in which the radiosensitivity of mature and maturing oocytes after radiation with the same dose rates was investigated, have led to the result that the mutation frequency dropped about fourfold in the adult (Russell and Kelly 1977). These data show very convincingly that the repair capacity especially in the oocytes is very high and has the lowest mutation frequency after radiation exposures with low dose rates.

In conclusion of these genetic investigations all the data studying mutation frequencies after exposures to ionising radiation show that a linear dose-response was found in the experiments. There was a very strong dependence of the mutation frequency on dose rate in the range down to around 8 mGy min^{-1} after low-LET radiation. This phenomenon was not observed after radiation exposures to high-LET radiation. In all cases no threshold doses have been reported.

Studies which have been performed with children from parents, who were subject to a radiation exposure, resulted in no significant increases of genetic mutations until now (especially data from the atomic bomb survivors in Hiroshima und Nagasaki). However, a trend with increasing radiation doses was observed and the evaluation of such trends will not result in smaller doubling doses as obtained from the animal studies. It may be higher. Therefore it can be concluded that a doubling dose of 1 Gy for genetic diseases is a conservative judgement for the hereditary risk of humans (UNSCEAR 2001).

3.6
Conclusions

For risk estimates the knowledge of the shape of dose-response curves after exposures to toxic agents is essential. This is also the case for the evaluation of radiation risk. For a number of radiation effects it has been shown that the dose-effect curve has a threshold and these radiation responses are only seen after radiation doses above the threshold doses which are in the range 100 mSv and higher. The underlying mechanisms are multicellular processes in which many cells have to be damaged for the development of these health effects. Usually death of stem cells is responsible for these effects.

However, for the induction of hereditary defects and of cancers it is assumed that dose-response curves exist without a threshold dose and this appears also to be the case for certain developmental effects in cases where a genetic predisposition for such developmental effects exists. The observed radiation effects e. g. the radiation-induced cancers do not have specific features. They cannot be distinguished by their clinical appearance or molecular changes from the "spontaneous" health effects, which develop without any radiation exposure. This means: Exposures to ionising radiation increase health effects, which are also found without these exposures. Radiation only increases the frequencies of these "spontaneous" events. Therefore the radiation effects must be higher than the scatter of the spontaneous damage in order to be determined significantly. This is usually the case only after radiation doses of around 100 mSv of low-LET radiation. It is also not very surprising that no increased cancer rates are observed in regions with high background radiation under these circumstances. Besides other conditions (size of group number, biasing factors etc.), the radiation dose rates are low and the doses are not high enough in order to cause a significant increase of cancer rates in the populations living in areas with high background radiation.

The most important question for risk evaluation is: Are any effects induced below these radiation doses with measurable effects, which cannot be seen only because they are within the variation of the baseline of cancer or do no radiation effects, like mutations or cancers, exist in the low-dose range? In order to answer the questions, the mechanisms for the development of the radiation effects must be evaluated. These so-called stochastic effects develop through several steps before a manifestation of health effects like cancer occurs (Fig. 3.3). The initial steps are apparently connected to radiation damage induced in the DNA, although some more recent data demonstrate that changes of signal transduction from the cell membrane to the nucleus and regulation processes for gene expression may also be important. In principle, similar DNA damages (polynucleotide breaks, base damage) can be caused by endogenous processes cf. induced through metabolically formed oxygen radicals. However, ionising radiation leads to clusters of such damaging events, whereas with oxygen radicals isolated damaged sites are most frequent. In the low-dose range, the dose distribution between individual cells shows a wide variability and cells will be found which have not been hit while neighbour cells experience comparatively high doses. In that way exposure to ionising radiation differs fundamentally with respect to the distribution of damaged sites in the DNA from exposures to chemicals. These microdosimetric considerations are important for the risk evaluation in the low-dose range.

Dose-response curves for chromosomal aberrations, malignant cell transformations and gene mutations have been studied after exposures to a wide variety of radiation qualities in many cellular and molecular systems *in vitro*. The irradiation has taken place in single cell cultures. The results from these experiments show that generally the data can be described by dose-effect curves without a threshold, but curves with a threshold dose cannot be excluded. In the low-dose range and with low dose rates a linear dose-response is usually dominating. In the medium and higher dose range a linear quadratic dose-effect curve has usually been observed with low-LET radiation. In contrast, with high-LET radiation the dose-response remains linear over a wider dose range. With low-LET radiation the dose-response becomes shallower when the dose rate decreases. The uncertainties in the low-dose range are larger for low-LET radiation than for high-LET radiation. In the latter case the dose-response curve can be fairly well and safely described by linearity without a threshold.

The experimental studies on cancer in animals give more conflicting results with respect to the shape of dose-response curves. In a number of cases the fit for the dose-response was better with a threshold, however, in many cases the data could be fitted in a more adequate way with dose-response curves without a threshold. This was especially the case for cancers with an endocrine regulation like mammary and ovary cancers, while bone cancers usually showed dose-effect curves with a threshold. Such a situation has also frequently been observed for skin cancers. With respect to hereditary effects it appears that in all cases the dose-response has no threshold.

The dose-response can be modified by a number of factors as time and duration of exposure, DNA repair, adaptive response, apoptosis, genetic instability and genetic disposition. With low-LET radiation the biological effects are smaller when the radiation exposure takes place with a low dose rate than with high rates in the range of 0.5 to 1.0 Gy min^{-1}. It seems that dose rates in the range of 1–10 mGy min^{-1} lead to the highest reduction of the radiation effects. In addition, fractionation of the radiation dose diminishes the effects. DNA repair is mainly responsible for these reductions. It is apparently possible to induce DNA repair by small radiation doses. However, this adaptive response only occurs under very limited conditions, it is dependent on radiation dose and dose rate. It is only seen to a small degree or not at all with high-LET radiation. Adaptive response cannot be observed in all tissue systems, in the prenatally developing organism and not with all individuals. The situation is very similar with respect to apoptosis, a biological mechanism for cell killing (suicide of the cell), which is triggered by signal transduction. In principle, it is possible through this mechanism that damaged cells, including malignant cells, can be eliminated and can therefore reduce the manifestation of health effects. All these modifying factors vary strongly between different cell systems, tissues and individuals.

With high-LET radiation these modifications of the dose-response are very small or do not exist at all. With long lasting exposures other mechanisms, like cell repopulation or a stimulation of the immune system, may be important, however, the data about such effects are comparatively limited. The development of cancer is a very complex process and not only the cellular phenomenon has to be evaluated. The influence of tissue response as well as of the immune system is important, however, these processes are little understood until now.

Clinical experiences as well as studies with cells from patients have shown that certain genetic syndromes exist with a strong hypersensitivity. These individuals fall clearly outside the normal distribution of radiosensitivity within our populations. Investigations of the mechanisms have demonstrated that very frequently deficiencies in certain DNA repair pathways and disturbances in the regulation of the cell proliferation cycle occur in these persons. An increase of acute and late radiation damages have been observed in persons with a genetic predisposition for an increased radiosensitivity. In addition, carcinogenesis is enhanced, however, there are no observations after small radiation doses available to date. This is probably because the numbers of individuals with such defects are small and therefore relevant epidemiological studies are very difficult if not impossible.

For practical radioprotection, where prospective estimates of risk have to be made and for risk evaluation as the basis for regulatory purposes, which are based on such estimates, it seems to be appropriate further to use the linear dose-response without threshold doses for stochastic radiation effects although the uncertainties are high and there is no solid scientific proof for such a shape of the dose-effect curve. This may be an overestimation of risk under certain conditions which is justified by taking precautionary principles into account.

3.7
References

Albertini RJ, Clark LS, Nicklas JA O'Neill P, Hui TE, Jostes R (1997) Radiation quality affects the efficiency of induction and the molecular spectrum of HPRT mutations in human T cells. Radiat Res 148: 76–86

Azzam EI, de Toledo SM, Gooding T, Little JB (1998) Intercellular communication is involved in the bystander regulation of gene expression in human cells exposed to very low fluences of α-particles. Radiat Res 150: 497–504

Baria K, Warren C, Eden OB, Roberts SA, West CM, Scott D (2002) Chromosomal radiosensitivity in young cancer patients: possible evidence of genetic predisposition. Int J Radiat Biol 78(5): 341–346

Cardis, E, ES Gilbert, L Carpenter Howe G, Kato I, Armstrong BK, Beral V, Cowper G, Douglas A, Fix J, Fry SA, Kaldor J, Lavé C, Salmon L, Smith PG, Voelz GL, Wiggs LD (1995) Effects of low doses and low dose rates of external ionizing radiation: cancer mortality among nuclear industry workers in three countries. Radiat Res 142: 117–132

Committee on the Biological Effects of Ionizing Radiations [BEIR IV] (1988) Health risks of radon and other internally deposited α-emitters. Natl Acad Sci USA, Natl Res Council. National Academy Press, Washington, DC

Committee on the Biological Effects of Ionizing Radiations [BEIR V] (1990) Health effects of exposure to low levels of ionizing radiation. Natl Acad Sci USA, Natl Res Council. National Academy Press, Washington, DC

Cregan SP, Brown DL, Mitchel RE (1999) Apoptosis and the adaptive response in human lymphocytes. Int J Radiat Biol 75(9): 1087–1094

Cross FT (1992) A review of experimental animal radon health effects data. In: Chapman JD, Dewey WC, Whitmore GF (eds) Radiation research: a twentieth-century perspective, Vol. II. Academic Press, San Diego, CA, pp 476–481

Ehling UH (1987) Quantifizierung des strahlengenetischen Risikos. Strahlenther Onkol 163: 283–291

Fearon ERB, Vogelstein (1990) A genetic model for colorectal tumourgenesis. Cell 61: 759–767

Feinendegen LE, Neumann RD (1999) Cellular responses to low doses of ionizing radiation. Workshop Report. United States Department of Energy and the National Institutes of Health, Washington DC

Goodhead DT (1994) Initial events in the cellular effects of ionizing radiations: clustered damage in DNA. Int J Radiat Biol 65: 7–17

Grahn D, Sacher GA (1968) Fractionation and protraction factors and the late effects of radiation in small mammals. In: Brown G, Cragle R, Noonan T (eds) Dose rate in mammalian radiation biology, CONF-680410. US Atomic Energy Commission, Oak Ridge, TN, pp 2.1–2.27

Greenblatt MS, Bennett W P, Hollstein M, Harris CC (1994) Mutations in the *p53* tumor suppressor gene: clues to cancer etiology and molecular pathogenesis. Cancer Res 54: 4855–4878

Hall E J (1994) Radiobiology for the radiologist. 4th Edition, J B Lippincott Company, Philadelphia, PA

Hendry JH, Potten CS, Merritt A (1995) Apoptosis induced by high- and low-LET radiations. Radiat Environ Biophys 34: 59–62

International Commission on Radiation Units and Measurements [ICRU] (1993) Microdosimetry. Report 36. ICRU Publications, Bethesda

International Commission on Radiological Protection [ICRP] (1977) ICRP Publication 26. Recommendations of the International Commission on Radiological Protection. Annals of the ICRP 1(3), Pergamon Press, Oxford, reprinted with additions 1987

International Commission on Radiological Protection [ICRP] (1990) ICRP Publication 60. Recommendations of the International Commission on Radiological Protection. Annals of the ICRP 21(1–3), Pergamon Press, Oxford, 1991

International Commission on Radiological Protection [ICRP] (1998) ICRP Publication 79 Genetic susceptibility to cancer. Annals of the ICRP 28(1–2), Pergamon Press, Oxford, 1999

Jacob P, Kenigsberg Y, Zvonova I, Goulko G, Buglova E, Heidenreich WF, Golovneva A, Bratilova AA, Drozdovitch V, Kruk J, Pochtennaja GT, Balonov M, Demidchik EP, Paretzke HG (1999) Childhood exposure due to the Chernobyl accident and thyroid cancer risk in contaminated areas of Belarus and Russia. Br J Cancer 80(9): 1461–1469

Joiner M, Marples B, Lambin P, Short SC, Turesson I (2001) Low-dose hypersensitivity: current status and possible mechanisms. Int J Radiat Oncol Biol Phys 49: 379–389

Kastan MB, Onyekwere O, Sidransky D, Vogelstein B, Craig RW (1991) Participation of *p53* protein in the cellular response to DNA damage. Cancer Res 51: 6304–6311

Kelly EM (1982) Mutation frequencies in male mice and the estimation of genetic hazards of radiation in man. Proc Natl Acad Sci USA 79: 542–544

Kondo S (1998) Apoptotic repair of genotoxic tissue damage and the role of *p53* gene. Mutat Res 402: 311–319

Little JB (1994) Failla Memorial Lecture: Changing views of cellular radiosensitivity. Radiat Res 140: 299–311

Little JB (1998) Radiation-induced genomic instability. Int J Radiat Biol 6 : 663–671

Little JB (2000) Radiation carcinogenesis. Carcinogenesis 21: 397–404

Lloyd DC, Edwards AA, Leonard A, Deknudt GL, Verschaeve L, Natarajan AT, Darroudi F, Obe G, Palitti F, Tanzarella C, Tawn EJ (1992) Chromosomal aberrations in human lymphocytes induced *in vitro* by very low doses of X-rays. Int J Radiat Biol 62: 53–63

Lubin JH, Boice JD Jr., Edling C Hornung RW, Howe GR, Kunz E, Kusiak RA, Morrison HI, Radford EP, Samet JM, Tirmarche M, Woodward A, Yao SX, Pierce DA (1995) Lung cancer risk in radon-exposed miners and estimation of risk from indoor exposure. J Natl Cancer Inst 87: 817–827

Lyon MF, Phillips RJS, Bailey HJ (1972) Mutagenic effect of repeated small radiation doses to mouse spermatogonia. I. Specific locus mutation rates. Mutat Res 15: 185–190

Maisin JR, Wambersie A, Gerber GB, Mattelin G, Lambiet-Collier M, De Coster B, Gueulette J (1983) The effects of a fractionated γ-irradiation on life shortening and disease incidence in BALB/c mice. Radiat Res 94: 359– 373

Mill AJ, Frankenberg D, Bettega D, Hieber L, Saran A, Allen LA, Calzolari P, Frankenberg-Schwager M, Lehane MM, Morgan GR, Pariset L, Pazzaglia S, Roberts CJ, Tallone L (1998) Transformation of C3H 10T? cells by low doses of ionising radiation: a collaborative study by six European laboratories strongly supporting a linear dose-response relationship. J Radiol Prot 18: 79–100

Miller RC, Randers-Pehrson G, Geard CR, Hall EJ, Brenner DJ (1999) The oncogenic transforming potential of the passage of single α-particles through mammalian cell nuclei. Proc Natl Acad Sci USA 96: 19–22

Mole RH (1992) Expectation of malformation after irradiation of the developing human in utero: the experimental basis for predictions. Adv Radiat Biol 15: 217–301

Molls M, Streffer C (1984) The influence of G2 progression on X-ray sensitivity of two-cell mouse embryos. Int J Radiat Biol 46: 355–65

Morlier JP, Morin M, Monchaux G, Fritsch P, Pineau JF, Chameaud J, Lafuma J, Masse R (1994) Lung cancer incidence after exposure of rats to low doses of radon: influence of dose-rate. Radiat Prot Dosim 56(1–4): 93–97

Morlier JP, Morin M, Chameaud J, Masse R, Bottard S, Lafuma J (1992) Importance du rôle du débit de dose sur l'apparition des cancers chez le rat après inhalation de radon. C R Acad Sci, Ser 3, 315: 436–466

Muirhead CR, Goodill AA, Haylock RGE, Vokes J, Little MP, Jackson DA, O'Hagan JA, Thomas JM, Kendall GM, Silk TJ, Bingham D, Berridge GLC (1999) Occupational radiation exposure and mortality: second analysis of the National Registry for Radiation Workers. J Radiol Prot 19: 3–26

Müller WU, Streffer C, Pampfer S (1994) The question of threshold doses for radiation damage: malformations induced by radiation exposure of unicellular or multicellular preimplantation stages of the mouse. Radiat Environm Biophys 33: 63–68

Müller WU, Bauch T, Stuben G, Sack H, Streffer C (2001) Radiation sensitivity of lymphocytes from healthy individuals and cancer patients as measured by the comet assay. Radiat Environ Biophys 40: 83–89

Norimura T, Nomoto S, Katsuki M, Gondo Y, Kondo S (1996) P53 dependent apoptosis suppresses radiation-induced teratogenesis. Nat Med 2: 577–580

Oya N, Zolzer F, Werner F, Streffer C (2003) Effects of serum starvation on proliferation and apoptosis in four human tumor cell lines with different *p53* status. Strahlenther Onkol 179(2): 99–106

Pampfer S, Streffer C (1989) Increased chromosome aberration levels in cell form mouse fetuses after zygote x-irradiation. Int J Radiat Biol 55(1): 85–92

Paretzke, HG (1989) Physical aspects of radiation quality. In: Baverstock KF, Stather IW (eds), Low dose radiation: biological bases of risk assessment. Taylor and Francis, London, pp 514–522

Pils S, Müller WU, Streffer C (1999) Lethal and teratogenic effects in two successive generations of the HLG mouse strain after radiation exposure of zygotes – association with genomic instability? Mutat Res 429: 85–92

Raabe OG, Culbertson MR, White RG, Parks NJ, Spangler WS, Samuels SJ (1995) Lifetime radiation effects in beagles injected with ^{226}Ra as young adults. In: van Kaick G, Karaoglou A, Kellerer AM (eds), Health effects of internationally deposited radionuclides: emphasis on radium and thorium. World Scientific, London, pp 313–318

Russell WL, Kelly EM (1977) Mutation frequencies in female mice and the estimation of genetic hazards or radiation in women. Proc Natl Acad Sci USA 74: 3523–3527

Russell WL, Russell LB, Kelly EM (1958) Radiation dose rate and mutation frequency. Science 128: 1546–1550

Scherer E, Streffer C, Trott K-R (eds) (1991) Radiopathology of organs and tissues. Springer Verlag, Berlin

Schiestl RH, Khogali F, Cals N (1994) Reversion of the mouse pink-eyed unstable mutation induced by low doses of X-rays. Science 266: 1573–1576

Schofield PN (1998) Impact of genomic imprinting on genomic instability and radiation-induced mutation. Int J Radiat Biol 74(6): 705–710

Searle AG (1974) Mutation induction in mice. Adv Radiat Biol 4: 131–207

Sparrow AH, Underbrink AG, Rossi HH (1972) Mutations induced in tradescantia by small doses of X-rays and neutrons: analysis of dose response curves. Science 176: 916–918

Streffer C (1991) Stochastische und nicht-stochastische Strahlenwirkungen. Nucl Med 30: 198–205

Streffer C (1997) Threshold dose for carcinogenesis: what is the evidence? In: Goodhead DT, O'Neill P, Menzel HG (eds) Microdosimetry. An interdisciplinary approach. Proc. 12th Symp on Microdosimetry. The Royal Society of Chemistry, Cambridge, pp 217–224

Streffer C (2002) Adaptive response after exposure to ionising radiation. Proc 4th Int Conf Health Effects of low-level radiation. British Nuclear Energy Society, Oxford

Streffer C, Bücker J, Cansier A, Cansier D, Gethmann CF, Guderian R, Hanekamp G, Henschler D, Pöch G, Rehbinder E, Renn O, Slesina M, Wuttke K (2000) Umweltstandards. Kombinierte Expositionen und ihre Auswirkungen auf die Umwelt. Wissenschaftsethik und Technikfolgenbeurteilung, Bd. 5. Springer, Berlin

Trott KR, Streffer C (1990) Occupational Radiation Carcinogenesis. In: Scherer E, Streffer C, Trott KR (eds) Radiation exposure and occupational risks. (Medical Radiology – Diagnostic Imaging and Radiation Oncology). Springer Verlag, Berlin, pp 61–74

Ullrich RL, Bowles ND, Satterfield LC, Davis CM (1996) Strain-dependent susceptibility to radiation-induced mammary cancer is a result of differences in epithelial cell sensitivity to transformation. Radiat Res 146: 353–355

Ullrich RL, Storer JB (1979a) Influence of γ-irradiation on the development of neoplastic disease in mice. I. Reticular tissue tumours. Radiat Res 80: 303–316

Ullrich RL, Storer JB (1979b) Influence of γ-irradiation on the development of neoplastic disease in mice. II. Solid tumours. Radiat Res 80: 317–324

Ullrich RL, Storer JB (1979c) Influence of γ-irradiation on the development of neoplastic disease in mice. III Dose-rate effects. Radiat Res 80: 325–342

United Nations Scientific Committee on the Effects of Atomic Radiation [UNSCEAR] (1977) Report to the General Assembly, with 10 annexes. United Nations, New York

United Nations Scientific Committee on the Effects of Atomic Radiation [UNSCEAR] (1982) Report to the General Assembly, with 12 annexes. United Nations, New York

United Nations Scientific Committee on the Effects of Atomic Radiation [UNSCEAR] (1988) Report to the General Assembly, with 7 annexes. United Nations, New York

United Nations Scientific Committee on the Effects of Atomic Radiation [UNSCEAR] (1993) Report to the General Assembly, with 9 scientific annexes. United Nations, New York

United Nations Scientific Committee on the Effects of Atomic Radiation [UNSCEAR] (1994) Report to the General Assembly, with 2 scientific annexes. United Nations, New York

United Nations Scientific Committee on the Effects of Atomic Radiation [UNSCEAR] (2000) Report to the General Assembly, with 10 scientific annexes. United Nations, New York

United Nations Scientific Committee on the Effects of Atomic Radiation [UNSCEAR] (2001). Report to the General Assembly, with 1 scientific annex. United Nations, New York

United States Department of Energy and the National Institutes of Health [NIEHS] (1999) Cellular Responses to Low Doses of Ionizing Radiation.Workshop Report, United Nations, New York

Warenius HM, Down JD (1995) *RBE* of fast neutrons for apoptosis in mouse thymocytes. Int J Radiat Biol 68: 625–629

Wei L, Sugahara T, Tao Z (eds) (1997) High Levels of Natural Radiation. Elsevier, Amsterdam

Weissenborn U, Streffer C (1988) Analysis of structural and numerical chromosomal anamalies at the first, second and third mitosis after irradiation of one-cell mouse embryos with X-rays or neutrons. Int J Radiat Biol 54: 381–394

Wojcik A, Streffer C (1994) Adaptive response to ionizing radiation in mammalian cells: a review. Biol Zent Bl 113: 417–434

Wright EG (1999) Inherited and inducible chromosomal instability. A fragile bridge between genome integrity mechanisms and tumorigenesis. J Pathol 187:19–27

4 Toxicology of Chemical Carcinogens

4.1
Introduction

The basic toxicodynamic principles of chemically and of radiation-induced carcinogenesis are largely identical. However, toxicokinetic factors of uptake, distribution, metabolism and elimination of chemical compounds lead to further complications. A chemical carcinogen may either act in the form of the parent compound, or, very frequently, in the form of an ultimately active metabolite. Hence, a
complicated interplay of activating and inactivating enzyme systems must be considered in performing low-dose extrapolations. This chapter provides some outstanding examples of typical ways of argumentation.

4.2
General concepts of chemical carcinogenesis and
dose-response relationships

General principles of the sequence of events that finally lead to cancer after exposure to genotoxic carcinogens have become increasingly clear (Fig. 4.1). The first
critical step in carcinogenesis often is metabolic activation of a procarcinogen
that by itself does not interact with DNA, resulting in a genotoxic DNA binding
substance (Fig. 4.1). Such ultimately carcinogenic metabolites, mostly of electrophilic nature, bind covalently to DNA structures (normally the nucleotide
bases), resulting in formation of "DNA adducts". This resulting DNA damage can
lead to DNA mutations. Formation of mutations from DNA adducts requires cell
proliferation. A typical example is methylation of guanine leading to the adduct
O^6-methylguanine. Due to DNA repair mechanisms this miscoding methylation
product may be removed. However, during DNA replication O^6-methylguanine
may lead to formation of a stable mispair with thymine, producing a GC→AT
transition. While the DNA adduct, for instance O^6-methylguanine is still
reversible, DNA mutations, such as the transition from GC to AT, usually represent irreversible DNA alterations. When mutations are induced into critical genes,
such as oncogenes or tumour suppressor genes, the cell cycle control may be disturbed resulting in proliferation (Fig. 4.1). This initiates a dynamic multistep
process leading to activation of further oncogenes and inactivation of tumour suppressor genes. Usually the multistep process of carcinogenesis takes several years
of "latency time" until a tumour becomes manifest.

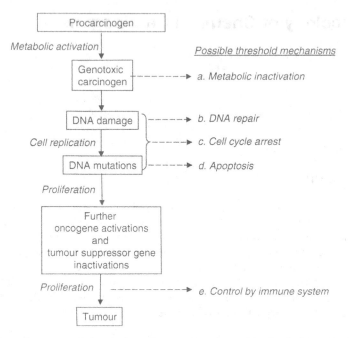

Fig. 4.1 General concept of chemical carcinogenesis and possible threshold mechanisms of genotoxic carcinogens

4.3
Modifiers of dose-response of the curve

Critical steps that finally lead to cancer have been summarised in the previous chapter. Several protective mechanisms (Fig. 4.1 a–e) may inhibit the multistep process of carcinogenesis. It is obvious that these parameters can influence the shape of the dose-effect curve for carcinogenesis and, in principle, can lead to thresholds. Lack of metabolic activation of a procarcinogen in a certain animal species (Fig. 4.1) would result in a mechanism for a complete preclusion of its carcinogenic action in that species. In general, the capacity of the activating metabolism influences the shape of the dose-effect curve. For instance, the slope of the dose-effect curve of induced liver tumour incidence in rats, elicited by the N,N-diethylnitrosamine or the tobacco-specific nitrosoketone NNK decreases at high doses due to saturation of cytochrome P450 2E1, which activates N,N-diethylnitrosamine and NNK to genotoxic intermediates (Bolt et al. 2003; see also sections 4.5.3 and 4.5.4).

Different types of mechanisms may lead to situations for a given chemical where a threshold-type dose-response might be at least scientifically plausible. Such cases may be addressed as those with a "practical threshold" (for differentiation of different types of threshold, see also Kirsch-Volders et al. 2000). One outstanding example is that of the aromatic hydrocarbon styrene, an industrial compound of wide use (DFG 1999).

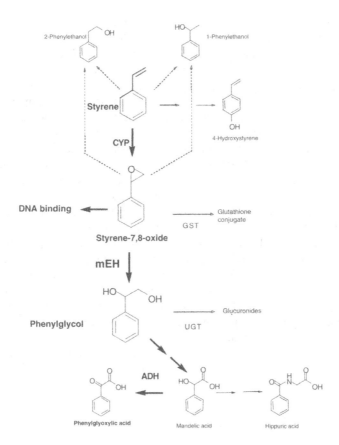

Fig. 4.2 Metabolism of styrene. After inhalation styrene is activated to the DNA binding agent styrene-7,8-oxide (STO) in a cytochrome P450 dependent reaction. STO can be detoxified by microsomal epoxide hydrolase (mEH) forming phenylglycol as an intermediate that is further conjugated by glucuronosyl transferases (UGT). The resulting glucuronides can easily be excreted in urine. Further metabolic pathways of phenylglycol involve formation of phenylglyoxylic acid by aldehyde dehydrogenase (ADH) from mandelic acid or formation of hippuric acid. An alternative metabolic pathway for detoxification of STO is catalysed by glutathione S-transferases (GST) resulting in non-toxic conjugates that can be excreted in urine (Herrero et al. 1997).

A pivotal step on the path to cancer that influences the shape of the dose-effect curve is metabolic inactivation of genotoxic carcinogens (a. in Fig 4.1). As a typical example, the influence of microsomal epoxide hydrolase (mEH) is presented that influences the dose-effect curve in a way that can be regarded as characteristic for the influence of metabolic inactivation. Styrene is activated to styrene-7,8-oxide by cytochrome P450, in first instance CYP2E1 (Fig. 4.2). Styrene oxide may bind to DNA, predominantly to the N^7 positon of guanine, which may lead to DNA strand breaks. Microsomal epoxide hydrolase (mEH) inactivates styrene-7,8-oxide by hydrolysis. To examine whether this detoxifying mechanism can introduce a threshold into the dose-response curve human mEH was transfected into a Chinese hamster cell line (V79 cells) that, before transfectio,n expresses only very low activities of this enzyme (Fig. 4.3a and b; Herrero et al. 1997). In such cells with low mEH, styrene oxide leads to the formation of DNA strand breaks in a dose-dependent manner with no observable threshold (Fig. 4.3c; Oesch et al. 2000; Herrero et al. 1997). Chinese hamster cells, genetically engineered to express human mEH at similar activities as observed in human liver, are protected from measurable genotoxic effects of styrene oxide up to 100 µM. Of course, for the technical reason of detection limit it is not possible to differentiate whether transfection of mEH introduced a "practical threshold" (characterised by a very low level of DNA damage, not measurably exceeding the background in absence of styrene oxide) or a "real threshold" (characterised by a dose-effect curve crossing the abscissa at 100 µM styrene oxide, i. e. representing a truly zero effect up to a concentration of 100 µM styrene oxide). Although a decision cannot be obtained experimentally, the model of a "practical threshold" appears favourable, since it seems improbable that a detoxifying enzyme can exclude that even very small numbers of genotoxic molecules still reach the DNA. If the assumption that detoxification does not guarantee completeness is accepted, it seems permissible to conclude that high activities of metabolic inactivation can introduce "practical", albeit not "perfect" thresholds. Whether a detoxifying enzyme activity qualifies as basis for a practical threshold depends on the speed and capacity of removal of the reactive species from the system, compared with the speed of the translocation of the reactive species from the

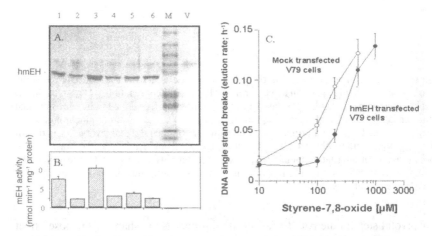

Fig. 4.3 Recombinant expression of human microsomal epoxide hydrolase (mEH) protects V79 Chinese hamster cells (V79 cells) from styrene-7,8-oxide induced DNA single strand breaks. **A.** Western Blot analysis of mEH transfected (lanes 1–6) and mock transfected (lane V) V79 cells. S9 fractions of the different clones were analysed with a polyclonal antibody against purified human mEH raised in rabbits. M: Pre-stained molecular weight markers (195, 112, 84, 63, 52, 35, and 32 *kD*). **B.** mEH activity of the same clones using styrene-7,8-oxide as a substrate. Values are expressed as nmol styrene glycol min^{-1} mg^{-1} protein. **C.** Effect of styrene-7,8-oxide on DNA single strand breaks using mEH transfected clone no. 3 (from A. and B.) and mock transfected V79 cells (Herrero et al. 1997).

site of its generation to the nucleus and reaction with the DNA. In the case of styrene oxide, a two-step mechanism of the inactivating microsomal epoxide hydrolase leads to the observation of a "practical threshold" (Fig. 4.3c). The mechanism consists of a very fast first step of removal of the epoxide, by covalent interaction with the enzyme followed by much slower hydrolysis of the complex. The high capacity of this mechanism is provided by an unusually high amount of enzyme. Thus, microsomal epoxide hydrolase sucks up the genotoxic epoxide like a sponge, up to a concentration of epoxide, which has titrated out the high amount of enzyme (Herrero et al. 1997: Oesch et al. 2000). Mechanisms of rapid inactivation vs. activation have also been discussed for other carcinogens, e. g. trichloroethylene (Brüning and Bolt 2000).

DNA repair protects cells from permanent fixation of DNA damage upon DNA replication, thereby generating the situation of a competition between repair and proliferation-dependent DNA synthesis (b in Fig. 4.1). Effective DNA repair can reduce the tumour outcome. However, again it seems unlikely that DNA repair mechanisms are perfect. If an extremely low dose of a carcinogen would induce only a single DNA adduct in a cell, this adduct has a probability to persist unrepaired until DNA replication takes place which may lead to "fixation" of the DNA adduct as a heritable mutation. Although this probability may be very low, it is not zero. If this assumption is accepted, the influence of DNA repair may be regarded similar to the consideration given above concerning metabolic inactivation: "practical", but not "perfect" thresholds may be introduced.

A carcinogen may induce DNA repair enzymes. For instance, the DNA repair protein O^6-methylguanine-DANN-methyltransferase (MGMT), responsible for direct reversal of the miscoding DNA lesion O^6-methylguanine, has been reported to be inducible in mammalian tissues (Vielhauer et al. 2001). Intraperitoneal injection of diethylnitrosamine in rats led to a 1.3-fold and 2.6-fold increase in MGMT mRNA expression in periportal and perivenous hepatocytes, respectively. Similarly, the carcinogen ethylnitrosourea induced mRNA levels in rat hepatocytes, up to 1.6-fold (Vielhauer et al. 2001). Overexpressing MGMT in transgenic mice has been shown to reduce the occurrence of several malignomas, e. g. age-related hepatocellular carcinomas (Walter et al. 2001; Zhou et al. 2001), and lymphomas induced by methylnitrosourea (Reese et al. 2001). Thus, one can speculate that under certain circumstances exposure to carcinogens might induce levels of DNA repair enzymes and therefore reduce levels of DNA damage below baseline levels (i. e. below the levels of "spontaneous" DNA lesions), leading to a tumour incidence even lower than in the control group. Although treatment with low doses of some carcinogens in animal studies has indeed been shown to lead to a lower yield of some tumour entities than in the control group, to our knowledge this theoretical possibility with respect to induction of DNA repair enzymes in humans has not yet been supported by tumour incidence data in epidemiological studies.

Cell cycle arrest can be induced as a consequence of DNA damage or interference with signal transduction in target cells (Fig. 4.1c). Low levels of a carcinogen may even decrease cell cycle progression below baseline rates (Lutz 1998; Lutz and Kopp-Schneider 1999). Since under specific circumstances the protective influence of decreased cell division can be stronger than the deleterious influence of increased DNA damage, the combination of both effects may result in a decrease in tumour incidence. Higher levels of the same substance will increase cell cycle progression due to cytotoxicity and regenerative cell proliferation, resulting in an increased tumour incidence. As a consequence, a J-shaped dose-effect curve would result. This mechanism has been observed for non-genotoxic carcinogens, such as TCDD (Kitchin et al. 1994) or caffeic acid (Lutz et al. 1997; reviewed in Lutz 1998), but has been postulated also for genotoxic carcinogens, such as 2-acetylaminofluorene and ionising radiation. Of course a "benefit" of the first, decreasing part of a J-shaped dose-effect curve must be interpreted with caution, since a decrease in cell proliferation below baseline might interfere with normal tissue regeneration. In addition, the protective influence may be tissue- or cell- type specific. Cell cycle delay may be induced in one cell type, whereas other more sensitive cell types may respond with cell proliferation. Nevertheless, cell cycle progression and cell proliferation probably represent the most relevant key-parameters concerning threshold mechanisms. Due to the lack of fixation of DNA damage as a stable mutation in a newly synthesised daughter strand of DNA a genotoxic substance will not be able to induce tumours in tissues that do not proliferate. For instance, the extremely low proliferative capacity of cardiomyocytes is likely to contribute to the high resistance of this cell type against neoplastic transformation. On the other hand, a genotoxic substance will cause tumours with an extremely high probability, when both, DNA damage and cell proliferation, will be induced in target tissues. The situation becomes difficult, when a given dose of a substance induces DNA damage, but not cytotoxicity and proliferative cell regeneration. In this case the result may depend on baseline proliferation of the relevant tissue. Rapidly proliferating cells in bone marrow or the crypt cells of the colon will be at high risk for neoplastic transformation. On the other hand, for cells with a relatively low baseline proliferation such as the olfactory epithelium, the latency period for carcinogenesis may exceed life expectancy (Lutz 2000). Of high interest are genotoxic substances for which both, induction of regenerative proliferation and genotoxicity act through a threshold process. For instance, paracetamol, formaldehyde and vinyl acetate appear to belong to this class of substances (see below).

Apoptosis (d in Fig. 4.1) and the control of neoplastically transformed cells by the immune system (e. in Fig. 4.1) are further mechanisms influencing the shape of the dose-effect curve. Cells may undergo apoptosis as a consequence of DNA damage induced by relatively high doses. There is no doubt that this process can reduce tumour rates. However, little is known about the efficiency of apoptotic mechanisms at low doses and whether such mechanisms can lead to thresholds of carcinogenesis.

Considering the complexity of mechanisms that may introduce "practical" or "perfect" thresholds, it becomes clear that evaluation of the risk of genotoxic substances is difficult, albeit not impossible. In order to underpin these theoretical considerations the dose-effect curves of some well-investigated carcinogens will be discussed to analyse the influence of the factors modifying the dose-effect relation. This comprises explicitly examples of genotoxic carcinogens, with and without threshold mechanisms.

4.4
Endogenous chemical carcinogens

Electrophilic metabolites that attach to DNA and thereby form DNA adducts are not only resulting from foreign compounds entering the body, but are also produced, to some extent, endogenously (Bolt 1997).

The existence of "endogenous" DNA adducts is now generally accepted (Nath et al. 1996). The discussion has been predominantely focussed on so-called "I-compounds" (indigenous compounds), detected by the method of "^{32}P-postlabelling" of DNA adducts (Randerath et al. 1993). Part of these adducts are apparently formed by natural products of lipid peroxidation (e. g. malondialdehyde or hydroxynonenal), but the structure and genesis of most of these "I-compounds" is not clear. Physiological background levels differ between tissues and are dependent on gender, age and nutritional factors. Typically, orders of magnitude of 1 adduct per 1 million DNA bases have been reported (Nath et al. 1996). Interestingly, etheno-bridged exocyclic DNA adducts, namely 1,N^6-ethenodeoxyguanine and 3,N^4-ethenodeoxycytidine, have been demonstrated at low and variable background levels in hepatic DNA of untreated rodents and in humans (Nair et al. 1995). Such adducts are associated with mutational consequences upon cell replication (Hang et al. 1998), and are typical products of the hepatic chemical carcinogens vinyl chloride and vinyl carbamate (Laib and Bolt 1980; Nair et al. 1995; Swenberg et al. 2000). The specific background of these promutagenic DNA lesions is thought to arise from endogenously formed lipid peroxidation products (Nair et al. 1995).

A chemically defined endogenous carcinogen is isoprene which, in physiologically activated form, serves to generate isoprenoids, such as cholesterol, steroids, bile acid, and the side chain of K vitamins (Peter et al. 1987). Isoprene is metabolised into a di-epoxide (2-methyl-2,2'-bi-oxirane) with mutagenic and genotoxic properties. In long-term bioassays on mice, isoprene is clearly carcinogenic, with a potency of about 1/10 that of 1,3-butadiene (Placke et al. 1996).

The quantitative role of endogenous ethylene oxide (ethene oxide, oxirane) has been more thoroughly investigated (Bolt 1996; Thier and Bolt 2000). Ethyl oxide is a directly reactive and genotoxic chemical. Long-term inhalation studies with this compound in mice and rats, at exposure levels between 10 and 100 ppm, have resulted in formation of malignant tumours at multiple sites (IARC 1982, 1987). In blood cells of humans occupationally exposed to ethylene oxide increased counts of chromosomal aberrations, of sister chromatid exchanges, and micronuclei have been found (IARC 1982).

In laboratory animals (Ehrenberg et al. 1977; Filser and Bolt 1984) as well as in humans (Törnqvist et al. 1989a) ethylene oxide is formed as a metabolite from ethylene by oxidative metabolism. The endogenous formation of ethylene has been repeatedly demonstrated by its exhalation in untreated rats (Frank et al. 1980; Sagai and Ichinose 1980; Shen et al. 1989) and in healthy humans (Ram Chandra and Spencer 1963; Shen et al. 1989). Consequently, also the metabolite ethylene oxide must be viewed as a natural body constituent (Filser et al. 1992; Swenberg et al. 2000).

According to current knowledge, possible endogenous sources of ethene and ethylene oxide are:

- lipid peroxidation (Liebermann and Mapson 1964; Frank et al. 1980; Sagai and Ichinose 1980; Törnqvist et al. 1989b);
- oxidation of amino acids (Lieberman et al. 1965; Clemens et al. 1983; Kessler and Remmer 1990);
- metabolism of intestinal bacteria (Törnqvist et al. 1989b).

These endogenous sources of ethylene and, in consequence, of ethylene oxide, contribute to a background alkylation (2-hydroxyethylation) of DNA in humans of 2–8 pmol mg^{-1} DNA (5-20 adducts per million bases). This appears even higher than the quantity represented by "I-compounds" as discussed above (Föst et al. 1989; Bolt and Leutbecher 1993; Cushnir et al. 1993). Therefore, macromolecular adduct loads caused by exogenous ethylene oxide may well be compared with this endogenous background. Such comparisons have been used to provide arguments towards regulation of ethylene oxide residues in consumer products (Filser et al. 1994). On this scientific background, it has even been concluded that "it is not expected that long-term occupational exposure to airborne concentrations at or below 1 ppm ethylene oxide produces an unacceptable increased risk in man" (van Sittert et al. 2000; further discussed by Thier and Bolt 2000). Such a conclusion, however, is not fully supported by studies of workers exposed to low levels of ethylene oxide, using the method of "alkaline elution" to detect DNA single strand breaks as biomarkers for genotoxicity (Fuchs et al. 1994). However, all available experimental data are generally consistent with this view (Swenberg et al. 2000).

4.5
Examples of carcinogens and modifiers leading to different dose-response types

4.5.1
Vinyl chloride

Vinyl chloride (VC) is a high-volume industrial chemical that is used as monomer in the production of polyvinyl chloride (PVC). It is of very low acute toxicity, characterised by prenarcotic and narcotic effects at airborne concentrations of 1 % (10,000 ppm) and higher. Following recognition of its carcinogenicity in humans, strict regulations have been applied in all industrialised countries.

Low levels of vinyl chloride, arising from degradation of chloroethylene solvents (trichloroethylene, perchloroethylene), are of environmental concern at polluted sites (Kielhorn et al. 2000). Close to landfills and areas contaminated with chlorinated hydrocarbons, dechlorination processes may lead to significant levels of VC in ambient air, leachate and groundwater. Low levels of VC have also been identified in tobacco smoke (cigarettes and small cigars). For an extended discussion of aspects of exposure see IPCS-WHO (1999).

4.5.1.1
General background of carcinogenicity of vinyl chloride

Toxicology and carcinogenicity of VC have been widely studied during the last 25 years, and a number of assessments of carcinogenic risk have been carried out, based on both, occupational epidemiology and animal experimental data. Vinyl chloride is classified as a human carcinogen by the European Union (EU) (category 1) and the International Agency for Research on Cancer (IARC) (group 1). The established carcinogenicity is the preponderant toxicological effect of VC.

As in the case of other chemical carcinogens, the assessments of risks concerning the environmental exposure and exposures in the occupational environments follow lines that are often considerably different. A number of available carcinogenic risk evaluations for the general (not professionally exposed) population are based on epidemiological studies on workers (as for instance, in the case of WHO Air Quality Guidelines 1987, 1999 and 2000). Recent risk evaluations also include the consideration of toxicokinetics in experimental animals and humans, by using experimental data and physiologically based pharmacokinetic models (PBPK models), in order to account for interspecies differences in physiology and VC metabolism (see below).

4.5.1.2
Toxicokinetics and metabolism

The main metabolic route of VC, after inhalation or ingestion, is by oxidation via cytochrome P450 (CYP2E1) that leads to the formation of chloroethylene oxide. This is a highly reactive epoxide, which rapidly rearranges to chloroacetaldehyde (Bolt 1978; Bartsch et al. 1979; Bolt et al. 1981). The detoxification of these two reactive metabolites, chloroethylene oxide and chloroacetaldehyde, is through conjugation with glutathione, catalysed by glutathione *S*-transferase(s).

Chloroethylene oxide and chloroacetaldehyde react with nucleic acid bases, forming DNA adducts. The *in vivo* formation of four etheno-DNA adducts has been experimentally demonstrated. These appear highly persistent and can cause defective transcription (miscoding). Targets for alkylation in nucleic acids are adenine, guanine, and cytosine moieties. Among others, the biologically relevant "etheno" adducts, $1,N^6$-ethenodeoxyadenosine and $3,N^4$-ethenodeoxycytidine, are formed (Laib and Bolt 1980). The metabolism of VC is a dose-dependent, saturable process (Bolt and Filser 1983). Inhalation toxicokinetics of vinyl chloride and vinyl bromide show linearity within a lower dose range, but saturation of the metabolism at inhalation concentrations of 500 ppm (Gehring et al. 1978; Bolt and Filser 1983). This leads to non-linearity of tumour response with inhalation concentration at high doses of vinyl chloride in inhalation bioassays (Gehring et al. 1978, 1979; see Fig. 4.4).

Studies of VC metabolism and the interpretation of carcinogenic dose-response in experimental animals represent one of the first integrations of metabolic data into carcinogenic risk assessment (Gehring et al. 1978; Bolt et al. 1980). A first attempt at extrapolating human cancer risks due to vinyl chloride from experimental animals to humans has been put forward by Gehring et al. (1979). However, at that time the authors did not yet incorporate the metabolic differences between the

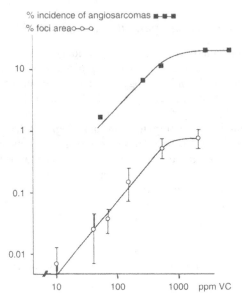

Fig. 4.4 Incidence of liver angiosarcomas (top curve), according to Maltoni (1977), and quantities of ATPase-deficient foci induced in female Wistar rats (bottom curve), according to Laib et al. (1985), versus vinyl chloride inhalation exposure concentration (log/log scale). Incidences of angiosarcomas and induced foci area increase linearly with increasing vinyl chloride exposure concentration until a plateau is reached, due to metabolic saturation.

species. In fact, humans metabolise vinyl chloride more slowly than rats, when expressed relative to body weight (Buchter et al. 1978). The use of the "physiologically-based pharmacokinetic" (PBPK) models is now the main tool to consider metabolic data. VC is a typical example of an industrial chemical of which the risk assessment requires such an approach (Clewell et al. 1995; Reitz et al. 1996; IPCS-WHO 1999).

4.5.1.3
Selection of the type of low-dose extrapolation

Although vinyl chloride is an established cause of human liver angiosarcoma, the evidence in humans is inconclusive as to whether it also causes other neoplastic and non-neoplastic chronic liver diseases as well as neoplasms in other (extrahepatic) organs. From available epidemiological studies, no clear statement as to the shape of the dose-tumour response curve for humans can be inferred (Ward et al. 2001).

From animal experimental data (long-term bioassays in rats) effective doses of the reactive metabolite(s) were calculated using the PBPK model of Clewell et al. (1995). The initial VC metabolism was hypothesised to occur via two saturable pathways, one representing low-capacity, high-affinity oxidation by cytochrome P450 2E1, and the other representing higher capacity, lower affinity oxidation by other isozymes of cytochrome P450. This calculation showed that the model was linear up to nearly 100 mg m^{-3}, and the calculated equivalence factor was later used

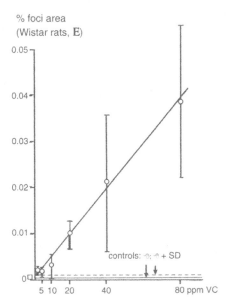

% foci area
(Wistar rats, E)

controls: ○; ◑ + SD

5 10 20   40   80 ppm VC

Fig. 4.5 Dose dependent induction of hepatocellular ATPase-deficient foci by vinyl chloride in female Wistar rats. The dose-response relationships can be described, within the accuracy of the method, by linear dose-response curves, which run through the origin (Laib et al. 1985).

to convert the risk from the inhalation experiments in animals (in the units of the dose metric) to human risk values.

Another line of research investigated the biological response of rat liver to low vinyl chloride doses. Investigations of cancer preliminary stages, as biomarkers, may provide a tool to study dose-response relationships, down to ranges where the carcinogenic risk caused by the substance can no longer be distinguished from non-exposed controls in long-term cancer bioassays.

Such a parameter is the development of pre-neoplastic enzyme-altered foci in the liver (see also section 6.5). Adenosine triphosphatase (ATPase) deficient foci were quantified by Laib et al. (1985) as an endpoint of VC effect in the "rat liver foci bioassay". The quantity of foci area follows dose-time response relationships identical to those observed for the induction of liver tumours (Kunz et al. 1983). In order to study dose-dependence, hepatocellular ATPase-deficient foci were evaluated after subchronic exposure of newborn rats to VC. Wistar rats were exposed from day 1 after birth over 10 weeks to 10, 40, 70, 150, 500 and 2,000 ppm VC (8 h d^{-1}; 5 d wk^{-1}). One week after cessation of exposure hepatic ATPase-deficient foci were quantified (Fig. 4.4). In a subsequent investigation at a lower dose range, groups of female and male Wistar and Sprague-Dawley rats were exposed (8 h d^{-1}; 5 d wk^{-1}) to 2.5, 5, 10, 20, 40 and 80 ppm VC. Exposure started at day 3 of life and lasted for 3 weeks. After cessation of exposure the animals were maintained for 10 weeks without further treatment until ATPase-deficient foci were quantified (Fig. 4.5 and 4.6). Below the range of saturation of the metabolism of VC (i. e. below 500 ppm exposure, see Fig. 4.4), both sets of experiments revealed a straight linear rela-

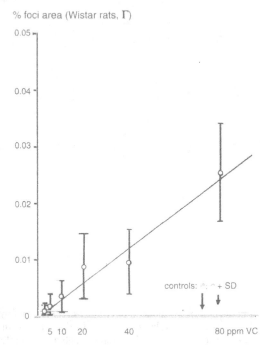

Fig. 4.6 Dose dependent induction of hepatocellular ATPase-deficient foci by vinyl chloride in male Wistar rats. The dose-response relationships can be described, within the accuracy of the method, by linear dose-response curves, which run through the origin (Laib et al. 1985).

tionship between the dose of VC and the % foci area (related to liver area) which was induced. Within the dose range investigated, i. e. down to 2.5 ppm airborne exposure, the dose dependence was linear, and no obvious threshold for the induction of pre-neoplastic foci by VC was apparent (Fig. 4.5 and 4.6; Laib et al. 1985). This exemplifies that, down to low doses of VC, no "threshold" dose exists. Investigations on covalent binding of (^{14}C)-VC to cellular macromolecules are supportive of this view, and the linearity of the dose-response curves disproves the hypothesis of a "threshold" for the carcinogenic action of VC in rats (Schumann et al. 1982).

It has been stated that vinyl chloride cancer risk assessments based on animal studies tend to overestimate the human risk of angiosarcoma of the liver (Kielborn et al. 2000). However, in the absence of convincing data on the quantitative nature of carcinogenic dose-response of VC in humans, and considering the above mentioned experimental data, a linear (no-threshold) dose-response for the assessment of human cancer risks associated with vinyl chloride exposures appears as the most appropriate practical procedure. This is backed by current procedures of official bodies. For instance, the US Environmental Protection Agency has issued guidelines for carcinogenic risk assessment (US EPA 1996). According to these guidelines, a default assumption of carcinogenicity based on linearity is appropriate when the scientific evidence supports a mode of action anticipated to be linear (e. g. DNA reactivity). This is the case for vinyl chloride.

4.5.1.4
Available carcinogenic risk evaluations of VC (low-dose linear extrapolation)

- The World Health Organisation (1987, 1999, 2000) has re-confirmed an earlier assessment of 1987. Angiosarcomas of the liver were considered, together with other tumours. The estimate was based on epidemiological data, and of relative risks related to working environments. A "unit risk" for lifetime exposure to 1 μg m-3 VC was assessed as 1×10^{-6} 1 μg^{-1} m^{-3}, or 2.59×10^{-3} for 1 ppm lifetime exposure.
- Clewell et al. (1995) have used epidemiologically derived relative risks, together with a PBPK model for taking into account aspects of metabolism of VC. The risk assessment was limited to liver angiosarcomas. The authors have produced three independent evaluations, based on three different epidemiological studies, with the following results:

1. based on Fox and Colliers (1977): $0.71 - 4.22 \times 10^{-3}$, for a lifetime exposure to 1 ppm;
2. based on Jones et al. (1988): $0.97 - 3.60 \times 10^{-3}$, for a lifetime exposure to 1 ppm;
3. based on Simonato et al. (1991): $0.40 - 0.79 \times 10^{-3}$, for a lifetime exposure to 1 ppm.

All three estimates by Clewell et al. (1995) are very close to each other. The average of the mean values of the confidence intervals of the three risk estimates is 1.8×10^{-6}, for lifetime exposure to 1 ppb. This is again consistent with the estimate of WHO (1999, 2000).

4.5.1.5
Risk estimates based on experimental data (low-dose linear extrapolation)

Risk estimates based on experimental data were mostly performed in view of workplace situations. As current occupational exposure limits for VC in various countries are set within the range between 1 and 5 ppm, the endpoint of extrapolation was mostly an assumed lifetime exposure to 1 ppm. For further relation to total (maximal) working time, the total working time may be considered as 14 % of lifetime. The generally used experimental protocols of long-term studies of chemical carcinogenicity are different from those used for radiation carcinogenicity studies. The assessment at a time of terminal sacrifice (mostly after 2 years of the study) leads to a combined consideration of tumour morbidity and mortality.

- Clewell et al. (1995) used the linearised multistage model, together with a PBPK model (in order to take into account the dose-related metabolic rates in experimental animals and humans, to provide a suitable interspecies extrapolation). The experimental data included two experimental studies by Maltoni et al. (1981, 1984) with inhalation of VC by rats and mice, another study by Maltoni et al. (1981, 1984) using gavage treatment of rats, and a study by Feron et al. (1981) with oral exposure of rats through the diet. The risk assessment was limited to liver angiosarcoma. For the present purpose, only the risk estimates based on inhalation experiments (Maltoni et al. 1981, 1984) are considered, which seems appropriate in view of human inhalation exposure. As an average of upper confidence limits calculated for humans on the basis of all inhalation studies in mice and rats, the authors arrived at a lifetime risk for 1 ppm VC inhalation of 3.0×10^{-3}. The physiologically based pharmacokinetic (PBPK) model for VC was later improved by Clewell et al. (2001). The initial metabolism of VC was described as occurring via two saturable pathways, one representing low capacity-high affinity oxidation by CYP2E1 (see Bolt et al. 2003) and the other representing higher capacity-lower affinity oxidation by other isozymes of P450, producing in both cases chloroethylene oxide and chloroacetaldehyde as intermediate reactive products. Depletion of glutathione by reaction with both reactive metabolites was also considered. Animal-based risk estimates for human inhalation exposure to VC, using total metabolism estimates from the new PBPK model, were found consistent with the previous risk estimates.

– Reitz et al. (1996) also used the linearised multistage model, together with PBPK modelling. The experimental data analysed included two experimental studies by Maltoni et al. (1974, 1981, 1984) in rats and mice. The upper confidence limit of lifetime risk at 1 ppm VC was calculated as 1.5×10^{-3}.

– The US Environmental Protection Agency (US EPA 2000) has used the linear "default method" as well as the linearised multistage model (with practically the same result), together with a PBPK model (in order to take into account the dose-dependent metabolic rates in experimental animal and humans and to provide a suitable interspecies extrapolation). The experimental data analysed were again those of Maltoni et al. (1974, 1981, 1984) in female rats. The risk assessment considered liver angiosarcomas, together with hepatomas and neoplastic nodules. The risk estimate presented by the agency, as upper confidence limit, was for continuos lifetime exposure during adulthood 11.4×10^{-3} for 1 ppm VC.

All these risk estimates, based on different sets and categories of data (animal experiments, epidemiological studies), lead to risk estimates that appear basically consistent, if the variability and uncertainty of such estimates is considered.

In essence, vinyl chloride is an established carcinogen, both in humans and in experimental animals. The primary target of its carcinogenicity is the liver, although there is clear experimental and some suggestive human evidence that it also acts at extrahepatic sites. The primary and most typical liver tumour is angiosarcoma (hemangioepithelioma), but experimental data also demonstrate formation of hepatocellular carcinomas. For experimental induction of liver tumours, there is experimental evidence that a linear low-dose extrapolation is justified. On this basis, quantitative risk assessments are available, including those based on human epidemiological data and those based on extrapolation from animal data by means of PBPK modelling. The different approaches are basically consistent with each other. Risk data derived from animal experiments and using PBPK modelling point to a similar order of magnitude and thereby confirm this approach.

There is some parallelism between genotoxicity induced by vinyl chloride on one hand and by ionising radiation on the other hand (see chapters 3 and 5): A linear dose-genotoxicity response, down to lower dose ranges, is experimentally verified by assessment of sensitive biomarkers, e. g. for vinyl chloride: enzyme-altered hepatic foci (Laib et al. 1985). There is a natural background of genotoxicity with endpoints characteristic for vinyl chloride. This refers to the presence of the promutagenic DNA "etheno" adducts $1,N^6$-ethenodeoxyadenosine and $3,N^4$-ethenodeoxycytidine (Swenberg et al. 2000), as well as to a natural background of enzyme-altered hepatic foci (Laib et al. 1985).

4.5.2
Aflatoxin B1

Aflatoxin B_1 (AFB$_1$) is a fungal metabolite produced by *Aspergillus flavus* that contaminates food supplies especially of improperly stored rice, corn and peanuts. It is one of the most potent human hepatocarcinogens known so far. It represents a liver cell type-specific toxin, because it induces the formation of tumours developing from parenchymal and bile duct epithelial liver cells, but not from other cell types present in the liver, such as Kupffer or endothelial cells (Steinberg et al. 1996). AFB$_1$ is a typical representative of carcinogens showing a linear dose-response relationship in the low-dose range. For instance, a 24 month study with male Fischer 344 (F344) rats exposed to five doses between 1 and 50 ppb in the drinking water resulted in a linear dose-response curve with the liver tumour incidence being 80 % at the highest dose (Fig. 4.7; Wogan et al. 1974; review: Poirier and Beland 1992). The linear increase in liver tumours fits well the also linear induction of the main DNA adduct (dG-N^7-AFB$_1$) formed by AFB$_1$ (Fig. 4.7; Buss et al. 1990). DNA adducts were measured after 8 weeks continuous administration of AFB$_1$ in the drinking water. After this period a steady state results between adduct formation and removal. A linear low-dose response to AFB$_1$ has also been shown in other species. For instance, tumour incidence in rainbow trout exposed to 50–250 ppb AFB$_1$

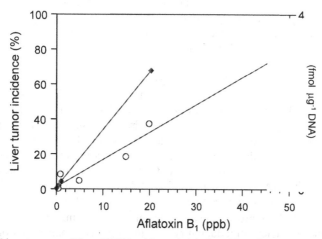

Fig. 4.7 Liver tumours (O) and DNA adducts (♦) induced by aflatoxin B$_1$. Male Fisher rats were exposed to five doses of aflatoxin B$_1$ between 1 and 50 ppb in the drinking water for 24 months for analysis of tumour incidence. DNA adducts were measured after 8 weeks continuous administration of AFB$_1$ in the drinking water (Wogan et al. 1974; Buss et al. 1990; Poirier and Beland 1992).

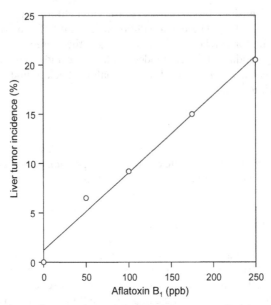

Fig. 4.8 Aflatoxin B1 liver carcinogenicity in rainbow trouts. Rainbow trout embryos were exposed to aflatoxin B1 for 30 min. At 11 months of age trouts were sacrificed and analysed for liver tumours. The number of fish were 200, 370, 249, 400 and 400 in the 0, 50, 100, 175 and 250 ppb exposure groups (Oganesian et al. 1999).

as embryos increased without an obvious threshold (Fig. 4.8; Oganesian et al. 1999). It should be mentioned that large inter-species differences in AFB_1 susceptibility are known (Hengstler et al. 1998). Rats represent a very sensitive species, whereas mice are much more resistant. It is important to use the rat model when extrapolating to human carcinogenicity, since both rats and man – compared to mice – are relatively poor conjugators of activated AFB_1. The molecular mechanisms of these interspecies differences have been reviewed in detail (Hengstler et al. 1998).

AFB_1 requires metabolic conversion to AFB_1 *exo*-8,9-epoxide in order to cause DNA damage (Essigmann et al. 1977; Swenson et al. 1977; Guengerich et al. 1998). In humans AFB_1 is activated primarily by CYP3A4 and 1A2. The AFB_1 epoxide reacts with guanine resulting in 8,9-dihydro-8-(N^7guanyl)-9-hydroxyaflatoxin B_1 (AFB_1-N^7-Gua) as a main DNA adduct. The positively charged imidazole ring of the resulting molecule facilitates depurination, leading to an apurinic site. Alternatively, the imidazole ring of AFB_1-N^7-Gua opens to form the stable AFB_1 form amidopyrimidine (AFB_1-FAPY). The initial AFB_1-N^7-Gua, AFB_1FAPY and the apurinic site represent likely precursors to the mutagenic effects of AFB_1.

Bailey et al. (1996) have presented strong evidence that the initial AFB_1-N^7-Gua adduct is extremely efficient in inducing point mutations. Two main mutations are induced (Fig. 4.9):

G→T transversions are targeted to the original site of the adduct (Fig. 4.9; upper panel). Such G→T transversions have been identified in the *p53* tumour suppressor gene in the third position of codon 249 (AGG) in approximately half of

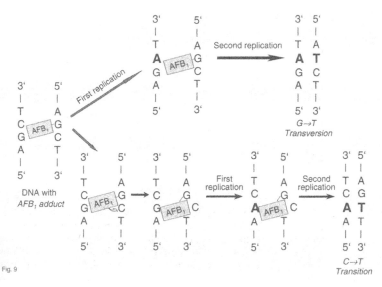

Fig. 9

Fig. 4.9 Two types of mutation (G→T transversion and C→T transition) resulting from the AFB_1-N^7-guanine adduct (8,9-dihydro-8-(N^7 guanyl)-9-hydroxyaflatoxin) B_1.

all examined hepatocellular carcinomas of humans exposed to AFB_1 (Bressac et al. 1991; Hsu et al. 1991).

The C→T transition occurs on the 3' face of the modified guanine (Fig. 4.9, lower panel). The AFB_1 moiety of the AFB_1-N^7-Gua adduct intercalates on the 5' face of guanine. As a consequence, the base 5' of the adduct (a cytosine in Fig. 4.9) may rotate out of the helix leading to the insertion of adenine across from the AFB_1 adduct, finally resulting in a C→T transition (Bailey et al. 1996). Such C→T transitions have been identified in codon 12 of the *c-ki-ras* oncogene of rat hepatocellular carcinomas (McMahon et al. 1990; Soman et al. 1993).

It can be assumed that the linear low-dose-response relationship between AFB_1 exposure and tumour incidence as well as the direct relationship between AFB_1-DNA adducts and tumour incidence require several preconditions: In the case of AFB_1 none of the possible mechanisms a–e in Fig. 4.1 seems to be effective enough to introduce a measurable threshold. Of course, it cannot be excluded that a threshold could be observed for lower doses than those used in the experiments shown in Fig. 4.7. Such experiments would be difficult, since extremely high numbers of animals would be required. However, unless more data are available the linear dose-response extrapolation model with no assumed threshold should be used for AFB_1 risk evaluation.

4.5.3
N,N-Diethylnitrosamine (DEN)

Nitrosamines may be formed in the stomach and intestine by the reaction of amines derived from protein and exogenous nitrites or nitrites formed endogenously from nitrates by the action of the colon microflora. Similar as AFB_1, several

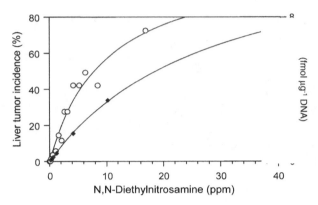

Fig. 4.10 Liver tumour incidence (O) in a large lifetime study including 1,140 male Wistar rats. Fifteen different doses of *N,N*-diethylnitrosamine ranging between 0.033 and 16.9 ppm were given in the drinking water (Peto et al. 1991). Data for O^4-ethylthymidine (♦) the major promutagenic DNA adduct responsible for induction of liver tumours were obtained from another study with male Fisher rats (Boucheron et al. 1987) since adduct data from Wistar rats are not available.

nitrosamines, for instance *N,N*-diethylnitrosamine (DEN), induce cancer when administered continuously in low doses. In a large lifetime carcinogenesis study, comprising 1140 male Wistar rats, 15 different doses ranging between 0.033 and 16.9 ppm were given in the drinking water (Peto et al. 1991). No threshold was observed in the low-dose range (Fig. 4.10). At high doses the dose-response curve was not linear but approached a plateau. This non-linearity at high doses is due to saturation of cytochrome P450 2E1, the enzyme responsible for activation of DEN to a genotoxic substance. O^4-Ethylthymidine (O^4-Et-dT) was considered to represent the major promutagenic DNA adduct responsible for induction of liver

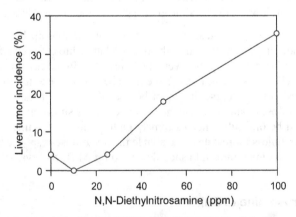

Fig. 4.11 Liver tumour induction by *N,N*-diethylnitrosamine in fish. Japanese Medaka were exposed to *N,N*-diethylnitrosamine for 48 h at 14 days post hatch. At 6 months of age fish were sacrificed and analysed for liver tumours. The number of fish were 49, 44, 49, 45 and 48 in the 0, 10, 25, 50, and 100 ppb exposure groups (Brown-Peterson et al. 1999).

tumours (Boucheron et al. 1987). Although DNA adducts were analysed in another rat strain (male Fisher rats) a similar shape of the dose-effect curve was obtained as for liver tumour incidence (Fig. 4.10). Thus, diethylnitrosamine represents a carcinogen showing a non-linear dose-response without threshold and with a very good correlation between DNA adduct induction and tumour incidence when administered continuously in the drinking water. However, different results were obtained in studies using short term or intermittent instead of chronic lifetime exposures. Seven hundred and fifty 14-day post-hatch Medaka were divided into 10 groups of 75 fish each and replicate groups were exposed to 0, 10, 25, 50 and 100 ppm DEN for a short term (48 h) (Brown-Peterson et al. 1999). No increase in hepatic adenomas or carcinomas was observed for the two lowest doses of 10 and 25 ppm DEN, whereas an increase was reported for 50 and 100 ppm (Fig. 4.11). Of course, the different dose-response curves may also be due to interspecies differences. Others observed non-linearity in several studies with male Fisher 344 rats (Williams et al. 1993, 1996, 1999, 2000). The data shown in Fig. 4.12 were obtained with five doses ranging from a cumulative total of 0.5 to 4 mmol DEN kg^{-1} body weight that were given intermittently as weekly i. p. injections for 10 weeks. Up to an exposure of 1 mmol kg^{-1} no liver carcinomas were observed. Exposure to 2 mmol kg^{-1} and greater caused liver carcinomas in almost all of the 12 exposed rats (Fig. 4.12; Williams et al. 1996). Whilst these data were obtained by the i. p. administration route, later the authors obtained similar results by intragastric instillation once weekly (Williams et al. 1999). Thus, the type of exposure, continuously lifetime versus intermittent or short-term exposure, appears to strongly influence the low-dose-response curve. The cumulative exposure of 2 mmol kg^{-1} yielded a 92 % liver cancer incidence (Williams et al. 1996). This effect can be compared best to that of the highest cumulative exposure achieved in the continuous lifetime study of

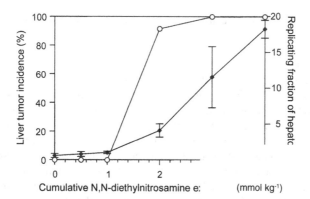

Fig. 4.12 Incidence of liver carcinomas (O) in male Fisher 344 rats (Williams et al. 1996). Five doses ranging from a cumulative total of 0.5 to 4 mmol N,N-diethylnitrosamine kg^{-1} body weight were given as weekly i. p. injections for 10 weeks. Twelve rats were examined per dose group. For determination of the replicating fraction (♦) of hepatocytes also male Fisher 344 rats and the same administration schedule were applied. Values are means and standard deviations of five rats. The replicating fraction was determined by immuno-histochemical analysis of bromodeoxyuridine incorporation.

Peto et al. (1991) that was calculated to be approximately 10 mmol kg^{-1} cumulative exposure resulting in a 78 % incidence of liver tumours. Thus, the cumulative exposure in the studies of Williams et al. (1993, 1996, 1999, 2000) was even smaller compared to that of the Peto et al. (1991) study. Of course, the single doses in the studies of Williams et al. were much higher. Williams et al. (1996, 1999) examined the mechanisms responsible for the observed non-linearity. Interestingly, induction of DNA adducts does not explain the observed non-linearity, since also the lowest exposures produced a clear level of DNA ethylation in the liver, even when given as single dose only (data not shown). However, cytotoxicity and cell proliferation in rat liver correlated well with tumour incidence. Whereas the non-tumourigenic cumulative doses of 0.5 and 1 mmol kg^{-1} did not cause a relevant increase in the replicating fraction of hepatocytes, a strong increase was observed for higher doses (Fig. 4.12). It is tempting to speculate that the differences between the studies of Peto et al. and Williams et al. are due to differences in toxicity in the low-dose range (with an expected linear toxicity dose-response in the Peto et al. study), but presently respective data are not available.

In conclusion, the low-dose-response relationship for DEN appears to depend on the type of administration of the test substance resulting in a linear relationship for continuous lifetime exposure but a non-linear relationship with a practical threshold at 1 mmol DEN kg^{-1} for intermittent weekly administration. Due to the remaining uncertainties the linear model with no threshold assumption is most appropriate for extrapolation of human cancer risk due to DEN until additional clarifying data become available. In addition, for extrapolations the lower linear part of the curve with doses < 5 ppm should be used. In this dose range the activating enzyme cytochrome P450 2E1 is not saturated, which should also be the case under conditions of human exposure.

4.5.4
NNK (4-[N-methyl-N-nitrosoamino]-1-[3-pyridyl]-1-butanone)

The tobacco-specific nitrosoketone NNK is a nicotine derivative forming the pro-mutagenic O^6-methylguanine and pyridyloxobutyl DNA adducts (Belinsky et al. 1986,1987; Staretz et al. 1997;) that are viewed in conjunction with the development of lung cancer. NNK is relevant also for non-smokers, since non-smoking women exposed to environmental tobacco smoke have been reported to take up and metabolise NNK, which could increase their risk of lung cancer (Anderson et al. 2001; review: Poirier and Beland 1992). The example of NNK demonstrates an example of an extrahepatic no-threshold carcinogen. As described for N,N-diethyl-nitrosamine a good correlation between DNA adducts (O^6-methylguanine) and tumour incidence has been observed (Fig. 4.13). Both, tumour incidence and DNA adducts show a non-linear dose-response curve without a threshold.

4.5.5
2-Acetylaminofluorene (2-AAF)

2-AAF is reasonably anticipated to be a human carcinogen. Occupational exposures were reported for chemists, chemical stockroom workers and biomedical

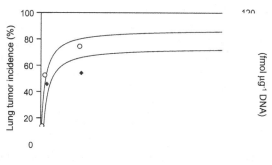

Fig. 4.13 Lung tumour incidence (O) was determined in male Fischer rats (experimental data and modelling of dose dependence). The N-nitrosoketone was administered as subcutaneous injections three times a week for 20 weeks. Rats were sacrificed and analysed for lung tumours after 31 months. O^6-Methyldeoxyguanosine (O^6-Me-dG) was measured in lung Clara cells at 4 weeks in male Fisher rats given subcutaneous injections three times a week (Belinsky et al. 1986, 1990).

researchers. In contrast, for the general population exposure to 2-AAF is very small. Nevertheless, 2-AAF is of high scientific interest since probably the largest tumourigenicity study ever done ("mega-mouse study") was conducted with this heterocyclic aromatic amide (Staffa and Mehlman 1979). Approximately 24,000 female BALB/c mice were given continuous administration of seven concentrations of 2-AAF in the diet, ranging from 30–150 ppm. Tumour incidence after 24 months is shown in Fig. 4.14. In the bladder, tumour incidence was not increased for 30, 34 and 45 ppm, but it increased weakly for 75 ppm, followed by a steep increase at doses of 100 and 150 ppm (Fig. 4.14.a). In contrast to bladder tumour incidence, the level of DNA adducts in bladder tissue increased linearly in the dose range between 15 and 150 ppm (Fig. 4.14.a). The tumourigenic effects of 2-AAF in mouse target organs have been reported to be associated with the formation of only one DNA

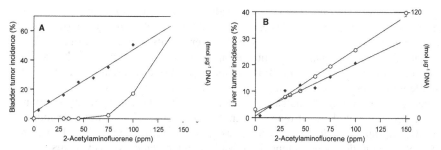

Fig. 4.14 Tumour incidence in female BALB/c mice after administration of 2-acetylamino-fluorene (ppm) in the diet (Staffa et al. 1979). Mice were analysed for tumours after 24 months. In the similarly conducted DNA adduct study N-(deoxyguanosin-8-yl)-2-aminoflu-orene (dG-C8-AF) was measured in bladder and in liver tissue of mice (Poirier et al. 1991). A: Bladder tumour incidence (O), N-(deoxyguanosin-8-yl)-2-aminofluorene (dG-C8-AF) in bladder tissue (♦). B: Liver tumour incidence (O).

adduct, N-(deoxyguanosine-8-yl)-2-aminofluorene (dG-C8-A7) (Poirier et al. 1991). However, a different scenario was observed in the liver (Fig. 4.14b). Liver tumour incidence and DNA adducts increased linearly with dose. Thus, the scenario observed with 2-AAF in liver is similar to that of AFB_1. The example of 2-AAF shows that low-dose-effect relationships do not only depend on the nature of the substance tested, but may also be tissue specific.

4.5.6
Arsenic

Very recent documentation on toxicity and carcinogenicity of inorganic arsenic is available (DFG 2002; EPA 2003). There is general agreement that inorganic arsenicals (except hydrogen arsenide) are human carcinogens. Also, neither experimental studies nor epidemiological investigations allow to establish a NOAEL (DFG 2002).

4.5.6.1
Experimental studies

Furst (1983) has reviewed animal carcinogenicity testing studies of nine inorganic arsenic compounds in over nine strains of mice, five strains of rats, in dogs, rabbits, swine and chicken. Testing was by oral, dermal, inhalation, and parental routes. All oxidation states of arsenic were tested. No study could clearly demonstrate that inorganic arsenic was carcinogenic in animals. Dimethylarsonic acid (DMA), the end metabolite predominant in humans and animals, has been tested for carcinogenicity in two strains of mice and was also not found positive (Innes et al. 1969). The meaning of non-positive experimental data for carcinogenicity of inorganic arsenic is uncertain, the mechanism of action in causing human cancer is not known, and probably rodents may not be a good model for arsenic carcinogenicity testing. It has been repeatedly concluded by official bodies that there has not been consistent demonstration of carcinogenicity in test animals, for various chemical forms of arsenic administered by different routes to several species (IARC 1980, DFG 2002, EPA 2003).

Some data indicate that arsenic may produce animal lung tumours if retention time in the lung can be increased (Pershagen et al. 1982, 1984). In short-term assays on generation of enzyme-altered foci in the liver of rats treated with dimethylnitrosamine, arsenite, arsenate, or combinations of dimethylnitrosamine with arsenite or arsenate, Laib and Moritz (1988) found indications of a cocarcinogenic effect of arsenite with dimethylnitrosamine, but no effect of arsenicals alone.

Jacobson-Kram and Montalbano (1985), reviewing the mutagenicity of inorganic arsenic, concluded that inorganic arsenic is inactive or very weak for induction of gene mutations *in vitro* but it is clastogenic with trivalent arsenic being an order of magnitude more potent than pentavalent arsenic. Moreover, arsenate, arsenite and monomethylarsonic acid (MMA) were considered clastogenic but the aberration response with dimethylarsonic acid (DMA) was insufficient to consider it a clastogen. Since arsenic exerts its genotoxicity by causing chromosomal effects, it has been suggested that it may act in a later stage of carcinogenesis of tumour progression, rather than as a classical initiator or promoter (Moore et al. 1997). A finding which supports this process is that arsenate (8–16 μM) and arsenite (3 μM) have been shown to induce 2–10-fold amplification of the dihydrofolate reductase gene in culture in methotrexate resistant 3T6 mouse cells (Lee et al. 1988). Although the mechanism of induction in rodent cells is not known, gene amplification of onco-genes is observed in many human tumours. Sodium arsenite (As^{+3}) induces DNA strand breaks which are associated with DNA-protein cross-links in cultured human fibroblasts at 3 mM but not 10 mM (Dong and Luo 1993), and it appears that arsen-ite inhibits the DNA repair process by inhibiting both excision and ligation (Laib and Moritz 1989; Jha et al. 1992; Lee-Chen et al. 1993). This is seen in relation to comutagenic and cocarcinogenic effects of arsenic (Hartwig and Schwerdtle 2002). *In vivo* studies in rodents have shown that oral exposure of rats to arsenate (As^{+5}) for 2–3 weeks resulted in major chromosomal abnormalities in bone marrow (Datta et al. 1986) and exposure of mice to As^{+3} in drinking water for 4 weeks (250 mg As l^{-1} as arsenic trioxide) caused chromosomal aberrations in bone marrow cells but not spermatogonia (Poma et al. 1987).

4.5.6.2
Epidemiology

Epidemiological studies on arsenic have been published indicating the origination of lung cancer after inhalation and urinary bladder, renal, skin and lung cancer after oral (drinking water) uptake (DFG 2002). In discussions on quantitative risk assess-ment, the following studies were considered in detail (EPA 2003).

A cross-sectional study of 40,000 Taiwanese exposed to arsenic in drinking water found significant excess skin cancer prevalence by comparison to 7,500 residents of Taiwan and Matsu who consumed relatively arsenic-free water (Tseng et al. 1968; Tseng 1977). Although this study demonstrated an association between arsenic expo-sure and development of skin cancer, several weaknesses and uncertainties, including poor nutritional status of the exposed populations, their genetic susceptibility and their exposure to inorganic arsenic from non-water sources, limit the study's useful-ness in risk estimation (EPA 2003). Dietary inorganic arsenic was not considered nor was the potential confounding by contaminants other than arsenic in drinking water. There may have been bias of examiners in the original study since no skin cancer or preneoplastic lesions were seen in 7,500 controls; prevalence rates rather than mortal-ity rates are the endpoint; and furthermore there is concern of the applicability of extrapolating data from Taiwanese to the US population because of different back-ground rates of cancer, possibly genetically determined, and differences in diet other than arsenic, e. g. low protein and fat and high carbohydrate (EPA 1988).

Several follow-up studies of the Taiwanese population exposed to inorganic arsenic in drinking water showed an increase in fatal internal organ cancers as well as an increase in skin cancer. Chen et al. (1985) found that the standard mortality ratios (*SMR*) and cumulative mortality rates for cancers of the bladder, kidney, skin, lung and liver were significantly greater in the Blackfoot disease endemic area of Taiwan when compared with the age adjusted rates for the general population of Taiwan. Blackfoot disease (BFD, an endemic peripheral artery disease) and these cancers were all associated with high levels of arsenic in drinking water. In the endemic area, *SMR*s were greater in villages that used only artesian well water (high in arsenic) compared with villages that partially or completely used surface well water (low in arsenic). However, dose-response data were not developed (Chen et al. 1985).

Chen et al. (1992) conducted a subsequent analysis of cancer mortality data from the arsenic-exposed population to compare risks of various internal cancers and also to compare risks between males and females. The study area and population had been described by Wu et al. (1989). It was limited to 42 southwestern coastal villages where residents have used water high in arsenic from deep artesian wells for more than 70 years. Arsenic levels in drinking water ranged from 0.010 to 1,752 ppm. The study population had 898,806 person-years of observation and 202 liver cancer, 304 lung cancer, 202 bladder cancer and 64 kidney cancer deaths. The study population was stratified into four groups according to median arsenic level in well water (< 0.10 ppm, 0.10–0.29 ppm, 0.30–0.59 ppm and 60+ ppm), and also stratified into four age groups (< 30 years, 30–49 years, 50–69 years and 70+ years). Mortality rates were found to increase significantly with age for all cancers and significant dose-response relationships were observed between arsenic level and mortality from cancer of the liver, lung, bladder and kidney in most age groups of both males and females. The data generated by Chen et al. (1992) were considered to provide evidence for an association of the levels of arsenic in drinking water and duration of exposure with the rate of mortality from cancers of the liver, lung, bladder, and kidney (US EPA 2003). Dose-response relationships were apparent from the tabulated data of Chen et al. (1992). Other studies summarised by US EPA (1988) showed a similar association in the same Taiwanese population with the prevalence of skin cancers.

4.5.6.3
Discussion of dose-response

The US EPA (1988) has performed an assessment of the carcinogenicity risk associated with ingestion of inorganic arsenic. The data provided by Tseng et al. (1968) and Tseng (1977) on about 40,000 persons exposed to arsenic in drinking water and 7,500 relatively unexposed controls were used to develop dose-response data. The number of persons at risk over three dose intervals and four time intervals of exposure, for males and females separately, were estimated from the reported prevalence rates as percentages. It was assumed that the Taiwanese persons had a constant exposure from birth, and that males consumed 3.5 l of drinking water per day and females consumed 2.0 l d^{-1}.

The multistage model with time was used to predict dose-specific and age-specific skin cancer prevalence rates associated with ingestion of inorganic arsenic; both linear and quadratic model fitting of the data were conducted. The maximum likelihood estimate (*MLE*) of skin cancer risk for a 70 kg person drinking 2 l of water per day ranged from 10^{-3} to 2×10^{-3} for an arsenic intake of 1 μg kg^{-1} d^{-1}. The cancer unit risk for drinking water was calculated to be 5×10^{-5} μg^{-1} l^{-1}. For details of the assessment, see US EPA (1988). Dose-response data had not been developed for internal cancers for the Taiwanese population; the data of Chen et al. (1992) are considered inadequate for such an evaluation (US EPA 2003).

The US EPA (Eastern Research Group, 1997) convened an Expert Panel on Arsenic Carcinogenicity. It was stated that it was clear from epidemiological studies that arsenic is a human carcinogen via the oral and inhalation routes. Moreover, it was concluded that "one important mode of action is unlikely to be operative for arsenic". It was agreed that arsenic and its metabolites do not appear to directly interact with DNA and that for each of the modes of action regarded as plausible, the dose-response would either show a threshold or would be non-linear. The panel agreed, however:

> ... that the dose-response for arsenic at low doses would likely be truly non-linear, i.e., with a decreasing slope as the dose decreased. However, at very low doses such a curve might be linear but with a very shallow slope, probably indistinguishable from a threshold.

In essence, it follows that the mechanism of carcinogenicity of inorganic arsenic are not sufficiently well understood, although there are indications that indirect mechanisms of action are operative which would lead to non-linearity of the dose-response at low doses. The evidence of carcinogenicity in humans is well established, although animal carcinogenicity data are considered inadequate for a valid assessment (US EPA 2003). In this situation, for regulatory matters, the default assumption of a linear dose-response appears appropriate at the present time, being in line with the present risk assessment of the US EPA (2003).

4.5.7
4-Aminobiphenyl (4-ABP)

4-Aminobiphenyl (4-ABP) occurs as a contaminant in 2-aminobiphenyl, which has been used in the manufacture of dyes. Ingesting foods with food additives containing trace amounts of 4-aminobiphenyl as a contaminant could possibly expose consumers. Also, mainstream cigarette smoke is reported to contain 4.6 ng cigarette[-1] of 4-aminobiphenyl, while sidestream smoke contains 140 ng cigarette[-1] of the chemical. Exposure of mice to 4-ABP has been shown to result primarily in the formation of one adduct, N-(deoxyguanosin-8-yl)-4-aminobiphenyl (dG-C8-ABP) (Beland et al. 1992). DNA adducts resulted in a linear dose-response relationship in liver and bladder tissue of male BALB/c mice 28 days after administration of six doses between 7 and 220 ppm in the drinking water (Fig. 4.15a and b; Schieferstein et al. 1985; Beland et al. 1992). In contrast, tumour incidence was not increased for 7, 14 and 28 ppm in the bladder, and no increase in liver tumours was observed in male BALB/c mice. The situation becomes more complex by a gender difference concerning susceptibility to 4-ABP. In contrast to male BALB/c mice (Fig. 4.15a) a linear low-dose-effect relationship for hepatic DNA adducts and liver tumour incidence was observed for female mice (Fig. 4.15c). In contrast to male mice only relatively low bladder tumour incidences could be induced in female animals (Fig. 4.15d). The example of 4-ABP shows that in rodents occurrence of thresholds can be sex- and tissue- dependent.

4.5.8
Formaldehyde

All carcinogens discussed so far did not exhibit thresholds, at least not in all organs or not under all treatment modes. Since the same seems to be the case for the major-

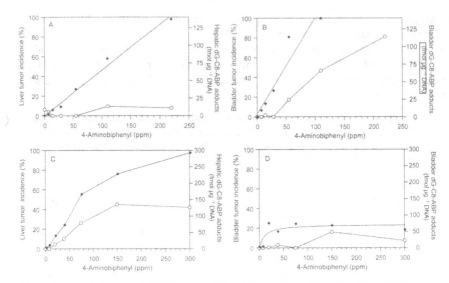

Fig. 4.15 4-Aminobiphenyl (ABP) induced liver (A. and C.) and bladder (B. and D.) tumours in relation to the DNA adduct N-(deoxyguanosin-8-yl)-4-aminobiphenyl (dG-C8-ABP) in male (A. and B.) and female (C. and D.) BALB/c mice. ABP was administered chronically in the drinking water. Tumour incidence was examined after 24 months, whereas DNA adducts were measured after 28 days of exposure (Schieferstein et al. 1985; Beland et al. 1992).

ity of genotoxic carcinogens it has been proposed that a no-threshold dose-response relationship should always be assumed. However, such a generalisation must be questioned. At least some genotoxic carcinogens induce tumours via mechanisms that lead to highly non-linear dose-response curves, from which practical thresholds may be inferred. Thoroughly studied examples are formaldehyde and vinyl acetate, and these will be illustrated here.

Toxicological effects of formaldehyde have been continuously discussed for the last 20 years (BMJFG 1984; Imbus 1988; Gilli et al. 2001). The primary focus was on the pathogenesis and interpretation of nasal tumours found in long-term rodent bioassays. Arguments on which regulatory discussions has focused were specifically related to the shape of the dose-response curve at low concentrations.

Nasal tumours in rodents were also a key finding for vinyl acetate. This has initiated research programmes on the underlying mechanisms. Both compounds, formaldehyde and vinyl acetate, are now very well studied. From these data, arguments have been inferred of non-linear carcinogenic dose-responses.

One of the main reasons for the common use of formaldehyde is its antimicrobial activity. This is intrinsically linked with its reactivity with functional groups of macromolecules, and also explains its damaging effect on cells and tissues, its irritating effects, sensitising properties, and genotoxicity. These different biological effects are therefore interconnected (Bolt 1987, Gilli et al. 2001).

More than 30 epidemiological studies concerning formaldehyde have been published. Possible causal links of formaldehyde have been claimed with cancers of the nasopharynx and, to a lesser extent, the nasal cavities. Inconsistencies have been seen in assigning a causal role of formaldehyde for human lung cancer (Blair et al. 1990). Against this background, experimental effects on the nasal epithelium induced by low doses of formaldehyde are pivotal for discussions on formaldehyde risks to humans under realistic exposure conditions (Morgan 1997).

4.5.8.1
Mutagenicity

The genetic toxicology of formaldehyde has been extensively reviewed (Ma and Harris 1988). In general, the available data show that formaldehyde is mutagenic in different test systems, especially when high concentrations act directly on cells (gene and chromosome mutations). Also, positive cell transformation assays have been reported *in vitro*. After inhalation of the compound, local DNA adducts were observed in rats, without systemic genetic effects (Casanova-Schmitz et al. 1984b).

Experiments on induction of sister chromatid exchange in human lymphocyte cultures (Kreiger and Garry 1983) have demonstrated no significant response below an apparent "threshold" of 5 µg ml^{-1} culture medium. In accordance with this, the cytotoxic dose-response curve showed a biphasic pattern with a marked increase in slope at about 10 µg formaldehyde ml^{-1}. Studies on UDS (unscheduled DNA synthesis) in primary cultures of human bronchial epithelial cells (Doolittle et al. 1985) did not demonstrate a response to formaldehyde in this system. Formaldehyde damage induced in DNA of different human cell culture systems comprises DNA-protein cross-links and DNA single-strand breaks; these lesions undergo efficient repair by complex mechanisms (Grafstrom et al. 1984).

4.5.8.2
Carcinogenicity

The "CIIT bioassay" (Swenberg et al. 1980, 1983a; Kerns et al. 1983; Morgan et al. 1986b) on which discussions have focused was performed on male and female F344 rats and B6C3F$_1$ mice that were exposed to 2.0, 5.6 or 14.3 ppm formaldehyde for 6 h d^{-1} and 5 d wk^{-1} over 24 months. Of main concern was the induction of squamous cell carcinomas in the nasal passages, occurring primarily in rats exposed to the highest concentration. A second lesion of concern was polypoid adenomas observed in treated and control rats. The results were basically confirmed by a second study (Albert et al. 1982; Sellakumar et al. 1985), with Sprague-Dawley rats exposed to 14 ppm formaldehyde.

In the "CIIT bioassay" mice showed no morphological changes at 2 ppm, whereas 5.6 and especially 14.3 ppm led to rhinitis, and to dysplasia and metaplasia of the nasal epithelium. Two (male) mice of the highest dose group developed squamous cell carcinoma (not statistically significant). The histological changes of dysplasia and metaplasia were more marked in rats. Changes were also observed at 2 ppm in this species. At 5.6 ppm one male and one female rat, but at 14.3 ppm nearly half of all rats developed squamous cell carcinomas. At the highest dose dysplasia and metaplasia of the trachea were also noticed.

Formal criticism of the study has been raised over its general outline (BMJFG 1984). In the rat study, the upper two doses involved severe toxic and lethal effects (20 % mortality at 18 months on 14.3 ppm). Severe rhinitis often led to dysplasia and subsequent death. It has been argued that low doses ought to have been selected. A study with hamsters (Dalbey 1982) with 10 ppm formaldehyde was negative in terms of tumour formation.

Subsequently, Appelman et al. (1988) have described a one-year inhalation study with male Wistar rats, (6 h d^{-1}, 5 d wk^{-1}) with 0 (controls), 0.1, 1 and 10 ppm formaldehyde using "naive" rats and rats in which the nasal epithelium had been pre-damaged by electrocoagulation. It was concluded that damaged rat nasal epithelia were more susceptible to the cytotoxicity of formaldehyde, and that 1 ppm formaldehyde did not visibly affect the intact nasal epithelium.

Monticello et al. (1996) conducted a formaldehyde inhalation carcinogenicity study (male F344 rats, 2 years inhalation, 6 h d^{-1}, 5 d wk^{-1}, at 0, 0.7, 2, 6, 10, or 15 ppm) in which a major endpoint was correlation of cell proliferation indices with origination of formaldehyde-induced nasal squamous cell carcinomas. From the study it was concluded, that target cell population size and sustained increases of cell proliferation in these populations, much determined by differences in regional airflow-driven formaldehyde dose to these sites, were decisive. Coupled with known non-linear kinetics of formaldehyde binding to DNA (v. i.), this was seen to account for the non-linearity and site specificity of formaldehyde-induced nasal squamous cell carcinomas in rats.

4.5.8.3
Formaldehyde-DNA interaction

Initially, data on formaldehyde-DNA interaction used for interpretation of the bioassay data (Casanova-Schmitz et al. 1984b) have been controversially debated (Cohn et al. 1985). This has led to an in-depth investigation of the quantitative coherence of primary DNA-formaldehyde interaction, resulting DNA-protein cross-links (DPX), cell replication and local tumour formation at the nasal epithelium.

No experimental evidence for formation by formaldehyde of DNA-DNA cross-links was obtained with mammalian cells (Ross and Shipley 1980; Bedford and Fox 1981; Ross et al. 1981; Harris et al. 1983), and steric considerations were put forward as likely reason (Bedford and Fox 1981). Against this background, the formation of DNA-protein cross-links in rodent nasal respiratory mucosa (Casanova-Schmitz and Heck 1983) has been regarded as an important primary biochemical lesion in the pathogenesis of formaldehyde-induced tumours (Lam et al. 1986). It is noteworthy that remarkable differences between different nucleo-proteins exist in terms of "cross-linkable" DNA-protein contacts (Solomon and Varshavsky 1985).

In essence, the general paradigm has been experimentally supported on different levels of the toxicological chain of events:

– The covalent binding of labelled formaldehyde to DNA of respiratory rat nasal mucosa was not linear with the exposure concentration, but increased steeply between 2 and 6 ppm (Casanova-Schmitz et al. 1984b). This has been put forward as an important argument against the applicability of linear (or quasi-linear) risk estimates for formaldehyde.
– The dose-response of incorporation of ^{14}C-formaldehyde into DPX of rat nasal mucosa has clearly demonstrated a non-linearity at formaldehyde doses below 5

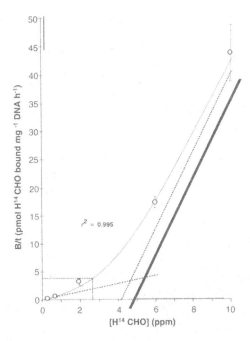

Fig. 4.16 DNA binding of ^{14}C-labelled formaldehyde (ordinate: inhaled by rats *in vivo*; abscissa: ppm).

ppm (Casanova et al. 1989, see Fig. 4.16). Effects of pre-exposures of rats on nasal DPX formations were visible, but were observed at high doses only and therefore considered of only marginal importance for risk extrapolations (Casanova et al. 1994).
– Formaldehyde exposure led to an increase in the rate of cell turnover in the respiratory mucosa (the tissue from where the nasal squamous cell carcinomas originate). This increase is non-linear with formaldehyde concentration, being undetectable at 2 ppm, highly significant at 6 ppm, then decreasing at 15 ppm due to cytotoxicity (Swenberg et al. 1983b). Such increased cell replication would substantially increase the number of non double-stranded DNA sites available for reaction with formaldehyde (Casanova et al. 1994).

4.5.8.4
Conclusions on risk extrapolation

Possible human risk estimates on the basis of the animal experiments have been discussed. The high incidence of experimental tumours in rats at 14.3 ppm, together with the mutagenic properties and the possibility of DNA interaction, have been main arguments in favour of a substantial human carcinogenic risk which ought to be calculated by conventional risk extrapolation models. On the other hand, already in 1984 the following arguments have been used against a linear risk extrapolation to humans (BMJFG 1984):

- species differences in target tissue doses due to differences in respiration physiology,
- the highest dose in the CIIT rat experiment having produced excessive toxicity and mortality,
- local cell and tissue lesions as being necessary precursor stages before tumours develop,
- dose-response of cell proliferation leading to higher proportions of single-stranded DNA which is susceptible to the generation of DNA-protein cross-links,
- non linear dose-response of covalent formaldehyde-DNA interaction,
  endogenous formation of formaldehyde and physiological formaldehyde levels,
- rapid detoxification of formaldehyde, if present in doses which are not excessive,
- formaldehyde seems not to act systemically.

Recommendations of various official bodies, on national and international levels, have generally considered these arguments. In particular, the concept that humans very probably are less susceptible than test animals, especially rats, has widely been accepted (Squire and Cameron 1984). Basing tumour data on the actually delivered dose (which is species-dependent because of differences in respiration physiology) has a substantial impact on the outcome of mathematical risk assessments (Starr and Buck 1984). Genotoxicity of formaldehyde was thought to be confined mainly to higher dose levels where increased cytotoxicity and regenerative cell division are found (Lutz 1986).

As a result of the experimental studies on formaldehyde, it was recommended that low concentration ($\leq$ 2 ppm airborne exposure) extrapolation, where no tissue damage is observed, be uncoupled from the responses at high concentrations ($\geq$ 6 ppm). At high concentrations epithelial degeneration, regenerative cell replication, and inflammation appear as the essential driving forces in formaldehyde carcinogenesis (Morgan 1997).

Thus, formaldehyde represents a well investigated example of a genotoxic chemical with a non-linear carcinogenic dose-response, implying a practical threshold.

4.5.9
Vinyl acetate

For two reasons the case of vinyl acetate will be discussed in more detail than all other substances presented so far (Hengstler et al. 2003):

1. Vinyl acetate is presently being debated intensively by several regulatory institutions and much research has been performed.
2. A relatively complete set of data has been generated that allows classification of vinyl acetate as substance acting by a threshold mechanism concerning carcinogenesis.

Thus, vinyl acetate is interesting for both, practical regulatory purposes and theoretical considerations concerning threshold mechanisms. Vinyl acetate is genotoxic since it induces chromosomal aberrations, DNA-protein cross-links and sister chromatid exchanges. Bioassay data show that vinyl acetate is carcinogenic in rats and mice by the oral route and in rats by the inhalation route. However, all carcinogenic responses are expressed at very high-dose levels that exceed standard definitions of Maximum Tolerated Dose (*MTD*). In this chapter we show that the dose-response curve of vinyl acetate has at least a practical threshold (for definition of a "practical threshold" see 4.2).

4.5.9.1
Carcinogenicity in rats and mice

Oral exposure. Vinyl acetate was clearly carcinogenic in mice and rats by oral ingestion. When the data are plotted collectively, a dose-response appears as depicted in Fig. 4.17. Responsiveness of rats and mice are similar when the dose is expressed in a mg kg⁻¹ d⁻¹ format. Because carcinogenicity only occurs at dose levels that exceed an *MTD* and are beyond what is expected of a limit test of 1,000 mg kg⁻¹ d⁻¹ according to OECD guidelines (1981, 1998), the respective studies must be considered excessive. Regarding multiple sites of carcinogenicity, the only tumours clearly associated with vinyl acetate exposure were of the upper digestive tract. Tumours were located in the oral cavity, oesophagus, and forestomach. All of these tissues are lined with squamous epithelium and do not display marked differences in histological make-up of the lining epithelium. Not surprisingly, all of the tumours are of the same histogenic origin. Generally, multiple-site carcinogenesis is considered for agents such as nitrosamines that affect a variety of systemic organs and different cell types. Nitrosamines are carcinogenic to oesophagus and brain, tissues of entirely different histology (Bartsch et al. 1987). It is not appropriate to consider the oral cavity, oesophagus, and forestomach as "multiple organs" when the tumours all

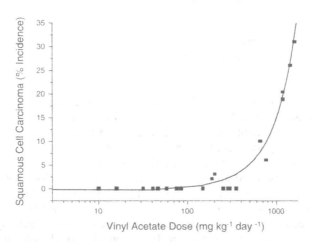

Fig. 4.17 Composite dose response data for rat and mouse drinking water bioassays. Vinyl acetate induced squamous cell carcinoma of the oral cavity, oesophagus and forestomach. The tumour incidence was greatest for the oral cavity. These data, which include male and female rats and mice, illustrate the sharp break in the dose-response curve with clear evidence for a practical threshold. Data are from Bogdanffy et al. (1993), Maltoni et al. (1997), and JBRC (1998).

appear to be derived from squamous epithelial lining and all of the tumours are histologically similar. This argument is consistent with published guidelines for combining neoplasms for evaluation in carcinogenesis studies (McConnell et al. 1986). There is, therefore, no evidence for multiplicity of tumour sites. Furthermore, the tumour incidence decreases from the oral cavity to the forestomach. This is a characteristic that would seem to be more appropriately associated with a site-of-contact carcinogen.

Inhalation exposure. Vinyl acetate was clearly carcinogenic by the inhalation route in rats, but not mice. While the Maltoni et al. (1974) study is considered inadequate for assessing carcinogenicity, the study of Bogdanffy et al. (1994) appears sufficient. Although vinyl acetate was clearly carcinogenic in rats carcinogenicity was only expressed at high exposure levels (600 ppm). The Maximal tolerated dose (*MTD*, defined as leading to 10 % retardation in weight gain) was exceeded in rats exposed to 600 ppm. There was no evidence for systemic carcinogenicity.

4.5.9.2
Epidemiological studies of vinyl acetate carcinogenicity

Epidemiological studies of vinyl acetate carcinogenicity are described in the epidemiological chapter. In essence, evaluation of epidemiological data on a possible carcinogenic effect of vinyl acetate is difficult, since most individuals in the existing epidemiological studies were exposed to several chemicals. Nevertheless, the existing data do *not* support a carcinogenic effect of vinyl acetate in humans.

4.5.9.3
Metabolism and genotoxicity of vinyl acetate

Exposure of tissues to vinyl acetate, at the site of contact, results in metabolic conversion to acetic acid and acetaldehyde. The histochemical localisation of carboxylesterase and aldehyde dehydrogenase in nasal tissue have been described in detail (Bogdanffy et al. 1986,1987; Olson et al. 1993; Lewis et al. 1994). These enzymes rapidly and almost completely convert vinyl acetate to acetic acid and acetaldehyde in nasal tissue. Acetaldehyde, at high concentrations, induces DNA protein cross-links that finally lead to chromosomal aberrations. Formation of DNA protein cross-links is facilitated by low intracellular pH (pH_i). Low pH microenvironment is caused by acetic acid formation from both vinyl acetate hydrolysis and acetaldehyde oxidation to acetic acid and liberation of protons (Bogdanffy 2002). Acetaldehyde, as discussed above, is a known clastogen but does not appear to induce point mutations. In fact, the profiles of genotoxic activity for acetaldehyde and vinyl acetate are almost identical and vinyl acetate is not active as a clastogen without a source of carboxylesterase added. Thus the clastogenic activity of vinyl acetate must be attributed to metabolic formation of acetaldehyde. It has been reported that acetic acid, formed intracellularly from vinyl acetate hydrolysis, contributes to the genotoxic activity (Morita et al. 1990; Sipi et al. 1992). The confounding effect that low pH can have on genetic toxicity tests that use mammalian cells is well known (Brusick 1986). Low pH has also been shown to induce cellular transformation of Syrian hamster embryo cells (LeBoeuf et al. 1992). In fact, acetic acid induces chromosomal aberrations in Chinese hamster ovary (CHO) cells. It is likely that the genotoxic activity of acetaldehyde is attributable at least in part to intracellular acidification since two protons are released when acetaldehyde is oxidised to acetic acid in the presence of aldehyde dehydrogenase and NAD^+. Intracellular acidification has been reported to facilitate acetaldehyde induced genotoxicity (Kuykendall and Bogdanffy 1992a). The authors used a model system for measuring DNA-protein cross-links (the initial event finally leading to chromosomal breaks) involving incubations of calf thymus histone protein with plasmid DNA and measurement of covalently bound DNA-histone protein complexes. Cross-links appeared to be between DNA and amino acid residues, guanosine and lysine, respectively. In their studies, it is shown first that DNA-protein complex formation requires a carboxylesterase-dependent metabolism of vinyl acetate to acetaldehyde and acetic acid. Next it is shown that the formation of acetaldehyde-induced DNA-protein cross-links is increased in the presence of increasing concentrations of acetic acid. Finally they demonstrate that acetaldehyde-induced DNA-protein cross-links are increased with simple reduction in pH. The mechanism for this increase is proposed to be due to ionisation to positively charged amino acid groups of histone proteins resulting in a higher affinity for the negatively charged DNA. The resulting tight association of histone protein with DNA might be a prerequisite for the formation of DNA-protein cross-links by acetaldehyde which, in addition, is increased by the higher electrophilicity of the carbonyl carbon upon protonation of acetaldehyde. In conclusion, although further studies are required in this field, strong evidence has been presented showing that intracellular acidification is a prerequisite for the genotoxic activity of vinyl acetate and acetaldehyde. In addition, the acetaldehyde DNA-protein cross-link was found to be very unstable with a half life of approximately 6 hours (Kuykendall and Bogdanffy 1992b, 1994).

4.5.9.4
Mechanistic data support a practical threshold for vinyl acetate

Tissues can accommodate exposure to acetic acid and acetaldehyde without adverse effect up to a certain level of exposure (Kuykendall et al. 1993). This is consistent with the known exposure of tissues to endogenous acetic acid and acetaldehyde. It is known that acetaldehyde is a natural constituent in the body and a metabolic by-product of threonine metabolism (Halvorson 1993; Lehninger et al. 1993). Background levels of acetaldehyde of approximately 0.3 µg ml^{-1} exist in blood. Therefore, it would seem reasonable that exposures to vinyl acetate that do not raise tissue acetaldehyde levels beyond the range of natural background levels in blood or tissues would also be below biological thresholds. Using a physiologically based pharmacokinetic (PBPK) model it is possible to predict tissue exposure to acetaldehyde that results from inhalation exposure to vinyl acetate. Fig. 4.18 shows the predicted basal cell acetaldehyde levels in humans during exposure to 1 ppm vinyl acetate. When critical levels of vinyl acetate are achieved, thresholds are exceeded and five critical steps in the mechanism that ultimately leads to cancer become active (Fig. 4.19). The threshold for pH$_i$ reduction in neuronal cells that does not induce cytotoxicity *in vitro* is 0.15 pH unit (Nedargaard et al. 1991). The lowest concentration of acetaldehyde that has been shown *in vitro* to induce sister chromatid exchanges (SCE) in Chinese

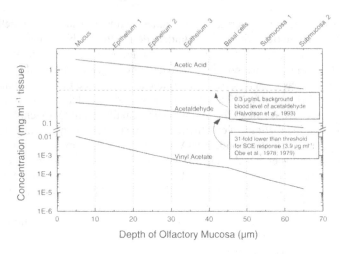

Depth of Olfactory Mucosa (μm)

Fig. 4.18 Dosimetry in human epithelium at an exposure to 1 ppm vinyl acetate. Predicted steady state concentrations of acetaldehyde, acetic acid, and vinyl acetate throughout the olfactory nasal mucosa of humans exposed continuously to 1 ppm vinyl acetate. Concentrations of acetaldehyde at the basal cell layer are critical for consideration because basal cells are the progenitor cells for the epithelium and are the target cell for carcinogenesis. The figure illustrates that at 1 ppm vinyl acetate, basal cell acetaldehyde concentrations are predicted to be approximately 3 times lower than background blood acetaldehyde levels and more than 3 times lower than the lowest concentration shown to induce sister chromatid exchanges (SCE). SCE are a sensitive marker of genetic damage and of questionable relevance. The margin of safety relative to the more appropriate endpoint of chromosomal aberrations is much greater. Obe et al. (1979) reported the lowest level to induce SCE in normal human lymphocytes or lymphocytes from Fanconi's anaemia patients to be 15.6 μg ml^{-1} or 7.8 μg ml^{-1} acetaldehyde, respectively. The margin of safety below the chromosomal aberrations endpoint is 124-fold.

hamster ovary (CHO) cells is 3.9 μg ml^{-1} (Obe et al. 1979). Sister chromatid exchanges are not considered to be a valid marker of mutagenic damage and are generally overly sensitive. A more appropriate and more widely accepted benchmark genetic toxicity endpoint would be chromosomal aberrations. Chromosomal aberrations are also mechanistically consistent with the data suggesting that acetaldehyde induces DNA-protein cross-links. However, since SCE occur already at lower concentrations than chromosomal aberrations SCE were chosen as an endpoint in order not to under- but rather overestimate the risk due to vinyl acetate exposure.

When critical levels of exposure to acetic acid and acetaldehyde are achieved, thresholds are exceeded and further critical steps in the mechanism that ultimately leads to cancer become active. These steps are illustrated for olfactory epithelium in Fig. 4.19. The PBPK model predicts that in rat nasal olfactory tissue exposure to 50 ppm vinyl acetate causes a 0.08 unit reduction in pH$_i$ and a basal cell acetaldehyde concentration of 1.7 μg ml^{-1}. Fifty ppm vinyl acetate is a NOAEL and the pH$_i$ reduction and basal cell acetaldehyde levels are below their thresholds. As the dose level increases to 200 ppm, pH$_i$ is predicted to be reduced by 0.25 pH units, a value slightly above the threshold, and cytotoxicity such as olfactory degeneration occurs

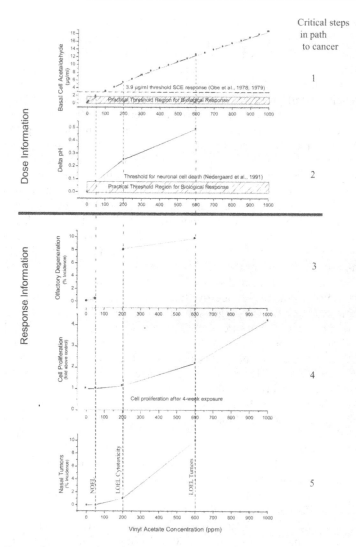

Fig. 4.19 Composite presentation of dose and response data for the 5 critical steps on the pathway to carcinogenesis in nasal olfactory epithelium. Panel 1 shows the predicted steady state concentration of acetaldehyde at the basal cells, the progenitor cells of nasal cancer, in relation to *in vitro* doses that produce sister chromatide exchanges. Panel 2 shows predicted pH_i changes in olfactory epithelium of the rat in relation to changes in pH_i that are cytotoxic to neuronal cells *in vitro*; pH_i reduction is proposed to be the critical step leading to cytotoxicity. Panel 3 shows olfactory degeneration in rats as a cytotoxic endpoint. Basal cell proliferation and the incidence of nasal tumours in rats is presented in panels 4 and 5. Olfactory degeneration (cytotoxicity) is observed at 200 ppm. Because acetaldehyde levels are only slightly above thresholds, there is no significant tumour response. At 600 ppm all thresholds are exceeded, cell proliferation is significantly enhanced and a significant incidence of nasal tumours is observed. The mechanism of action for nasal respiratory and oral cavity tumours is similar with the exception that the cause of the proliferative response in oral cavity may be only subtly related to cytotoxicity and is more likely the result of known mitogenic effects due to reduced pH_i.

at an incidence between 8 and 10 % (Fig. 4.19; step 3). However, the cell proliferation response at 200 ppm is weak (step 4). Levels of acetaldehyde of 5.4 µg ml^{-1} are slightly in excess of the threshold for genotoxicity. Thus at 200 ppm there is minimal exposure above threshold levels of acetaldehyde (step 1), minimal pH$_i$ reduction above the threshold (step 2), enhanced olfactory degeneration (step 3) and slightly enhanced cell proliferation (step 4). At 200 ppm, one nasal tumour was observed which was not statistically significant (step 5). At 600 ppm acetaldehyde levels are predicted to be markedly above threshold at 12.4 µg ml^{-1} (step 1), pH$_i$ is predicted to be markedly reduced by 0.49 pH units (step 2), olfactory degeneration is strongly enhanced (step 3) and basal cell proliferation is more than 2-fold above control (step 4). At 600 ppm, all of the critical steps in the mechanism of carcinogenesis are active and tumours now appear at a statistically significantly increased rate (step 5). This sequence of events and the physiological modelling suggest that only when critical exposure concentrations, threshold levels for pH$_i$ reduction and cellular proliferation (induced by cytotoxicity in olfactory epithelium) are achieved, are the conditions necessary for a complete carcinogenic mechanism all in place.

There is another area of mechanistic work, which deserves discussion regarding the generality of the mechanism discussed above for both, the olfactory and respiratory tissue as well as the upper digestive tract. The mechanism of action described above for olfactory epithelium suggests that cytotoxicity is the first adverse cellular response to vinyl acetate exposure of both respiratory and olfactory tissue. Events such as cell proliferation and tumour formation become significant only at higher concentrations. The data to support this mechanism are clear for the olfactory tissue but require careful analysis for the respiratory tissue. Also in respiratory tissue there are several pieces of information that support cytotoxicity as the first step in carcinogenesis. These are the observations of respiratory epithelial degeneration and cell proliferation in rats exposed for one or five days to 1,000 ppm (Bogdanffy et al. 1997) and the *in vitro* cytotoxicity studies which show acid phosphatase release from nasal turbinates in culture (Kuykendall and Bogdanffy 1993). The cell proliferation responses, which were significant also at 600 ppm most likely represent subtle cytotoxic responses that are repaired or not evident microscopically. The lack of a more pronounced response in respiratory tissue has been recognised as an unresolved question (Bogdanffy et al. 1994; Plowchalk et al. 1997).

An alternative hypothesis for the cell proliferation stimulus in respiratory epithelium has recently emerged (Bogdanffy, 2002) that might also clarify the mode of action in the upper gastrointestinal tract. Literature reports suggest that reductions in pHi can also induce mitogenesis. Alterations in pH$_i$ are involved in stimulation of cell growth and transformation. For instance, Syrian Hamster Embryo cells, cultured at pH 6.7, show a marked increase in life span, compared to those cultured at pH 7.3, as measured by the number population doublings that occur before cellular senescence (Kerckaert et al. 1996). The higher proton burden of the intracellular environment has been shown to displace Ca^{2+} from intracellular binding sites (Batlle et al. 1993). Ca^{2+} displaced from the growth and differentiation factor (GDF) protein blocks the intracellular signalling that leads to differentiation (Isfort et al. 1993). Blockage of the differentiation pathway could promote sustained proliferation, expansion of the undifferentiated cell population, and clonal expansion of spontaneous- or chemical-induced mutants. Although substantiation of the hypothesis that intracellular acidification is mitogenic in nasal or oral cavity mucosal cells suggests further experimentation, the proposal is supported by the literature and could provide a fundamental linkage to many tumour promotion mechanisms. Regardless of the mechanism for induction it is clear that cell proliferation is induced in nasal respiratory and upper gastrointestinal tract epithelial tissues and that this step is critical to the complete expression of the carcinogenic potential of vinyl acetate.

Since olfactory tissue is the most susceptible tissue to vinyl acetate toxicity and is the focus for risk assessment, this unresolved question can be considered minor. It has been argued that respiratory and olfactory tissue are not likely to respond similarly to the same exposure to acetic acid, or reduced pH_i and that respiratory and olfactory tissues may have different biochemical capacities for responding to alterations in pH_i. There is precedent in the literature to support this position (Stott and McKenna 1985; Frighi et al. 1991; Tobey et al. 1992). Similar to the nasal cavity the oral cavity possesses carboxylesterase and the activity has been localised to squamous epithelium (Yamahara and Lee 1993; Reed and Robinson 1998). The activity in oral mucosa of rats and mice were similar and were correlated to regions shown to be active by histochemistry (Morris 2000, as presented in Sarangapani et al. 2000). However, the carboxylesterase activity of the rat oral mucosa was about 100 times lower than that of nasal tissue. Recently cell proliferation has been measured in the oral mucosa of rats and mice where vinyl acetate was administered in drinking water (Valentine et al. 2002). The oral cavity is lined with squamous epithelium. Rats and mice were exposed to concentrations of up to 24,000 ppm for 92 days. Less than 2-fold but significant increases in oral mucosal basal cell proliferation were observed in rats evaluated on days 29 and 92. In mice, the responses were more pronounced with approximately 2.4- and 3.4-fold increases being observed at day 92 in the 10,000 and 24,000 ppm groups, respectively. The greatest proliferative response observed in mice was in the lower jaw, which was also the region of greatest tumour formation observed in the JBRC study (Japanese Bioassay Research Centre 1992). In conclusion, although the support for the proposed mechanism of action of vinyl acetate on oral cavity mucosa is not as robust as for the nasal cavity the research to date provides a parallel picture in which enhanced epithelial cell proliferation is induced when critical thresholds are exceeded. The data further suggest an even higher level of a practical threshold in these tissues for which also the formation of some tumours have been reported after exposure to very high doses of vinyl acetate. The implication of the discussed five-step mechanism is that practical thresholds of exposure exist below which there is no substance related increased risk for cancer.

4.5.9.5
Practical thresholds in relation to human vinyl acetate exposure

As shown above there is clear evidence for a practical threshold for vinyl acetate induced carcinogenesis: an exposure ranging between 50 and 200 ppm vinyl acetate was shown to cause cytotoxicity, but the proliferation response was only weak. In this concentration range no significant increase in carcinogenicity was observed in experimental animals, but started only at 600 ppm. An even more conservative practical threshold represents 50 ppm. Below 50 ppm no cytotoxicity and no cell proliferation could be induced by vinyl acetate. Consequently, no carcinogenesis was observed in experimental animals in this dose range. Besides these studies based on experimental animals, PBPK modelling predicts that exposure to 1 ppm vinyl acetate leads to a basal cell acetaldehyde level that is about three times lower than the endogenous concentration of acetaldehyde *in vivo*. Thus, a concentration of 1 ppm can be expected to be far below concentrations that could cause toxic effects.

Such practical thresholds can be compared to existing occupational vinyl acetate exposures. Concentrations of 0.07–0.57 ppm vinyl acetate were reported in ambient air in an area, where several vinyl acetate manufacturers were located (US Agency for Toxic Substances and Disease Registry, ATSDR 1992). An ambient air concentration of 0.14×10^{-3} ppm was detected near a chemical waste disposal site (Pellizzari 1982). Although most studies published 1990 or later reported relatively low vinyl acetate exposures, for instance < 0.22 ppm for polyvinyl acetate painters (International Technology Corporation 1992) or < 9.9 ppm in various Finnish industries (Finnish Institute of Occupational Health 1994), much higher occupational exposures have been reported in earlier studies. For instance, in 1969 a maximal exposure of 49 ppm has been reported in vinyl acetate production and polymerisation industries in the USA (Deese and Joyner 1969). As a consequence of the practical threshold for vinyl acetate it seems to be very important to avoid extreme exposures, since at concentrations of 600 ppm, albeit for a lifetime exposure, carcinogenesis was observed. On the other hand, to our knowledge, almost all occupational exposures reported in the past decade were much lower. Exposure limits for vinyl acetate are different in several countries. Examples for current national occupational exposure limits (time-weighted average) are 10 ppm in the USA and Germany, 8.5 ppm in France and 2.8 ppm in Poland and the Russian Federation (ACGIH 2001; DFG 2001). Regarding ambient lifetime exposures, a limit of 0.4–1.0 ppm has been recommended (Bogdanffy et al. 1999). Lower concentrations are below a practical threshold, where the prerequisites for vinyl acetate carcinogenesis, namely cytotoxicity and regenerative cell proliferation are not observed, whilst significant vinyl acetate-induced carcinogenesis itself is not experimentally observed below 600 ppm.

4.5.10
Non DNA-reactive genotoxins

The classification of carcinogens and of germ cell mutagens is in a state of present discussion. In Germany, the Senate Commission of the DFG for the Investigation of Health Hazards in the Work Area (MAK-Commission) has issued new recommendations to distinguish between 5 groups of proven and suspected carcinogens (Neumann et al. 1998), instead of 3 groups, as provided by the Labelling Guide of the European Union (see Box 4.1 on page 130; categories 1, 2, 3; in Germany: GefStoffV, Anhang I, no. 1.4.2.1). The proposal (see Box 4.2 on page 131) has resulted from a continuing discussion over 10 years (Bolt et al. 1988). It includes as Category 5:

> Substances with carcinogenic and genotoxic potential, the potency of which is considered so low that, provided that the MAK-value is observed, no significant contribution to human cancer risk is to be expected.

In consequence, it means that a genotoxic potential, which is low, might be negligible, at low doses, for human cancer risk. An example for this category (based on effective biological inactivation mechanisms) is styrene which has been discussed above (see section 4.3) in some detail.

Moreover, there should not only be a distinction between "genotoxic" and "non-genotoxic" carcinogens, but within the group of "genotoxic" carcinogens also a differentiation between those characterised by "threshold" and "non-threshold" effects should be attempted.

The following chapter characterises "non-mutational" genotoxic mechanisms as those where the primary event is interaction with proteins or protein systems leading to changes on the chromosome level, not a direct interaction with DNA leading to a mutation. Such "indirect" mechanisms of genotoxicity are expected to be accompanied by practical thresholds (Crebelli et al. 2000; Parry et al. 2000). In this respect, topoisomerase and motor protein interactions have been studied in some detail.

4.5.10.1
Topoisomerase poisons

DNA topoisomerases are nuclear enzymes, which induce transient breaks in the DNA allowing DNA strands or double helices to pass through each other (Hengstler et al. 1999b; Toonen and Hande 2001; Wang 2002). By this action topoisomerases solve topological problems of DNA in replication, transcription, recombination and chromosome condensation as well as decondensation. DNA topoisomerases fall into two major classes: the type I enzymes that induce single stranded cuts in DNA, and the type II enzymes that cut and pass double stranded DNA.

Recent discussion on dietary flavonoids. Due to the periods of rapid cell turnover it is likely that fetal tissues or tissues of children are more susceptible to some environmental or dietary genotoxic agents (Hengstler et al. 1998, 1999a; Von Mach, 2002). Recently, some studies suggested a causal relationship between infant leukemia induced *in utero* and maternal exposure to dietary bioflavonoids. These compounds are being discussed as endocrine disruptors (Degen and Bolt 2000), and as topoisomerase II-poisons (Ross 2000; Strick et al. 2000; Alexander et al. 2001; McDonald et al. 2001). In contrast to infant leukemia no role of dietary topoisomerase II-poisons has been observed in pathogenesis of adult leukemia. The majority of studies reported beneficial effects of dietary bioflavonoids for adult individuals. Several epidemiological studies observed a decreased risk for prostate, breast, uterus, colon and lung cancer for adults consuming high levels of bioflavonoids. Numerous mechanisms have been reported that may explain these protective effects including inhibition of tyrosine kinases, antioestrogenic effects, release of transforming growth factor b, induction of apoptosis and antioxidative effects. One of the main sources of dietary flavonoids are soy beans and soy products (Tab. 4.1). Soy products, including for instance soy burgers, soy hot digs, soy cheese, etc., have proliferated due to the labelling of soy as a food that reduces the risk of some tumour types and also the risk of heart disease. Exposures to dietary isoflavones have been reported to range between 50 and 100 mg d^{-1} in East Asian populations resulting in plasma concentrations of 40–240 ng ml^{-1} daidzein and genistein (combined) (Bolt et al. 2000). In European populations dietary exposures to isoflavones are much lower usually not exceeding 1 mg d^{-1}. However, the consequences of exposure to bioflavonoids may differ between adults and trans-placentally exposed embryos. Due to the rapid cell proliferation, the aspect of topoisomerase II inhibition by bioflavonoids may be much more critical for embryos than for adults.

Mechanism of action. The mechanism of action of topoisomerase II can be dissected into a series of steps initiated by the binding of DNA to both subunits (S1 and S2) of the enzyme (Fig. 4.20a). In the next step, topoisomerase II cleaves the DNA, forming a phosphotyrosine linkage between each DNA single strand break product and the catalytic tyrosines of both topoisomerase II subunits (Fig. 4.20b). The latter step depends on the presence of magnesium. In a next ATP-dependent step, topoisomerase II traps a second DNA double strand (Fig. 4.20c). Trapping of the second DNA duplex is achieved by a conformational change of the N-terminal domains of both subunits, leading to a closed clamp. After trapping of the second DNA double strand (termed trans- or T-strand) it will pass the gap of the first DNA double strand (termed gap- or G-strand) (Fig. 4.20d). As soon as the T-strand has passed the G-strand the carboxy terminal portion of the enzyme undergoes a conformational change (Fig. 4.20e) that allows the exit of the T-strand (Fig. 4.20f). In a next step topoisomerase II reverses the cleavage reaction of the G-strand (Fig. 4.20g). After religation of the G-strand hydrolysis of ATP leads to a conformational change of the N-terminal part of both subunits of topoisomerase II that allows dissociation of the G-strand (Fig. 4.20h).

"Classical" topoisomerase II targeting substances act by trapping the cleaved G-strand-enzyme intermediate, thus blocking religation and enzyme release, leaving the DNA with a permanent double strand break. These substances that lead to

Tab. 4.1 Topoisomerase II-poisons and/or catalytic inhibitors in environment and food

Substance	Sources of human exposure
Flavonoids	
Quercetin	Fruits and vegetables
Genistein	Soy
TTIG-2535	Investigational soy drug product
Catechins	Tea, wine, chocolate
Caffeine	Coffee, tea
Thiram	Agricultural fungicide
Benzene metabolites	Gasoline
p-Benzoquinone	
Hydroquinone	
Catechol	
4,4'-Biphenol	
Senna	Anthraquinone laxative
Dipyrone, Baygon, Thiram	Insecticides

higher levels of covalent topoisomerase II-DNA complexes have been termed topoisomerase II-"poisons", whereas substances that inhibit the enzyme during other steps are referred to as "catalytic inhibitors" (Walker and Nitiss 2002). The best known topoisomerase II-poisons belong to two classes of antineoplastic agents, the epipodophyllotoxins (e. g. etoposide and teniposide) and the anthracyclines (e. g. doxorubicin). Catalytic inhibitors include for instance derivatives of coumarin antibiotics, such as the coumermycins and novobiocin, the thiobarbiturate merbarone and the bisdioxopiperazines (Walker and Nitiss 2002). However, apart from drugs there is a relatively large number of natural and synthetic products present in the environment and in food that act as topoisomerase II-poisons and/or catalytic inhibitors (Tab. 4.1).

Carcinogenicity of topoisomerase II-poisons. DNA double strand breaks induced by topoisomerase II-poisons can induce apoptosis of tumour cells contributing to the therapeutic effects of epipodophyllotoxins and anthracyclins (Walker and Nitiss 2002). In addition, the presence of covalent topoisomerase II-DNA complexes arrests the replication fork, which also contributes to the antineoplastic effects. However, if healthy cells survive exposure to topoisomerase II-poisons, DNA double strands may lead to chromosomal aberrations. The best studied chromosomal aberrations induced by topoisomerase II-inhibitors in humans are translocations involving the *MLL* gene located at chromosome 11q23. After exposure to etoposide or doxorubicin *MLL* is rearranged with partner genes in at least 40 different translocations, the most common being translocations with chromosomes 4, 6 and 9 (Strick et al. 2000). The latter translocations are frequently observed in patients that have been treated with etoposide and teniposide and developed therapy-related acute myeloid leukemias (AML). There is no doubt that the topoisomerase II-poison etoposide is carcinogenic in humans. AML develops relatively early after

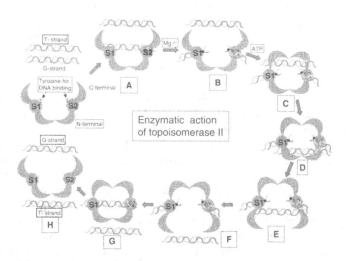

Fig. 4.20 Enzymatic action of topoisomerase II. The "classical" topoisomerase II-poisons act by trapping the G-strand-enzyme intermediate (step f. in this figure), thus blocking religation and enzyme release, leaving the DNA with a permanent double strand break.

etoposide therapy (2–3 years). Cumulative etoposide doses of 2 to 3 g m^{-2} body surface are associated with a cumulated 6-year risk for development of secondary leukemia of 2.2 % (Toonen and Hande 2001). For patients receiving less than 2 g m^{-2} etoposide a 0.4–0.6 % risk was found.

In conclusion, exposures to high doses of topoisomerase II-poisons in cancer therapy have clearly shown that topoisomerase II-poisons induce chromosomal translocations and leukemia.

Dietary exposure to topoisomerase II-poisons: role in infant leukemia? The carcinogenicity of high doses of etoposide and the fact that humans are exposed to low doses of several topoisomerase II-poisons in food (Tab. 4.1) led to the concern that exposure to these substances may contribute to carcinogenesis in the general population. In this context, the question whether carcinogenic topoisomerase II-poisons act by threshold or non-threshold mechanisms is of high relevance (Hengstler et al. 2003). On one hand, one might expect that carcinogens that act by inhibition of enzymes have thresholds (Von Mach et al. 2002, 2003). This may be true for inhibitors of detoxifying enzymes or some DNA repair enzymes, where low doses of inhibitors will be without consequences as long as they do not reduce enzyme activity to a relevant extent. However, the situation for topoisomerase II can be expected to be more critical: in principle a single molecule of a topoisomerase II-poison might trap a topoisomerase II protein at a step where DNA strands have been cleaved. This could lead to a single DNA double strand break, induce a single chromosome translocation and initiate a series of steps finally leading to neoplastic transformation. However, the probability for neoplastic transformation resulting from a single DNA double strand break seems to be extremely low. Hence, from a very theoretical point of view it remains open whether some topoisomerase II-poisons are to be expected to evoke deterministic or stochastic effects. From a more practical point of view the concept of a "practical threshold" could be favoured.

Recently, some alarming studies have been published suggesting that exposure to low doses of topoisomerase II-poisons may indeed be relevant (Ross 2000; Strick et al. 2000; Alexander et al. 2001; McDonald et al. 2001): Approximately 80 % of infants (younger than 1 year of age) with acute myelogenous leukemia (AML) and acute lymphoblastic leukemia (ALL) have chromosome translocations involving the *MLL* gene. The above mentioned leukemias (AML) associated with etoposide chemotherapy manifested identical chromosomal translocations involving the *MLL* gene (Ross 1998). The MLL cleavage site induced by etoposide colocalised with those observed in infant leukemia (Strick et al. 2000). In addition, molecular studies have suggested that leukemias in the first postnatal year have been initiated *in utero*. This led to the hypothesis that maternal exposure during pregnancy to environmental substances that inhibit topoisomerase II may cause development of leukemia in infants. This hypothesis has been examined in several studies (Ross 2000; Strick et al. 2000; Alexander et al. 2001). Since topoisomerase II-inhibitors have been found in soy, wine, coffee, tea, cocoa, pesticides, medications, specific fruits and vegetables, mothers of infant cases and matched controls were interviewed for potential exposure to topoisomerase II-inhibitors (Ross 1998). An approximately 10-fold higher risk of infant AML with increasing maternal consumption of topoisomerase II-poison containing foods was reported in this study. Based on 20 bioflavonoids tested, Strick et al.

(2000) identified a common structure essential for topoisomerase II-induced DNA cleavage, suggesting that maternal ingestion of bioflavonoids induce *MLL* breaks leading to infant leukemia in utero. In addition, at least some bioflavonoids that act as topoisomerase II-poisons have been reported to be able to cross the rat placenta and reach fetal tissues (Schroder-van der Elst et al. 1998; Degen et al. 2002). In addition, the incidence of infant leukemia is almost 2-fold higher in several Asian cities, for example Hong Kong or Osaka, than in Western countries. This difference might be explained by the high food intake of bioflavonoids in many regions of Asia, especially by consumption of soybeans and soybean products (Strick et al. 2000). The observation that topoisomerase II associated chromosomal aberrations occur in infant leukemia leads to the question whether *MLL*-breaks induced by topoisomerase II-poisons are important also for leukemias of adults. However, *MLL* abnormalities in adult leukemia that is not associated with topoisomerase II-therapy are less frequent than 5 % in contrast to approximately 80 % in infant leukemia. In addition, no epidemiological data support a central role of diet in pathogenesis of adult leukemia. Differences in the number of proliferating cells between adult and fetal bone marrow might serve as explanation. It is known that the α-isoform of topoisomerase II is expressed at higher levels in proliferation compared to quiescent cells. Due to the periods of rapid cell turnover it is likely that fetal tissues may be more susceptible to dietary topoisomerase II-poisons than adult tissues.

However, it should also be considered that a large number of studies reported health benefits of bioflavonoids. Several epidemiological studies of human cancers, including prostate, colon and lung, demonstrated a decreased risk associated with consumption of foods containing high levels of flavonoids. It has been hypothesised that high plasma concentrations of isoflavones could act like "natural chemotherapeutic" agents (Record et al. 1997; Strick et al. 2000). The example of topoisomerase II-poisons shows that the same substances may be beneficial or harmful probably depending on target cell types and their proliferation status during exposure.

In conclusion, it is presently being discussed whether dietary topoisomerase II-poisons contribute to infant leukemia. If this will be confirmed, soy products and other foods containing high levels of topoisomerase II-poisons should not be consumed during pregnancy. Based on the nature of the interaction of topoisomerase with a toxicant, the existence of a "practical" threshold of a resulting genotoxic effect seems at least plausible.

4.5.10.2
Motor protein inhibitors

Motor proteins are basically involved in biological mechanics. Specifically, in the process of cell division they provide the driving force for the chromosomal segregation. Hence, interaction of chemical compounds with motor protein systems is a type of protein interaction that may lead to genotoxic effects on a chromosomal level such as aneuploidy. Because the primary event is not a mutation, "traditional" dose-effect relations are to be expected, with the implication of thresholds (Kirkland and Müller 2000; Parry et al. 2000).

Microtubule cytoskeleton and motor proteins. The cytoskeleton basically consists of microtubules, actin-microfilaments and intermediate filaments. Filaments and bundles assemble, together with some kinds of motor proteins (kinesins, dyneins,and myosins) and other structural and regulatory components, into functional cellular units such as myofibrils, the spindle apparatus, flagella or cilia which are cooperatively involved in diverse cell physiological functions. Kinesin is involved in chromosome segregation, cytokinesis and the cytoplasmic transport of vesicles, and some other processes depending on force generation. Kinesin molecules are able to move over long distances along microtubules using a protofilament as track (Sosa et al. 1997).

The internal molecular structure of the cytoskeleton and the function of its associated proteins enable morphogenetic or force- and movement-generating processes. Microtubules are obligate proteinaceous elements of eukaryotic organisms. In functional cooperation with other components of the cytoskeleton, mainly microfilaments (actin) and intermediate filaments, they are essentially involved in numerous diverse cell physiological events mentioned above. Disturbances in microtubule functions indispensably cause serious changes in cell structuring, cell proliferation, and cell motility; permanent influences can finally lead to cell death. Through the electron microscope, microtubules appear as long hollow cylinders with outer and inner diameters of about 24 and 14 nm, respectively. Their lengths vary remarkably between one and several hundreds of micrometers, depending on cell type and cell function.

Microtubules consist mainly of α- and β-tubulin. Both forms have a molecular mass of approximately 50 kD and a diameter of about 4–5 nm and form heterodimers, which constitute the basic building blocks of the microtubules. Especially when whole-mount samples are viewed by electron microscopy, the microtubules reveal longitudinal striations, representing the tubulin protofilaments, which can be described as chain-like association products of $\alpha\beta$-dimers. Because of the dimeric character of tubulin and the strong dimer alternation, at one end of the protofilament an α-subunit and at the opposite one β-subunit is situated, making the protofilament polar. Within the microtubule, the protofilaments are associated laterally with the same polarity. Therefore, the microtubule also appears as a polar structure with a "plus" and a "minus" end. Under steady state conditions, at the "plus" end more dimers are added than lost and at the "minus" end more dimers are released than new ones re-associate. Polarity is a very important feature for microtubule function. It is the basic property for realisation of direction-dependent cellular events, e. g. vesicle transport. Microtubules of mammalian cells are commonly known to be sensitive to cold. That means that they disintegrate at temperatures below 10 °C. On the other hand, microtubules can be reconstituted from tubulin (Shelanski et al. 1973; Weingarten et al. 1975).

Beside numerous natural as well as synthetic microtubule-acting drugs (vinblastine, colchicine, taxol, and nocodazole), applied as cell proliferation inhibitors in modern medicine, publications deal with toxic actions of non-pharmaceutical compounds on the microtubule cytoskeleton. As a result, a number of organic compounds have been found to inhibit microtubule formation and/or to disrupt microtubules. This includes important inorganic compounds (e. g. mercury, zinc, cobalt, nickel and cadmium), as well as a number of organic compounds (e. g. methyl mercury, triethyl lead, 2,5-hexanedione, acrylamide, acrylonitrile, formaldehyde, acetaldehyde and other aldehydes). Especially metal ions, e. g. divalent mercury, are known to change the assembly behaviour of tubulin or stability of tubulin assembly products. Other metal ions are capable of changing the structure of the assembly products.

Using purified tubulin (with or without MAPs), it is possible to follow microtubule formation *in vitro* by means of turbidity measurements at 340 to 360 nm. Assembly is accompanied by an increase of light scattering (turbidity) within the assembly solution. Upon incubation, turbidity reaches a plateau level, which correlates directly to the amount of assembly products formed. In this way, microtubule formation can be quantitatively analysed under *in vitro* conditions.

Members of the kinesin family are integrated in mitotic spindle processes and in the distribution of chromosomes (Gelfrand and Scholey 1992; Wordeman and Mitchison 1995). Kinesin is a tetramer of two heavy and two light chains. The genes of the heavy chain of different species, including humans, have been isolated and sequenced. An approximately 340 amino acid long globular domain, termed heavy chain, is functional as the "motor domain" that is highly conserved among different organisms. It comprises both the ATP and the microtubule-binding site. A central stalk region links a smaller tail domain (light chain) containing binding sequences for loads (Kozielski et al. 1997, 1998).

The carboxy terminus of the heavy chain of conventional kinesin contains positively charged groups, which are thought to be responsible for the interaction with the light chains, the cellular loads, and regulatory activities of the motor proteins. Microtubules are polar structures that give the basis for the unidirectional motion of mechanochemical proteins. It has been shown that the working direction of kinesins is decisively determined by the motor domain position inside the molecule in relation to its carboxy- or amino-terminus (Hirokawa 1996). Kinesins with an amino-terminally situated motor domain (Henningsen and Schliwa 1997) transport their loads typically towards the plus end of the microtubules, i. e. away from the microtubule-organising centre (MTOC), and, vice versa, carboxy-terminally situated motor domains transport in the opposite direction.

Intracellular movement and transport can be mimicked *in vitro* using purified kinesin and taxol-stabilised microtubules in a so-called "microtubule gliding assay" (Ray et al. 1993). The velocity of kinesin-generated motility *in vitro* is about 0.6 μm s^{-1}, dependent on experimental conditions, e. g. Mg^{2+} concentration, pH, temperature, ionic strength, and ATP concentration. Differential interference contrast (DIC) video microscopy is routinely used to follow kinesin-driven microtubule-motility *in vitro* (Fig. 4.21a).

4.5.10.3
Examples: lead and mercury

Questions of carcinogenicity and genotoxicity of inorganic lead and mercury are now in a state of discussion, in connection with setting Occupational Exposure Limits (DFG 1999, 2000). In this context, the influence of organic lead and mercury salts on genotoxicity (endpoint: micronuclei in V79 cells) and on tubulin assembly has been studied in detail (Thier et al. 2003a). Lead and mercury salts induced micronuclei are showing a classical dose dependence with no-effect-concentrations established. Lead chloride, acetate, nitrate and $HgCl_2$ also interfered dose-dependently with the tubulin assembly *in vitro*, and no-effect-concentrations

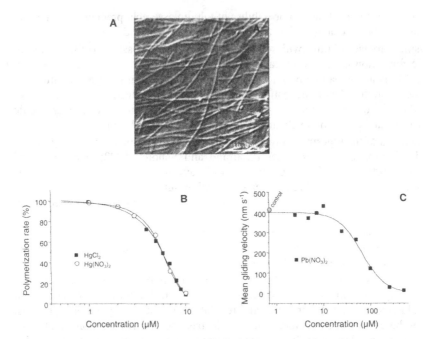

Fig. 4.21a Visualisation of microtubules by DIC phase contrast video microscopy (courtesy of E Ungerand and KJ Böhm).
Fig. 4.21b Tubulin assembly: dose-response for mercury salts (Thier et al. 2003a).
Fig. 4.21c Gliding Assay: dose-response curve for lead(II) nitrate (Thier et al. 2003a).

for these salts were established. Inhibition of the tubulin assembly by $HgCl_2$ started at 2 μM. The gliding velocity of microtubules along immobilised kinesin molecules was affected by 25 μM $Pb(NO_3)_2$ and 0.1 μM $HgCl_2$, again both in a dose-dependent manner (Fig. 4.21 b and c).

In essence, such data are compatible with the view that compounds inducing chromosomal genotoxicity via interaction with cytoskeletal proteins and, in consequence, disturbance of chromosome segregation, follow a conventional dose-response. Such non-mutational types of genotoxicity, based on chemical-protein interactions, should involve a threshold of genotoxicity.

4.5.10.4
Acrylamide

Recent discussions of the carcinogenicity of acrylamide have been focussed on findings of traces of acrylamide in food, originating during processes of cooking and frying (Tareke et al. 2000; see section 1.3).

The toxicological literature on acrylamide has recently been reviewed and assessed by the European Union within the risk assessment procedure for existing chemicals (European Union 2002). With regard to mutagenicity, it was demonstrated that a substantial body of information was available covering an array of genotoxicity end points. Although acrylamide is not mutagenic in bacteria, its genotoxic potential on the chromosomal level is clear-cut in mammalian systems *in vitro*. Acrylamide is metabolised through the cytochrome P450 CYP2E1 to *a potentially genotoxic and reactive metabolite*, glycidamide (Bolt et al. 2003), and has therefore been viewed as a direct-acting mutagen, although the relevance of this metabolite for the genotoxicity of acrylamide has recently been questioned (Park et al. 2002). Acrylamide is also genotoxic *in vivo* on a chromosomal level to both somatic cells and germ cells. In the case of germ cells, acrylamide has been experimentally demonstrated to induce heritable mutations. Acrylamide affects the kinesin-based microtubule functions and therefore acts as a *motor protein inhibitor* (Sickles et al. 1996).

Acrylamide is clearly carcinogenic in rats (Johnson et al. 1986; Friedman et al. 1995) producing increased incidences of a number of benign and malignant tumours in a variety of organs (e. g. thyroid, adrenals, tunica vaginalis of the testes). The tumour types observed show a possible relationship with disturbed endocrine functions and raise the possibility of a hormonal mechanism (KS Crump Group 1999a,b). The carcinogenicity of acrylonitrile in humans has not been thoroughly investigated so far, and no firm conclusions can be drawn from these studies (European Union 2002). However, in view of the high neurotoxicity potential of acrylamide the safety precautions for acrylamide-exposed workers have always been very strict in the past; a low carcinogenic potential is therefore not expected to be visible from epidemiological studies of exposed cohorts. It has been claimed (KS Crump Group 1999a) that, while a genotoxic mechanism cannot be ruled out, evidence for non-linearity in the slope of the dose-response curve has accumulated, and that this strongly suggests that at the bioassay doses (Johnson et al. 1986; Friedman et al. 1995) acrylamide would be at best a weak, opportunistic genotoxin in cells that would require additional promotional signals to progress.

Although the European Union (2003) has addressed the possibility of a hormonal mechanism (v. s.) it was stated that, given the genotoxicity profile of acrylamide, genotoxicity cannot be discounted from contributing to the tumour formation, and that there were no mechanistic arguments to indicate that the tumour findings would be restricted to animals and not humans. This implies also that more research on the mechanisms of tumour formation by acrylamide is needed.

In essence, there are arguments in the direction of acrylamide being a non-DNA-reactive genotoxin, acting on the chromosomal level. This is supported by interference with motor protein functions (Sickles et al. 1996) and would argue in favour of existence of a threshold of genotoxicity. However, the genotoxic mechanisms seem to be more complex and are not sufficiently well understood so far (Park et al. 2002). Therefore, until more scientific information on the mechanisms of acrylonitrile carcinogenicity and genotoxicity are available, it appears prudent for regulatory matters to use the default assumption of a linear dose-response at the present time (as in the calculation given in section 1.4). However, it appears likely that this will over-estimate the carcinogenic risk of acrylamide at low doses.

4.5.11
Non-genotoxic carcinogens

As already mentioned (see section 4.5.9), the Senate Commission of the Deutsche Forschungsgemeinschaft for the Investigation of Health Hazards in the Work Area

(MAK-Commission) has issued new recommendations to distinguish between 5 groups of proven and suspected carcinogens (Neumann et al. 1998), instead of 3 groups, as provided by the Labelling Guide of the European Union (see box 4.1). The distinction of new categories intends that the classification of carcinogens, in future, should be based much more on mechanisms by which carcinogenic effects are elicited.

To consider non-genotoxic mechanisms, the new proposal (see box 4.2) includes as Category 4: Substances with carcinogenic potential for which genotoxicity plays no or at most a minor role. No significant contribution to human cancer risk is expected, provided that the MAK value is observed.

In general, distinguishing between "genotoxic" and "non-genotoxic" carcinogens is important in the characterisation of "threshold" vs. "non-threshold" effects. The classification is supported especially by the evidence, that increases in cell proliferation or changes in cellular differentiation are important in the mode of action. To characterise the cancer risk, the manifold mechanisms contributing to carcinogenesis and their characteristic dose-time-response relationships are taken into consideration.

Besides typical organ toxicants (e. g. chloroform, carbon tetrachloride) also peroxisome proliferators have been included into category 4 of "non-genotoxic carcinogens" which implies that a dose threshold may be defined below which no cancer risk is to be expected. Peroxisome proliferators exemplify compounds leading to experimental carcinogenicity based on a receptor-mediated effect. The most prominent example for a non-genotoxic, receptor-based carcinogen, however, is TCDD. Because of its importance, this merits exemplification.

4.5.11.1
2,3,7,8-Tetrachlorodibenzo-p-dioxin (TCDD)

2,3,7,8-Tetrachlorodibenzo-p-dioxin (TCDD) is a combustion product of chlorine-containing wastes and a contaminant in certain organohalogenated compounds, such as the herbicide 2,4,5-trichlorophenol. TCDD has received considerable attention due to its presence as a trace contaminant in food (especially in fish and meat), water and soil and the extremely high acute toxicity observed in some experimental animals, e. g. guinea pigs (LD_{50}: 1 µg kg^{-1}) (Mc Connell et al. 1978).

In 1997, the International Agency for Research on Cancer (IARC) has classified TCDD as a human carcinogen. TCDD is a multisite carcinogen in all species and strains of laboratory animals tested. Tumours were observed in the liver, thyroid, respiratory tract and other organs. The most seriously affected organ is liver in female rodents (Kociba et al. 1978; Portier et al. 1984). TCDD is a non-genotoxic carcinogen. It has been reported to selectively promote the clonal proliferation of populations of initiated cells by epigenetic mechanisms (see section 6.5.2), such as reduction of cell proliferation in normal hepatocytes, reduction of intrafocal apoptosis and increased intrafocal cell proliferation (Schulte-Hermann et al. 1992; Buchmann et al. 1994; Baumann et al. 1995). One of the molecular mechanisms through which TCDD modulates these processes appears to be oestrogen-dependent (Lucier et al. 1991). In this context, the antioestrogenic properties of TCDD might explain reduction of proliferation in females observed at low doses of dioxins (Hengstler et al. 1998).

> **Box 4.1: Classification of carcinogens[1] and suspected carcinogens**
>
> ## European Union (Commission of the European Communities 1987)
>
> For the purpose of classification and labelling, and having regard to the current state of knowledge, such substances are divided into three categories:
>
> *Category 1*
>
> Substances known to be carcinogenic to man. There is sufficient evidence to establish a causal association between human exposure to a substance and the development of cancer.
>
> *Category 2*
>
> Substances which should be regarded as if they are carcinogenic to man. There is sufficient evidence to provide a strong presumption that human exposure to a substance may result in the development of cancer, generally on the basis of:
>
> – appropriate long-term animal studies,
> – other relevant information.
>
> *Category 3*
>
> Substances that cause concern for man owing to possible carcinogenic effects but in respect of which the available information is not adequate for making a satisfactory assessment. There is some evidence from appropriate animal studies, but this is insufficient to place the substance in category 2.

Three large epidemiological studies of human occupational exposure report a TCDD-mediated increase in all cancers and suggest that the human lung is a sensitive target for TCDD (Zober et al. 1990; Fingerhut et al. 1991; Manz et al. 1991; review: NIH Draft Report: www.niehs.nih.gov). Human data suggest that 1 % of exposed humans get an additional tumour (ED_{01}) based on body burden of 6–80 ng kg^{-1} body weight for all cancers combined and in the range of 36–250 ng kg^{-1} for lung cancer. Based on TCDD daily intake of 1.0 pg TCDD kg^{-1} body weight, Steenland et al. (2001) estimated an excess lifetime (75 years) risk for dying of cancer of 0.05–0.9 % above the background lifetime risk of cancer death.

Several studies suggested a non-linear J- or U-shaped dose-response curve for TCDD induced carcinogenesis in animals, suggesting a practical threshold mechanism. Examples are J-shaped dose-response curves for liver tumour induction in rats (Kociba 1991) or formation of rat liver foci using an initiation/promotion protocol with TCDD (Kitchin et al. 1994; Lutz 1998). A TCDD induced cell cycle delay is the most probable explanation for this phenomenon. For instance, Teeguarden et al. (1999) have shown a TCDD-induced reduction of cell proliferation in female Sprague Dawley rats. In this study a two-stage model of hepatocarcinogenesis was applied. After initiation with partial hepatectomy combined with administration of 10 mg diethylnitrosamine (DEN) kg^{-1} rats were administered TCDD at

---

[1] For a general comparison of carcinogenicity classification criteria of different organisations within Europe, see Seeley et al. (2001).

Box 4.2: Classification of carcinogens and suspected carcinogens

Deutsche Forschungsgemeinschaft (Commission for the Investigation of Health Hazards of Chemical Compounds in the Work Area; Neumann et al. 1998)

Category 1

Substances that cause cancer in man and can be assumed to make a significant contribution to cancer risk. Epidemiological studies provide adequate evidence of a positive correlation between the exposure of humans and the occurrence of cancer. Limited epidemiological data can be substantiated by evidence that the substance causes cancer by a mode of action that is relevant to man.

Category 2

Substances that are considered to be carcinogenic for man because sufficient data from long-term animal studies substantiated by evidence from epidemiological studies indicate that they can make a significant contribution to cancer risk. Limited data from animal studies can be supported by evidence that the substance causes cancer by a mode of action that is relevant to man and by results of in vitro tests and short-term animal studies.

Category 3

Substances that cause concern that they could be carcinogenic for man but cannot be assessed conclusively because of lack of data. The classification in Category 3 is provisional.
3A. Substances for which the criteria for classification in Category 4 or 5 are fulfilled but for which the database is insufficient for the establishment of a MAK value.
3B. Substances for which in vitro or animal studies have yielded evidence of carcinogenic effects that are not sufficient for classification of the substance in one of the other categories. Further studies are required before a final decision can be made. A MAK or BAT value can be established provided no genotoxic effects have been detected.

Category 4

Substances with carcinogenic potential for which genotoxicity plays no or at most a minor part. No significant contribution to human cancer risk is expected provided the MAK value is observed. The classification is supported especially by evidence that increases in cellular proliferation or changes in cellular differentiation are important in the mode of action. To characterise the cancer risk, the manifold mechanism contributing to carcinogenesis and their characteristic dose-time-response relationships are taken into consideration.

Category 5

Substances with carcinogenic and genotoxic effects, the potency of which is considered so low that, provided the MAK and BAT values are observed, no significant contributions to human cancer risk is to be expected. The classification is supported by information on the mode of action, dose-dependence and toxicokinetic data pertinent to species comparison.

0, 0.01, 0.1, 1.0 or 10 ng kg^{-1} d^{-1}. After 1 and 3 months 0.1 ng kg^{-1} d^{-1}TCDD reduced proliferation of hepatocytes in DEN-treated rats.

Several studies modelling dose-response relationships of cancer endpoints in animals suggested threshold mechanisms for TCDD. Nevertheless, the situation remains controversial. In a draft paper about dose-response relationships for dose-response modelling of dioxins, the NIH reported that only half of the cancer endpoints observed in published animal studies were not linear (www.niehs.nih.gov). Non-cancer endpoints had a greater degree of non-linearity with only 40 % of the observed responses being linear. The human situation is even more difficult to assess. In most epidemiological studies data were not adequate to model the dose-response relationship. In one study where dose-response modelling has been applied, the estimated shape of the dose-response curve was non-linear (Becher et al. 1996). These data did not indicate the existence of a threshold value.

In conclusion, relevant animal studies suggest that TCDD induced carcinogenesis follows a receptor-based threshold mechanism. However epidemiological data a, presently available re not adequate to address the exact nature of the low-dose-response relationship in humans. Notably, the quantity of the carcinogenic risk in humans seems to be much lower than that predicted on the basis of animal experimentation (see section 4.6.3).

4.5.11.2
Hormones

The link of hormonal action with chemical carcinogenesis has been reviewed by the International Agency for Research on Cancer, in conjunction with hormonal contraception and post-menopausal hormonal therapy (IARC 1999) and thyrotropic agents (IARC 2001). It has been exemplified for oestrogens and gestagens that receptor-mediated mechanisms can explain a majority of responses, but non-receptor processes may also exist (Yager and Liehr 1996). Cell proliferation is viewed as the most important receptor-mediated mechanism by which hormonally active compounds act in carcinogenesis at hormone-sensitive target tissues. Cell proliferation is fundamental to the process of carcinogenesis; it is an essential (co)-factor and enhances cancer incidence (i. e. tumour promotion) by preferentially stimulating the growth of genetically altered and preneoplastic cells (Preston-Marin et al. 1990; IARC 1999).

Taking experimental thyroid neoplasia as an example, IARC (2001) has specifically postulated that mechanistic evidence should be applied in the evaluation of chemicals that cause thyroid tumours in experimental animals. Criteria for such a use of mechanistic data have been described by Capen et al. (1999) in the following way:

- Agents that lead to the development of thyroid neoplasia through an adaptive physiological mechanism belong to a different category from those that lead to neoplasia through genotoxic mechanisms or through mechanisms involving pathological responses with necrosis and repair.
- Agents that cause thyroid follicular-cell neoplasia in rodents solely through hormonal imbalance can be identified on the basis of the following criteria:

- – no genotoxic activity (agent and/or metabolite) found in an overall evaluation of the results of tests *in vivo* and *in vitro*,
- – hormone imbalance demonstrated under the conditions of the assay for carcinogenicity,
- – the mechanism whereby the agent leads to hormone imbalance defined.
- – When tumours are observed both, in the thyroid and at other sites, they should be evaluated separately on the basis of the modes of action of the agent.
- – Agents that induce thyroid follicular-cell tumours in rodents by interfering with thyroid hormone homeostasis can, with some exceptions, notably the sulfonamides, also interfere with thyroid hormone homeostasis in humans if given at a sufficient dose for a sufficient time. Such agents can be assumed not to be carcinogenic in humans at concentrations that do not lead to alterations in thyroid hormone homeostasis.

IARC (2001) has specified that in these cases hormonal assays and morphological evaluation of the thyroid gland should be carried out in animals of the same species, and preferably the same strain, as were used in the bioassay in which thyroid tumours developed. Evidence for hormonal imbalance could include measurements of serum thyroid hormone and TSH and of morphological changes characteristic of increased TSH stimulation, including increased thyroid gland weight and diffuse follicular-cell hyperplasia and/or hypertrophy. In this conjunction, a detailed statement regarding the distinction of genotoxicity and non-genotoxicity has been issued to which reference can be made (IARC 2001).

In essence, this means that the induction of tumours, which arise in consequence of hormonal imbalance, implies a threshold mechanism, and that no tumour formation should be expected below the no-hormonal-effect-level.

4.6
Interspecies extrapolation

4.6.1
Introduction

In the previous chapter several chemical carcinogens have been discussed with respect to their low-dose-response relationships and possible threshold mechanisms concerning neoplastic transformation. Relevant data were obtained from experimental animals. It is well known that animal data may not correctly predict the human situation. Thus, interspecies extrapolation is required. Extrapolation from laboratory animals to man represents one of the most complex challenges to the toxicologist. For this purpose, species differences in uptake, distribution, metabolism, site of action, elimination and accumulation must be taken into account. Differences in the metabolism of xenobiotics represent probably the most frequent explanation for observed qualitative and quantitative differences in toxic effects among animal species.

Ideally, species, which metabolise the respective test substance like humans, should be used for toxicological tests. Generally this ideal is not attainable as metabolism differs widely among species. However, it is possible to identify the species with the metabolism for a specific compound closest to man. For this purpose isolated hepatocytes have been used successfully, since a good correlation between the metabolism of a xenobiotic *in vivo* and isolated hepatocytes of the same species *in vitro* has been observed (Hengstler et al. 2000). A reasonable interspecies extrapolation from animals to humans and safety assessment can in most cases be performed, if we have (i) an adequate understanding of the metabolism and mechanism of toxicity of a given compound, (ii) data on its metabolism and toxicity (e. g. on DNA adducts organotoxicity endpoints) in primary hepatocytes and other relevant cell types of humans and of experimental animal species and (iii) toxicity data, including carcinogenicity, in these experimental animal species. In the present chapter interspecies extrapolations will be described for aflatoxin B_1 (linear dose-response) and TCDD (threshold response), two very different substances that have been analysed for possible threshold mechanisms in the previous section. In addition, heterocyclic amines will be discussed since metabolic pathways explaining interspecies differences have been intensively examined.

Consideration should be given to the fact that there are examples where interspecies differences in toxicity are a consequence of differences in target mechanisms, for instance the $\alpha2\mu$-globulin nephropathy leading to renal carcinogenesis, which has been recognised as a male rat-specific problem, and for peroxisome proliferation and associated hepatocarcinogenesis, which is often observed in rats and mice (Caldwell 1992). However, keeping such exceptions in mind, mechanisms of toxicity are similar in animals and man for the majority of substances. In this chapter it will be shown that – although for a large number of substances interspecies differences can be explained by differences in metabolism – there is no general mechanism that applies to species-specificity. Thus, extrapolation from animals to man most probably will never be possible by one or a small number of routine experiments, but requires a diligent evaluation.

4.6.2
The classical example: contraceptive hormones and the Beagle dog dilemma

With considerable publicity, in 1975 a series of contraceptives (17?-hydroxyprogesterone derivatives), containing the gestagen *megestrol acetate*, were withdrawn from the market in Germany and some other European countries (e.quens®, Oraconal®, Planovin®, Tri-Ervonum®). The reason was induction of mammary tumours in Beagle bitches upon 7 years of continuous experimental treatment with megestrol acetate (Bolt 1976). This was a repetition of a very similar situation concerning contraceptives containing the gestagens *anagestone acetate* and *chlormadinone acetate*, which had occurred earlier (in 1968–1970) leading to withdrawal of contraceptives such as Ne-Novum®, Aconcen®, Eunomin®, Estirona®, Disut® and Gestacliman®.

At this time, international requirements for experimental safety testing of contraceptives (Neumann and Elger 1971) included long-tern carcinogenicity studies in mice (1.5 years of treatment), rats (2 years), dogs (7 years) and monkeys (10 years). Doses to be tested in these studies were the 1-, 10- and 25-fold of the human therapeutic dose, based on mg kg^{-1} body weight (Nelson et al. 1973). The rationale for these experimental requirements was precaution in view of the potential long-term use of oral contraception in large parts of the healthy female population.

The dilemma, which occurred in consequence of recognition of the development of mammary tumours in female Beagle dogs, induced very detailed studies on the relevance of these tumours for humans. As result of these investigations, the following arguments were put forward, indicative of a pathogenesis of these specific experimental tumours which may not be translated to the situation in human therapy (Kewitz 1971; Neumann and Elger 1971):

– profound species-specific differences in reproductive physiology (no luteolytic factor, leading to pseudo-pregnancy in dogs),
– species-specific chronic over-stimulation of the mammary gland in dogs by gestagens, involving tubular and lobulo-alveolar growth,
– metabolic elimination of gestagens of the 17α-hydroxyprogesterone series being much slower in dogs than in humans, leading to much higher target doses in dogs,
– high spontaneous incidences of mammary tumours in Beagle bitches (e. g. 50 % of animals aged 5 years bearing tumours, 90 % of tumour-bearing animals at an age of 10 years).

As a consequence, it was later generally accepted that the occurrence of mammary tumours in female dogs, upon long-term and high-dose administration of gestagen-containing contraceptives, was not indicative of a similar cancer risk in humans, indicating that a species extrapolation should not be made. Later, oral contraceptives containing gestagens of the 17α-hydroxyprogesterone series were again admitted to the market (Bolt 1976).

In essence, this provides a historical and practically important case where inter-species extrapolation is not feasible, mainly based on marked physiological species differences.

4.6.3
2,3,7,8-Tetrachlorodibenzo-p-dioxin (TCDD) and related compounds

Most but not all of the toxic effects of TCDD and other polyhalogenated aromatic hydrocarbons are mediated by the aryl hydrocarbon receptor (review: Hankinson et al. 1996; see also section 6.5.2). The aryl hydrocarbon receptor protein evolved about 450 million years ago, early in vertebrate evolution. Ligands for the aryl hydrocarbon receptor, particularly TCDD, cause several toxic effects, including cancer, progressive weight loss, toxicity to the immune system, fetal toxicity, birth defects, dysregulation of endocrine (thyroid, androgen, oestrogen and growth factor) homeostasis and decreases in male and female reproductive performance. TCDD has proven to be the most potent tumour promotor analysed, and it also acts as a complete carcinogen. Unlike polycyclic aromatic hydrocarbons or heterocyclic amines, TCDD is not activated to a mutagen and does not bind to DNA.

The unligated aryl hydrocarbon receptor is located in the cytoplasm, where it is associated with two molecules of the 90 kDa heat shock protein and another protein of about 43 kDa (review: Hankinson et al. 1996). This complex is termed the "unligated aryl hydrocarbon receptor complex". After binding of TCDD or other ligands, the aryl hydrocarbon receptor is released from the 90 kDa heat shock proteins and the 43 kDa protein and associates with the aryl hydrocarbon receptor nuclear translocator protein (ARNT) (Fig. 4.22). The latter complex binds to specific recognition motifs of DNA (termed dioxin-responsive element, which has a consensus sequence [T(T/A)GCGTG] upstream of the *CYP1A1* gene), resulting in enhanced transcription. Genes which have been shown to exhibit increased transcription rates by this mechanism include *CYP1A1, CYP1A2*, and *CYP1B1*, the glutathione *S*-transferase subunit *GSTYa*, the UDP-glucuronosyltransferase *UGT1*O^6*, the aldehyde dehydrogenase *ALDH3c* and the NAD(P)H:quinone reductase *NQO$_1$*.

Contrary to what its name suggests, ARNT is not involved in translocating the aryl hydrocarbon receptor to the nucleus, since TCDD also induced nuclear translocation of the liganded aryl hydrocarbon receptor in ARNT-deficient mouse Hepa-1 cells (review: Hankinson et al. 1996). Nevertheless, ARNT is essential for the aryl hydrocarbon receptor to bind to DNA.

Homozygous knockout mice for the aryl hydrocarbon receptor nuclear translocator (ARNT) died *in utero* between 9.5 and 10.5 days of gestation (Kozak et al. 1997). The primary cause of lethality was failure of the embryo to vascularise and form the spongiotrophoblast. This is in agreement with the known role of ARNT in induction of angiogenesis.

Further insight was obtained by generation of aryl hydrocarbon receptor knockout mice (Fernandez-Salguero et al. 1995). Almost half of the knockout mice (Ahr -/-) died shortly after birth, whereas survivors were fertile, but showed deficiencies in liver function (50 % reduction in size; decrease in retinoic acid metabolism; decreased CYP1A2 levels) and in the immune system (Fernandez-Salguero et al. 1996; Andreola et al. 1997). The aryl hydrocarbon receptor-deficient mice are relatively resistant to doses of TCDD (2,000 µg kg^{-1}) 10-fold higher than those found to induce severe toxicity in littermates expressing a functional aryl hydrocarbon receptor (Fernandez-Salguera et al. 1996). However, at higher doses of TCDD the aryl hydrocarbon receptor-deficient mice displayed single cell necrosis and vasculi-

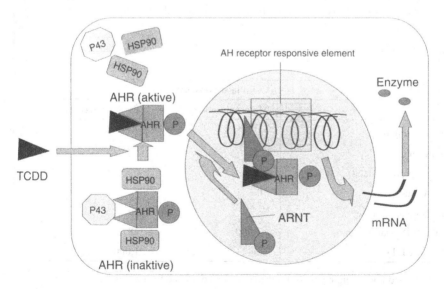

Fig. 4.22 Transformation of the aryl hydrocarbon receptor (Hankinson et al. 1996) by 2,3,7,8-tetrachloro-*p*-dibenzodioxin (TCDD).

tis in their livers and lungs, showing the existence of aryl hydrocarbon receptor-independent pathways of TCDD-induced toxicity.

Extremely large interspecies differences in TCDD-induced toxicity are known. The guinea pig is the most TCDD-susceptible mammal known, with an LD_{50} in the range of 1–2 µg TCDD kg^{-1} (Schwetz et al. 1973). In contrast, the hamster is the most TCDD-resistant species concerning acute toxicity with an LD_{50} greater than 3,000 µg kg^{-1} (Olson et al. 1980).

Exposure to TCDD during pregnancy causes prenatal mortality in all mammals examined (Peterson et al. 1993). The rank order of susceptibility from the most to the least sensitive species was reported to be monkey = guinea pig > rabbit = rat = hamster > mouse (Tab. 4.2). *Adult* hamsters are about three orders of magnitude more resistant than adult guinea pigs. Yet, the difference in *prenatal* mortality between these species was only about one order of magnitude. Interpretation of the prenatal mortality data is quite complex, since species-specific differences in the most sensitive periods during pregnancy have to be considered. In most laboratory mammals, gestational exposure to TCDD caused a common pattern of fetotoxic responses. These common responses include decreased fetal growth, subcutaneous oedema and thymic hypoplasia. In addition to these common effects highly species-specific malformations have been described. Examples of TCDD effects that occur only in a single susceptible species are cleft palate formation in the mouse, intestinal haemorrhage in the rat, and the formation of extra ribs in the rabbit (Peterson et al. 1993).

TCDD acts as a complete carcinogen in chronic animal studies with doses as low as 0.001 µg kg^{-1} d^{-1} in rats, leading to the formation of neoplasms in the lung, oral/nasal cavity, thyroid, adrenal glands and liver of rats. In mice, TCDD induced neoplasms in the liver, subcutaneous tissue, thyroid gland, and thymic lymphomas. In hamsters, squamous cell carcinomas of the facial skin were observed after

Tab. 4.2 Interspecies differences in prenatal toxicity and toxicity to adult female animals (maternal toxicity) of 2,3,7,8-tetrachlorodibenzo-*p*-dioxin (data from Peterson et al. 1993).

Species	Cumulative doses (µg kg^{-1})	
	Maternal toxicity[a]	Prenatal toxicity[b]
Rhesus monkey	1	1
Guinea pig (Hartley)	1.5	1.5
Rabbit (New Zealand)	2.5	2.5
Rat (Wistar)	10	5
Rat (Sprague-Dawley)	5	20
Hamster (Syrian-Golden)	>3,000	18
Mouse (CD1)	2,000	1,000

[a] Decreased body weight gain or marked oedema compared to controls.
[b] Cumulative doses are given, which caused an at least 2-fold increase in the percentage of absorption plus late gestational deaths over controls.

intraperitoneal as well as subcutaneous injections of relatively high doses of TCDD (total dose: 600 µg kg^{-1}). Based on these data, carcinogenesis is one of the primary concerns associated with human exposure to TCDD. Scientists who had carefully reviewed the data have come to different conclusions regarding the susceptibility of humans to the carcinogenic effects of TCDD: the majority position is that humans are less susceptible than most laboratory animals to the carcinogenic effects of TCDD, while others conclude that humans and susceptible rodents respond similarly (Hays et al. 1997). The TCDD-related carcinogenic response in rats was compared to that in humans (Hays et al. 1997). For this purpose the internal biological dose, measured as blood lipid or adipose tissue TCDD levels, was determined for rats and humans (NIOSH cohort) exposed to TCDD. Although workers exposed to TCDD (NIOSH cohort) experienced internal exposures similar or higher than rats treated with TCDD, rats exhibited a greater tumour response. At comparable peak serum lipid TCDD levels (about 7,000 ppt) the tumour response in rats was more than 9-fold higher than the human response. It should be considered that the average area under the blood concentration vs. time curve (AUC) for TCDD in the NIOSH cohort was about 1,000-fold greater than the average background in the general population. It was concluded that humans are less susceptible to TCDD induced carcinogenesis than rats and that human exposure to background levels of TCDD (about 5 ppt serum lipid) is not likely to cause an incremental cancer risk (Hays et al. 1997).

Although the interspecies differences regarding TCDD-toxicity cover more than 3 orders of magnitude, only little is known about the mechanisms responsible for these differences. It seems probable that a universal mechanism determining TCDD-susceptibility does not exist. Possibly, the basic mechanisms responsible for specific effects differ between species.

Some differences in TCDD-susceptibility have been shown to be due to differences in the affinity of the aryl hydrocarbon receptor for TCDD (Gielen et al. 1972; Poland et al. 1994), although this explanation seems to be limited to some mouse strains. The murine *Ah* locus was originally defined as a difference in susceptibility of mouse strains to polycyclic aromatic hydrocarbons. Later, the *Ah* locus was shown to encode the aryl hydrocarbon receptor, which binds planar aromatic ligands and mediates most of their effects. For instance, C57BL/6 mice treated with 3-methylcholanthrene show an induction of CYP1A1, whereas DBA/2 mice fail to respond. Administration of TCDD induces CYP1A1 activity in both mouse strains, however, a larger dose is required for DBA/2 mice. Genetic crosses and backcrosses between these mouse strains showed that responsiveness is inherited as an autosomal dominant trait. The allele associated with responsiveness was termed Ah^b, whereas the non-responsive allele was termed Ah^d. The later allele expresses a protein with diminished binding affinity for ligands, resulting in a diminished sensitivity to TCDD (Gielen et al. 1972; Poland and Glover 1980; Poland et al. 1994). The cDNAs of four murine aryl hydrocarbon alleles have been cloned and sequenced (Poland et al. 1994). Three alleles with the higher ligand binding affinity (Ah^{b-1}, Ah^{b-2}, and Ah^{b-3}), that differ only by a few point mutations, and one allele with the low binding affinity (Ah^d) have been described. The latter allele is most appropriately compared with the Ah^{b-2} allele, in that both express proteins of the same size (~104 kDa) that differ in only two amino acids. The lower ligand affinity of the Ah^d receptor has been shown to be caused by an alanine → valine substitution at position 375 of the Ah^{b-2} receptor. In similar genetic crosses other effects of TCDD, such as thymic atrophy (Poland and Glover 1980), hepatic porphyria (Jones and Sweeney 1980), and immunosuppressive effects (Vecchi et al. 1983) have also been shown to segregate with the *Ah* locus. In addition, developmental toxicity was related to the *Ah* locus in mice (review: Peterson et al. 1993). In five mouse strains with low affinity receptors there was only a 0 to 3 % incidence of cleft palate formation, whereas four of five strains with high-affinity aryl hydrocarbon receptors developed a ≥ 50 % incidence (Poland and Glover 1980). However, one strain with the high-affinity receptor was resistant, showing that alternative mechanisms protecting from developmental toxicity must exist.

Recently physicochemical differences in the *Ah* receptors of the most TCDD-susceptible and the most TCDD resistant rat strains have been observed (Pohjanvirta et al. 1999). Long-Evans rats are more than 1,000-fold more sensitive to the acute lethal effects than are Han/Wistar rats (LD_{50}s: 10 µg kg^{-1}versus > 9,600 µg kg^{-1}). However, this difference is highly end point dependent, since the difference between both rat strains is negligible for CYP1A1 induction. Immunoblotting revealed that the *Ah* receptor of the sensitive Long Evans rats has a significantly higher molecular mass (~ 106 kDa) than the resistant Han/Wistar rats (~ 98 kDa) (Pohjanvirta et al. 1999). In addition, the sensitive Long Evans rats contained about twofold higher levels of binding sites for [^{3}H]TCDD and threefold higher hepatic concentrations of ARNT than the resistant Han/Wistar rats. In another study the sensitive and resistant rat strains were crossed and the resistant alleles of genes determining TCDD susceptibility were identified (Tuomisto et al. 1999). Susceptibility was determined by two genes, the *Ah* receptor gene and a novel gene designated "*B*". The "Han/Wistar-type allele" of both genes strongly increased resistance to TCDD acute lethal effects (Tuomisto et al. 1999). Although the data convincingly show that the affinity of the aryl hydrocarbon receptor for TCDD correlates with TCDD susceptibility, it seems impossible to extrapolate interspecies toxicity by receptor affinity. The time course of association of [^{3}H]-TCDD with hepatic cytosol from hamsters, mice, rats, gerbils and guinea pigs showed significant interspecies-differences (Nakai et al. 1994). However, their rank order of affinity did not correlate with the rank ordering of their toxic potency. In another study, binding of [^{3}H]-TCDD to cytosol as well as TCDD induced binding of the aryl hydrocarbon receptor to the dioxin responsive element (determined by gel retardation analysis) was compared for various species (Bank et al. 1992). These experiments indicate that the TCDD-resistant hamster shows a similar [^{3}H]-TCDD binding to cytosolic protein as the sensitive guinea pig (Tab. 4.3). On the other hand, binding to the dioxin responsive element was greater for the guinea pig compared to hamster. However, the relatively small difference between hamsters and guinea pigs is unlikely to explain the large interspecies differences shown in Tab. 4.2. An additional argument against the thesis that aryl hydrocarbon receptor affinity determines TCDD-susceptibility as a common mechanism was given by a study comparing Han/Wistar to Long-Evans rats (Rozman 1989). LD_{50} of Han/Wistar rats was > 3,000 µg TCDD kg^{-1}, compared to only 12 µg TCDD kg^{-1}for Long-Evans rats. This interstrain difference in TCDD toxicity did not match receptor-binding affinity, which was almost identical in both strains.

A typical sign of acute TCDD intoxication is the wasting syndrome, which is defined as decreased feed intake as well as body weight. In Sprague-Dawley rats

Tab. 4.3 Interspecies differences in specific binding of [³H]-TCDD to hepatic cytosol and binding of transformed aryl hydrocarbon receptor to the dioxin responsive element can not sufficiently explain interspecies differences in toxicity shown in Tab. 4.2 (data from Bank et al. 1992)

Species	Binding of [³H]-TCDD to hepatic cytosol (fmol mg⁻¹ protein)	Binding to the dioxin responsive element (relative amount)
Human LS180 cells	260.0 ± 3.0	+++
Guinea pig (Hartley)	43.2 ± 4.5	++++
Rabbit (New Zealand)	75.1 ± 3.5	+++
Rat (Sprague-Dawley)	51.0 ± 2.0	+++
Hamster (Syrian-Golden)	50.4 ± 5.4	++
Mouse (C57BL/6N)	41.2 ± 1.7	+
Chicken	41.3 ± 2.3	+
Rainbow Trout	5.1 ± 2.6	–

TCDD has been shown to inhibit gluconeogenesis. It has been suggested that the progressive hypoglycemia observed in TCDD-treated rats causes the wasting syndrome (Rozman 1989; Weber et al. 1991). However, in another study plasma glucose remained unaltered in TCDD-treated hamsters and guinea pigs (Unkila et al. 1995). In addition, no decrease in gluconeogenesis due to suppression of phosphoenolpyruvate carboxykinase was observed in TCDD-treated guinea pigs. Thus, dysregulation of glucose homeostasis and inhibition of gluconeogenesis cannot constitute a general mechanism for the wasting syndrome (Unkila et al. 1995). An alternative to explain the mechanism of the wasting syndrome is that TCDD leads to an increase of tryptophan levels in plasma, based on the observation that in susceptible Long-Evans rats TCDD caused a dose dependent increase in free tryptophan in plasma, whereas TCDD-resistant Han/Wistar rats did not exhibit this change (Unkila et al. 1994). It was suggested that increased tryptophan levels in the brain might mediate TCDD-anorexia. However, treatment of guinea pigs with doses of TCDD, which caused wasting syndrome, did not affect brain tryptophan levels or plasma total and free tryptophan concentrations (Unkila et al. 1995). Thus, it seems probable that the mechanisms responsible for the wasting syndrome – in spite of similar clinical symptoms – differ between species.

It has been suggested that the adipose tissue serves as a protective reservoir against the toxic effects of TCDD (Geyer et al. 1997). In the adipose tissue persistent lipophilic compounds, such as TCDD, may accumulate, so that only a small fraction of the incorporated dose can reach the target organs and exert toxic effects. A relationship between the oral 30-day LD_{50} (μg kg⁻¹) of TCDD in different mammalian species and their total body fat content (TBF %) was reported: $LD_{50} = 6.03 \times 10^{-4}$ (TBF %)$^{5.30}$ (Geyer et al. 1997). This equation obviously suggests "survival of the fattest". It would predict an LD_{50} of about 6,000 μg kg⁻¹ for an adult man of 70 kg body weight. In this context human new-borns and embryos would be much more sensitive. However, several aspects that could be concluded from the body-fat-model still have to be confirmed. For instance – despite similar TCDD exposure – some individuals develop chloracne, whereas others do not. It would be interesting to examine whether this difference in susceptibility is associated with differences in body fat content.

In addition, species variation in TCDD susceptibility could be due to other physiological characteristics, including pharmacokinetics and metabolism. The relation between the external dose of TCDD and resulting TCDD concentrations in liver and adipose tissue of humans, rats and mice varies by as much as 700-fold. It is known that at the same external doses, internal concentrations increase with body weight, an effect that is not specific to TCDD. Thus, at the same external dose (calculated for 100 pg TCDD kg d^{-1} by Lawrence and Gobas 1997), the internal concentration of TCDD in the organism will be about 7-fold higher for humans than mice and 4-fold higher than for rats, due to the smaller elimination rate constant in humans.

Marked interspecies differences have also been described for TCDD metabolism, which renders TCDD more water soluble and increases excretion. Primary hepatocytes from untreated rats, having an LD_{50} 25-fold greater than guinea pigs, metabolised TCDD 2.8-fold faster then primary hepatocytes from untreated guinea pigs (Wroblewski and Olson 1985). Pre-treatment with TCDD (5 µg kg^{-1}; i. p.) 72 h prior to hepatocyte isolation increased the metabolic rate of TCDD 3.2-fold for rats, whereas no increase was observed in guinea pigs, resulting in 9-fold greater metabolic rates of treated rats versus treated guinea pigs. Thus, constitutive TCDD metabolism as well as the species-specific ability of TCDD to induce its own rate of metabolism may contribute to the varying susceptibility of species to TCDD.

In conclusion, extremely large interspecies differences in TCDD-induced toxicity are known. The guinea pig is the most susceptible mammal known that is more than three orders of magnitude more sensitive than the hamster, which is the most resistant mammal. Humans appear to be less sensitive than most laboratory animals. Environmental exposure to current background levels of TCDD is not likely to cause in increase in human cancer risk.

4.6.4
Aflatoxin B1

Striking interspecies variations in susceptibility to AFB_1 carcinogenesis have been observed, with rats representing the most sensitive and mice the most resistant species, with refractory to dietary levels 3 orders of magnitude higher than for rats (Busby and Wogan 1984). TD_{50} values (statistical estimate of the dose required to result in 50 % incidence of tumours) from lifetime feeding studies were 1.3, 5.8, and > 70 µg AFB_1 kg^{-1} d^{-1} in male Fischer rats, male Wistar rats and male C57BL mice, respectively (review: Wild et al. 1996). In another study the susceptibilities of male Fischer rats and male Syrian golden hamsters were compared. Male Fisher rats treated with 1 mg AFB_1 kg^{-1} d^{-1} (5 d wk^{-1} for 6 weeks) developed hepatocellular cancer within 46 weeks, whereas only one hamster (treated with 2 mg kg^{-1} d^{-1}; 5 d wk^{-1} for 6 weeks) developed a hepatocellular carcinoma after 78 weeks.

The level of DNA adducts in livers of AFB_1 treated animals was at least qualitatively associated with species specific susceptibilities to AFB_1 hepatocarcinogenesis (Wild et al. 1996). The levels of AFB_1-DNA adducts in liver were in the order:

rat ≥ guinea pig > hamster > mouse, suggesting that guinea pigs – for which to our knowledge no carcinogenicity studies are available – are more susceptible than mice and hamsters and almost as sensitive as rats (Wild et al. 1996). These data are consistent with other studies, which reported that AFB_1-DNA adduct levels in rats were 1.5-fold higher than in guinea pigs (Ueno et al. 1980), 3-fold higher than in hamsters (Garner and Wright 1975; Lotlikar et al. 1984) and between 40–600-fold higher than in mice (Lutz et al. 1980; Croy and Wogan 1981).

Thus, for risk assessment of humans one would ideally want to measure AFB_1-N^7-guanine adducts in the DNA of hepatocytes, the target cell for carcinogenesis. However, such measurements in human liver have been rare, because of difficulties in obtaining tissue samples including samples from appropriate subjects. As an alternative, determination of AFB_1-albumin adducts (expressed as pg AFB_1-lysine equivalent mg^{-1} albumin) in peripheral blood has been established (Wild et al. 1996). The amount of AFB_1-albumin adducts in peripheral blood has been shown to correlate with AFB_1-DNA adduct levels in the liver of rodents and to be at least qualitatively associated with species susceptibility to AFB_1 hepatocarcinogenesis in rats, hamsters and mice (Wild et al. 1996). The level of AFB_1-albumin adducts formed as a function of a single dose of AFB_1 in rodents was compared to humans from Gambia and Southern China with an estimated exposure of about 850 ng AFB_1 kg^{-1} d^{-1}. This cross-species comparison resulted in values for Sprague-Dawley, Fischer and Wistar rats of 0.3–0.5 pg AFB_1-lysine equivalent mg^{-1} albumin per 1 µg AFB_1 kg^{-1} body weight, values of < 0.025 for the mouse, whereas 1.56 was estimated for exposed humans (Fig. 4.23). This suggests that exposure to AFB_1 results in even higher adduct levels in humans than in rats, whereby it should be taken into account that rats already represent a species with high susceptibility to

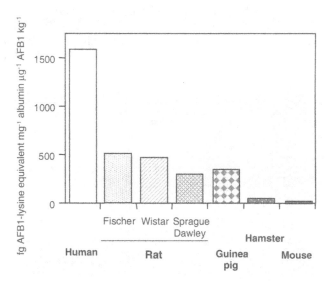

Fig. 4.23 Interspecies comparison of aflatoxin B_1-albumin adducts (expressed as fg AFB_1-lysine equivalent mg^{-1} albumin) in rodents and humans for a given intake of aflatoxin B_1 (from Wild et al. 1996).

AFB$_1$-carcinogenesis. However, cross-species extrapolation was made on a µg kg^{-1} body weight basis. In most instances (e. g. in clinical practice of cytostatic drug application) the use of body surface is considered to be a more appropriate basis. If the cross-species comparison in Fig. 4.24 is made on the basis of surface area, using a mean surface area of 1.6 m^2 for humans (60 kg) and 0.02 m^2 for rats with 100 g body weight, the adduct range for rats would be 0.06–0.07 pg AFB$_1$-lysine equivalent per mg albumin per 1 µg aflatoxin per m^2 surface area, whereas the value for humans would be 0.04. Thus, the susceptibility of humans seems to be similar to that of rats, which means that humans have to be considered as a species with relatively high sensitivity to AFB$_1$. This is in agreement with another study, which compared DNA binding of [^{3}H] AFB$_1$ to DNA of primary hepatocytes of humans, rats and mice (Cole et al. 1988). Similar DNA binding was observed in hepatocytes of humans and female rats, whereas DNA binding in hepatocytes of mice was much lower (Fig. 4.24). The constellation of mice representing the most resistant and rats a relatively susceptible species for AFB$_1$ genotoxicity was also confirmed by determination of chromosomal aberration and micronuclei in bone marrow (Anwar et al. 1994). In addition, induction of unscheduled DNA synthesis (UDS) by AFB$_1$ in primary hepatocyte cultures was strongest for human hepatocytes and somewhat weaker in hepatocyte cultures of male Fischer 344 rats, whereas mouse hepatocyte cultures were the least responsive to AFB$_1$ exposure (Steinmetz et al. 1988). Hepatocyte cultures from male Cynomolgus monkeys yielded a similarly weak response as mouse hepatocytes. This may be due to the lack of constitutive CYP1A2 expression which, as in the case of some heterocyclic amines (see below), renders the Cynomolgus monkey an inadequate species for extrapolation of human risk.

Numerous studies were performed to examine whether interspecies variation in susceptibility to AFB$_1$ carcinogenesis and AFB$_1$-adducts can be explained by interspecies differences in metabolism. Phase I-metabolism of AFB$_1$ by mammalian liver microsomes results in 4 main metabolites. The AFB$_1$-*exo*-8,9-epoxide represents the most relevant metabolite, which forms DNA adducts – mainly at the N^7 position of guanine – much more efficiently than the AFB$_1$-*endo*-8,9-epoxide

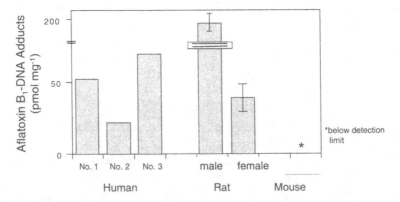

Fig. 4.24 Amount of [^{3}H]aflatoxin B$_1$ bound to DNA isolated from cultured primary hepatocytes from human, rat (Sprague-Dawley) and mouse (CD1 Swiss) after exposure to 200 nM aflatoxin B$_1$ for 24 h (from Cole et al. 1988).

(Guengerich 1996). AFM1 and AFQ1 are considered as detoxification products. AFQ1 is the predominant product produced by microsomes of humans and monkeys (Ramsdell and Eaton 1990), whereas the main metabolite formed by rat and mouse microsomes is AFB_1-8,9-epoxide. In addition, in mouse and rat microsomes a 2-fold higher rate of AFB_1-8,9-epoxide formation was observed, suggesting that toxification of AFB_1 by phase I-metabolism is more efficient in rodent than in human liver microsomes. This was confirmed by a study that examined induction of sister chromatid exchanges (SCEs) by AFB_1 in human mononuclear leukocytes,

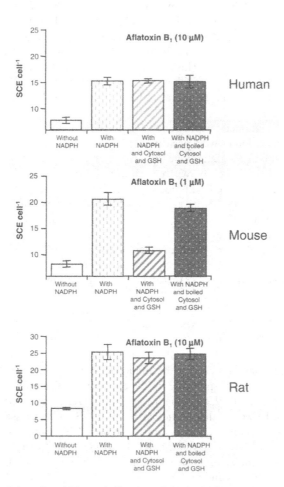

Fig. 4.25 Effect of phase I and II-metabolism on aflatoxin B1 genotoxicity as evidenced by induction of sister chromatid exchanges in human lymphocytes after incubation of human venous blood with aflatoxin B1. Incubations were performed using microsomes from human, mouse and rat livers. Incubations were performed with and without addition of NADPH as a cofactor of cytochrome P450 to analyse the influence of members from the latter enzyme family. To examine the influence of glutathione S-transferases liver cytosol supplemented with glutathione (GSH) was added in an additional incubation. As a negative control liver cytosol was boiled before the incubation (data from Wilson et al. 1997).

with addition of human, rat and mouse liver microsomes as a metabolising system (Wilson et al. 1997). Mouse liver microsomes activated AFB_1 to a greater extent than rat or human liver microsomes, although the differences were rather moderate (Fig. 4.25). In humans, interindividual differences in the potential of liver microsomes to bioactivate AFB_1 to an SCE-inducing metabolite were observed, which significantly correlated with CYP1A2 activity of the liver, as determined by tacrine 1-hydroxylation (Wilson et al. 1997). Besides CYP1A2 also CYP3A4 can activate AFB_1 in human liver. However, different kinetic characteristics were reported, with a relatively low K_m for CYP1A2 compared to the relatively high K_m of CYP3A4, the main cytochrome P450 in human liver (Gallagher et al. 1994). Thus, at the low AFB_1 concentrations expected in human liver *in vivo*, metabolic activation is expected to be catalysed predominantly by CYP1A2 in most individuals (Gallagher et al. 1994; Buetler et al. 1996). However, after exposure of humans with low levels of CYP1A2 and high levels of CYP3A4 to high amounts of AFB_1, significant amounts of the mycotoxin might also be converted to AFB_1-8,9-epoxide by CYP3A4.

If the above mentioned incubations of human mononuclear blood cells with AFB_1 and liver microsomes were performed in the presence of mouse, rat or human liver cytosol plus reduced glutathione (at a concentration of 2.5 mM, which is high enough to support GST activity, whereas almost no GST activity is present in the microsomal incubations due to dilution of the cofactor), the rate of SCE-induction was reduced by almost 90 % by mouse cytosol, whereas cytosol of humans and rats showed no significant effect (Fig. 4.25). This suggests that efficient conjugation with glutathione confers resistance to mice, but not to humans or rats. Indeed, hepatic GST-mediated conjugation of microsomally generated AFB_1-8,9-epoxide was 7,080, 930 and 140 pmol min^{-1} mg^{-1} cytosolic protein for the mouse, hamster and rat, respectively, whereas GST-mediated conjugation by human liver cytosol was below the detection limit (Slone et al. 1995). The high AFB_1-8,9-epoxide conjugating activity of mouse liver cytosol is due to an in principle expressed α-class GST isoform (mYc) with an extremely high activity towards AFB_1-8,9-epoxide (Quinn et al. 1990; Ramsdell and Eaton 1990; Hayes et al. 1991; Buetler et al. 1992; Slone et al. 1995). A homologous GST isoform (Yc2) with high activity towards AFB_1-8,9-epoxide was observed to be inducible in rats by antioxidants such as ethoxyquin and is mainly responsible for the chemoprotective effect of a variety of natural dietary compounds that act via the *antioxidant response element* (Eaton and Gallagher 1994). However, this particular form of GST is expressed in principle only at very low levels in rats. In contrast to mouse GSTs, human α-class GSTs exhibit relatively little activity towards AFB_1-8,9-epoxide (Slone et al. 1995). In conclusion, phase I-metabolism provides a greater extent of AFB_1 activation in mice compared to humans and rats. However, an extremely efficient conjugation of AFB_1-8,9-epoxide with GSH by GST mYc confers resistance to mice, whereas humans and not-induced male rats represent species with high susceptibility to AFB_1 hepatocarcinogenesis, due to the lack of such an effective phase II-detoxification pathway as in mice.

In addition to the differences between species in covalent binding of AFB_1 to liver DNA, Fig. 4.24 shows a 5 to 6-fold greater binding of [^{3}H] AFB_1 to DNA from male compared to female hepatocytes (Cole et al. 1988). Similarly, binding of AFB_1

in vivo was shown to be greater in male compared to female rats (Gurtoo and Moty-cka 1976) and liver microsomes from males were about three times more active than microsomes from females in metabolising AFB_1 to DNA binding metabolites. In addition, male rats are more susceptible than female rats to the hepatocarcino-genic effects of AFB_1 (Eaton and Gallagher 1994). This sex difference in suscepti-bility might be explained by the lack of expression of CYP3A2 (an orthologue of CYP3A4) in female rats and by an about 10-fold greater expression of GSTYc2 in livers of adult female rats compared to adult male rats, whereby the latter in princi-ple express only very low levels of GSTYc2 (Hayes et al. 1994; Buetler et al. 1996). Interestingly, the frequency of hepatocellular carcinoma in AFB_1-exposed individu-als in Thailand has been shown to be greater for men than for women (Shank et al. 1972; Peers and Linsell 1977). This might be due to the fact that CYP1A2 activity, determined by caffeine metabolism, is significantly lower in women than in men as reported by Rasmussen and Brosen (1996). It might be of relevance that among women, those taking oral contraceptives have lower CYP1A2 activities (Horn et al. 1995). In addition, nulliparous women had lower CYP1A2 activities than multi-parous women. Since activation of AFB_1 by human liver microsomes significantly correlated with the *CYP1A2* phenotype (Wilson et al. 1997), it might be worthwhile to examine whether oral contraceptives – besides their well known protection against ovarian cancer – also protect AFB_1-exposed women from hepatocellular cancer.

Oltipraz, an antischistosomal agent, has been shown to protect rats from the hepa-tocarcinogenic effect of AFB_1 (Bolton et al. 1993). Based on the anticarcinogenic effect of oltipraz in rats, phase II clinical trials have been initiated in the Qidong region in China, where individuals are highly exposed to AFB_1. Most probably the rat is protected against AFB_1-induced hepatocarcinogenesis by oltipraz treatment, because this compound induces GSTs including GSTYc2, which possesses an espe-cially high conjugating activity for AFB_1-8,9-epoxide (Buetler et al. 1996). However, up to the present time there is no evidence to suggest the existence of a human GST isoenzyme with a high AFB_1-8,9-epoxide-conjugating activity relevant to the *in vivo* situation. For this reason, clinical trials with oltipraz as a protecting agent against hepatocarcinogenesis have been criticised (Buetler et al. 1996). Although this argu-ment may be correct for GSTs, oltipraz has also been reported to inhibit AFB_1 acti-vation by decreasing CYP1A2 and CYP3A4 activities in cultured primary human hepatocytes (Langouet et al. 1995; Morel et al. 1997). In addition, oltipraz inhibited human recombinant *CYP1A2* and *3A4*. Thus, humans may also be protected by oltipraz (although to our knowledge this has not yet been confirmed by the still ongoing clinical studies), whereby chemoprotection of humans may be due to an inhibition of AFB_1 activation, in contrast to rats, which seem to be protected by an enhancement of AFB_1-8,9-epoxide conjugation with glutathione.

In conclusion, large interspecies differences in susceptibility to aflatoxin B_1 car-cinogenesis are known, with rats representing the most sensitive and mice the most resistant species, the latter being refractory to dietary aflatoxin B_1 levels three orders of magnitude higher than for rats. An extremely efficient conjugation of the major reactive metabolite of AFB_1 confers resistance to mice. Since humans as well as rats lack a similarly effective detoxification pathway, both represent species with a relatively high susceptibility to aflatoxin B_1 carcinogenesis.

4.6.5
Heterocyclic amines in foods

Up to now at least twenty different heterocyclic amines have been isolated from cooked foods, primarily fish, chicken, and pork, and several additional heterocyclic amines have been reported, but their structures have not been elucidated (Adamson et al. 1996). Formation of heterocyclic amines results from the reaction of muscle creatine with amino acids at customary cooking temperature. Since all known heterocyclic amines have planar structures they are able to intercalate into GC pair-rich stretches of DNA. 2-Amino-3-methylimidazo[4,5-*f*]quinoline (IQ), 2-amino-3,8-dimethylimidazo[4,5-*f*]quinoxaline (MeIQx) and 2-amino-1-methyl-6-phenyl-imidazo[4,5-*b*]pyridine (PhIP) represent three of the most prevalent heterocyclic amines found in the diet of Western countries.

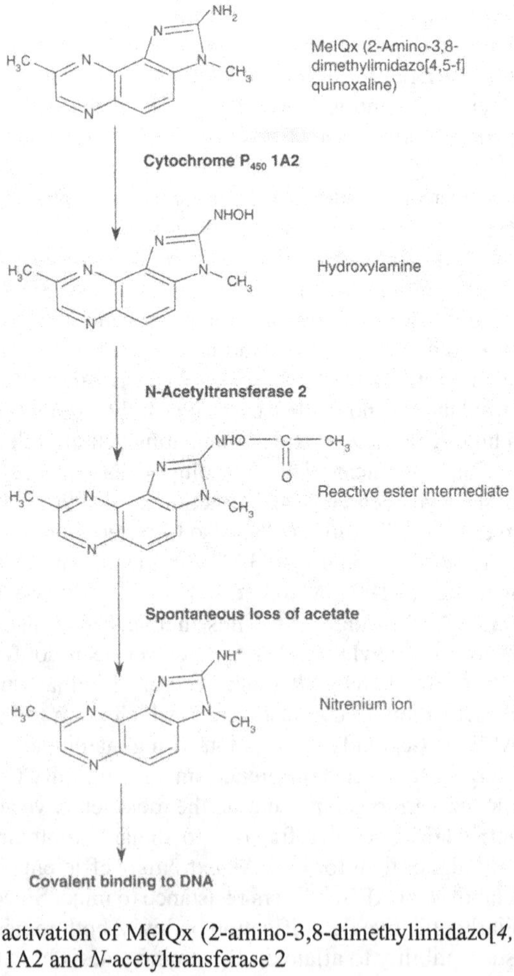

Fig. 4.26 Metabolic activation of MeIQx (2-amino-3,8-dimethylimidazo[*4,5-f*]quinoxaline) by cytochrome P450 1A2 and *N*-acetyltransferase 2

The main route of metabolic activation of heterocyclic amines includes *N*-oxidation to hydroxylamines, primarily by CYP1A2 but also by other CYP isoforms, and their subsequent *O*-esterification by acetyltransferase or sulphotransferase to reactive ester-intermediates, which can spontaneously lose acetate or sulphate to form a nitrenium ion that can covalently bind to DNA (Fig. 4.26). Hydroxylation of exocyclic amino groups of heterocyclic amines is believed to be an initial activation step, whereas ring-hydroxylation and subsequent hydroxylations are predominantly detoxification pathways (Buonarati et al. 1990). Some metabolites of heterocyclic amines, e. g. *N*-hydroxy-PhIP or *N*-acetoxy-PhIP, can be transported from the liver to extrahepatic tissues (Kadlubar et al. 1995). In addition, glucuronides derived from heterocyclic amines are secreted into the bile, due to their relatively high molecular weight. In the colon the heterocyclic hydroxylamines may then be released from the conjugates due to the action of bacterial glucuronidases. In contrast to the original *non*-hydroxylated heterocyclic amines, the hydroxylamines represent good substrates for NAT2, which is expressed in relatively high levels in colon epithelial cells. This mechanism may explain the colon-specific carcinogenic effect of some heterocyclic amines.

Large interspecies-differences in metabolism, mutagenicity and carcinogenicity of heterocyclic amines have been observed. At least 10 heterocyclic amines, including IQ, MeIQx, and PhIP, have been shown to be carcinogenic in rats and mice (Adamson et al. 1996). In addition IQ has been shown to be a potent hepatocarcinogen in Cynomolgus monkeys, whereas MeIQx lacks the potency of IQ to induce hepatocellular carcinoma after a 5-year dosing period (Snyderwine et al. 1997). The carcinogenicity of PhIP in the Cynomolgus monkey is not yet known. In order to examine interspecies differences, liver microsomes from Cynomolgus monkeys, rats and humans were added to *Salmonella typhimurium* TA 98 in the Ames test as an activating system (Davis et al. 1993). Differences between species were most pronounced when MeIQx was tested. Incubations with 0.5 μg MeIQx resulted in 1243 ± 48, 1070 ± 75, 695 ± 35 and 129 ± 19 revertant colonies plate[-1] for human, male rat, female rat and Cynomolgus monkey microsomes, respectively (Fig. 4.27). Similarly, DNA

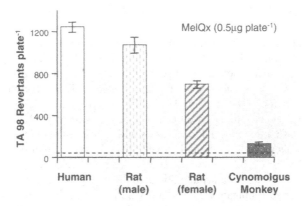

Fig. 4.27 Activation of 2-amino-3,8-dimethylimidazo[4,5-*f*]quinoxaline (MeIQx) to a bacterial mutagen by microsomes from humans, male and female rats and Cynomolgus monkeys (untreated animals). The line gives the number of spontaneous TA 98 revertants (data from Davis et al. 1993).

adducts after oral application of MeIQx were highest in male rats, followed by female rats and were much lower in Cynomolgus monkeys (Davis et al. 1993). Both, rats and humans have constitutive levels of hepatic CYP1A2 (Sesardic et al. 1989, 1990), whereas Cynomolgus monkeys do not (Sadrieh and Snyderwine 1995). The results suggest that lack of constitutive CYP1A2 expression in Cynomolgus monkeys is responsible for the low capacity to activate MeIQx, which protects this species from the genotoxic and carcinogenic effects of this compound. In contrast to the Cynomolgus monkey, the Marmoset monkey shows constitutive hepatic expression of CYP1A2 similar to that in human liver (Edwards et al. 1994). Liver microsomes from the Marmoset monkey activated MeIQx to a bacterial mutagen to a similar extent as human liver microsomes (Edwards et al. 1994). Thus, the Marmoset is a more suitable model than the Cynomolgus monkey for carcinogenicity studies with MeIQx.

Comparing human and rat CYP1A2 it was suggested that the rat enzyme is a less efficient catalyst than the human enzyme with respect to the activation of MeIQx via N-hydroxylation (review: Boobis et al. 1996). In addition, the almost exclusive N-hydroxylation of MeIQx by human liver microsomes, in contrast to the competing detoxifying ring hydroxylation reactions by rat liver microsomes (Turesky et al. 1988; Alexander et al. 1989), may explain the higher efficiency of human microsomes to activate MeIQx to mutagenic metabolites.

The higher capacity of male compared to female rat liver microsomes to activate MeIQx to mutagenic metabolites might be explained by CYP2C11, an isoform that exists in male but not in female rats (Yamazoe et al. 1988; Davis et al. 1993). The difference between male and female rats cannot be explained by CYP1A2, since about 2-fold higher levels of CYP1A2 have been observed in female compared to male rats (Yamazoe et al. 1988).

For IQ a completely different constellation has been reported. The N-hydroxylation of IQ appears to be carried out largely by hepatic CYP3A4 and/or CYP2C9/10 (Snyderwine et al. 1997). Vice versa, CYP3A4 and CYP2C9/10 were unable to N-hydroxylate MeIQx.

Western blot analysis showed that Cynomolgus monkey hepatic microsomes in principle express CYP isoforms immunologically related to the human CYP3A, CYP2C, and very low levels of CYP1A1 (Sadrieh and Snyderwine 1995). CYP3A constitutes about 20 % of the total hepatic CYP in Cynomolgus monkeys and is at least 90 % homologous to human CYP3A4 (Ohmori et al. 1993). This might explain the fact that the mutagenic activation of IQ by human and Cynomolgus liver microsomes is very similar, with only slightly higher numbers of TA98 revertants per plate for human microsomes (Davis et al. 1993). The differences in CYP isoenzymes involved in the initial activation of MeIQx and IQ have also been confirmed by differential induction of cytochrome P450 (Sadrieh and Snyderwine 1995): treatment of Cynomolgus monkeys with rifampicin induced hepatic microsomal proteins related to human CYP3A and CYP2C and was accompanied by a 3-fold increase in the mutagenic activation of IQ, without any alterations in the mutagenic activation of MeIQx. In addition, human recombinant CYP3A4 and CYP2C9 were shown to activate IQ by mutation but not MeIQx (Aoyama et al. 1990). Treatment of Cynomolgus monkeys with TCDD significantly increased mutagenic activation of both, MeIQx and IQ, which is consistent with an induction of CYP1A isozymes and suggests that TCDD-inducible CYP1A enzymes N-hydroxylate both substrates without any selectivity.

In an extensive study considering phase I and II-metabolism, species differences in the bioactivation of PhIP between human, rat and mouse were examined (Lin et al. 1995). Human hepatic microsomes had the highest capacity to catalyse the initial activation step to N-hydroxy-PhIP, which was 1.8- and 1.4-fold higher than in rats and mice, respectively. In addition, the ratio of the activating N-hydroxylation to the detoxifying ring-hydroxylation resulting in 4′-hydroxy-PhIP was 97 : 1 for human hepatic microsomes, which was much greater than that of rat (33:1) or mouse (1.7:1). The high ratio of N- to 4′-hydroxylation by human microsomes may be due to the extremely low CYP1A1 expression in human liver, which is the principal enzyme responsible for the formation of 4′-hydroxy-PhIP in rodents (Wallin et al. 1990; Lin et al. 1995). To examine the role of acetyltransferase and sulfotransferase, calf thymus DNA was incubated with [^{3}H]N-hydroxy-PhIP and liver cytosol of humans, mice and rats with and without addition of the cofactors acetyl CoA and 3′-phosphoadenosine-5′-phosphosulfate (PAPS). Subsequently, PhIP-DNA binding was determined. Acetyl coenzyme A-dependent DNA binding of N-hydroxy-PhIP with rat cytosol was similar to that in human rapid acetylators, but was 2.6-fold higher than that in human slow acetylators. The O-acetyltransferase activity for PhIP was lowest in mice, that is only 11 % of that in rats, 12 % of that in human rapid acetylators and 29 % of that in the slow ones. In contrast, mouse hepatic cytosols exhibited the highest sulfotransferase activity for PhIP activation, which was 4.9- and 2.3-fold higher than that in rats and humans. GST inhibitors (triethyltin bromide for human and triphenyltin chloride for rodent glutathione S-transferases) increased significantly PhIP-DNA binding in human and rat cytosols, but not in mouse cytosols.

In conclusion, humans may be more susceptible to the carcinogenic effect of heterocyclic amines than monkeys, rats or mice. Some individuals, namely those with high CYP1A2 and CYP3A4 activities and the rapid acetylator phenotype can be expected to be at an especially high cancer risk. The polymorphism of NAT2 further complicates the situation of heterocyclic amines (review: Hengstler et al. 1998). The rapid acetylator phenotype, assessed by the rate of orally given sulfamethazine, a substrate specific to NAT2, was shown to be associated with colorectal cancer (odds ratio: 1.8; 95 %, CI: 1.0–3.3; number of patients: 110; Roberts-Thomson et al. 1996). In this study the highest risk was observed in the youngest third (< 64 years) of patients (odds ratio: 8.9; 95 %, CI: 2.6–30.4) and increased with increasing intake of meat in rapid, but not in slow acetylators. In addition, the risk of colorectal cancer was higher for individuals consuming cooked well done meat, which is known to contain higher levels of heterocyclic amines, compared to individuals usually eating meat less well done (Schiffmann and Felton 1990; Gerhardsson de Verdier et al. 1991; Lang et al. 1994). Although these studies have not directly measured the intake of heterocyclic amines, the whole constellation suggests that heterocyclic amines consumed with meat contributes to human carcinogenesis.

In addition to rapid acetylators individuals possessing high CYP1A2 activities are also expected to have a higher risk for heterocyclic amine-induced colon cancer. It is known that cigarette smoking (and other hydrocarbon-containing mixtures) induces CYP1A2 in human liver (Boobis et al. 1996). Thus, it should be expected that, for individuals regularly eating meat, cigarette smoking represents a risk factor for colon cancer. At a first glance it seems intriguing that epidemiological studies suggest the opposite (Giovannuci et al. 1994). Although smoking itself is a positive risk factor, smokers eating meat were at reduced risk of colorectal cancer compared with non-smokers with similar meat consumption. However, this is not the only report showing that cigarette smoking – although representing a risk factor by itself – may protect from additional genotoxic exposures, e. g. to *N*-nitrosodiethanolamine or ethene oxide (Oesch et al. 1994). It has been speculated that cigarette smoking, besides increasing CYP1A2 levels, leads to reduced geno-toxicity of these compounds through parallel induction of glutathione *S*-trans-ferase A1-1 (GSTA1), which may serve to detoxify heterocyclic amines in the colon (Boobis et al. 1996). Indeed, it has been shown that GSTA1 can be induced in human hepatocytes, but inducibility is only observed in a subgroup of individu-als (Morel et al. 1993; Schrenk et al. 1995). Whether this reflects interindividual variation in protection against heterocyclic amines remains to be examined.

Since heterocyclic amines most probably contribute to human carcinogenesis, strategies to minimise their intake and genotoxicity should be suggested. In this con-text it seems important that heterocyclic amines induce their own activation in humans (Sinha et al. 1994). In non-smokers, consumption of pan-fried meat cooked at high temperature (250 °C) for 7 days (which was shown to contain MeIQx, PhIP and DiMeIQx) significantly increased CYP1A2 activity, whereas meat cooked at only 100 °C did not. In addition, chronic treatment of Cynomolgus monkeys with IQ (10 mg kg^{-1}) increased the capacity of liver microsomes to activate MeIQx to a bacterial mutagen (Sadrieh and Snyderwine 1995). Indeed, at a substrate concentration of 5 μg MeIQx, microsomes from IQ-treated versus control monkeys led to 430 versus 40 revertants per plate, respectively. Thus, heterocyclic amine genotoxicity might be reduced if meat consumption is followed by a period without meat-consumption to avoid having the next heterocyclic amine exposure when CYP1A2 still is induced. In addition, exposure to heterocyclic amines might be reduced by cooking meat at a low temperature (about 100 °C) and eating beef "medium" instead of "well done". Meth-ods to minimise the intake of heterocyclic amines (Adamson et al. 1996) – the use of microwaves, wrapping meat or fish in aluminium foil to prevent contact with an open flame, using bacon grease sparingly for cooking, removing the delicious meat juice and blackened parts of meat –may be mentioned for completeness, but some of these may be unacceptable from a gourmet's point of view.

In conclusion, humans seem to be more susceptible to the carcinogenic effect of heterocyclic amines than Cynomolgus monkeys, rats and mice. Especially human individuals with high CYP1A2 and CYP3A4 activities and of the rapid acetylator phenotype may be expected to have an increased risk of colon cancers.

4.7
Interindividual variations of toxicological response in the human population

Differences between individuals in response to toxic challenges are a matter of cur-rent concern. Specifically, the recognition of "vulnerable" subpopulations has led to discussions, how these could be adequately protected by regulatory action. For instance, with respect to occupational exposure limits (OEL) practical approaches have been pragmatic, being located between the paradigms of personal selection for specific workplaces and of OEL standard settings oriented towards the most susceptible subgroup within a workforce (Bolt 1998). At present, most widely discussed are individual influences of age, specifically regarding children and adolescents, and of genetically determined individual susceptibility which ought to have an influence on risk assessment. Major avenues of argumentation will be outlined in the following.

4.7.1
Age (children)

Risks of chemical exposures especially to the health of children have become a current policy issue, both in the United States (Landrigan 1999) and the European Union (Schwenk et al. 2003). Primary attention has mostly been given to exposure of infants and small children through consumer products, by drugs and pesticides. As a specific example, it has been noted that derivations of the Acceptable Daily Intake (*ADI*) of food contaminants must give special consideration regarding the use of food additives to apply to infants below the age of 12 weeks, who depend entirely on infant formula for nutrition (Östergaard and Knudsen 1998).

The general proposal to introduce an additional 10-fold "uncertainty factor" (safety factor) for pesticides when exposure of infants and children is anticipated has been opposed because of lack of a scientific rationale (Renwick et al. 2000), but it was considered necessary to view premature neonates, infants, and children of different ages as separate risk groups (Östergaard and Knudsen 1998).

Toxicokinetic differences in the course of human development have been taken into the argumentation of children's susceptibility, yet the very complexity of the situation has been noted (Faustman et al. 2000). A review of age-relations of available pharmacokinetic data on selected drugs has shown that the ratio of internal doses (AUC) in children vs. adults for the same relative drug dose (in mg kg^{-1}) was surprisingly constant. A well known exception was neonates and very young infants, especially in relation to their glucuronidation capacity (Renwick et al. 2000). Hence, it must be noted that the generalisation of infants and children being more susceptible to toxic effects than adults is founded more on assumptions than on scientific facts.

Immature physiological functions of the fetus and young child theoretically make these groups more vulnerable to toxicants, at least up to one year of age (Östergaard and Knudsen 1998). In general, for a particular compound, children may be more sensitive than adults, or they may be less sensitive (Faustman et al. 2000).

It also has been suggested to consider adolescents as a separate group vulnerable to chemical toxicants, because adolescence is the second most rapid period of growth and development, after infancy. However, under public health aspects for this age group the individual exposure situation is much more relevant than theoretical differences in toxicological susceptibility. A recent compilation of surveys of exposure situations of adolescents due to occupational, environmental and lifestyle factors has revealed that the overwhelming problem of exposure in this age group is due to lifestyle: tobacco, alcohol, cannabis and designer drugs (Bolt 2002). This problem is of global nature and does not touch questions of low-dose extrapolations.

4.7.2
Relevance of polymorphisms for low-dose exposures

4.7.2.1
Introduction and definitions

It has become clear that interindividual differences in human susceptibility to carcinogenic substances may depend on the individual's genetic makeup. Polymorphisms are defined as less frequent phenotypes, which occur in at least 1 % of a population. If a polymorphism, by definition, requires an aberrant phenotype, a point mutation that does not lead to an amino acid exchange would not be classified as a polymorphism. In this regard there is some confusion as to the meaning of polymorphism. Therefore, we suggest to differentiate between silent genetic polymorphisms (for instance single nucleotide exchanges that do not lead to amino acid exchanges), polymorphisms that cause alterations in amino acid sequence but do not lead to (known) functional alterations and – certainly the most relevant type – the functional polymorphism. The Human Gene Guidelines (Ingelman-Sundberg et al. 2000) suggest the following nomenclature: the abbreviated name of the gene in capital letters (for instance: GST for glutathione *S*-transferase) followed by "*" and a number that defines a specific polymorphic gene. Usually these numbers are in the order of discovery of individual polymorphisms. Deletions are usually indicated by "*0".

It is important to differentiate between polymorphisms and rare variants of genes. In contrast to the relatively frequent polymorphic genes that, by definition, occur in more than 1 % of individuals in a population, rare variants are present in less than 1 %. Some rare variants are extremely relevant, since almost all carriers will develop cancer. Well-known examples are a rare variant of the *p53* gene leading to Li-Fraumeni syndrome or a variant of the *APC* gene that causes adenomatosis/polyposis coli, a disease leading to colon cancer in all affected individuals. Thus, some rare variants are characterised by their extremely low frequency in combination with a very high penetrance concerning cancer risk. In these aspects polymorphisms differ from rare variants. Some of them occur quite frequently. For instance *GSTM1*0*, a large gene deletion of glutathione *S*-transferase can be found in approximately 50 % of Europeans ("Caucasians"; Fig. 4.28; reviews: Hengstler et al. 1998; Thier et al. 2003). However, the risk to die from cancer is usually only slightly increased for carriers of polymorphic high risk alleles. In cases where polymorphisms are associated with cancer risk, odds ratios usually range between 1.4 and 3.0. To our knowledge no polymorphism is consistently associated with an odds ratio higher than 5.0; although such high odds ratios were reported in individual studies, they were not confirmed by other investigators. For instance, a cigarette smoker who carries a high risk allele of glutathione *S*-transferase M1, namely *GSTM1*0*, has an approximately 1.3 to 1.4-fold increased risk to die from lung cancer compared to a smoker carrying the respective wild-type allele of *GSTM1*. Cigarette smoking itself is already associated with a 30- to 40-fold increased risk to die from lung cancer. Thus, single high-risk alleles seem of minor relevance considering an individual's cancer risk. However, considering the high frequency of polymorphic alleles it is clear that despite of the low individual risk world wide much more cancer cases are due to polymorphisms than to rare variants.

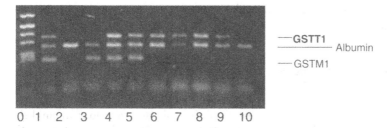

Fig. 4.28 Polymerase chain reaction (PCR) analysis of polymorphisms of glutathione *S*-transferase (GST) M1 and T1. Both polymorphisms are due to gene deletions (*GSTM1*0* and *GSTT1*0*) that result in complete absence of the respective enzyme activities. The primers were constructed to amplify DNA fragments that are located in the deleted regions (Arand et al. 1996). Primers for human albumin serve as positive controls. Lanes 1 and 3–5 show positive amplification with *GSTM1* primers indicating carriers of wild type *GSTM1*, whereas lanes 2 and 6–10 show results from individuals with the deleted *GSTM1* gene (*GSTM1*0*). Compared to individuals with the *GSTM1* wild type carriers of the GSTM1*0 allele have an approximately 1.4-fold increased risk to die from lung cancer (review: Hengstler et al. 1998; figure from: Hengstler et al. 1998). Lane 0: *Alu*I-restricted pBR322 DNA size marker.

This overview does not give a complete review of all known polymorphisms possibly relevant for cancer susceptibility, since this has already been presented elsewhere (Hengstler et al. 1998, 2000; Thier et al. 2003). It is intended to discuss relevant principles of polymorphic high risk alleles, with respect to low-dose exposures.

4.7.2.2
Frequency of polymorphisms in the human genome

After sequencing of the human genome of single individuals one of the next milestones will be to compare the genomes of larger populations to identify polymorphisms. Nevertheless, based on the number of single nucleotide polymorphisms (SNPs, that can be silent or functional), the total number of polymorphisms in the human genome can already be estimated. Such SNPs occur approximately every 500 base pairs. The average size of a gene is approximately 20,000 base pairs including 5' and 3'-flanking regions and introns. Thus, on average, already a single gene contains approximately 40 SNPs. If the human genome contains 30,000 genes the resulting number of SNPs is 1.2 million. This estimation is based exclusively on SNPs. SNPs are certainly the most frequent type of polymorphism. Nevertheless, other genetic alterations, such as deletions, insertions or amplifications also contribute to polymorphisms. Thus, the total number of polymorphisms is probably even higher than 1.2 million. Of course, only a minority of them can be expected to be functional polymorphisms. On the assumption that 1 % of all SNPs lead to functional consequences the number of functional polymorphisms is 12,000 and – of course – only a small fraction of all functional polymorphisms will be associated with cancer risk. Nevertheless, these calculations show that the specific polymorphisms reviewed in the following paragraph possibly represent only a small fraction of all polymorphisms that are relevant for cancer susceptibility.

Most probably, many polymorphisms relevant for susceptibility to chemical carcinogens are still unknown.

4.7.2.3
Polymorphisms relevant for cancer risk

Most polymorphisms can be expected to be irrelevant for human cancer risk. For instance, polymorphic alleles that determine hair and eye colour are unlikely to be relevant for an individual susceptibility to chemical carcinogens. Gene categories that are critically involved in carcinogenesis have been summarised in Tab. 4.4. Polymorphisms that lead to a decreased activity of tumour suppressor genes, for instance a single amino acid exchange of *p53* that facilitates degradation, may be associated with an increased cancer risk (review: Hengstler et al. 2000). Similarly, alleles of DNA repair enzymes that lead to a decreased repair activity are likely to influence susceptibility to carcinogens. The risk for hormone-dependent tumours certainly depends on enzymes and factors that regulate hormone concentrations in blood. For instance, high activity alleles of enzymes that contribute to biosynthesis as well as low activity alleles of enzymes that are involved in biodegradation of critical hormones represent candidates for high risk alleles. One of the best documented example is the polymorphism of cytochrome P450 19 (CYP19, chromosomal location: 15q21.1), the cytochrome P450 subfamily responsible for aromatisation of androgens, and thereby the biosynthesis of oestrogens (review: Dunning et al. 1999). Importantly, transformation of androgens to oestrogens may not only take place in the gonads, but also in adipose tissue (Bolt and Göbel 1972). Three independent case control studies came to the conclusion that the *CYP19 (TTTA)$_{10}$*-allele is associated with an increased risk for breast cancer. Another category of genes (Tab. 4.4) is xenobiotics transporters, such as *P*-glycoproteins (MDR1) and MRP2. The first genetic polymorphisms have recently been identified (Brockmöller et al. 2000). Besides their well-documented role in anticancer resistance, transporters involved in cellular clearance of carcinogens are likely to have great impact as cancer susceptibility genes.

Certainly the most intensively studied category of genes with respect to cancer susceptibility are the drug metabolising enzymes (reviews: Hengstler et al. 1998, 2000; Brockmöller et al. 2000; Thier et al. 2003). Drug metabolism can be divided into functionalisation (phase I-metabolism) and conjugation (phase II-metabolism)

Tab. 4.4 Categories of cancer susceptibility genes[a]

Oncogenes and tumour suppressor genes
DNA repair genes
Enzymes of hormone synthesis and metabolism and hormone receptors
Xenobiotics transporters
Carcinogen metabolising enzymes: phase I and phase II metabolising enzymes

[a] Functional polymorphisms of cancer susceptibility genes are likely to influence an individuals cancer risk.

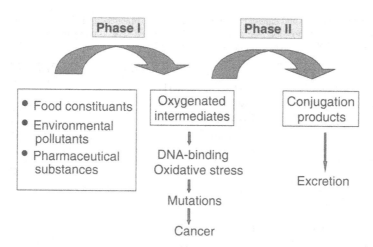

Fig. 4.29 Phase I- and phase II metabolism of xenobiotics. Polymorphic high activity alleles of phase I metabolising enzymes are often associated with an increased cancer risk. In contrast, high activity alleles of phase II-metabolising enzymes are usually associated with a decreased cancer risk. Exceptions exist, where phase II-enzymes are involved in activation of procarcinogens.

see Fig. 4.29. Phase I-metabolism often results in generation of reactive intermediates (Fig. 4.29). Thus, high activity alleles of phase I-metabolising enzymes are often associated with an increased cancer risk. In contrast, most conjugating reactions lead to detoxification of reactive intermediates. Thus, high activity alleles of phase II-metabolising enzymes are often associated with decreased cancer risk. Exceptions exist, since in some cases phase II-enzymes are involved in activation of procarcinogens.

Tab. 4.5 summarises the most important polymorphisms of phase I (a) and phase II (b) metabolising enzymes. If an association of a specific polymorphism with cancer risk has been reported by at least two independent groups this has been indicated by "+" for increased risk associated with high activity alleles or by "-" in case of a decreased cancer risk. If major uncertainties could not be excluded, "+" or "-" were put in parentheses. However, for most polymorphisms clear conclusions concerning the relevance for cancer susceptibility can not yet be drawn, due to missing or inconsistent data. Polymorphisms of drug metabolising enzymes that have been shown to influence cancer susceptibility and that were confirmed by independent studies include (i) CYP1A1*2A and CYP1A1*2C and increased risk for lung cancer, (ii) high CYP1A2 activity and increased risk for bladder cancer, (iii) high CYP2D6 activity and increased risk for lung cancer, (iv) NAT2 (N-acetyltransferase 2) rapid alleles and increased risk for colorectal but a decreased risk for bladder cancer (the discrepancy is well understood, since NAT2 contributes to activation of well known colon carcinogens such as MeIQx (3-methyl-3,8-dimethylimidazo[4,5-f]quinoxaline), but predominantly inactivates some bladder carcinogens, such as 4-aminobiphenyl, (v) GSTM1*0 and increased lung, breast and bladder cancer risk and (vi) GSTP1 that most probably is associated with lung cancer (Hengstler et al. 1998).

Tab. 4.5 Polymorphisms and cancer risk. a: Phase I-polymorphisms. b: Phase II-polymorphisms. +: increased risk; (+): possibly increased risk; -: decreased risk; (-):possibly decreased risk; 0: no detectable influence; nd: data not sufficient

a	CYP1A1			CYP1A2	CYP2D6	CYP2E1
	MSPI (3'-flanking) *CYP1A1*2A*	I462V (exon7) *CYP1A1*2C*	AHH inducibility			
Lung	+ (Asian) 0 (Caucasian)	+ (Asian) 0 (Caucasian)	+	nd	+	(0)
Colorectal	(+)	nd	nd	(+)	nd	nd
Breast	(+)	(+)	nd	nd	0	0
Bladder	0	0	nd	+	0	0
Liver	nd	nd	nd	nd	(+)	0
Endometrium	(+)	(+)	nd	nd	nd	nd
Skin (BCC)	0	(+)	nd	nd	0	nd
Brain	nd	nd	nd	nd	(-)	nd

b	NAT1	NAT2	GSTM1*0	GSTT1*0	GSTP1	mEH	
	rapid	rapid			Codon 104 (low activity)	Exon 3 low	Exon 4 low
Lung	nd	(0?)	+	(0)	(+)	(0)	nd
Colorectal	(+)	+	(+)	(+)	0	nd	nd
Breast	nd	(-)	+	nd	0	nd	nd
Bladder	nd	-	+	(+)	(+)	nd	nd
Liver	nd	(-)	nd	nd	nd	(+?)	nd
Endometrium	nd	nd	(0)	nd	nd	nd	nd
Skin (BCC)	nd	nd	nd	(+)	nd	nd	nd
Brain	nd	nd	nd	(+)	nd	nd	nd

4.7.2.4
Practical consequences

Establishment of threshold values. In Germany, the Commission for the Investigation of Health Hazards of Chemical Compounds in the Work Area (MAK Commission) sets maximum workplace concentration values for single substances (Bartsch et al. 1998). The MAK (Maximale Arbeitsplatzkonzentration) is the maximum concentration of a substance in the ambient air in the workplace, which has no adverse effect on the worker's health. For substances with a genotoxic and carcinogenic potential, the commission considered it scientifically inappropriate to establish a safe threshold. Therefore, TRK-values (Technische Richtkonzentrationen) have been established. The TRK-value gives the concentration of a substance (usually in the ambient air in the workplace) that can be achieved using technically available measures. Once established, both types of limits are legally binding in Germany and have to be observed in the workplace.

With respect to polymorphisms, it should be considered whether polymorphic genes influence susceptibility to an individual substance. If this is the case care should be taken to protect the more susceptible subpopulation.

Epidemiological studies. Knowledge about human carcinogens often stems from epidemiological studies. A limitation of epidemiological studies is their relatively low sensitivity, even if large cohorts of individuals are examined. The low sensitivity in due to the high baseline cancer rate (therefore carcinogens that induce rare cancer types can be identified with a better sensitivity) and the relatively low exposure of humans (low at least, when exposures are compared to exposures in animal studies). If polymorphic gene products control a substance, sensitivity of epidemiological studies could be increased by integration of carriers of high-susceptibility alleles.

Identification of individuals at higher cancer risk. Perhaps one of the most relevant questions concerning polymorphisms and low-dose exposure to carcinogens is, whether polymorphisms help to identify individuals that have a strongly increased cancer risk. This is relevant, since specific strategies of cancer prevention could be applied for individuals at high risk. For instance early diagnosis of lung cancer by magnetic resonance imaging (MRI) and early treatment represents the best method for improving survival rates (Goldberg and Unger 2000; Leutner and Schild 2001). However, MRI is too expensive to be extended to the total population, but could be performed for a high-risk subpopulation. However, presently, the question whether such individuals can be identified by polymorphisms must be denied. As explained above, odds ratios for cancer risk of individual high-risk alleles range between 1.3 and 3.0. This seems too low to justify extensive prevention programs. In our opinion such programs would only make sense if risk factors exceed values of 20. Single high-risk alleles are not associated with such a high risk. However, several high-risk alleles may interact with each other. For instance, such an interaction has been observed for cytochrome P4501A2 and NAT2, both of which are involved in activation of heterocyclic aromatic amines (review: Hengstler et al. 1998). In case of an interaction of several high-risk alleles individual risk factors may often behave in a multiplicative way. Thus, carriers of two high-risk alleles, each associated with a risk factor of 2.0 have a combined risk of 4.0. Hypothetically, carriers of five interacting high-risk alleles would have a risk of 32, which would be high enough to justify extensive prevention strategies. However, presently it is not known whether as many as five high-risk alleles can interact. Such studies would require analysis of very high case numbers ranging between 1,000 and 5,000 individuals. Although technically possible such studies have not yet been performed, mainly due to problems in funding such large studies.

4.8
Conclusion

Predominant results of the case studies of compounds have been presented in this chapter. Criteria for the selection of compounds to be discussed in this chapter were representations of characteristics relevant for risk assessment, and the availability of scientific data supporting different mechanisms. The overall results of these case studies are summarised in Tab. 4.6.

There is almost general agreement that distinction should be made between *genotoxic and non-genotoxic* chemicals when conducting assessments of cancer risk to humans. Although the risk assessment approach used for non-genotoxic chemicals is fairly similar among different regulatory bodies, risk assessment approaches for genotoxic carcinogens vary widely (Seeley et al. 2001).

Non-genotoxic carcinogens (examples: TCDD and hormones) are characterised by a "conventional" dose-response which allows derivation of a No-Observed-Adverse-Effect-Level (NOAEL). Insertion of an uncertainty (or safety) factor will lead to derivation of permissible exposure levels at which no relevant human cancer risk is anticipated.

For *genotoxic carcinogens*, the case studies of chemicals demonstrate a whole array of different possibilities. Positive data of chromosomal effects only, e. g. aneugenicity or clastogenicity, in the absence of mutagenicity, may support the characterisation of a compound that produces carcinogenic effects only at high, toxic doses (Schoeny 1996). Examples of *non-DNA-reactive genotoxins*, affecting topoisomerase or motor proteins of the spindle apparatus, have been presented. In these cases, relevant arguments are in favour of the existence of "practical" thresholds (Crebelli 2000; Parry et al. 2000). Moreover, genotoxicity (especially when of local nature) may be relevant only under conditions of sustained local tissue damage and associated increased cell proliferation. Formaldehyde and vinyl acetate were highlighted as examples. Also in such cases, the derivation of practical thresholds and, in consequence, of health-based exposure limits appears sufficiently justified. A further discussion of the notion of practical threshold is given in section 8.2.2.

The assumption of a linear dose-carcinogenicity response without threshold is supported in a plausible way for a number of carcinogens, e. g. for vinyl chloride and the tobacco smoke-specific nitrosamine NNK. Other carcinogens may behave differently, but the precise nature of the dose-response, at low doses, has not been sufficiently established. Also, differences may occur for the same compound between different tissues. In such cases, precautionary considerations will mostly lead to application of the conservative approach of a linear low-dose extrapolation.

In essence, more aspects of mechanisms involved should be considered within regulatory assessments of carcinogenic risks of chemicals (Parry et al. 2000; Wiltse and Dellarco 2000).

Tab. 4.6 Essential features of the case studies presented in chapter 4.
EXP = Animal experiments, HUM = Human epidemiology, MEC = Mechanisms, PBPK = PBPK model, CELL = Cell culture studies

Substance	Effect	Linearity	Threshold	Mechanism behind threshold/ non-linearity	Basis of evidence	Sensitivity; interspecies comparison
DNA-reactive, genotoxic carcinogens						
Vinyl chloride	Human liver angiosarcoma	Yes	No		EXP HUM MEC PBPK	Animals > Humans
Aflatoxin B1	DNA adducts Liver tumour	Yes Yes	– No		EXP MEC (?)	Mice > Rats
N,N-Diethylnitrosamine (DEN)	Liver cancer	Saturation at high doses	No (?) "Practical" for chronic exposure	High-dose plateau due to enzyme saturation Low-dose non-linearity uncertain Stimulation of DNA repair enzymes (?)	EXP MEC	
4-[N-Methyl-N-nitroso-amino]-1-[3-pyridyl]-1--butanone (NNK)	Lung cancer	Yes	No		EXP	
2-Acetylaminofluorene (2-AAF)	DNA adducts Bladder tumour Liver tumour	Yes No Yes			EXP	
4-Aminobiphenyl (4-ABP)	Female mice DNA adducts Liver tumour Bladder tumour Male mice DNA adducts Liver tumour Bladder tumour	Yes Yes No (low) Yes No No	Sex and tissue specific		EXP	

Tab. 4.6 continued

Substance	Effect	Linearity	Threshold	Mechanism behind threshold/ non-linearity	Basis of evidence	Sensitivity: interspecies comparison
Formaldehyde	Mutagenicity DNA adducts; DPX; DNA (ssb) Carcinogenicity (nasopharynx and nasal cavities; lung ?)	No No No	Yes (?) Yes (?) Yes (?)	DNA repair Cell proliferation, formaldehyde DNA binding kinetics (non-linear) Increased cell replication increases ssb DNA sites available for reaction with formaldehyde	EXP HUM (nasal tumours good; lung inconsistent) MEC CELL	Probably humans < rats (differences in respiration physiology)
Vinyl acetate	Mutagenicity DPX, Chromosome aberrations; SCE Carcinogenicity (Upper digestive tract and nasal tumours)	No No No	Yes Yes Yes ("practical")	Metabolic activation to acetic acid and acetaldehyde. Confounding effect of low pH Unstable DPX; Background levels of acetaldehyde Cell proliferation	EXP HUM no support of carcinogenicity MEC (?) PBPK	Nasal tumours in rats but not mice
Non DNA-reactive genotoxins						
Motor protein inhibitors (e. g. Pb, Hg)	Micronuclei	No	Yes ("practical")	Disturbance of chromosomal segregation inhibit microtubule formation Cell proliferation inhibition	EXP HUM (Pb) MEC	
Topoisomerase poisons	Transient DNA breaks Chromosome aberrations Infant leukemia	No	No ? (DNA break) Yes ? (enzyme inhibition)		MEC HUM (??)	
Non-genotoxic carcinogens						
TCDD	Carcinogenicity	?	Yes	Receptor-based threshold mechanism; Reduction of cell proliferation	EXP HUM MEC (?)	Humans less sensitive (?)

4.9
References

Abelson PH (1994) Risk assessment of low-level exposures. Science 265: 1507

Adamson RH, Thorgeirsson UP, Sugimura T (1996) Extrapolation of heterocyclic amine carcinogenesis data from rodents and nonhuman primates to humans. Arch Toxicol Suppl 18: 303–318

Adler ID, Andrae U, Kreis P, Neumann HG, Thier R, Wild D (2000) Recommendations for the categorization of germ cell mutagens. Int Arch Occup Environ Health 73: 428–432

Agency for Toxic Substances and Disease Registry – US Dept. of Health and Human Services [ATSDR] (1997) Toxicological Profile for Vinyl Chloride. ATSDR, Atlanta, GA

Agency for Toxic Substances and Disease Registry (ATSDR) (1992) Toxicological profile for vinyl acetate. U.S. Public Health Service, U.S. Department of Health and Human Services, Atlanta, GA

Albert RE, Sellakumar AR, Lashkin S, Kuschner M, Nelson M, Snyder CA (1982) Gaseous formaldehyde and hydrogen chloride induction of nasal cancer in the rat. J Natl Cancer Inst 68: 597–602

Alexander FE, Patheal SL, Biondi A, Brandalise S, Cabrera ME, Chan LC, Chen Z, Cimino G, Cordoba JC, Gu LJ, Hussein H, Ishii E, Kamel AM, Labra S, Magalhaes IQ, Mizutani S, Petridou E, de Oliveira MP, Yuen P, Wiemels JL, Greaves MF (2001) Transplacental chemical exposure and risk of infant leukemia with MLL gene fusion. Cancer Res 61: 2542–2546

Alexander J, Wallin H Home JA, Becher G (1989) 4-(2-amino-1-methylimidazo[4,5-*b*]pyrid-6-yl)phenyl sulphate – a major metabolite of the food mutagen 2-amino-1-methyl-6-phenyl-imidazo[4,5-*b*]pyridine (PhIP) in the rat. Carcinogenesis 10: 1543–1547

American Conference of Governmental Industrial Hygienists [ACGIH] (2001) TLVs and BEIs. [Documentation on the Threshold Limit Values and Biological Exposure Indices], 7th edition. American Conference of Governmental Industrial Hygienists, Cincinnati, OH

Anderson KE, Carmella SG, Ye M, Bliss RL, Le C, Murphy L, Hecht SS (2001) Metabolites of a tobaccospecific lung carcinogen in nonsmoking women exposed to environmental tobacco smoke. J Natl Cancer Inst 93: 378–381

Andreola F, Fernandez-Salguero PM, Chiantore MV, Petkovich MP, Gonzalez FJ, De-Luca LM (1997) Aryl hydrocarbon receptor knockout mice (*AHR-/-*) exhibit liver retinoid accumulation and reduced retinoic acid metabolism. Cancer Res 57: 2835–2838

Anwar WA, Khalil MM, Wild CP (1994) Micronuclei, chromosomal aberrations and aflatoxin-albumin adducts in experimental animals after exposure to aflatoxin B1. Mutat Res 322: 61–67

Aoyama T, Gelboin HV, Gonzalez FJ (1990) Mutagenic activation of 2-amino-3-methyl-imidazo[4,5-*f*]quinoline by complementary DNA expressed human liver P450. Cancer Res 50: 2060–2063

Appelman LM, Woutersen RA, Zwart A, Falke HE, Feron V (1988) One-year inhalation toxicity study of formaldehyde in male rats with a damaged or undamaged nasal mucosa. J Appl Toxicol 8: 85–90

Arand M, Mühlbauer R, Hengstler JG, Jäger E, Fuchs J, Winkler L, Oesch F, (1996) A multiplex polymerase chain reaction protocol for the simultaneous analysis of the glutathione *S*-transferase *GSTM1* and *GSTT1* polymorphisms. Anal Biochem 236: 184–186

Argyris TS (1985) Promotion of epidermal carcinogenesis by repeated damage to mouse skin. Am J Indust Med 8: 29–37

Bailey EA, Iyer RS Stone MP, Harris TM, Essigmann JM (1996) Mutational properties of the primary aflatoxin B₁-DNA adduct. Proc Natl Acad Sci USA 93: 1535–1539

Bamberger J (2000) Vinyl acetate: Subchronic toxicity 90-day drinking water in rats and mice. In: E I du Pont de Nemours & Company, Haskell Laboratory for Toxicology and Industrial Medicine, DuPont-3978, Newark, DE

Bank PA, Yao EF, Phelps CL, Harper PA, Denison MS (1992) Species-specific binding of transformed *Ah* receptor to a dioxin responsive transcriptional enhancer. Eur J Biochem 228: 85–94

Bartsch H, Malaveille C, Barbin A, Planche G (1979) Mutagenic and alkylating metabolites of haloethylenes, chlorobutadienes, and dichlorobutenes produced by rodent or human liver tissues. Arch Toxicol 41: 249–277

Bartsch H, O'Neill IK, Schulte-Hermann R (1987) Relevance of *N*-nitroso compounds to human cancer: overview. In: Relevance of *N*-nitroso compounds for human cancer: exposures and mechanism, IARC Sci. Publ. 84: 5–10

Bartsch R, Förderkunz S, Reuter U, Sterzl-Eckert H, Greim H (1998) Maximum workplace concentration values and carcinogenicity classification for mixtures. Environ Health Perspect 106 Suppl 6: 1291–1293

Batlle DC, Peces R, LaPointe MS, Ye M, Daugirdas JT (1993) Cytosolic free calcium regulation in response to acute changes in intracellular pH in vascular smooth muscle. Am J Physiol Cell Physiol 264: 932–943

Bauman JW, Goldsworthy TL, Dunn CS, Fox TR (1995) Inhibitory effects of 2,3,7,8-tetrachlorodibenzo-p-dioxin on rat hepatocyte proliferation induced by 2/3 partial hepatectomy. Cell Prolif 28: 437–451

Becher H, Steindorf K, Flesch-Janys D (1998) Quantitative cancer risk assessment for dioxins using an occupational cohort. Environ Health Perspect 106 Suppl 2: 663–670

Bedford P, Fox BW (1981) The role of formaldehyde in methylene dimethanesulfonate-induced DNA cross-links and its relevance to cytotoxicity. Chem Biol Interact 38: 119–126

Beland FA, Fullerton NF, Smith BA, Poirier MC (1992) DNA adduct formation and aromatic amine tumourigenesis. Prog Clin Biol Res 37479–37492

Belinsky SA, Foley JF, White CM, Anderson MW, Maronpot RR (1990) Dose-response relationship between O^6-methylguanine formation in Clara cells and induction of pulmonary neoplasia in the rat by 4-(methylnitrosamino)-1-(3-pyridyl)-1-butanone. Cancer Res 50: 3772–37780

Belinsky SA, Walker VE, Maronpot RR, Swenberg JA, Anderson MW (1987) Molecular dosimetry of DNA adduct formation and cell toxicity in rat nasal mucosa following exposure to the tobacco specific nitrosamine 4-(N-methyl-N-nitrosamino)-1-(3-pyridyl)-1-butanone and their relationship to induction of neoplasia. Cancer Res 47: 6058–6065

Belinsky SA, White CM, Boucheron JA, Richardson FC, Swenberg JA, Anderson M (1986) Accumulation and persistence of DNA adducts in respiratory tissue of rats following multiple administrations of the tobacco specific carcinogen 4-(N-methyl-N-nitrosamino)-1-(3-pyridyl)-1-butanone. Cancer Res 46: 1280–1284

Bergman K, Mueller L, Teigen SW (1996) The genotoxicity and carcinogenicity of paracetamol: a regulatory (re)view. In: Current issues in mutagenesis and carcinogenesis, No. 65. Mutat Res 349: 263–288

Blair A Saracci R, Stewart PA, Hayes RB, Shy C (1990) Epidemiologic evidence on the relationship between formaldehyde and cancer. Scand J Work Environ Health 16: 381–393

Bogdanffy MS, Dreef-van der Meulen HC, Beems RB, Feron VJ, Cascieri TC, Tyler TR, Vinegar MB, Rickard RW (1994) Chronic toxicity and oncogenicity inhalation study with vinyl acetate in the rat and mouse. Fundam Appl Toxicol23: 215–229

Bogdanffy MS, Gladnick LN, Kegelman T, Frame SR (1997) Four week inhalation cell proliferation study of the effects of vinyl acetate on rat nasal epithelium. Inhal Toxicol 9: 331–350

Bogdanffy MS, Plowchalk DR, Sarangapani R, Starr TB, Andersen ME (200) Mode of action-based dosimeters for interspecies extrapolation of vinyl acetate inhalation risk. Inhal Toxicol 13: 377–396

Bogdanffy MS, Randall HW, Morgan KT (1986) Histochemical localization of aldehyde dehydrogenase in the respiratory tract of the Fischer 344 rat. Toxicol Appl Pharmacol 82: 560–567

Bogdanffy MS, Randall HW, Morgan KT (1987) Biochemical quantitation and histochemical localization of carboxylesterase in the nasal passages of the Fischer 344 rat and B6C3F1 mouse. Toxicol Appl Pharmacol 88: 183–194

Bogdanffy MS, Sarangapani R, Kimbell JS, Frame SR, Plowchalk DR (1998) Analysis of vinyl acetate metabolism in rat and human nasal tissues by an in $vitro$ gas uptake technique. Toxicol Sci 46: 235–246

Bogdanffy MS, Sarangapani R, Plowchalk DR, Jarabek AM, Andersen ME (1999) A biologically-based risk assessment for vinyl acetate-induced cancer and noncancer toxicity. Toxicol Sci 51: 19–35

Bogdanffy MS, Taylor ML (1993) Kinetics of carboxylesterase-mediated metabolism of vinyl acetate. Drug Metab Dispos 21: 1107–1111

Bogdanffy MS, Tyler TR, Vinegar MB, Rickard RW, Carpanini FMB, Cascieri TC (1993) Chronic toxicity and oncogenicity study with vinyl acetate in the rat: In utero exposure in drinking water. Fund Appl Toxicol 23: 206–214

Bolt HM (1976) Probleme der Toxikologie von Sexualhormonen. Fortschr Med 94: 731–734

Bolt HM (1978) Metabolic activation of halogenated ethylenes. In: Remmer H, Bolt HM, Bannasch P, Popper H (eds) Primary Liver Tumours. MTP Press, Lancaster, pp 285–294

Bolt HM (1987) Experimental toxicology of formaldehyde. J Cancer Res Clin Oncol 113: 305–309

Bolt HM (1996) Quantification of endogenous carcinogens. The ethylene oxide paradox. Biochem Pharmacol 52: 1–5

Bolt HM (1997) Risiken durch Stoffe am Arbeitsplatz. Möglichkeiten der Quantifizierung des endogenen Risikos. Gefahrstoffe – Reinhaltung der Luft 57: 241–242

Bolt HM (1998) Toxicological defence and susceptibility in relation to risk assessment: limit values in occupational risk management. EUROTOX Newsl 21(1): 17–20

Bolt HM (2002) Occupational versus environmental and lifestyle exposures of children and adolescents in the European Union. Toxicol Lett 127: 121–126

Bolt HM, Filser J, Laib RJ, Ottenwälder H (1980) Binding kinetics of vinyl chloride and vinyl bromide at very low doses. Quantitative aspects of risk assessment in chemical carcinogenesis. Arch Toxicol Suppl 3: 129–142

Bolt HM, Filser JG (1983) Quantitative Aspekte der Kanzerogenität von Vinylbromid. Verh Dtsch Ges Arbeitsmedizin 23: 433–437

Bolt HM, Gelbke HP, Kimmerle G, Laib RJ, Neumann HG, Norpoth KH, Pott F, Steinhoff D, Wardenbach P (1988) Stoffe mit begründetem Verdacht auf krebserzeugendes Potential (Abschnitt III, Gruppe B der MAK-Werte-Liste): Probleme und Lösungsmöglichkeiten. Arbeitsmed Sozialmed Präventivmed 23: 139–144

Bolt HM, Göbel P (1972) Formation of oestrogens from androgens by human subcutaneous adipose tissue *in vitro*. Horm Metab Res 4: 312–313

Bolt HM, Janning P, Michna H, Degen GH (2001) Comparative assessment of endocrine modulators with oestrogenic activity: I. Definition of a hygiene-based margin of safety (HBMOS) for xeno-oestrogens against the background of european developments. Arch Toxicol 74: 649–662

Bolt HM, Laib R, Filser JG, Ottenwälder H, Buchter A (1981) Vinyl chloride and related compounds: mechanisms of action on the liver. In: Berk PD, Chalmers TC (eds) Frontiers in liver disease. Thieme-Stratton, New York, pp 80–92

Bolt HM, Leutbecher M (1997) A note on the physiological background of the ethylene oxide adduct 7(2-hydroxyethyl)guanine in DNA from human blood. Arch Toxicol 71: 719–721

Bolt HM, Roos PH, Thier R (2003) The cytochrome P450 isoenzyme CYP2E1 in the biological processing of industrial chemicals: consequences for occupational and environmental medicine. Int Arch Occup Environ Health 76: 174–185

Bolton MG, Munoz A, Jacobson LP, Groopman JD, Maxuitenko YY, Roebuck BD, Kensler TW (1993) Transient intervention with oltipraz protects against aflatoxin-induced hepatic tumourigenesis. Cancer Res 53: 3499–3504

Boobis AR, Gooderham NJ, Edwards RJ, Murray S, Lynch AM, Yadollahi-Farsani M, Davies DS (1996) Enzymic and interindividual differences in the human metabolism of heterocyclic amines. Arch Toxicol Suppl 18: 286–302

Boucheron JA, Richardson FC, Morgan PH, Swenberg JA (1987) Molecular dosimetry of O^4-ethyldeoxythymidine in rats continuously exposed to diethylnitrosamine. Cancer Res 47: 1577–1581

Bressac B, Kew M, Wands J, Ozturk M (1991) Selective G to T mutations of $p53$ gene in hepatocellular carcinoma from southern Africa. Nature 350: 429–431

Brown HR, Leininger JR (1992) Pathobiology of the aging rat, Vol. 1. ILSI Press, Washington, DC, pp 377–87

Brown-Peterson NJ, Krol RM, Zhu Y, Hawkins WE (1999) *N*-nitrosodiethylamine initiation of carcinogenesis in Japanese medaka (Oryzias latipes): hepatocellular proliferation, toxicity, and neoplastic lesions resulting from short-term, low-level exposure. Toxicol Sci50: 186–194

Brüning T, Bolt HM (2000) Renal toxicity and carcinogenicity of trichloroethylene: key results, mechanisms, and controversies. Crit Rev Toxicol 30: 253–285

Brusick D (1986) Genotoxic effects in cultured mammalian cells produced by low pH treatment conditions and increased ion concentrations. Environ Mutagen 8: 879–886

Buchmann A, Stinchcombe S, Korner W, Hagenmaier H, Bock KW (1994) Effects of 2,3,7,8-tetrachloro- and 1,2,3,4,6,7,8-heptachlorodibenzo-*p*-dioxin on the proliferation of preneoplastic liver cells in the rat. Carcinogenesis 15: 1143–1150

Buchter A, Bolt HM, Filser J (1978) Pharmakokinetik und Karzinogenese von Vinylchlorid. Arbeitsmedizinische Risikobeurteilung. Verh Dtsch Ges Arbeitsmed 18: 111–124

Buetler TM, Bammler TK, Hayes JD, Eaton DL (1996) Oltipraz-mediated changes in aflatoxin B_1 biotransformation in rat liver: Implications for human chemointervention. Cancer Res 56: 2306–2313

Buetler TM, Slone D, Eaton DL (1992) Comparison of the aflatoxin B_1-8,9-epoxide conjugating activities of two bacterially expressed alpha class glutathione *S*-transferase isozymes from mouse and rat. Biochem Biophys Res Commun 188: 597–603

Bundesanstalt für Arbeitsschutz und Arbeitsmedizin (BAuA) (1997) Verordnung zum Schutz vor gefährlichen Stoffen – Gefahrstoffverordnung und Anhänge. Schriftenreihe der Bundesanstalt für Arbeitsschutz und Arbeitsmedizin – Regelwerk Rw 14, Wirtschaftsverlag NW, Bremerhaven

Bundesministerium für Jugend, Familie und Gesundheit [BMJFG] (1984) Formaldehyd. Ein gemeinsamer Bericht des Bundesgesundheitsamtes, der Bundesanstalt für Arbeitsschutz und des Umweltbundesamtes. Kohlhammer, Stuttgart

Buonarati M, Felton JS (1990) Activation of 2-amino-1-methyl-6-phenylimidazo[4,5-*b*]pyridine (PhIP) to mutagenic metabolites. Carcinogenesis 11: 1133–1138

Busby WF, Wogan GN (1984) Aflatoxins. In: Chemical Carcinogens, Vol. 2. Am Chem Soc Monogr 182: 945–1136

Buss P, Caviezel M, Lutz WK (1990) Linear dose-response relationship for DNA adducts in rat liver from chronic exposure to aflatoxin B_1. Carcinogenesis 11: 2133–2135

Butterworth BE, Bogdanffy MS (1999) A comprehensive approach for integration of toxicity and cancer risk assessments. Reg Toxicol Pharmacol. 29: 23–36

Caldwell J (1992) Problems and opportunities in toxicity testing arising from species differences in xenobiotic metabolism. Toxicol Lett 64/65: 651–659

Capen CC (1997) Mechanistic data and risk assessment of selected toxic end points of the thyroid gland. Toxicol Pathol 25: 39–48

Casanova M, Deyo DF, Heck HD'A (1989) Covalent binding of inhaled formaldehyde to DNA in the nasal mucosa of Fischer 344 rats: analysis of formaldehyde and DNA by high-performance liquid chromatography and provisional pharmacokinetic interpretation. Fund Appl Toxicol 12: 397–417

Casanova M, Morgan KT, Gross EA, Moss OR, Heck, HD'A (1994) DNA-protein crosslinks and cell replication at specific sites in the nose of F344 rats exposed subchronically to formaldehyde. Fund Appl Toxicol 23: 525–536

Casanova-Schmitz M, Heck HD'A (1983) Metabolism of formaldehyde in the rat nasal mucosa *in vivo*. Toxicol Appl Pharmacol 70: 239–253

Casanova-Schmitz M, Starr TB, Heck HD'A (1984) Differentiation between metabolic incorporation and covalent binding in the labeling of macromolecules in the rat nasal mucosa and bone marrow by inhaled ^{3}H- and ^{14}C-formaldehyde. Toxicol Appl Pharmacol 76: 26–44

Chen CJ, Chen CW, Wu MM, Kuo TL (1992) Cancer potential in liver, lung bladder and kidney due to ingested inorganic arsenic in drinking water. Br J Cancer 66: 888–892

Chen CJ, Chuang YC, Lin TM, Wu HY (1985) Malignant neoplasms among residents of a Black-foot disease-endemic area in Taiwan: High-arsenic artesian well water and cancers. Cancer Res 45: 5895–5899

Clemens MR, Einsele H, Frank H, Remmer H, Waller HD (1983) Volatile hydrocarbons from hydrogen peroxide-induced lipid peroxidation of erythrocytes and their cell compounds. Biochem Pharmacol 32: 3877–3878

Clewell HJ, Gentry PR, Gearhart JM, Allen BC, Andersen ME (2001) Comparison of cancer risk estimates for vinyl chloride using animal and human data with a PBPK model. Sci Total Environ 274: 37–66

Clewell HJ, Gentry PR, Gearhart JM, Allen BC, Anderson ME (1995) Considering pharmacokinetic and mechanistic information in cancer risk assessments for environmental contaminants: examples with vinyl chloride and trichloroethylene Chemosphere 31: 2561–2578

Cohn MS, DiCarlo FJ, Turturro A, Ulsamer AG (1985) Letter to the Editor. Toxicol Appl Pharmacol 77: 363–364

Cole KE, Jones TW, Lipsky MM, Trump BF, Hsu IC (1988) *In vitro* binding of aflatoxin B_1 and 2-acetylaminofluorene to rat, mouse and human hepatocyte DNA: the relationship of DNA binding to carcinogenicity. Carcinogenesis 9: 711–716

Commission of the European Communities (1987) Legislation on dangerous substances. Classification and labelling in the European Communities. Vol. I. Graham & Trotman, London

Crebelli R (2000) Threshold-mediated mechanisms in mutagenesis: implications in the classification and regulation of chemical mutagens. Mutat Res 464: 129–135

Croy RG, Wogan GN (1981) Quantitative comparison of covalent aflatoxin-DNA adducts formed in rat and mouse livers and kidneys. J Natl Cancer Inst 66: 761–768

Cushnir JR, Naylor S, Lamb JH, Farmer PB (1993) Tandem mass spectrometric approaches for the analysis of alkylguanines in human urine. Org Mass Spectrom 28: 552–558

Dalbey WE (1982) Formaldehyde and tumours in hamster respiratory tract. Toxicology 24: 9–14

Datta S, Talukder G, Sharma A (1986) Cytotoxic effects of arsenic in dietary oil primed rats. Sci Culture 52: 196–198

Davis CD, Adamson RH, Snyderwine EG (1993) Studies on the mutagenic activation of hetero-cyclic amines by Cynomolgus monkey, rat and human microsomes show that Cynomolgus monkeys have a low capacity to N-oxidize the quinoxaline-type heterocyclic amines. Cancer Lett 73: 95–104

Deese DE, Joyner RE (1969) Vinyl acetate: a study of chronic human exposure. Am Ind Hyg Assoc J 30: 449–57

Degen GH, Bolt HM (2000) Endocrine disruptors: update on xenoestrogens. Int Arch Occup Env-iron Health 73: 433–41

Degen GH, Janning P, Diel P, Michna H, Bolt HM (2002) Transplacental transfer of the phytoe-strogen daidzein in DA/Han rats. Arch Toxicol 76: 23–9

Deutsche Forschungsgemeinschaft [DFG] (1999a) Quecksilber und organische Quecksil-berverbindungen. In: Greim H (ed) Gesundheitsschädliche Arbeitsstoffe, toxikologisch-arbeitsmedizinische Begründungen von MAK-Werten, 28. Lieferung. Wiley-VCH, Weinheim

Deutsche Forschungsgemeinschaft [DFG] (1999b) Styrol, Nachtrag 1998. In: Greim H (ed) Gesundheitsschädliche Arbeitsstoffe, toxikologisch-arbeitsmedizinische Begründungen von MAK-Werten, 28. Lieferung. Wiley-VCH, Weinheim

Deutsche Forschungsgemeinschaft [DFG] (2000) Blei und seine anorganischen Verbindungen, außer Bleiarsenat und Bleichromat In: Greim H (ed) Gesundheitsschädliche Arbeitsstoffe, toxikologisch-arbeitsmedizinische Begründungen von MAK-Werten, 31. Lieferung. Wiley-VCH, Weinheim

Deutsche Forschungsgemeinschaft [DFG] (2001) List of MAK and BAT values 2001. Maximum concentrations and biological tolerance values at the workplace, Report No. 37. Wiley-VCH, Weinheim

Deutsche Forschungsgemeinschaft [DFG] (2002) Arsen und anorganische Arsenverbindungen. In: Greim H (ed) Gesundheitsschädliche Arbeitsstoffe: toxikologisch-arbeitsmedizinische Begrün-dungen von MAK-Werten, 35. Lieferung. Wiley-VCH, Weinheim, pp 1–50

Doolittle DJ, Furlong JW, Butterworth BE (1985) Assessment of chemically induced DNA repair in primary cultures of human bronchial epithelial cells. Toxicol Appl Pharmacol 79: 28–38

Dunning AM, Healey CS, Pharoah PD, Teare MD, Ponder BA, Easton DF (1999) A systematic review of genetic polymorphisms and breast cancer risk. Cancer Epidemiol Biomarkers Prev (10): 843–854

Eastern Research Group (1997) Report on the expert panel on arsenic carcinogenicity: review and workshop. Prepared by Eastern Research Group, Lexington, MA, for the National Center for Environmental Assessment, Washington, DC, EPA contract no. 68-C6-0041

Eaton DL, Gallagher EP (1994) Mechanism of aflatoxin carcinogenesis. Annu Rev Pharmacol Toxicol 34: 135–172

Eberhard DC (1991) Species differences in the toxicity and cytochrome P450 IIIA-dependent metabolism of digitoxin. Mol Pharmacol 40: 859–867

Edwards RJ, Murray BP, Murray S, Schultz T, Neubert D, Gant TW, Thorgeirsson SS, Boobis AR, Davies DS (1994) Contribution of CYP1A1 and CYP1A2 to the activation of heterocyclic amines in monkeys and human. Carcinogenesis 15: 829–836

Ehrenberg L, Osterman-Golkar S, Segerbäck D, Svensson K, Calleman CJ (1977) Evaluation of genetic risks of alkylating agents. III. Alkylation of hemoglobin after metabolic conversion of ethene to ethene oxide *in vivo*. Mutat Res 45: 175–184

Essigmann JM, Croy RG, Nadzan AM, Busby WF, Reinhold VN, Buchi G, Wogan GN (1977) Structural identification of the major DNA adduct formed by Aflatoxin B$_1$ *in vitro*. Proc Natl Acad Sci USA 74: 1870–1874

European Union (2002) European Risk Assessment Report. Acrylamide. CAS No: 79-06-1, EINECS No: 201-173-7, Office for Official Publications of the European Communities, Luxembourg

Faustman EM, Silbernagel SM, Fenske RA, Burbacher TM, Ponce RA (2000) Mechanisms underlying children's susceptibility to environmental toxicants. Environ Health Perspect 108 Suppl 1: 13–21

Fedtke N, Wiegand HJ (1990) Hydrolysis of vinyl acetate in human blood. Arch Toxicol 64: 428–429

Fernandez-Salguero P, Pineau T, Hilbert DM, McPhail T, Lee SS, Kimura S, Nebert DW, Rudikoff S, Ward JM, Gonzalez FJ (1995) Immune system impairment and hepatic fibrosis in mice lack-ing the dioxin-binding Ah receptor. Science 268: 722–726

Fernandez-Salguero PM, Hilbert DM, Rudikoff S, Ward JM, Gonzalez FJ (1996) Aryl-hydrocar-bon receptor-deficient mice are resistant to 2,3,7,8-tetrachlorodibenzo-p-dioxin-induced toxic-ity. Toxicol Appl Pharmacol 140: 173–179

Feron VJ, Hendriksen CFM, Speek AJ, Til HP, Spit BJ (1981) Lifespan oral toxicity study of vinyl chloride in rats. Food Cosmet Toxicol 19: 317–333

Filser JG (1992) The closed chamber technique – uptake, endogenous production, excretion, steady-state kinetics and rates of metabolism of gases and vapors. Arch Toxicol 66: 1–10

Filser JG, Bolt HM (1984) Inhalation toxicokinetics based on gas uptake studies. VI. Comparative evaluation of ethylene oxide and butadiene monoxide as exhaled reactive metabolites of ethylene and 1,3-butadiene in rats. Arch Toxicol 55: 219–223

Filser JG, Kreuzer PE, Greim H, Bolt HM (1994) New scientific arguments for regulation of ethylene oxide residues in skin-case products. Arch Toxicol 68: 401–405

Fingerhut MA, Halperin WE, Marlow DA, Piacitelli LA, Honchar PA, Sweeney MH, Greife AL, Dill PA, Steenland K, Suruda AJ (1991) Cancer mortality in workers exposed to 2,3,7,8-tetrachlorodibenzo-p-dioxin. N Engl J Med 324: 212–218

Finnish Institute of Occupational Health (1995) Työhygieenisten mittausten rekisteri [Registry of Industrial Hygiene Measurements] (Fin.). In: European Foundation for the Improvement of Living and Working Conditions (ed) The European health and safety database (HASTE) – Summaries of descriptions of systems for monitoring health and safety at work. Office for Official Publications of the European Communities, Luxemburg. Also available at http://192.58.80.9/e/eu/haste/index.htm

Föst U, Marczynski M, Kasemann R, Peter H (1989) Determination of 7-(2-hydroxyethyl)guanine with gas chromatograph/mass spectrometry as a parameter for genotoxicity of ethylene oxide. Arch Toxicol Suppl 13: 250–253

Foster PL, Eisenstadt E, Miller JH (1983) Base substitution mutations induced by metabolically activated Aflatoxin B_1. Proc Natl Acad Sci USA 80: 2695–2698

Fox AJ, Collier PF (1977) Mortality experience of workers exposure to vinyl chloride monomer in the manufacture of polyvinyl chloride in Great Britain. Br J Ind Med 34: 1–10

Frank H, Hintze T, Remmer H (1980) Volatile hydrocarbons in breath, an indication for peroxidative degradation of lipids. In: Kolb B (ed) Applied head-space gas chromatography. Heyden, London, pp 155–164

Fraser I (2001) Butadiene – progress under the European Union existing substances regulation. Chem Biol Interact 136: 103–108

Friedman M, Dulak L, Stedham M (1995) A lifetime oncogenicity study in rats with acrylamide. Fundam Appl Toxicol 27: 95–105

Frighi V, Ng LL, Lewis A, Dhar H (1991) Na^+/H^+ antiport and buffering capacity in human polymorphonuclear and mononuclear leukocytes. Clin Sci 80: 95–99

Fuchs J, Wullenweber U, Hengstler JG, Bienfait HG, Hiltl G, Oesch F (1994) Genotoxic risk for humans due to work place exposure to ethylene oxide: remarkable individual differences in susceptibility. Arch Toxicol 68: 343–348

Furst A (1983) A new look at arsenic carcinogenesis. In: Lederer W, Fensterheim R (eds) Arsenic: industrial, biomedical, and environmental perspective. Van Nostrand Reinhold, New York, pp 151–163

Gallagher EP, Wienkers LC, Stapleton PL, Kunze KL, Eaton DL (1994) Role of human microsomal and human complementary DNA-expressed cytochrome P450 1A2 and cytochrome P450 3A4 in the bioactivation of aflatoxin B_1. Cancer Res 54: 101–108

Garner RC, Wright CM (1975) Binding of [^{14}C]aflatoxin B_1 to cellular macromolecules in the rat and hamster. Chem Biol Interact 11: 123–131

Gehring PJ, Watanabe PG, Park CN (1978) Resolution of dose-response toxicity data for chemicals requiring metabolic activation. Example: vinyl chloride. Toxicol Appl Phramacol 44: 581–591

Gehring PJ, Watanabe PG, Park CN (1979) Risk of angiosarcoma in workers exposed to vinyl chloride as predicted from studies in rats. Toxicol Appl Pharmacol 49: 15–21

Gelfand VI, Scholey JM (1992) Cell biology. Every motion has its motor. Nature 359(6395): 480–482

Gerhardsson de Verdier M, Hagman U, Peters RK, Steineck G, Overvik E (1991) Meat cooking methods and colorectal cancer: a case- referent study in Stockholm. Int J Cancer 49: 520–525

Geyer HJ, Schramm K-W, Scheunert I, Schughart K, Buters J, Wurst W, Greim H, Kluge R, Steinberg CEW, Kettrup A, Madhukar B, Olso JR, Gallo MA (1997) Considerations on genetic and environmental factors that contribute to resistance or sensitivity of mammals including humans to toxicity of 2,3,7,8-tetrachlorodibenzo-p-dioxin (TCDD) and related compounds. Ecotoxicol Environ Safety 36: 213–230

Gielen JE, Goujon FM, Nebert DW (1972) Genetic regulation of aryl hydrocarbon hydroxylase induction. II. Simple Mendelian expression in mouse tissues *in vivo*. J Biol Chem 247: 1125–1137

Gilli G, Bugliosi EH, Schiliro T (2001) Human immunological response against the molecular adduct F-HSA as biomarker of occupational and environmental exposure to formaldehyde. EPI Marker 5(3): 1–4

Giovannucci E, Rimm EB, Stampfer MJ, Colditz GA, Ascherio A, Willett WC (1994) Intake of fat, meat and fiber in relation to risk of colon cancer in men. Cancer Res 54: 2390–2397

Goldberg M, Unger M (2000) Lung cancer. Diagnostic tools. Chest Surg Clin N Am 10(4): 763–79

Goldman M (1996) Cancer risk of low-level exposure. Science 271: 1821–1822

Grafstrom RC, Fornace A, Harris CC (1984) Repair of DNA damage caused by formaldehyde in human cells. Cancer Res 44: 4323–4327

Guengerich FP (1996) Metabolic control of carcinogens. In: Hengstler JG, Oesch F (eds) Control mechanisms of carcinogens. Publishing House of the Editors, Mainz

Guengerich FP, Johnson WW, Shimada T, Ueng YF, Yamazaki H, Langouet S (1998) Activation and detoxication of Aflatoxin B$_1$. Mutat Res 402: 121–128

Gupta RS, Chopra A, Stetsko DK (1986) Cellular basis for the species differences in sensitivity to cardiac glycosides (digitalis). J Cell Physiol. 127: 197–206

Gurtoo HL, Motycka L (1976) Effect of sex differences on the *in vitro* and *in vivo* metabolism of aflatoxin B$_1$ by the rat. Cancer Res 36: 4663–4671

Halvorson MR, Noffsinger JK, Peterson CM (1993) Studies of whole blood-associated acetaldehyde levels in teetotallers. Alcohol 10: 409–413

Hang B, Medina M, Fraenkel-Conrat H, Singer B (1998) A 55 kDa protein isolated from human cells shows DNA glycosylase activity toward 3,N^4-ethenocytidine and the G/T mismatch. Proc Natl Acad Sci USA 95: 13561–13566

Hankinson O, Bacsi SG, Fukunaga BN, Kozak KR, McNulty SE, Minehart E, Probst MR, Reisz-Porszasz S, Sun W, Zhang J (1996) Role of the aryl hydrocarbon receptor in carcinogenesis. In: Hengstler JG, Oesch F (eds) Control Mechanisms of Carcinogenesis. Publishing House of the Editors, Mainz

Harris CC, Grafstrom RC, Lechner JF, Autrup H (1983) In: Banbury Report No.12, Nitrosamines and human cancer. Cold Spring Harbor, NY, p 121

Hartwig A, Schwerdtle T (2000) Interactions by carcinogenic metal compounds with DNA repair processes: toxicological implications. Toxicol Lett 127: 47–54

Haseman JK, Haily JR (1997) An update of the National Toxicology Program database on nasal carcinogens. Mutat Res 380: 3–11

Hayes JD, Judah DJ, McLellan LI, Kerr LA, Peacock SD, Neal GE (1991) Ethoxyquin-induced resistance to aflatoxin B1 in the rat is associated with the expression of a novel alpha-class glutathione *S*-transferase subunit Yc2, which possesses high catalytic activity for aflatoxin B$_1$-8,9-epoxide. Biochem J 279: 385–398

Hayes JD, Nguyen T, Judah DJ, Petersson DG, Neal GE (1994) Cloning of cDNAs from fetal rat liver encoding glutathione *S*-transferase Yc polypeptides: the Yc2 subunit is expressed in adult rat liver resistant to the hepatocarcinogen aflatoxin B$_1$. J Biol Chem 269: 20707–20717

Hays SM, Aylward LL, Karch NJ, Paustenbach DJ (1997) The relative susceptibility of animals and humans to the carcinogenic hazard posed by exposure to 2,3,7,8-tetrachlorodibenzo-*p*-dioxin: an analysis using standard and internal measures of dose. Chemosphere 34: 1507–1522

Hazelton GA, Klaassen CD (1988) UDP-glucuronosyltransferase activity toward digitoxigenin-monodigitoxide. Differences in activation and induction properties in rat and mouse liver. Drug Metab Dispos 16: 30–36

Hengstler JG, Arand M, Herrero ME, Oesch F. (1998) Polymorphisms of *N*-acetyltransferases, glutathione *S*-transferases, microsomal epoxide hydrolase and sulfotransferases: influence on cancer susceptibility. Recent results. Cancer Res 154: 47–85

Hengstler JG, Bogdanffy MS, Bolt HM, Oesch F (2003) Challenging dogma: thresholds for genotoxic chemicals? The case of vinyl acetate. Annu Rev Pharmacol Toxicol 43: 485–520

Hengstler JG, Kett A, Arand M, Oesch F, Pilch H, Tanner B (1998) Glutathione *S*-transferase *T1* and *M1* gene defects in ovarian carcinoma. Cancer Lett 130: 43–48

Hengstler JG, Lange J, Kett A, Dornhöfer N, Meinert R, Arand M, Knapstein PG, Becker R, Oesch F, Tanner B (1999b) Contribution of *c-erbB-2* and topoisomerase IIα to chemoresistance in ovarian cancer. Cancer Res 59: 3206–3214

Hengstler JG, Utesch D, Steinberg P, Ringel M, Swales N, Biefang K, Platt KL, Diener B, Böttger T, Fischer T, Oesch F (2000) Cryopreserved primary hepatocytes as an *in vitro* model for the evaluation of drug metabolism and enzyme induction. Drug Metab Rev 32: 81–118

Hengstler JG, Van der Burg B, Steinberg P, Oesch F (1999) Interspecies differences in cancer susceptibility and toxicity. Drug Metab Rev 31: 917–970

Henningsen U, Schliwa M (1997) Reversal in the direction of movement of a molecular motor. Nature 389: 93–96

Herrero ME, Arand M, Hengstler JG, Oesch F (1997) Recombinant expression of human microsomal epoxide hydrolase protects V79 Chinese hamster cells from styrene oxide- but not from ethylene oxide-induced DNA strand breaks. Environ Mol Mutagen 30: 429–439

Hirokawa N (1996) The molecular mechanism of organelle transport along microtubules: the identification and characterization of KIFs (kinesin superfamily proteins). Cell Struct Funct 21: 357–367

Homann N, Karkkainen P, Koivisto T, Nosova T, Jokelainen K, Salaspuro M (1997) Effects of acetaldehyde on cell regeneration and differentiation of the upper gastrointestinal tract mucosa. J Natl Cancer Instit 89: 1692–1697

Horn EP, Tucker MA, Lambert G, Silverman D, Zametkin D, Sinha R, Hartge T, Landi MT, Caporaso NE (1995) A study of gender-based cytochrome P450 1A2 variability: a possible mechanism for the male excess of bladder cancer. Cancer Epidemiol Biomarkers Prev 4: 529–533

Hsu IC, Metcalf RA, Sun T, Welsh JA, Wang NJ, Harris CC (1991) Mutational hotspot in the *p53* gene in human hepatocellular carcinomas. Nature 350: 427–428

Imbus HR (1988) A review of regulatory risk assessment with formaldehyde as an example. Regul Toxicol Pharmacol 8: 356–366

Ingelman-Sundberg M, Daly AK, Oscarson M, Nebert DW (2000) Human cytochrome P450 (CYP) genes: recommendations for the nomenclature of alleles. Pharmacogenetics 10(1): 91–93

Innes JRM, Ulland BM, Valerio MG (1969) Bioassay of pesticides and industrial chemicals for tumorigenicity in mice: a preliminary note. J Natl Cancer Inst USA 42: 1101–1114

International Agency for Research on Cancer [IARC] (1980) Some metals and metallic compounds. IARC Monogr Eval Carcinog Risk Chem Hum 23: 1–415

International Agency for Research on Cancer [IARC] (1985) Allyl compounds, aldehydes, epoxides and peroxides. IARC Monogr Eval Carcinog Risks Hum 36

International Agency for Research on Cancer [IARC] (1987) Overall evaluation of carcinogenicity: An updating of IARC Monographs Volumes 1 to 42 IARC Monogr Eval Carcinog Risks Hum Suppl 7

International Agency for Research on Cancer [IARC] (1997) Polychlorinated dibenzo-*para*-dioxins and polychlorinated dibenzofurans. IARC Monogr Eval Carcinog Risks Hum 69 1–666

International Agency for Research on Cancer [IARC] (1999) Hormonal Contraception and postmenopausal hormonal therapy. IARC Monogr Eval Carcinog Risks Hum 72: 1–660

International Agency for Research on Cancer [IARC] (2001) Some Thyrotropic Agents. IARC Monogr Eval Carcinog Risks Hum 79: 1–763

International Programme on Chemical Safety, World Health Organisation [IPCS-WHO] (1999): Vinyl Chloride, IPCS Environmental Health Criteria 215, International Programme on Chemical Safety, World Health Organisation, Geneva

International Technology Corporation (1992) Exposure to volatile components of polyvinyl. acetate (PVA) emulsion paints during application and drying, National. Paint and Coating Association, Vinyl Acetate Toxicology Group, Washington, DC

Isfort RJ, Cody DB, Asuith TN, Ridder GM, Stuard SB, LeBoeuf RA (1993) Induction of protein phosphorylation, protein synthesis, immediate-early-gene expression and cellular proliferation by intracellular pH modulation. Eur J Biochem213: 349–357

Jacobson-Kram D, Montalbano D (1985) The reproductive effects assessment group's report on the mutagenicity of inorganic arsenic. Environ Mutagen 7: 787–804

Japanese Bioassay Research Center (1998) Summary data provided to BGVV by the Vinyl Acetate Toxicology Group, Inc.; in correspondence dated 2 December, 1998

Jha AN, Noditi M, Nilsson R, Natarajan AT (1992) Genotoxic effects of sodium arsenite on human cells. Mutat Res 284: 215–221

Johnson K, Gorzinski S, Bodner K (1986) Chronic toxicity and oncogenicity study on acrylamide incorporated in the drinking water of Fischer 344 rats. Toxicol Appl Pharmacol 85: 154–168

Jones KG, Sweeney GD (1980) Dependence of the porphyrogenic effect of 2,3,7,8-tetrachlorodibenzo-*p*-dioxin upon inheritance of aryl hydrocarbon hydroxylase responsiveness. Toxicol Appl Pharmacol 53: 42

Jones RW, Smith DM, Thomas PG (1988) A mortality study of vinyl chloride monomer workers employed in the United Kingdom in 1940–1974. Scand J Work Environ Health 14: 153–160

K.S.Crump Group Inc. (1999a) Consideration of the potency classification of acrylamide based on the incidence of tunica vaginalis mesotheliomas (TVMs) in male Fischer 344 rats. Prepared for the Acrylamide Monomer Producers Association (AMPA) Karlstr. 21, D-69200 Frankfurt/M

K.S.Crump Group Inc. (1999b) Medchanism of acrylamide induction of benign fibroadenomas in the aging female Fischer 344 rat: Relevance to human health risk assessment. Prepared for the Acrylamide Monomer Producers Association (AMPA) Karlstr. 21, D-69200 Frankfurt/M

Kadlubar FF, Kaderlik KR, Butler MA, Lin DX, Mulder GJ, Minchin RFM, Friesen MD, Bartsch H, Nago M, Esumi H, Sugimura T, Lang NP (1995) Metabolic activation of PhIP in rats, dogs, and humans. Princess Takamatsu Symp 1995 23: 207–213

Kerckaert GA, LeBoeuf RA, Isofort RJ (1996) pH effects on the lifespan and transformation frequency of Syrian hamster embryo (SHE) cells. Carcinogenesis 17: 1819–1824

Kerns WD, Pavkov KL, Donofio DJ, Gralla EJ, Swenberg JA (1983) Carcinogenicity of formaldehyde in rats and mice after long-term inhalation exposure. Cancer Res 43: 4382–4392

Kessler W, Remmer H (1990) Generation of volatile hydrocarbons from amino acids and proteins by an iron/ascorbate/GSH system. Biochem Pharmacol 39: 1347–1351

Kewitz H (1971) Nebenwirkungen contraceptiver Steroide. Westkreuz-Verlag, Berlin

Kielhorn J, Melber C, Wahnschaffe U, Aitio A, Mangelsdorf I (2000) Vinyl chloride: still a cause for concern. Environ Health Perspect 108: 579–588

Kirkland DJ, Müller L (2000) Interpretation of the biological relevance of genotoxicity test results: the importance of thresholds. Mutat Res 464: 137–147

Kirman CR, Hays SM, Kedderis GL, Gargas ML, Strother DE (2000) Improving cancer dose-response characterization by using physiologically based pharmacokinetic modelling: an analysis of pooled data for acrylonitrile-induced brain tumours to assess cancer potency in the rat. Risk Anal 20: 135–150

Kirsch-Volders M, Aardema M, Elhajounji A (2000) Concepts of threshold in mutagenesis and carcinogenesis. Mutat Res 464: 3–11

Kitchin KT, Brown JL, Setzer RW (1994) Dose-response relationship in multistage carcinogenesis: promoters. Environ Health Perspect 102: 255–64

Kociba RJ (1991) Rodent bioassays for assessing chronic toxicity and carcinogenic potential of TCDD. In: Gallo MA, Scheuplein RJ, Van der Heijden K (eds) Biological Bbasis for risk assessment of dioxins and related compounds. Cold Spring Harbor Laboratory Press, Cold Spring Harbor, NY, pp 3–11

Kociba RJ, Keyes DG, Beyer JE, Carreon RM, Wade CE, Dittenber DA, Kalnins RP, Frauson LE, Park CN, Barnard SD, Hummel RA, Humiston CG (1978) Results of a two-year chronic toxicity and oncogenicity study of 2,3,7,8-tetrachlorodibenzo-*p*-dioxin in rats. Toxicol Appl Pharmacol 46: 279–303

Kozak KR, Abbott B, Hankinson O (1997) ARNT-deficient mice and placental differentiation. Dev Biol 191: 287–305

Kozielski F, Arnal I, Wade RH (1998) A model of the microtubule-kinesin complex based on electron cryomicroscopy and X-ray crystallography. Curr Biol 8: 191–198

Kozielski F, Sack S, Marx A, Thormahlen M, Schonbrunn E, Biou V, Thompson A, Mandelkow EM, Mandelkow E (1997) The crystal structure of dimeric kinesin and implications for microtubule-dependent motility. Cell 91: 985–994

Kreiger RA, Garry VF (1983) Formaldehyde-induced cytotoxicity and sister-chromatid exchanges in human lymphocyte cultures. Mutat Res 120: 51–55

Kunz HW, Tennekes HA, Port RE, Schwarz M, Lorke D, Schaude G (1983) Quantitative aspects of chemical carcinogenesis and tumour promotion in liver. Environ Health Perspect 50: 113–122

Kuykendall JR, Bogdanffy MS (1992a) Reaction kinetics of DNA-histone crosslinking by vinyl acetate and acetaldehyde. Carcinogenesis 13: 2095–2100

Kuykendall JR, Bogdanffy MS (1992b) Efficiency of DNA-crosslinking induced by saturated and unsaturated aldehydes *in vitro*. Mutat Res 283: 131–36

Kuykendall JR, Bogdanffy MS (1994) Formation and stability of acetaldehyde-induced crosslinks between poly-lysine and poly-deoxyguanosine. Mutat Res 311: 49–56

Kuykendall JR, Taylor ML, Bogdanffy MS (1993) Cytotoxicity and DNA-protein crosslink formation in rat nasal tissues exposed vinyl acetate are carboxylesterase-mediated. Toxicol Appl Pharmacol 123: 283–292

Laib RJ, Bolt HM (1980) Trans-membrane-alkylation. A new method for studying irreversible binding of reactive metabolites to nucleic acids. Biochem Pharmacol 29: 449–452

Laib RJ, Bolt HM (1986) Vinyl acetate, a structural analog of vinyl carbamate, fails to induce enzyme-altered foci in rat liver. Carcinogenesis 7: 841–843

Laib RJ, Pellio T, Wünschel UM, Zimmermann N, Bolt HM (1985) The rat liver foci bioassay: II. Investigations on the dose-dependent induction of ATPase-deficient foci by vinyl chloride at very low doses. Carcinogenesis 6: 69–72

Lam CW, Casanova M, Heck HDA (1985) Depletion of nasal mucosal glutathione by acrolein and enhancement of formaldehyde-induced DNA-protein cross-linking by simultaneous exposure to acrolein. Arch Toxicol 58: 67–71

Landrigan PJ (1999) Risk assessment for children and other sensitive populations. Ann NY Acad Sci 895: 1–9

Lang NP, Butler MA, Massengill J, Lawson M, Stotts RC, Hauer-Jensen M, Kadlubar FF (1994) Rapid metabolic phenotypes for acetyltransferase and cytochrome P4501A2 and putative exposure to food-borne heterocyclic amines increase the risk for colorectal cancer or polyps. Cancer Epidemiol Biomarkers Prev 3: 675–682

Langouet S, Coles B, Morel F, Becquemont L, Beaune P, Guengerich FP, Ketterer B, Guillouzo A (1995) Inhibition of CYP1A2 and CYP3A4 by oltipraz results in reduction of aflatoxin B1 metabolism in human hepatocytes in primary culture. Cancer Res 55: 5574–5579

Lawrence GS, Gobas FA (1997) A pharmacokinetic analysis of interspecies extrapolation in dioxin risk assessment. Chemosphere, 35: 427–452

LeBoeuf RA, Lin P, Kerchaert G, Gruenstein E (1992) Intracellular acidification is associated with enhanced morphological transformation in Syrian hamster embryo cells. Cancer Res 32: 144–148

Lee TC, Tanaka N, Lamb PW, Gilmer TM, Barrett JC (1988) Induction of gene amplification by arsenic. Science 241: 79–81

Lee-Chen SF, Gurr JR, Lin IB, and Jan, KY (1993) Arsenite enhances DNA double-strand breaks and cell killing of methyl methanesulfonate-treated cells by inhibiting the excision of alkali-labile sites. Mutat Res 294: 21–28

Lehninger AL, Nelson DL, Cox MM (1993) Principles of biochemistry. Worth Publishers, New York, pp 506–541

Leone G, Voso MT, Sica S, Morosetti R, Pagano L (2001) Therapy related leukemias: susceptibility, prevention and treatment. Leuk Lymphoma 41: 255–276

Leutner C, Schild H (2001) MRI of the lung parenchyma. Rofo Fortschr Geb Rontgenstr Neuen Bildgeb Verfahr 173(3): 168–175

Lewis JL, Nikula KJ, Novak R, Dahl AR (1994) Comparative localization of carboxylesterase in F344 rat, beagle dog, and human nasal mucosa. Anat Rec 239: 55–64

Lieberman M, Kunishi AT, Mapson LW, Wardale DA (1965) Ethylene production from methionine. Biochem J 97: 449–459

Lieberman M, Mapson LW (1964) Genesis and biogenesis of ethylene. Nature 204: 343–345

Lijinksy W, Reuber MD (1983) Chronic toxicity studies of vinyl acetate in Fischer rats. Toxicol Appl Pharmacol 68: 43–53

Lin DX, Lang NP, Kadlubar FF (1995) Species differences in the biotransformation of the food-borne carcinogen 2-amino-1-methyl-6-phenylimidazo[4,5-b]pyridine by hepatic microsomes and cytosols from humans, rats, and mice. Drug Metab Dispos 23: 518–524

Lotlikar PD, Jhee EC, Insetta SM, Clearfield MS (1984) Modulation of microsome-mediated aflatoxin B_1 binding to exogenous and endogenous DNA by cytosolic glutathione S-transferase in rat liver. Carcinogenesis 5: 269–276

Lucier GW, Tritscher A, Goldsworthy T, Foley J, Clark G, Goldstein J, Maronpot R (1991) Ovarian hormones enhance 2,3,7,8-tetrachlorodibenzo-p-dioxin-mediated increases in cell proliferation and preneoplastic foci in a two-stage model for rat hepatocarcinogenesis. Cancer Res 51: 1391–1397

Lutz U, Lugli S, Bitsch A, Schlatter J, Lutz WK (1997) Dose-responseDose-response for the stimulation of cell division by caffeic acid in forestomach and kidney of the male F344 rat, Fund Appl Toxicol 39: 131–237

Lutz WK (1986) Quantitative evaluation of DNA binding data for risk estimation and for classification of direct and indirect carcinogens. J Cancer Res Clin Oncol 112: 85–91

Lutz WK (1998) Dose-response relationships in chemical carcinogenesis: superposition of different mechanisms of action, resulting in linear-nonlinear curves, practical thresholds, J-shapes. Mutat Res 405: 117–124

Lutz WK (2000) A true threshold dose in chemical carcinogenesis cannot be defined for a population, irrespective of the mode of action. Hum Exp Toxicol 19: 566–68

Lutz WK (2001) Susceptibility differences in chemical carcinogenesis linearize the dose-response relationship: threshold doses can be defined only for individuals. Mutat Res 482: 71–76

Lutz WK, Jaggi W, Luthy J, Sagelsdorff P, Schlatter C (1980) *In vivo* covalent binding of aflatoxin B_1 and aflatoxin M_1 to liver DNA of rat, mouse and pig. Chem Biol Interact32: 249–250

Lutz WK, Kopp-Schneider A (1999) Threshold dose-response for tumour induction by genotoxic carcinogens modelled via cell-cycle delay. Toxicol Sci 49: 110–115

Ma TH, Harris MM (1988) Review of the genotoxicity of formaldehyde. Mutat Res 196: 37–59

Maltoni C (1977) Recent findings on the carcinogenicity of chlorinated olefins, Environ Health Perspect 21: 1–5

Maltoni C, Ciliberti A, Lefemine G, Soffritti M (1997) Results of a long-term experimental study on the carcinogenicity of vinyl acetate monomer in mice. Ann NY Acad Sci 209–238

Maltoni C, Lefemine C, Chieco P, Carretta D (1974) Vinyl chloride carcinogenesis. Current results and perspectives. Med Lav 65: 421

Maltoni C, Lefemine G, Ciliberti A, Cotti G, Carretti D (1981) Carcinogenicity bioassay of vinyl chloride monomer: a model of risk assessment on an experimental basis. Environ Health Perspect 41: 2–29

Maltoni C, Lefemine G, Ciliberti A, Cotti G, Carretti D (1984) Experimental research on vinyl chloride carcinogenesis. In: Maltoni C, Mehlman M (eds) Archives of Research on Industrial Carcinogenesis, Vol. 2. Princeton Scientific Publishers., Princeton, NJ

Manz A, Berger J, Dwyer JH, Flesch-Janys D, Nagel S, Waltsgott H (1991) Cancer mortality among workers in chemical plant contaminated with dioxin. Lancet 338: 959–64

McConnell EE, Solleveld HA, Swenberg JA, Boorman GA (1986) Guidelines for combining neoplasms for evaluation of rodent carcinogenesis studies. J Natl Cancer Inst 76: 283–289

McDonald TA, Holland NT, Skibola C, Duramad P, Smith MT (2001) Hypothesis: phenol and hydroquinone derived mainly from diet and gastrointestinal flora activity are causal factors in leukemia. Leukemia 15: 10–20

McMahon G, Davis EF, Huber LJ, Kim Y, Wogan GN (1990) Characterization of *c-Ki-ras* and *N-ras* oncogenes in aflatoxinaflatoxin B_1-induced rat liver tumours. Proc Natl Acad Sci USA 8: 1104–1108

Monticello TM, Swenberg JA, Gross EA, Leininger JR, Kimbell JS, Seilkop S, Starr TB, Gibson JE, Morgan KT (1996) Correlation of regional and nonlinear formaldehyde-induced nasal cancer with proliferating populations of cells. Cancer Res 56: 1012–1022

Moolgavkar SH, Luebeck EG (1990) Two event model for carcinogenesis: biological, mathematical and statistical considerations. Risk Anal10: 323–341

Moolgavkar SH, Luebeck EG, Krewski D, Zieliski JM (1993) Radon, cigarette smoke, and lung cancer: a reanalysis of the Colorado uranium miners' data.. Am J Epidemiol 4: 207–217

Moore MM, Harrington-Brock K, Doerr CL (1997) Relative genotoxic potency of arsenic and its methylated metabolites. Mutat Res 386(3):279-290

Morel F, Fardel O, Meyer DJ, Langouet S, Gilmore KS, Meunier B, Tu CP, Kensler TW, Ketterer B, Guillouzo A (1993) Preferential increase of glutathione *S*-transferase class alpha transcripts in cultured human hepatocytes by phenobarbital, 3-methylcholanthrene, and dithiolethiones. Cancer Res 53: 231–234

Morel F, Langouet S, Maheo K, Guillouzo A (1997) The use of primary hepatocyte cultures for the evaluation of chemoprotective agents. Cell Biol Toxicol 13: 323–329

Morgan KT (1997) A brief review of formaldehyde carcinogenesis in relation to rat nasal pathology and human risk assessment. Toxicol Pathol 25: 291–307

Morgan KT, Jiang XZ, Starr TB, Kerns WD (1986) More precise localization of nasal tumours associated with chronic exposure of F344 rats to formaldehyde gas. Toxicol Appl Pharmacol 82: 264–271

Morita T (1995) Low pH leads to sister-chromatid exchanges and chromosomal aberrations, and its clastogenicity is S-dependent. Mutat Res. 334: 301–308

Moritz R (1989) Investigation of tumor initiating and/or cocarcinogenic properties of arsenite and arsenate with the rat liver foci bioassay. Exp Pathol 37: 231–233

Nair J, Barbin A, Guichard Y, Bartsch H (1995) 1,N^6-Ethenodeoxyadenosine and 3,N^4-ethenodeoxycytidine in liver DNA from humans and untreated rodents detected by immunoafdfinitry/^{32}P-postlabelling. Carcinogenesis 16: 613–617

Nakai JS, Winhall MJ, Bunce NJ (1994) Comparative kinetic study of the binding between 2,3,7,8-tetrachlorodibenzo-*p*-dioxin and related ligands with the hepatic *Ah* receptors from several rodent species. J Biochem Toxicol 9: 199–209

Nath RG, Randerath K, Donghui L, Chung FL (1996) Endogenous production of DNA adducts. Regulat Toxicol Pharmacol 23: 22–28

Nedergaard M, Goldman SA, Desai S, Pulsinelli WA (1991) Acid-induced death in neurons and glia. J Neurosci 11: 2489–2497

Nelson LW, Weikel JH, Reno FE (1973) Mammary nodules in dogs during four years' treatment with megestrol acetate or chlormadinone acetate. J Natl Cancer Inst 51: 1303

Neumann F, Elger W (1971) Kritische Überlegungen zu den biologischen Grundlagen von Toxizitätsstudien mit Sexualhormonen. In: Plotz EJ, Haller J (eds) Methodik der Steroidtoxikologie, Thieme, Stuttgart, p 7

Neumann HG, Vamvakas S, Thielmann HW, Gelbke HP, Filser JG, Reuter U, Greim H, Kappus H, Norpoth KH, Wardenbach P, Wichmann HE (1998) Changes in the classification of carcinogenic chemicals in the work area. Int Arch Occup Environ Health 71: 566–574

Obe G, Natarajan AT, Meyers M, Den Hertog A (1979) Induction of chromosomal aberrations in peripheral lymphocytes of human blood *in vitro*, and of SCEs in bone-marrow cells of mice *in vivo* by ethanol and its metabolite acetaldehyde. Mutat Res 68: 291–294

Oesch F, Arand M, Hengstler JG (2000) Enzympolymorphismen als Ursachen erhöhter Suszeptibilität in der chemischen Karzinogenese. In: Müller WU, Heinemann G, Fehringer F (eds) Strahlenbiologie und Strahlenschutz, individuelle Strahlenempfindlichkeit und Strahlenschutz. Publikationsreihe: Fortschritte im Strahlenschutz, Band II, Bad Kissingen, pp 374–393

Oesch F, Hengstler JG, Fuchs J (1994) Cigarette smoking protects mononuclear blood cells of carcinogen exposed workers from additional work-exposure induced DNA single strand breaks. Mutat Res 321: 175–185

Oesch F, Herrero ME, Hengstler JG, Lohmann M, Arand M (2000) Metabolic detoxification: implications for thresholds. Toxicol Pathol 28: 382–87

Oganesian A, Hendricks JD, Pereira CB, Orner GA, Bailey GS, Williams DE (1999) Potency of dietary indole-3-carbinol as a promoter of Aflatoxin B_1-initiated hepatocarcinogenesis: results from a 9000 animal tumour study. Carcinogenesis 20: 453–58

Ohmori S, Horie T, Guengerich FP, Kiuchi M, Kitada M (1993) Purification and characterization of two forms of hepatic microsomal cytochrome P450 from untreated Cynomolgus monkeys. Arch Biochem Biophys 305: 405–413

Olson JR, Holscher MA, Neal RA (1980) Toxicity of 2,3,7,8-tetrachlorodibenzo-*p*-dioxin in the golden Syrian hamster. Toxicol Appl Pharmacol 55: 67–78

Olson MJ, Martin JL, Larosa AC, Brady AN, Pohl LR (1993) Immunohistochemical localization of carboxylesterase in the nasal mucosa of rats. J Histochem Cytochem 41: 307–311

Organisation for Economic Co-operation and Development [OECD] (1981) Combined chronic toxicity/carcinogenicity studies. Test Nr. 453.OECD Guidelines for Testings of Chemicals. OECD, Paris

Organisation for Economic Co-operation and Development [OECD] (1998) Repeated dose 90-day oral toxicity study in rodents. Test Nr. 408. OECD Guidelines for Testings of Chemicals. OECD, Paris

Östergaard G, Knudsen I (1998) The applicability of the ADI (Acceptable Daily Intake) for food additives to infants and children. Food Addit Contam 15, Suppl: 63–74

Ott MG, Teta MJ, Greenberg HL (1989) Lymphatic and hematopoietic tissue cancer in a chemical manufacturing environment. Am J Ind Med 16: 631–643

Park J, Kamendulis LM, Friedman MA, Klaunig JE (2002) Acrylamide-induced cellular transformation. Toxicol Sci 65: 177–183

Parry JM (2000) Reflections on the implications of thresholds of mutagenic activity for the labelling of chemicals by the European Union. Mutat Res 464: 155–158

Parry JM, Jenkins GJS, Haddad F, Bourner R, Parry EM (2000) *In vitro* and *in vivo* extrapolations of genotoxin exposures: consideration of factors which influence dose-response thresholds. Mutat Res 464: 53–63

Peers FG, Linsell CA (1977) Dietary aflatoxins and human primary liver cancer. Ann Nutr Aliment 31: 1005–1018

Pellizzari ED (1982) Analysis for organic vapor emissions near industrial and chemical waste disposal sites. Environ Sci Technol 16: 781–85

Pershagen G, Lind B, Bjorklund NE (1982) Lung retention and toxicity of some inorganic arsenic compounds. Environ Res 29: 425–434

Pershagen G, Norberg G, Bjorklund NE (1984) Carcinomas of the respiratory tract in hamsters given arsenic trioxide and/or benxo(a)pyrene by the pulmonary route. Environ Res 34: 227–241

Peter H, Wiegand HJ, Bolt HM, Greim H, Walter G, Berg M, Filser JG (1987) Pharmacokinetics of isoprene in mice and rats. Toxicol Lett 36: 9–14

Peterson RE, Theobal HM, Kimmel GL (1993) Developmental and reproductive toxicity of dioxins and related compounds: cross-species comparisons. Crit Rev Toxicol 23: 283–335

Peto R, Gray R, Brantom P, Grasso P (1991) Effects on 4080 rats of chronic ingestion of N-nitrosodiethylamine or N-nitrosodimethylamine: a detailed dose-response study. Cancer Res 51: 6415–6451

Placke ME, Griffis L, Bird M, Bus M, Persing RL, Cox LA (1996) Chronic inhalation oncogenicity study of isoprene in B6C3F1 mice. Toxicology 110: 253–262

Plowchalk DR, Andersen ME, Bogdanffy MS (1997) Physiologically based modelling of vinyl acetate uptake, metabolism, and intracellular pH changes in the rat nasal cavity. Toxicol Appl Pharmacol 142: 386–400

Pohjanvirta R, Viluksela M, Tuomisto JT, Unkila M, Karasinska J, Franc MA, Holowenko M, Giannone JV, Harper PA, Tuomisto J, Okey AB (1999) Physicochemical differences in the AH receptors of the most TCDD-susceptible and the most TCDD-resistant rat strains. Toxicol Appl Pharmacol 155: 82–95

Poirier MC, Beland FA (1992) DNA adduct measurements and tumour incidence during chronic carcinogen exposure in animal models: implications for DNA adduct-based human cancer risk assessment. Chem Res Toxicol 5: 749–755

Poirier MC, Fullerton NF, Smith BA, Beland FA (1995) DNA adduct formation and tumourigenesis in mice during the chronic administration of 4-aminobiphenyl at multiple dose levels. Carcinogenesis 16: 2917–2921

Poland A, Glover E (1980) 2,3,7,8-tetrachlorodibenzo-p-dioxin: segregation of toxicity with the Ah locus. Mol Pharmacol 17: 86

Poland A, Palen D, Glover E (1994) Analysis of the four alleles of the murine aryl hydrocarbon receptor. Mol Pharmacol 46: 915–921

Poma K, Degraeve N, Susanne C (1987) Cytogenic effects in mice after chronic exposure to arsenic followed by a single dose of ethyl methanesulfonate. Cytologia 52: 445–450

Portier C, Hoel D, van Ryzin J (1984) Statistical analysis of the carcinogenesis bioassay data relating to the risks from exposure to 2,3,7,8-tetrachlorodibenzo-p-dioxin. In: Lowrance WW (ed) Public health risks of the dioxins. W Kaufmann, Los Altos, NM, pp 99–120

Preston-Marin S, Pike MC, Ross RK, Jones PA, Henderson RE (1990) Increased cell division as a cause of human cancer. Cancer Res 50: 7415–7421

Quinn BA, Crane TL, Kocal TE, Best SJ, Cameron RG, Rushmore TH, Faber E, Haye MA (1990) Protective activity of different hepatic cytosolic glutathione S-transferases against DNA-binding metabolites of aflatoxin B_1. Toxicol Appl Pharmacol 105: 351–363

Ramsdell HS, Eaton DL (1990) Species susceptibility to aflatoxin B_1 carcinogenesis: comparative kinetics of microsomal biotransformation. Cancer Res 50: 615–620

Randerath K, Putnam KL, Osterburg HH, Johnson SA, Morgan DG, Finch CE (1993) Age-dependent increase of DNA adducts (I-compounds) in human and rat brain DNA. Mutat Res 295: 11–18

Rasmussen BB, Brosen K (1996) Determination of urinary metabolites of caffeine for the assessment of cytochrome P450 1A2, xanthine oxidase, and N-acetyltransferase activity in humans. Ther Drug Monit 18: 254–262

Ray S, Meyhöfer E, Milligan RA, Howard J (1993) Kinesin follows the microtubule's protofilament axis. J Cell Biol 121: 1083–1093

Record IR, Broadbent JL, King RA, Dreosti IE, Head RJ, Tonkin AL (1997) Genistein inhibits growth of B16 melanoma cells *in vivo* and *in vitro* and promotes differentiation *in vitro*. Int J Cancer 72: 860–864

Reed CJ, Robinson DA (1998) Histochemical localisation and biochemical quantitation of carboxylesterase activity in rat and mouse oral cavity mucosa. Report to the Vinyl Acetate Toxicology Group, School of Biomolecular Sciences, John Moores University, Liverpool

Reese JS, Allay E, Gerson SL (2001) Overexpression of human O^6-alkylguanine DNA alkyltransferase (AGT) prevents MNU induced lymphomas in heterozygous $p53$ deficient mice. Oncogene 20: 5258–5263

Reitz RH, Gargas ML, Andersen ME, Provan WM, Green TL (1996) Predicting cancer risk from vinyl chloride exposure with a physiologically based pharmacokinetic model. Toxicol Appl Pharmacol 137: 253–267

Renwick AG, Dorne JL, Walton K (2000) An analysis of the need for an additional uncertainty factor for infants and children, Regul Toxicol Pharmacol 31: 286–296

Roberts-Thomson IC, Ryan P, Khoo KK, Hart WJ, McMichael AJ, Butler RN (1996) Diet, acetylator phenotype, and risk of colorectal neoplasia. Lancet 347: 1372–1374

Ross JA (1998) Maternal diet and infant leukemia: a role for DNA topoisomerase II inhibitors? Int J Cancer Suppl 11: 26–28

Ross JA (2000) Dietary flavonoids and the MLL gene: a pathway to infant leukemia? Proc Natl Acad Sci USA 97: 4411–4413

Ross WE, McMillan DR, Ross CF (1981) Comparison of DNA damage by methylmelanines and formaldehyde. J Natl Cancer Inst 67: 217–221

Ross WE, Shipley N (1980) Relationship between DNA damage and survival in formaldehyde-treated mouse cells. Mutat Res 79: 277–283

Rotstein JB, Slaga TJ (1988) Acetic acid, a potent agent of tumour progression in the multistage mouse skin model for chemical carcinogenesis. Cancer Lett 42: 87–90

Rozman K (1989) A critical review on the mechanisms of toxicity of 2,3,7,8-tetrachlorodibenzo-*p*-dioxin. Implications for human safety assessment. Derm Beruf Umwelt 38: 95–95

Sadrieh N, Snyderwine EG (1995) Cytochromes P450 in cynomolgus monkeys mutagenically activate 2-amino-3-methylimidazo[4,5-*f*]quinoline (IQ) but not 2-amino-3,8-dimethyl-imidazo[4,5-*f*]quinoxaline (MeIQx). Carcinogenesis 16: 1549–1555

Sagai M, Ichinose T (1980) Age-related changes in lipid peroxidation as measured by ethane, ethylene, butane and pentane in respired gases in rats. Life Sci 27: 731–738

Sarangapani R, Teeguarden J, Clewell HJ, Centry R, Covington T, Andersen ME (2000) Oral hazard identification and dose-response characterization for vinyl acetate. Report to the Vinyl Acetate Toxicology Group, KS Crump Group, ICF Consulting, Research Triangle Park, NC

Schieferstein GJ, Littlefield NA, Gaylor DW, Sheldon WG, Burger GT (1985) Carcinogenesis of 4-aminobiphenyl in BALB/cStCrlfC3Hf/Nctr mice. Eur J Cancer Clin Oncol 21: 865–873

Schiffmann MH, Felton JS (1990) Fried foods and the risk of colon cancer. Am J Epidemiol 131: 376–378

Schoeny R (1996) Use of genetic toxicology data in US EPA risk assessment: the mercury study report as an example. Environ Health Perspect 104, Suppl. 3: 663–673

Schrenk D, Stuven T, Goh G, Viebahn R, Bock KW (1995) Induction of CYP1A and glutathione *S*-transferase activities by 2,3,7,8-tetrachlorodibenzo-*p*-dioxin in human hepatocyte cultures. Carcinogenesis 16: 943–946

Schroder-van der Elst JP, van der Heide D, Rokos H, Morreale de Escobar G, Kohrle J (1998) Synthetic flavonoids cross the placenta in the rat and are found in fetal brain. Am J Physiol 274: 253–256

Schulte-Hermann R, Bursch W, Kraupp-Grasl B, Oberhammer F, Wagner A (1992) Programmed cell death and its protective role with particular reference to apoptosis. Toxicol Lett 64–65: 569–574

Schumann AM, Watanabe PG, Reitz RH, Gehring PJ (1982) The importance of pharmacokinetic and macromolecular events as they relate to mechanisms of tumourigenicity and risk assessment. In: Plaa G, Hewitt WR (eds) Toxicology of the Liver. Target Organ Toxicology Series, Raven Press, New York, pp 311–331

Schwenk M, Gundert-Remy U, Heinemeyer G, Olejniczak K, Stahlmann R, Kaufmann W, Bolt HM, Greim H, von Keutz E, Gelbke HP (2003) Children as a sensitive subgroup and their role in regulatory toxicology. Arch Toxicol 77: 2–6

Schwetz B, Norris J, Sparschu G, Rowe V, Gehring P, Emerson J, Gerbig C (1973) Toxicology of chlorinated dibenzo-*p*-dioxins. Environ Health Perspect 5: 87–99

Scott WJ, Beliles RP, Silverman HI (1971) The comparative acute toxicity of two cardiac glycosides in adult and newborn rats. Toxicol Appl Pharmacol 20: 599–601

Seeley MR, Tonner-Navarro, LE, Beck, BD, Deskin R, Feron VJ, Johanson G, Bolt HM (2001) Procedures of health risk assessment in Europe. Regul Toxicol Pharmacol 34: 153–169

Sellakumar AR, Snyder CA, Solomon JJ, Albert RE (1985) Carcinogenicity of formaldehyde and hydrogen chloride in rats. Toxicol Appl Pharmacol 81: 401–106

Sesardic D, Boobis AR, Edwards RJ, Davies DS (1989) A form of cytochrome P450 in man, orthologous to form d in the rat catalyzes the *O*-deethylation of phenacetin and is inducible by cigarette smoking. Br J Clin Pharmacol 26: 363–372

Sesardic D, Cole KJ, Edwards RJ, Davies DS, Thomas PE, Levin W, Boobis AR (1990) The inducibility and catalytic activity of cytochromes P450c (P4501A1) and P450d (P4501A2) in rat tissues. Biochem Pharmacol 39: 499–506

Shank RC, Bhamarapravati N, Gordon JE, Wogan GN (1972) Dietary aflatoxins and human liver cancer. IV Incidence of primary liver cancer in two municipal populations in Thailand. Food Cosmet Toxicol 10: 171.179

Shelanski ML, Gaskin F, Cantor CR (1973) Microtubule assembly in the absence of added nucleotides. Proc Natl Acad Sci USA 70: 765–768

Shen J, Kessler W, Denk B, Filser JG (1989) Metabolism and endogenous production of ethylene in rat and man. Arch Toxicol Suppl 13: 237–239

Sickles DW, Brady ST, Testino A, Friedman MA, Wrenn RW (1996) Direct effect of the neurotoxicant acrylamide on kinesin-based microtubule motility. J Neurosci Res 46: 7–17

Simon P, Filser JG, Bolt HM (1985) Metabolism and pharmacokinetics of vinyl acetate. Arch Toxicol 57: 191–95

Simon P, Ottenwälder H, Bolt HM (1985) Vinyl acetate: DNA-binding assay *in vivo*. Toxicol Lett 27: 115–120

Simonato L, L'Abbe KA, Andersen A, Belli S, Comba P, Engholm G, Ferro G, Hagmar L, Langard S, Lundberg I (1991) A collaborative study of cancer incidence and mortality among vinyl chloride workers. Scand J Work Environ Health 17: 159–169

Sinha R, Rothman N, Brown ED, Mark SD, Hoover RN, Caporaso NE, Levander OA, Knize MG, Lang NP, Kadlubar FF (1994) Pan-fried meat containing high levels of heterocyclic aromatic amines but low levels of polycyclic aromatic hydrocarbons induces cytochrome P450 1A2 activity in humans. Cancer Res 54: 6154–6159

Sipi P, Joarventaus H, Norppa H (1992) Sister-chromatid exchanges induced by vinyl esters and respective carboxylic acids in cultured human lymphocytes. Mutat Res 279: 75–83

Slone DH, Gallagher EP, Ramsdell HS, Rettie AE, Stapleton PL, Berlad LG, Eaton DL (1995) Human variability in hepatic glutathione *S*-transferase-mediated conjugation of aflatoxin B_1-epoxide and other chemicals. Pharmacogenetics 5: 224–233

Snyderwine EG, Turesky RJ, Turteltaub KW, Davis CD, Sadrieh N, Schut HA, Nagao M, Sugimura T, Thorgeirsson UP, Adamson RH, Thorgeirsson SS (1997) Metabolism of food-derived heterocyclic amines in nonhuman primates. Mutat Res 376: 203–210

Solleveld HA, Haseman JK, McConnell EE (1984) Natural history of body weight gain, survival and neoplasia in the F344 rat. J Natl Cancer Inst 72: 929–40

Solomon MJ, Varshavsky A (1985) Formaldehyde-mediated DNA-protein crosslinking: A probe for *in vivo* chromatin structures. Proc Natl Acad Sci USA 82: 6470–6474

Soman NR, Wogan GN (1993) Activation of the *c-Ki-ras* oncogene in Aflatoxin B_1-induced hepatocellular carcinoma and adenoma in the rat: detection by denaturing gradient gel electrophoresis. Proc Natl Acad Sci USA 90: 2045–2049

Sosa H, Hoenger A, Milligan RA (1997) Three different approaches for calculating the three-dimensional structure of microtubules decorated with kinesin motor domains. J Struct Biol 118: 149–158

Squire RA, Cameron LL (1984) An analysis of potential carcinogenic risk from formaldehyde. Regul Toxicol Pharmacol 4: 107–129

Staffa JA, Mehlman MA (1979) Innovations in cancer risk assessment (ED01 study). Environ Pathol Toxicol 3: 1–246

Staretz ME, Foiles PG, Miglietta LM, Hecht SS (1997) Evidence for an important role of DNA pyridyloxobutylation in rat lung carcinogenesis by 4-(methylnitrosamino)-1-(3-pyridyl)-1-butanone: effects of dose and phenethyl isothiocyanate. Cancer Res 57: 259–266

Starr TB, Buck RD (1984) The importance of delivered dose in estimating low-dose cancer risk from inhalation exposure to formaldehyde. Fund Appl Toxicol 4: 740–753

Steenland K, Deddens J, Piacitelli L (2001) Risk assessment of 2,3,7,8-tetrachlorodibenzo-*p*-dioxin (TCDD) based on an epidemiologic study. Am J Epidemiol 154: 451–458

Steinberg P, Jennings GS, Schlemper B, Oesch F (1996) Molecular mechanisms underlying the liver cell-type specific toxicity of aflatoxin B_1. In: Hengstler JG, Oesch F (eds) Control Mechanisms of Carcinogenesis, Druckerei Thieme, Mainz

Steinmetz KL, Green CE, Bakke JP, Spak DK, Mirsalis JC (1988) Induction of unscheduled DNA synthesis in primary cultures of rat, mouse, hamster, and human hepatocytes. Mutat Res 206: 91–102

Stone R (1995) A molecular approach to cancer risk. Science 268: 356–357

Storm JE, Rozman KK (1997) Evaluation of alternative methods for establishing safe levels of occupational exposure to vinyl halides. Regul Toxicol Pharmacol 25: 240–255

Stott WT, McKenna MJ (1985) Hydrolysis of several glycol ether acetates and acrylate esters by nasal mucosal carboxylesterase in vitro. Fundam Appl Toxicol5: 399–404

Strick R, Strissel PL, Borgers S, Smith SL, Rowley JD (2000) Dietary bioflavonoids induce cleavage in the MLL gene and may contribute to infant leukemia. Proc Natl Acad Sci USA 97: 4790–4795

Swenberg JA, Barrow CS, Boreiko CD, Heck HDA, Levine RJ, Morgan KT, Starr TB (1983b) Non-linear biological responses to formaldehyde and their implications for carcinogenic risk assessment. Carcinogenesis 4: 945–952

Swenberg JA, Gross EA, Randall HW, Barrow CS (1983a) In: Clary JJ, Gibson JE, Waritz RS, (eds) Formaldehyde: toxicology, epidemiology, mechanisms. Marcel Dekker, New York, pp 225–236

Swenberg JA, Ham A, Koc H, Morinello E, Ranasinghe A, Tretyakova N, Upton PB, Wu KY (2000) DNA adducts: effects of low exposure to ethylene oxide, vinyl chloride and butadiene. Mutat Res 464: 77–86

Swenberg JA, Kerns WD, Mitchell RE, Gralla EJ, Pavkov KL (1980) Induction of squameous cell carcinomas of the rat nasal cavity by inhalation exposure to formaldehyde vapor. Cancer Res 40: 3398–3402

Swenson DH, Lin JK, Miller EC, Miller JA (1977) Aflatoxin B₁-2,3-oxide as a probable intermediate in the covalent binding of aflatoxins B₁ and B₂ to rat liver DNA and ribosomal RNA in vivo. Cancer Res 37: 172–81

Tareke E, Rydberg P, Karlsson P, Eriksson S, Törnqvist M (2000) Acrylamide: a cooking carcinogen? Cherm Res Toxicol 13: 517–522

Teeguarden JG, Dragan YP, Singh J, Vaughan J, Xu YH, Goldsworthy T, Pitot HC (1999) Quantitative analysis of dose- and time-dependent promotion of four phenotypes of altered hepatic foci by 2,3,7,8-tetrachlorodibenzo-p-dioxin in female Sprague-Dawley rats. Toxicol Sci 51: 211–23

Thier R, Bolt HM (2000) Carcinogenicity and genotoxicity of ethylene oxide: new aspects and recent advances. Crit Rev Toxicol 30: 595–608

Thier R, Bonacker D, Stoiber T, Böhm KJ, Wang M, Unger E, Bolt HM, Degen G (2003a) Interaction of metal salts with cytoskeletal motorprotein systems. Toxicol Lett 140/141: 75–81

Thier R, Brüning T, Roos PH, Rihs HP, Golka K, Ko Y, Bolt HM (2003b) Markers of genetic susceptibility in human environmental hygiene and toxicology: the role of selected CYP, NAT and GSTgenes. Int J Hyg Environ Health 206: 149–171

Tobey NA, Reddy SP, Keku TO, Cragoe EJ, Orlando RC 1992. Studies of pHᵢ in rabbit esophageal basal and squamous epithelial cells in culture. Gastroenterology 103: 830–839

Toonen TR, Hande KR (2001) Topoisomerase II inhibitors. Cancer Chemother Biol Response Modif 19: 129–147

Törnqvist M, Almberg JG, Bergmark EN, Nilsson S, Osterman-Golkar S (1989a) Ethylene oxide doses in ethene-exposed fruit store workers. Scand J Work Environ Health 15: 436–438

Törnqvist M, Gustafsson B, Kautiainen A, Harms-Ringdahl M, Granath F, Ehrenberg L (1989b) Unsaturated lipids and intestinal bacteria as sources of endogenous production of ethene and ethylene oxide. Carcinogenesis 10: 39–41

Tseng WP (1977) Effects and dose-response relationships of skin cancer and Blackfoot disease with arsenic. Environ Health Perspect 19: 109–119

Tseng WP, Chu HM, How SW, Fong JM, Lin CS, Yen S (1968) Prevalence of skin cancer in an endemic area of chronic arsenicism in Taiwan. J Natl Cancer Inst USA 40: 453–463

Tuomisto JT, Viluksela M, Pohjanvirta R, Tuomisto J (1999) The AH receptor and a novel gene determine acute toxic responses to TCDD: segregation of the resistant alleles to different rat lines (1999) Toxicol Appl Pharmacol 155: 71–81

Turesky RJ, Aeschbacher HU, Würzner HP, Skipper PL, Tannenbaum SR (1988) Major routes of metabolism of the food-borne carcinogen 2-amino-3,8-dimethylimidazo[4,5-f]quinoxaline in the rat. Carcinogenesis: 9: 1043–1048

Ueno I, Friedman L, Stone CL (1980) Species difference in the binding of aflatoxin B₁ to hepatic macromolecules. Toxicol Appl Pharmacol 52: 177–180

Unkila M, Pohjanvirta R, MacDonald E, Tuomistro J (1994) Characterization of 2,3,7,8-tetra-chlorodibenzo-*p*-dioxin (TCDD)-induced brain serotonin metabolism in the rat. Eur J Pharmacol 270: 157–166

Unkila M, Ruotsalainen M, Pohjanvirta R, Viluksela M, MacDonald E, Tuomistro J, Rozman K, Tuomisto J (1995) Effect of 2,3,7,8-tetrachlorodibenzo-*p*-dioxin (TCDD) on tryptophan and glucose homeostasis in most TCDD-susceptible and the most TCDD-resistent species, guinea pigs and hamsters. Arch Toxicol 69: 677–683

US Environmental Protection Agency [EPA] (1988) Special report on ingested inorganic arsenic: skin cancer; nutritional essentiality. EPA Report, EPA/625/3-87/013. Risk Assessment Forum, Washington, DC

US Environmental Protection Agency [EPA] (1996) Proposed guidelines for carcinogen risk assessment. US Fed. Register 61, No.79, April 23, 1996. US Environmental Protection Agency, Washington, DC, available at http: //www.epa.gov./ORD/WebPubs/carcinogen/

US Environmental Protection Agency [EPA] (2000) Vinyl chloride, IRIS file online, US Environmental Protection Agency, Washington, DC; available at http: //www.epa.gov/iris/subst/l001.htm/

US Environmental Protection Agency [EPA] (2003) Arsenic, inorganic (CASRN 7440-38-2) In: U.S. Environmental Protection Agency Integrated Risk Information System (IRIS) http: //www.epa.gov/iris/subst/0278.htm

Van Sittert NJ, Boogaard PJ, Natarajan AT, Tates AD, Ehrenberg LG, Törnqvist MA (2000) Formation of DNA adducts and induction of mutagenic effects in rats following 4 weeks inhalation exposure to ethylene oxide as a basis for cancer risk assessment. Mutat Res 447: 27–48

Vecchi A, Sironi M, Antonia M, Recchia CM, Garattini S (1983) Immunosuppressive effects of 2,3,7,8-tetrachlorodibenzo-*p*-dioxin in strains of mice with different susceptibility. Proc Natl Acad Sci USA 87: 6917

Vielhauer V, Sarafoff M, Gais P, Rabes HM (2001) Cell type-specific induction of O^6-alkylguanine-DNA alkyltransferase mRNA expression in rat liver during regeneration, inflammation and preneoplasia. J Cancer Res Clin Oncol 127: 591–602

Von Mach MA (2002) Primary biliary cirrhosis in classmates: coincidence or enigmatic environmental influence? EXCLI Journal 1: 1–7

Von Mach MA, Lauterbach M, Kaes J, Hengstler JG, Weilemann JG (2003) Suizidale und para-suizidale Intoxikationen mit Paracetamol: eine Analyse von 1995 bis 2002. Dtsch Med Wochenschr 128: 15–19

Von Mach MA, Weilemann LS (2002) Zunehmende Bedeutung von Antidepressiva bei suizidalen und parasuizidalen Intoxikationen. Dtsch Med Wochenschr 127: 2053–2056

Walker JV, Nitiss JL (2002) DNA topoisomerase II as a target for cancer chemotherapy. Cancer Invest 20: 570–589

Wallin H, Milkalsen A, Guengerich FP, Ingelman-Sundberg M, Solberg KE, Rossland OJ, Alexander J (1990) Differential rates of metabolic activation and detoxication of the food mutagen 2-amino-1-methyl-6-phenylimidazo[4,5-*b*]pyridine by different cytochrome P450 enzymes. Carcinogenesis 11: 489–492

Walter CA, Zhou ZQ, Manguino D, Ikeno Y, Reddick R, Nelson J, Intano G, Herbert DC, McMahan CA, Hanes M (2001) Health span and life span in transgenic mice with modulated DNA repair. Ann NY Acad Sci 928: 132–140

Wang JC (2002) Cellular roles of DNA topoisomerases: a molecular perspective. Nat Rev Mol Cell Biol 3: 430–40

Ward E, Boffetta P, Andersen A, Colin D, Comba P, Deddens JA, De Santis M, Engholm G, Hagmar L, Langard S, Lundberg I, McElvenny D, Pirastu R, Sali D, Simonato L (2001) Update of the follow-up of mortality and cancer incidence among European workers employed in the vinyl chloride industry. Epidemiology 12: 710–718

Waxweiler RJ, Smith AH, Falk H, Tyroler HA (1981) Excess lung cancer risk in a synthetic chemicals plant. Environ Health Perspect 41: 159–165

Weber LDW, Lebofsky M, Stahl BU, Gorski JR, Muzi G, Rozman K (1991) Reduced activities of key enzymes of gluconeogenesis as a possible cause of acute toxicity of 2,3,7,8-tetra-chlorodibenzo-*p*-dioxin (TCDD) in rats. Toxicology 66: 133

Weingarten MD, Lockwood AH, Hwo SY, Kirschner MW (1975) A protein factor essential for microtubule assembly. Proc Natl Acad Sci USA 72: 1858–1862

Wild CP, Hasegawa R, Barraud L, Chutimataewin S, Chapot B, Nobuyuki I, Montesano R (1996) Aflatoxin-albumin adducts: a basis for comparative carcinogenesis between animals and humans. Cancer Epidemiol Biomarkers Prev 5: 179–189

Williams GM, Gebhardt R, Sirma H, Stenback F (1993) Non-linearity of neoplastic conversion induced in rat liver by low exposures to diethylnitrosamine. Carcinogenesis 14: 2149–2156

Williams GM, Iatropoulos MJ, Jeffrey AM, Luo FQ, Wang CX, Pittman B (1999) Diethylnitrosamine exposure-responses for DNA ethylation, hepatocellular proliferation, and initiation of carcinogenesis in rat liver display non-linearities and thresholds. Arch Toxicol 73: 394–402

Williams GM, Iatropoulos MJ, Jeffrey AM (2000) Mechanistic basis for nonlinearities and thresholds in rat liver carcinogenesis by the DNA-reactive carcinogens 2-acetylaminofluorene and diethylnitrosamine. Toxicol Pathol 28: 388–395

Williams GM, Iatropoulos MJ, Wang CX, Ali N, Rivenson A, Peterson LA, Schultz C, Gebhardt R (1996) Diethylnitrosamine exposure-responses for DNA damage, centrilobular cytotoxicity, cell proliferation and carcinogenesis in rat liver exhibit some non-linearities. Carcinogenesis 17: 2253–2258

Wilson AS, Williams DP, Davis CD, Tingle MD, Park BK (1997) Bioactivation and inactivation of aflatoxin B_1 by human, mouse and rat liver preparations: effect on SCE in human mononuclear leucocytes. Mutat Res 373: 257–264

Wilson JD (1997) So carcinogens have thresholds: how do we decide what exposure levels should be considered safe? Risk Anal 17: 1–3

Wiltse JA, Dellarco VL (2000) U.S. Environmental Protection Agency's revised giudelines for carcinogen risk assessment: evaluating a postulated mode of carcinogenic action in guiding dose-response extrapolation. Mutat Res 464: 105–115

Wogan GN, Paglialunga S, Newberne PM (1974) Carcinogenic effects of low dietary levels of Aflatoxin B_1 in rats. Food Cosmet Toxicol. 12: 681–685

Wordeman L, Mitchison TJ (1995) Identification and partial characterization of mitotic centromere-associated kinesin, a kinesin-related protein that associates with centromeres during mitosis. J Cell Biology 128: 95–105

World Health Organisation [WHO] (1987) Air quality guidelines for Europe. WHO Regional Publications, European Series No. 23, World Health Organisation (WHO) Regional Office, Copenhagen

World Health Organisation [WHO] (1999) Guidelines for air quality. World Health Organisation (WHO), Geneva

World Health Organisation [WHO] (2000) Air quality guidelines for Europe, 2nd edition. WHO Regional Publications, European Series, No. 91, World Health Organisation (WHO) Regional Office, Copenhagen

Wroblewski VJ, Olson JR (1985) Hepatic metabolism of 2,3,7,8-tetrachlorodibenzo-p-dioxin (TCDD) in the rat and guinea pig. Toxicol Appl Pharmacol 81: 231–240

Wu MM, Kuo TL, Hwang YH, Chen CJ (1989) Dose-response relation between arsenic concentration in well water and mortality from cancers and vascular diseases. Am J Epidemiol 130: 1123–1132

Yager JD, Liehr JG (1996) Molecular mechanisms of oestrogen carcinogenesis. Ann Rev Pharmacol Toxicol 36: 203–232

Yamahara H, Lee VHL 1993. Drug metabolism in the oral cavity. Adv Drug Deliv Rev 12: 25–39

Yamazoe Y, Abu-Zeid M, Manabe S, Toyama S, Kato R (1988) Metabolic activation of a protein pyrolysate promutagen 2-amino-3,8-dimethylimidazo[4,5-f]quinoxaline by rat liver microsomes and purified cytochrome P450. Carcinogenesis 9: 105–109

Zhou ZQ, Manguino D, Kewitt K, Intano GW, McMahan CA, Herbert DC, Hanes M, Reddick R, Ikeno Y, Walter CA (2001) Spontaneous hepatocellular carcinoma is reduced in transgenic mice overexpressing human O^6-methylguanine-DNA methyltransferase. Proc Natl Acad Sci USA 98: 12566–12571

Zober A, Messerer P, Huber P (1990) Thirty-four-year mortality follow-up of BASF employees exposed to 2,3,7,8-TCDD after the 1953 accident. Int Arch Occup Environ Health 62: 139–157

5 Epidemiological Perspectives on Low-Dose Exposure to Human Carcinogens

5.1
Introduction

5.1.1
Epidemiological considerations

The major strength of epidemiological studies is that human beings are under study, risk is thus not extrapolated from animal data or derived from molecular studies. It should be underlined that epidemiological studies are concerned with events which occur in populations. The primary unit is thus groups of people and not an individual. This is how epidemiology differs from clinical medicine. The clinician is interested in what is wrong with a patient and how to cure the patient, while the epidemiologists ask: "What caused the disease and how could we prevent it?". Thinking in epidemiological terms often seems strange to clinicians who are used to think of the problem of each single patient.

The epidemiologists focus on the distribution of disorders in a population, which types of people are at increased risk and which are not (at least as important). Has the disease frequency changed over time, sex, age, occupation, socio-economic class or area, these are typical questions of concern. The aim is to evaluate whether associations between a disease and an effect are causal or spurious. Laboratory scientists identify carcinogenic compounds, e. g. tobacco, that causes lung cancer in animals. However, the argument that cigarette smoke causes lung cancer in humans would remain unconvincing if it was not shown that the disease is more common in smokers than in non-smokers. Epidemiology is the most important source of direct scientific evidence about exposure and health effects in the human population.

Epidemiologists are often criticised that proof is impossible in epidemiological studies, with the implication that it is possible in other studies. In contrast to mathematics, which is axiomatic and seeks reasons based on logic, science is observational and seeks reasons based on experience. In its broadest meaning, science consists of making observations, seeking reasons for the observations, and offering explanations, that is, formulate a hypothesis where a cause is supposed to explain an effect.

A "sufficient cause" is defined as the minimal set of conditions that cause a disease. The minimal prerequisite implies that all of the conditions are necessary. A sufficient cause normally consists of a number of complementary causes. For biological effects most, if not all, of the component causes of a sufficient cause are unknown. Let us illustrate how the concept of sufficient, complementary, and necessary causes work on a common exposure such as tobacco. Smoking is not a sufficient cause of cancer since not everyone who smokes develops cancer. Approximately 90 % of all smokers never develop a lung cancer. Moreover, it is not a necessary cause because not everyone who develops cancer has a history of smoking. But the scientific evidence shows that cancer occurs more often among those who smoke. It follows, therefore, that among smokers who developed cancer, smoking was sufficient in the circumstances to cause cancer in some instances. That is, smoking was one of the complementary causes of lung cancer. Other complementary causes could be a genetic predisposition making the individual sensitive to the carcinogens in tobacco. Large efforts are today devoted to identifying these "susceptible" cases, the area of genetic epidemiology is growing fast.

There is a misbelieve that definitive scientific knowledge only stems from experimental studies. The view of some experimental scientists is that epidemiology gives nothing more than an association and that detailed laboratory studies give the final cause-effect relationship. Laboratory studies normally involve a degree of observer control that is not possible in epidemiological studies and this control, not the observation *per se*, can strengthen the inference from laboratory studies, but the control is not a guarantee against error. If only laboratory studies give definite knowledge we should not believe in the evolution of species or that ionising radiation causes cancer in humans. In addition, experiments do not give us full proof since they could be controversial, contradictory and irreproducible. David Hume's, and other philosophers, belief that proof is impossible in any empirical science is especially important to epidemiologists (for more details see chapter 8). However, whatever conclusion about proof is made there are different degrees of scientific criteria for evidence.

When interpreting the results it has to be kept in mind, however, that there are several weaknesses in epidemiological studies, especially studying low doses and their effects. Some of the more obvious limitations to epidemiology and biostatistics are discussed below. Epidemiological studies are observational rather than experimental in their design. This is particularly true for what is called etiological epidemiological studies. It would, for instance, probably be very hard convincing any ethical committee of a study where individuals are exposed to different levels of a compound if the compound is considered harmful, e. g. ionising radiation or smoking. On the other hand, people often, willingly or not, expose themselves to harmful factors and we thus have to rely on populations that have been exposed due to reasons beyond our control.

Therefore the epidemiological methods used have one main goal and that is to create an "experimental-like" situation, e. g. making exposed and non-exposed groups as comparable as possible. In order to do this, systematic and random errors have to be brought to a minimum (as done in any scientific study). The systematic error of the study is dealt with through study design and the random error or precision through statistical methods reflected by the width of the confidence intervals for each risk estimate. Any epidemiological result must be interpreted in the light of previous results, supporting experimental and animal evidence, a plausible and a possible dose-response relationship.

A specific feature of etiological epidemiological studies is that there is no possibility to tell if a specific carcinogen caused a cancer. Although molecular geneticists are trying to identify alterations that imply that a specific etiological agent caused a malignancy, however, no reliable tool is available today. We thus still have to depend on statistical differences between exposed and non-exposed populations. Models extrapolating risks for intermediate and high doses to low-dose situations are necessary because of the inability of epidemiological studies to evaluate small effects of the exposure variables. During the last 30–40 years much has been learnt about the genetic and hereditary features of cancer through mechanistic modelling of observational data. (This approach is described in detail in chapter 6).

Various types of biases could distort an epidemiological study. Three important problems, central to any epidemiological study – confounding, selection bias, and information bias will be addressed:

Confounding means that some risk factor, other than the one under study, is differently distributed among exposed and non-exposed. The effect of not controlling for confounding is that disease occurrence will differ independently of the effect under study. As an example, if an exposed group contains more men than women, this will lead to a difference in incidence of myocardial infarction (myocardial infarction being more common among men in the western part of the world). If the exposure under study increases the risk of myocardial infarction then the difference in gender strengthens the effect and the result is a combination of the exposure under study and the effect of being a man. This problem could be controlled for by stratification, i. e. the effect of exposure being studied separately for men and women. A most serious problem occurs if the end point under study, e. g. cause of death, varies with dose.

A selection bias occurs if the underlying cause of exposure influences the effect under study in populations medically exposed to ionising radiation. In a study of approximately 36,000 patients receiving ^{131}I as a diagnostic procedure the overall risk (standardised incidence ratio, SIR) of thyroid cancer was found to be 3.1 when the number of cancers was compared to what could be expected from the country as a whole (Dickman et al. 2003). However, when the patients were divided based on reason for referral and previous treatment the figures changed dramatically (Tab. 5.1).

Reason for referral could be suspicion of thyroid cancer or hyperthyroidism. For 1,792 patients who received radiotherapy towards the neck region for a benign disorder SIR was 14.20 (95 % confidence interval, CI, 8.64–19.77). But for 23,795 patients referred for other reasons than suspicion of a thyroid tumour a non-significant increased risk of 1.40 was seen (Tab. 5.1). This example shows the difficulty in using patients for risk estimations since previous therapy, underlying disorder and/or reason for referral has to be taken into consideration. There are, however, certain positive aspects of patient cohorts, the dosimetry and follow-up are most often better than in, for example, occupational cohorts.

Another major problem is individuals lost to follow up in a study. If not identified, these persons will continue to contribute person years at risk (e. g. still included in the study despite that they have died or migrated) without being at risk of developing a disease. Even the information on who has migrated can influence the results if the reason for migration is linked to exposure or effect.

One example is the Techa River cohort consisting of approximately 28,000 individuals exposed to radioactive discharge from the Mayak nuclear facility in the Southern Urals, Russia (Kossenko et al. 1997). Individuals in the Techa River cohort lived in the riverside villages during the period of highest releases, 1950–56. Before the fall of the Soviet Union there was little migration from the area that originally defined the study cohort. After 1992 migration has increased dramatically and if this was due to lack of health care in the areas along the river it could be a selection of individuals with life threatening disorders, such as cancer, that are leaving to seek medical attention. Another problem emerges if the follow-up is related to exposure level, e. g. patients supposed to be exposed are screened for the disease under study.

Tab. 5.1 Thyroid cancer risks for patients receiving diagnostic amounts of ^{131}I. Standardised incidence ratio (SIR) and 95 % Confidence interval (CI) (Hall, unpublished data)

No. of patients	Observed no. of patients	SIR	95 % CI
All patients			
36,443	124	3.11	2.56–3.65
Referred under the suspicion of a thyroid tumour			
10,856	64	4.89	3.69–6.08
Previous external radiotherapy			
1,792	25	14.20	8.64–19.77
Referred for other reasons			
23,795	35	1.40	0.94–1.86

Information bias is a problem when interpreting the recent findings of a sharply increased risk of thyroid cancer among children in the Chernobyl area. Screening programmes have increased the ascertainment of occult thyroid tumours through the use of ultrasound examination, a possibility discussed in one of the original reports (Baverstock et al. 1992). Thyroid screening was locally organised in the most contaminated areas after the accident, and large-scale screening with ultrasound examination, supported by the Sasakawa and IPHECA programmes did not start until 1991 and 1992 (Souchkevitch 1996; Yamashita and Shibata 1997). As can be seen in Tab. 5.2, an increased risk of thyroid cancer was noted in the Chernobyl area in 1990 (UNSCEAR 2000).

Tab. 5.2 Thyroid cancer incidence rates in children under 15 years of age at diagnosis in Belarus, Bryansk region in Russian Federation, and Ukraine (UNSCEAR 2000)

Country/ Region	No. of cases per 100,000 children												
	1986	-87	-88	-89	-90	-91	-92	-93	-94	-95	-96	-97	-98
Belarus	0.2	0.3	0.4	0.3	1.9	3.9	3.9	5.5	5.1	5.6	4.8	5.6	3.9
Bryansk	0.0	0.3	0.0	0.3	0.3	0.3	0.9	0.3	2.8	2.5	0.6	2.2	–
Ukraine	0.2	0.1	0.1	0.1	0.2	0.2	0.5	0.4	0.4	0.5	0.6	0.4	0.5

Approximately 1,800 thyroid cancer cases have been reported in Belarus, the Russian Federation and Ukraine in children and adolescents for the period 1990–1998. The influence of screening is difficult to estimate but taking the advanced stage of the tumours into consideration, it is likely that most of the tumours would have been detected sooner or later. Studies comparing thyroid cancer incidence with radioactive contamination have shown significantly increased risks in highly contaminated areas, even after screening was taken into consideration (Jacob et al. 1998). It is thus most likely that the vast majority of the diagnosed thyroid cancers are related to the Chernobyl accident.

5.1.2
Epidemiological methods

Incidence provides the clearest measure of disease burden and effect of a carcinogenic compound at the population level. Incidence rates give the effect of genetic and environmental causal and preventive factors. Different weights of these factors, additively synergism or antagonism between them are integrated in the measure. In contrast to mortality, incidence allows, in theory, a better comparison between population, areas and time periods, since mortality is also reflecting prognosis and, hence, treatment effect. In addition, cancer survival also depends on their stage at diagnosis, a function of public and professional awareness, as well as access to health care. On the other hand, incidence data are available only for a minority of the world population due to lack of population based cancer registries and census data stratified on age and gender. Increased diagnostic intensity, e. g. computed tomography and magnetic resonance imaging, might entail detection of non-lethal slow growing cancers. In contrast, decreasing autopsy rates decreases the number of cancers reported to cancer registries.

In epidemiology, three types of studies dominate: cohort studies, case-control studies and geographical correlation (or ecological) studies. A cohort study could in some instances be considered in analogy with an experiment, where exposed and non-exposed groups are compared. An alternative is to use a cohort with a wide range of exposure and use the lowest dose group for comparison. A case-control study employs an extra step of sampling controls. This makes the case-control study more efficient than a cohort study of a whole population and also allows for better control of confounders, but the sampling procedure increases the risk of a selection bias.

A cohort study consists of a population that is either defined and followed (prospective study) or constructed of a cohort of persons alive sometimes in the past (retrospective study) and followed forward in time. The cohort could consist of workers (e. g. nuclear power plant workers, Chernobyl recovery workers), patients or people living in certain areas. The exposure is given and the effect or outcome is measured. The weakness of a cohort study is that information, e. g. doses, vital status, cause of death, has to be gathered for all individuals which can be costly and time consuming. A cohort study is also highly dependent on reliable follow up possibilities, e. g. cause of death or cancer registries of uniform and high quality.

In a case-control study, cases with a specified disorder are identified and controls without the disease are selected. Explanatory variables among cases and controls are compared after collection of data. If cases and controls are selected from a previous well-defined cohort, the case-control study is said to be nested within the cohort – a nested case-control study. The crucial step is to adequately select the controls since they should be representative of the entire source population with the exception of the exposure under study. A case-control study could be used when it is difficult or too costly to obtain information on factors that might influence the result for the whole cohort or more detailed information is needed than available in the cohort setting.

The following is an example of difficulties encountered when follow up data differ with regard to exposure in a cohort study. Ivanov et al. (Ivanov 1996; Ivanov et al. 1997) have studied the cancer incidence in 142,000 Russian Chernobyl recovery operation workers. A significantly increased risk of leukaemia was found when the observed cases were compared with those expected from national incidence rates. However, the studies have been criticised for not using internal comparison because the increased medical surveillance and active follow-up of the emergency workers, coupled with underreporting in the general population, most likely influenced the results (Boice 1997; Boice and Holm 1997).

In contrast, the same investigators did not find an increased risk of leukaemia related to ionising radiation in a case-control setting, when controls were selected among fellow workers (Ivanov et al. 1997). These findings demonstrate that cancer incidence ascertainment in the exposed populations may differ from that in the general population. Future epidemiological investigations might be more informative if they are based on appropriate internal comparison groups.

An ecological correlation study examines the relationship between disease frequency and selected environmental factors, and place and time of residency are used as surrogates for actual exposure. This approach could be useful for generating

Tab. 5.3 Factors that complicate calculation of radiation risk, especially at low doses. Some of these factors are discussed in more detail in chapter 4.

Factor	Comment
Dose	Cell killing at high doses
Dose rate	Probably higher risk at brief exposure, time for repair at protracted exposure (only true for low-LET radiation)
Gender	Gender specific risks for different organs
Age	Higher risks for those exposed as young, dependent on type of tumour
Latency	Differ by age and tissue exposed
Smoking	Interacts with ionising radiation
Medical exposures	Chemotherapy induces leukaemia, confounding by indication
Outcome	Cancer incidence and mortality may differ
Background rates	Radiation risk varies with background rates
Tumour tissue	Differ in susceptibility
Heredity	Individual susceptibility to ionising radiation
Molecular factors	Extent of molecular repair at low doses is unknown

hypothesis regarding aetiology of a disorder but the use of non-individual exposure data and the simplification of complex relationships limits the studies of causal relationships. As long as individual dosimetry is not performed it will be difficult to identify causal relationships. In most instances, confounding factors are not taken into consideration and many risk factors, other than ionising radiation, produce a variation in cancer incidence and mortality.

Smoking habits, demographic characteristics such as ethnicity, urbanisation, socio-economical factors, migration and environmental factors are very seldom considered in ecological studies as has been pointed out elsewhere (Modan 1991). Other problems are: lack of adequate information on number of cancers, accuracy of cancer diagnosis, natural variability in base line cancer incidence and autopsy rates, and a diluting effect through migration. Factors, besides size of the study populations, that might influence the results of a radiation epidemiological study are listed and commented in Tab. 5.3. The different biases that arise from group-level studies have been given the general term 'ecological bias', and lead to what has been termed the ecological fallacy.

5.1.3
Relative and absolute risk estimates

Risk estimates are often presented as excess relative risk (ERR) or excess absolute risk (EAR). These terms represent the increased rates relative to an unexposed population, i. e. background rates, measured on proportional and absolute scales. For example, an *ERR* of 1 corresponds to a doubling of the cancer rate, while an *EAR* may be expressed as the extra annual number of cancers per 10,000 persons. Under the assumption of a linear dose-response relationship the risks may additionally be expressed as amounts per unit dose, e. g. *ERR* Sv^{-1}. The risks could also be given for a specific dose, e. g. *ERR* at 1 Sv. In contrast to the absolute risk, the relative risk is more influenced by the background incidence of the disease.

Since several factors influence the radiation-induced cancer risk one possibility is to present results as *ERR* and/or *EAR* specific to particular values for these factors, e. g. specific to gender, age at exposure or time since exposure. It has also become increasingly popular to present models that describe modifying effects. These models are generally empirical, in that they attempt to provide a good fit to the data.

5.1.4
Biostatistical considerations

From the Japanese atomic bomb survivor data it is approximated that exposure to 2 Sv would double the risk of dying from cancer, i. e. a relative risk of 2 (UNSCEAR 2000). The ability of an epidemiological study to detect such an increase is good, even a causal association after exposure to 1 Sv (increased risk of approximately 50 %) is possible to detect. In 1,000 persons, 180 individuals (18 %) are supposed to die of cancer in Sweden. If 1 Gy adds an additional 9 % to those dying from cancer, i. e. 90 individuals, an epidemiological study would probably have the ability to detect such an increase.

However, if we want to measure the risk after exposure to 50 mSv the additional increase in death due to cancer is 0.5 % or 5 cases, and we would not have the possibility to detect such an increase in 1,000 individuals. In order to have the statistical ability to detect an increase caused by 50 mSv (i. e. a 0.5 % increased mortality) we have to have a cohort of 57,000 exposed individuals if we assume a linear dose-response relationship and a 95 % confidence limit to our point estimate. In this exercise we have to keep in mind that the size of the study population is not the only prerequisite, an ability to control for confounding factors, consistent exposure data, and most importantly, reliable mortality registration and complete follow up are also needed.

5.2
Ionising radiation

5.2.1
Introduction

During the past decades extensive research on the long-term effects of ionising radiation has been conducted and epidemiological studies have contributed to our present knowledge. There are a number of populations under study and the most important source is the Japanese atomic bomb survivors. The data derived form this cohort has been used for determining exposure standards to protect the public and the workforce from harmful effects of ionising radiation. The standards were set by using modelling approaches to extrapolate from cancer risks observed following exposure to intermediate or high doses to predict changes in cancer frequency at low radiation doses. However, there are a number of difficulties and problems that influence the risk estimates at low doses and they will be discussed in the present chapter.

The major exposure to low dose and low dose rate radiation derives from medical tests, occupational, and environmental situations. The established model for determining carcinogenic effects at low doses in radiation protection is based on the hypothesis that the cancer incidence increases with radiation dose. Most national and international bodies have adopted a so-called linear model (ICRP 1991; UNSCEAR 2000). The major implication of the model for stochastic effects is that all doses, regardless of how low they are, must be considered potentially carcinogenic.

5.2.2
Contributions from epidemiology

5.2.2.1
Atomic bomb survivors

The Life Span Study (LSS) is the single most important study of radiation carcinogenesis in human populations. It is a well-defined cohort of people who has been followed from 1950 in order to determine the cause of death, cancer incidence as well as other outcomes. The cohort is large, includes men and women of all ages, the dose range is substantial, and the individual doses well characterised. Weaknesses are that the follow-up started in 1950, cancer incidence was not registered prior to 1958, and the follow-up was not complete. The effect of these shortcomings is not known.

When referring to the LSS it should be noted that the exposure was of high-dose rate but that the majority of the individuals in the cohort were actually exposed to low doses. Close to 73 % of the 86,572 individuals included was exposed to doses less than 50 mSv (weighted dose to the colon) and only 6 % received more than 500 mSv (UNSCEAR 1993). A statistically increased risk of dying from cancer was seen after exposure to < 50 mSv (Pierce et al. 1996). This finding was in conflict with previously published incidence data and it was suggested that the difference could partly be explained by misclassification of causes of death for survivors close to the hypocenter. The survivors close to the hypocenter could have had a higher likelihood of having a cancer recorded as the underlying cause of death.

However, in a recent study, Pierce and Preston (2000) clarified the issue focusing on survivors exposed to doses less than 500 mSv. That study was restricted to individuals exposed within 3 km from the hypocenter. A total of 7,000 cancers, diagnosed in approximately 50,000 survivors between 1958 and 1994, yielded useful risk estimates and a statistically significant increased risk of cancer and leukaemia was found. A linear dose-response model has previously seemed to fit the risk of solid tumours in the LSS best (Thompson et al. 1994; Pierce et al. 1996; Little and Muirhead 2000; Pierce and Preston 2000). In Fig. 5.1, taken from the publication by Pierce and Preston (2000), it can be seen that the degree of linearity below 500 mSv is high. The conclusion was that a linear dose-response model does not overestimate risks at low doses.

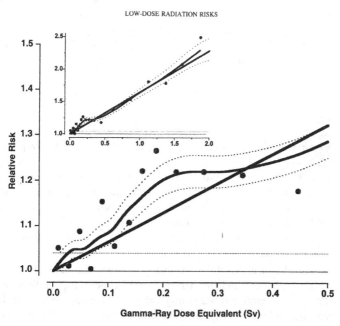

Fig. 5.1 Cancer incidence among atomic bomb survivors in the 3 km zone in relation to γ-ray equivalent dose (from Pierce and Preston 2000)

There has been considerable debate on the possibility of a dose below which there is no excess risk, that is a threshold dose. The threshold discussion is probably even more relevant after low dose rate exposure since protracted exposure might theoretically allow for molecular repair (see chapter 3). However, the latter issue is not possible to address in the LSS studies. The no-threshold model assumes the threshold to be 0 Sv. Fitting confidence intervals to their finding, Pierce and Preston (2000) found the upper limit to the confidence interval to be 60 mSv in the latest finding from the atomic bomb survivors.

Data on the dose-response relationship provided by the atomic bomb survivors gives no clear low-dose reduction factor for solid tumours or a factor very close to 1 (UNSCEAR 2000). When taking uncertainties of dose estimates into consideration, the dose-response for leukaemia fitted a linear-quadratic relationship and it was concluded that the best estimate of a reduction factor for leukaemia would be 2 (Little and Muirhead 2000; UNSCEAR 2000). Other studies regarding the dose-response relationship are females examined by repeated fluoroscopies where the risk of breast cancer was found to linearly relate to the absorbed breast tissue dose (Miller et al. 1989; Boice et al. 1991). Leukaemia risk in women given radiotherapy for cervical cancer was consistent with a linear dose-response relationship, although a quadratic term could not be excluded (Boice 1997).

The childhood thyroid gland is, besides red bone marrow and premenopausal female breast, one of the most radiosensitive organs in the body (UNSCEAR 2000). Studies of thyroid cancer risks are therefore of importance when examining risks at low doses. Age at exposure is the strongest modifier of risk; a decreasing risk with increasing age has been found in several studies (Thompson et al. 1994; Ron et al. 1995). Among survivors of the atomic bombings, the most pronounced risk of thyroid cancer was found among those exposed before the age of 10 years, and the highest risk was seen 15–29 years after exposure and was still increased 40 years after exposure (Thompson et al. 1994).

5.2.2.2
The Chernobyl accident

The accident at the Chernobyl nuclear power plant in 1986 was the most severe ever to have occurred in the nuclear industry. Within a few days or weeks, 28 power plant employees and firemen died as a consequence of radiation exposure. The catastrophe brought about the evacuation and relocation of about 350,000 people (UNSCEAR 2000). Vast territories were contaminated and deposition of released radionuclides was measurable in all countries of the northern hemisphere. In addition, about 240,000 workers were called upon in 1986 and 1987 to take part in the major mitigation activities at the reactor. All together, about 600,000 persons received the special status of so-called liquidator.

There can be no doubt about the relationship between the radioiodine released from the Chernobyl accident and the risk of thyroid cancer, especially in children (Astakhova et al. 1998). Among those younger than 15 years at the time of the accident, an increased incidence of thyroid cancer was seen (Tab. 5.2). No increased risk of leukaemia or solid tumours related to the releases from the accident has been proven among recovery operation workers or in residents of contaminated areas.

Several studies on adverse pregnancy outcomes related to the Chernobyl accident have been performed and so far, no increase in birth defects, congenital malformations, stillbirths, or premature births could be linked to radiation exposures caused by the accident (UNSCEAR 2000).

5.2.2.3
Occupational exposure

Workers employed in a number of industries face the potential for exposure to natural or man-made sources of ionising radiation. Significant internal exposures to natural radionuclides can occur in mining of radioactive ores (uranium, mineral sands). Other industries where exposure to radiation may still be significant are the nuclear fuel cycle, the production of nuclear weapons, and in nuclear medicine.

Several studies of nuclear industry workers have been conducted and one of the largest including 124,743 workers from the United Kingdom and a second analysis of the cohort was recently published (Muirhead et al. 1999). No increased risk of solid tumours was detected but a borderline significant risk of leukaemia (excluding chronic lymphatic leukaemia) resembling the central estimate of the LSS data (UNSCEAR 2000). It was concluded that the study provided some evidence of an elevated risk of leukaemia associated with occupational exposure to radiation and that the data was consistent with risk estimates of the ICRP Publication 60 (1991).

In a combined analysis of workers from Canada, United Kingdom and United States, including 95,673 workers, a total of 3,975 deaths due to cancer, and 119 due to leukaemia, were found (Cardis et al. 1995). The mean dose was 40.2 mSv but the distribution of doses was very skewed since close to 60 % of the workers had doses below 10 mSv. The risk of leukaemia, excluding chronic lymphatic leukaemia, did not reach significance using the 95 % confidence limit and it was concluded that the current radiation risk estimates for low-dose exposure was not appreciably in error.

5.2.2.4
Natural background radiation

The major exposure from ionising radiation is natural background radiation and it is an inescapable feature of life. Humans are exposed from two sources, high-energy cosmic ray particles from outer space and radioactive nuclides that originate from the earth crust. The magnitude of the exposure is dependent on residency, occupation, and life-style factors. Altitude above sea level determines the dose received from cosmic radiation, while radiation from the ground depends on the local geology, construction, and ventilation of houses. Living at a high altitude can lead to a 5-fold increased dose, while dose dependent on the local geology can vary with a factor of 100. The average annual effective dose from natural sources for all humans on earth is estimated to be 2.4 mSv, where 1.3 mSv derives from exposure to radon (UNSCEAR 1993 and Tab. 3.3).

It is difficult to estimate effects of natural background radiation since any effect is likely to be small in comparison to other causes of cancer. Due to the low effects anticipated, large populations have to be studied in order to obtain sufficient statistical certainty. To study risk at low doses within the frame of epidemiological studies can be difficult. One problem is to maintain the same standards of diagnosis and registration throughout large populations. Most studies include comparisons of cancer incidence or mortality among individuals living in areas with different background radiation.

As previously pointed out, an ecological correlation study examines the relationship between disease frequency and selected environmental factors, and place and time of residency are used as surrogates measures for actual exposure. With few exceptions, studies on background radiation are of the ecological type where association is taken as proof of causation. As long as individual dosimetry is not performed it will always be unclear whether the cancer observed is associated with ionising radiation or not. Smoking habits, demographic characteristics such as ethnicity, urbanisation, socio-economical factors, migration, and environmental factors including, among others, insulation, water quality, and maybe chemical pollution, are very seldom measured in ecological studies.

Darby (1991) calculated the fraction of malignancies in the US population that were associated with natural background radiation by using the models based on the Committee on the Biological Effects of Ionising Radiation (BEIR-V 1990). It was predicted that 11 % of the deaths from leukaemia and 4 % from solid tumours were caused by postnatal exposure to natural radiation other than radon. However, a number of studies during the years have failed to identify any increased risk of cancer or leukaemia due to background radiation (Court Brown et al. 1960; Allwright et al. 1983; Muirhead et al. 1991; Richardson et al. 1995). The correlation between cosmic radiation and cancer has been studied but no increased risks of leukaemia or cancer were found to be related to altitude (Mason and Miller 1974; Weinberg and Brown 1987).

In Yangjiang, a province in China, thorium-containing minerals have been washed down from the nearby heights and raised the background radiation level 3 times compared to adjacent areas of similar altitude. Approximately 80,000 individuals live in the high background areas and the annual dose to the red bone marrow was estimated to be 1.96 mSv compared to 0.72 mSv in the low-dose control area. When comparing overall cancer mortality and risk of dying of leukaemia, breast cancer and lung cancer, non-significant lower rates were seen in the high background areas (Tao and Wei 1986; Wei et al. 1990; Chen and Wei 1991; Wei and Wang 1994). A surprising finding was that chromosomal aberrations in lymphocytes were found to a higher extent in individuals residing in high-dose areas.

5.2.3
Environmental exposure to ionising radiation

In 1983, a local TV station announced a cluster of childhood leukaemia in Seascale, Great Britain, a few kilometres from the Sellafield nuclear fuel reprocessing plant in West Cumbria. Since then, many epidemiological studies have set out to analyse the risk of leukaemia near nuclear installations. Today, after nearly 18 years of accumulated results, the existence of an increased risk of leukaemia among young people living near a nuclear installation still remains highly controversial.

A large number of cluster studies have been performed near nuclear installations since 1984 (Heasman et al. 1983; Poole et al. 1988; Sharp et al. 1996). These studies were generally small, including only a few cases, and most of them show no excess of leukaemia among the young people living in the vicinity of these installations. An excess of leukaemia exists near some nuclear installations, at least for the reprocessing plants at Sellafield and Dounreay, Great Britain (Heasman et al. 1986; Ewings et al. 1989), and the nuclear power plant Krümmel, Germany (Westermeier and Michaelis 1995; Meinert et al. 1999). However, causal relationship between ionising radiation and leukaemia has not been identified. Excesses of leukaemia have also been identified far away from any nuclear installations, and the results of the multisite studies makes it less likely that the hypothesis of an increased risk of leukaemia is solely related to nuclear discharge. One explanation to clusters of leukaemia is the hypothesis of an infectious aetiology associated with population mixing. (Greaves 1988; Kinlen 1994; for more detailed discussion see chapter 8.)

5.2.3.1
Radon

Radon is an inert noble gas resulting from the decay of naturally occurring uranium-238. Uranium-238 is universally present in the environment and radon is an indoor air pollutant. It is also present in outdoor air, but at a far lower concentration. The α-particle emissions from radon progeny and not from radon itself are responsible for the critical dose of radiation delivered to the lung.

During the last 30 years health effects associated with uranium miners have received much attention. Although mortality rates are elevated for such causes as accidents and non-malignant respiratory disease, lung cancer caused by exposure to radon decay products is the primary hazard to underground uranium miners. Findings are frequently reviewed and summarised, most recently in a report from the National Research Council's Biological Effects of Ionising Radiation Committee (BEIR-VI, 1999). In general, the temporal patterns of excess risk following exposure are also similar among the studies; that is, risks change in a similar fashion with time since exposure and with age of the individual.

Initially, risks of indoor radon were estimated primarily by extrapolating the risks observed in the studies of underground miners to the exposures sustained by the general population indoors. To use the miner data is not done without certain precautions. The occupational exposure is relatively short-term compared to domestic exposure and concentrations are substantially above those typically found in homes, the extension of estimates from men, largely smokers, to the entire population, and differing dosimetry of radon progeny for the circumstances of exposures in mines and in homes. Linear non-threshold models were used for the extrapolation from higher to lower exposures, which invited the criticism that risks were overestimated.

The risk estimate derived from the eight completed indoor radon studies is indicative of an effect consistent with the risk extrapolations from the miner studies where no increased risk was found in the low-dose range (Fig. 5.2). There are, however, several problems to tackle when conducting indoor radon studies. Using building materials as crude measurements for radiation dose is likely to induce

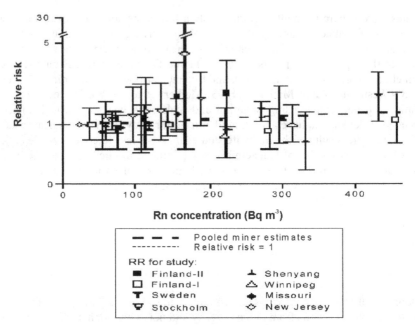

Fig. 5.2 Relative risks (RR) from 8 lung cancer case-control studies of indoor radon. - - - -, extrapolation of risk from miners (adopted from Lubin et al. 1994). . . . . ., relative risk of 1.

large errors since radon is in most cases the highest contributor and the concentration of radon daughters is highly dependent on ventilation, which in turn is dependent on climate and season. Not including exposure in all dwellings where a person lived will introduce problems since exposure during the first years of life could be the most important from a carcinogenic point of view, as shown in numerous studies (UNSCEAR 1994).

Correlating background radiation in the dwelling where the patient lived at the time of diagnosis is therefore misleading. It could be that in future the use of biomarkers for exposure, e. g. DNA adducts, could increase the accuracy of measurements (Poirier 1997). The largest problem emerges when the profound impact of smoking on lung cancer risk is not taken in to consideration. The number of attributable cancer deaths are far higher in smokers than in never-smokers and most studies have failed to show an increased risk of lung cancer in never-smokers. Risk projections imply that radon is the second leading cause of lung cancer after smoking.

In a Swedish case-control study (1980–84) of 586 women and 774 men diagnosed with lung cancer, radon was measured in 8,992 dwellings occupied by the study subjects at some time since 1947 (Pershagen et al. 1994). Information on smoking habits and other risk factors for lung cancer was obtained from questionnaires. The risk of lung cancer increased in relation to both estimated cumulative and time-weighted exposure to radon. The interaction between radon exposure and smoking with regard to lung cancer exceeded additivity and was closer to a multiplicative effect (Streffer et al. 2000).

5.2.4
Medical exposure

Medical use of ionising radiation could be divided into diagnostic examinations and therapy. Either external beam therapy or radionuclides are used. Radiotherapy procedures are not dealt with in this chapter since they cannot be considered being of low-dose exposure. Currently, with the exception of hyperthyroidism, polycytemia vera and possibly arthritis, ionising radiation is not used to treat benign disorders. Diagnostic X-rays are the largest man-made source of exposure to ionising radiation for the general population.

5.2.4.1
Nuclear medicine

The use of radiopharmaceuticals is widely practised throughout the world and nuclear medicine is the term used to describe medical diagnostic and therapeutic techniques based on the use of radionucleides. Radiopharmaceuticals have been used for diagnostic and therapeutic purposes for more than 60 years. The diagnostic procedures are the most common in nuclear medicine. Decreased or absent function of an organ can be inferred from reduced concentration of the radiopharmaceutical, as for instance caused by the presence of a tumour. Thyroid examinations, scans and uptake tests, constitute approximately 28 % of all nuclear medical procedures in the world, closely followed by bone scans (26 %), and cardiovascular examinations (15 %) (UNSCEAR 2000).

Studies of patients administered 131I for diagnostic purposes have not shown an increased risk of thyroid cancer related to ionising radiation (Hall et al. 1996). An explanation could be the striking finding of a profound age dependency, which implies that the thyroid gland is especially sensitive to ionising radiation during periods of rapid cell proliferation. Data from the atomic bomb survivors underline the strong modifying effect of age at exposure, with no excess risk seen in individuals older than 20 years (Thompson et al. 1994). Children are normally not exposed to radiopharmaceuticals.

5.2.4.2
Diagnostic X-ray examinations

The existence of a causal relation between prenatal exposure to ionising radiation and childhood leukaemia remains controversial. No radiation-related excess of leukaemia has been identified among the approximately 3,000 atomic bomb survivors exposed *in utero* (Yoshimoto et al. 1994, Delongchamp et al. 1997). Stewart et al. (1956) reported that prenatal radiation following diagnostic X-ray was associated with a subsequent increased risk of leukaemia and solid tumours during childhood. This report was followed by others and contributed to major changes in medical practice (MacMahon 1962; Oppenheim et al. 1975; Bithell 1988).

However, the presence of a causal relation still remains the subject of debate. Although major case-control studies, together with meta-analysis, consistently have shown a smaller increased risk for childhood leukaemia following a history of prenatal radiation (Salonen and Saxén 1975; Muirhead and Kneale 1989; Mole 1990) most cohort studies have not supported this association (Oppenheim et al. 1975). In a recent case-control study, where exposure information was retrieved from departments of radiology (minimising recall bias), no increased risk of childhood leukaemia was noted after *in utero* exposure (Naumburg et al. 2001).

5.3
Non-ionising radiation

5.3.1
UV-light

Exposure to UV-light could have, besides some beneficial effects (vitamin D production), many harmful effects on human health (Armstrong and Kricker 2001. Cancer development requires an accumulation of numerous genetic changes. UV-light produces undesirable DNA changes, which could lead to skin cancer (de Gruijl et al. 2001. UV-light has also been shown to suppress the immune system. However, it is still not known whether such immunomodulating effects may lead to an increase in the number and severity of certain tumours and/or infections in humans.

The incidence and mortality rates of cutaneous melanoma have risen dramatically during the past decades (Parkin et al. 1998). Sunlight exposure is suspected to be the main reason for the increase. However, the results from epidemiological studies are not consistent and the relationship between melanoma and UV-light is complex. The risk of malignant melanoma does not simply increase with accumulated exposure to UV-light since melanoma does not predominantly occur in body sites most exposed to the sun. To explain this paradox the "intermittent sunlight hypothesis" has been put forward (Walter et al. 1999). It is claimed that sunburns, especially in childhood, have an adverse effect while chronic exposure has a less carcinogenic or even protective effect.

Epidemiologic studies implicate sunlight as the principal environmental cause of cutaneous melanoma and the childhood has been suspected to be a period of particular susceptibility. The question of whether sunlight exposure during childhood confers higher risks than similar exposure at older ages is not only of intrinsic biological interest, but also has important implications for primary prevention. There is, however, no straightforward answer to the question how dangerous sunlight is. This is partly explained by the methodological problems inherited in these types of studies. A person's lifetime dose of solar ultraviolet radiation can be extremely difficult to quantify. Investigators have attempted to measure past sun exposure in several ways, which leads to strikingly different conclusions regarding the association between sun exposure at specific ages and risk of melanoma.

A somewhat less controversial area is the inter-individual sensitivity to UV-light and melanoma risk. In a meta-analyses of 20 studies it was found that fair-skinned people born and raised in environments of low solar irradiation have significantly lower risks of melanoma than people of similar complexion born and raised in sunny environments (Whiteman et al. 2001). No quantitative risk estimates were given due to the differences in measuring exposure.

In conclusion, UV-light has been causally related to malignant melanoma in numerous epidemiological studies. The relationship is, however, complex and the risks difficult to determine due to misclassification of exposure.

5.3.2
Mobile phones

Hand-held phones were introduced in the mid 1980s and became widely used 10 years later. By early 2000, the number of subscribers to cellular-telephone services had grown to 500 million world-wide (CTIA 2000). Given the widespread current use of mobile phones the matter is potentially of considerable public health importance. Concern has arisen about adverse health effects, especially the possibility that the low-power microwave-frequency signal transmitted by the antennas on handsets might cause brain tumours or accelerate the growth of subclinical tumours (Rothman et al. 1996; McKinlay 1997). The heating of brain tissue by cellular telephones is probably negligible and direct genotoxic effects are unlikely. Data concerning the risk of cancer associated with the exposure of humans to non-ionising radiation of the frequencies used by cellular telephones are limited.

In the largest study to date, 782 patients with CNS tumours, diagnosed between 1994 and 1998, were compared to 799 controls (Inskip et al. 2001). When comparing never, or sporadic, users to individuals who used a cellular telephone for more than 100 hours a relative risk of 0.9 for glioma, 0.7 for meningioma, 1.4 for acoustic neuroma and 1.0 for all types of tumours combined, were found. The authors concluded that there was no evidence of an increased risk of a CNS tumour for a person using a mobile phone for 60 or more minutes per day or regularly for five or more years. Tumours did not occur disproportionately often on the side of head on which the telephone was typically used.

In conclusion, several reports on the adverse effects of the use of mobile phones have been published and at present there are no firm results that indicate a risk of CNS tumours after use of mobile phones.

5.3.3
Other electro-magnetic fields

The potential carcinogenic effects of extremely low frequency electromagnetic fields (EMF) have been evaluated in epidemiological and experimental studies for over two decades. Because genotoxic effects of EMF have not been shown, most recent laboratory research has attempted to show biological effects that could be related to cancer promotion. Cancer epidemiology studies in relation to EMF have focused primarily on brain cancer and leukaemia both, from residential sources of exposure in children and adults and from occupational exposure in adult men.

Studies of residential exposure and childhood brain tumours have produced inconsistent results, regardless of the exposure metrics used. This outcome holds for both current and past estimates of EMF, whether based on wire codes, distance, or measured or calculated fields. Initially, it appeared that the EMF association was stronger with childhood brain tumours than with leukaemia (Ahlbom 1988), however, several of the most recent studies have found no association (Feychting and Ahlbom 1993, Tynes and Haldorsen 1997). Studies examining use of appliances by children, or by their mothers during pregnancy, have also found an inconsistent pattern of risk, and recent studies of parental occupational exposure and childhood brain tumours suggest a lack of an association (Wilkins and Hundley 1990).

Residential studies have found little or no association between electric and magnetic fields exposure and brain cancer in adults (Feychting et al. 1990). No evidence either epidemiological or experimental, has emerged to provide reasonable support for a causal role of EMF on brain cancer or leukaemia.

5.4
Smoking

5.4.1
Active smoking

Tobacco is one of the most studied human carcinogens. It is also the largest preventable risk factor for morbidity and mortality in industrialised countries. WHO estimates that tobacco will cause 8.4 million deaths annually by the year 2020. More than 50 carcinogens, among the most potent of which are polycyclic aromatic hydrocarbons (PAHs) and tobacco-specific nitrosamines (TSNs), are found in tobacco smoke (see section 4.4.3.4). The level of tar and nicotine in cigarettes has decreased, along with the level of PAHs, but the level of TSNs has increased.

It is difficult to determine the effect of tobacco since it not only includes the number of cigarettes smoked per day and type of cigarette but also carcinogen metabolism and DNA repair. Many studies have shown a relationship between tobacco smoke, carcinogen-DNA adduct formation, tumour specific mutations (e. g. *p53* mutational spectra), and cancer risk but data are conflicting. Newer methods can probably clarify the role of environmental tobacco smoke in carcinogenesis.

For many years the carcinogenic effects of tobacco were thought to be restricted to the lung, pancreas, bladder and kidney, and (synergistically with alcohol) larynx, oral cavity, pharynx, and oesophagus (Doll and Peto 1981). More recent evidence indicates that also stomach, liver, and (possibly) cervix cancer are also increased by smoking (Liu et al. 1998). The relative importance of smoking related cancers varies widely in different parts of the world and seems to multiply the effect of the background incidence rates due to other, known or unknown, risk factors.

Lung cancer continues to be the main adverse effect of tobacco smoking and is the leader in cancer deaths in the industrialised world. Tobacco smoking causes approximately 85–90 % of all lung cancer. The highest risk is seen in those that started smoking at young ages and continued smoking throughout their life. In the western world smoking became popular among males in the 1940s and the male use of tobacco reached its peak in the 1970s. In many industrialised countries the prevalence of male smokers has since then dropped and parallel to that, the lung cancer incidence among males. However, male lung cancer rates are still increasing in most developing countries and eastern Europe, where cigarette consumption remains high and in some areas even increasing.

5.4.2
Passive smoking

The risk of lung cancer among never-smokers living with a spouse who smokes has been extensively studied (Tab 5.4). An increased risk of lung cancer related to envi-

Tab. 5.4 Relative risk (RR) and 95 % confidence interval (CI) of lung cancer and exposure to environmental tobacco smoke in 39 epidemiological studies according to sex, geographical region, year of publication, and study design

	No. of lung cancers	Pooled RR	95 % CI
By sex			
Women	4,626	1.24	1.13–1.36
Men	274	1.34	0.97–1.84
Both	5,095	1.23	1.13–1.34
By geographical region			
USA	1,959	1.17	1.05–1.31
Europe	439	1.53	1.21–1.94
Japan	550	1.28	1.04–1.57
China and Hong Kong	1,678	1.22	0.99–1.50
By year of publication			
1981–85	809	1.29	1.06–1.58
1986–90	1,591	1.28	1.07–1.54
1991–97	2,226	1.19	1.06–1.33
By study design			
Case-control	4,115	1.24	1.12–1.38
Cohort	511	1.27	1.05–1.53

ronmental tobacco smoke was found in an analysis of 37 published epidemiological studies of the risk of lung cancer (4,626 cases) in non-smokers who did and did not live with a smoker (Hackshaw et al. 1997).

A significant 24 % increased risk of lung cancer was found in never smokers who lived with a smoker compared to those who did not (Hackshaw et al. 1997). A dose-response relation of the risk of lung cancer with both the numbers of cigarettes smoked by the spouse and the duration of exposure was significant. Tobacco specific biomarkers were also found in the blood and urine of never smokers exposed to environmental tobacco smoke. It was shown that relative risk estimates of lung cancer and exposure to environmental tobacco smoke did not significantly differ between men and women ($P = 0.31$), between geographical regions ($P = 0.26$), with year of publication ($P = 0.16$), or between cohort and case-control studies ($P = 0.53$).

For each study a linear regression analysis was performed between the relative risk (in logarithms) and the number of cigarettes smoked by the spouse (Fig. 5.3a). The summary estimate (allowing for the within and between study variation) shows a significant dose-response relation. Risk increases by 23 % (14 % to 32 %) for every 10 cigarettes smoked per day by the partner (88 % if he/she smoked 30 cigarettes daily). Fig. 5.3b shows the summary estimate of risk in relation to duration of exposure. Risk increases by 11 % (4 % to 17 %) for every 10 years of exposure (35 % for 30 years' exposure).

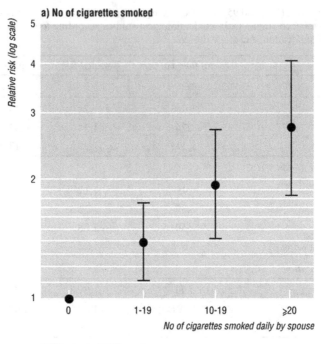

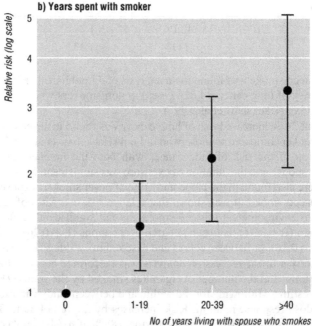

Fig. 5.3 Dose-response relation between the relative risk (95 % confidence interval) of lung cancer and (a) the number of cigarettes smoked daily by the spouse and (b) the number of years living with a spouse who smokes (Hackshaw et al. 1997)

The impact of lifetime residential and workplace environmental tobacco smoke has received little attention and case-control studies have failed to identify a relationship (Johnson et al. 2001).

5.5
Other potential carcinogenic compounds

The carcinogens covered in this section are discussed in detail in chapter 4. We have chosen the compounds with reliable epidemiological data. There are three major types of origin of potential carcinogenic substances that, through environmental contamination, could have an effect on humans. Environmentally persistent *insecticides* (dichloradiphenyltrichoroethane-DDT, hexachlorohexane-HCH) have been used for decades and are still extensively applied. *Chemicals* used in industrial processes which were not meant to be released into the environment but which have nevertheless reached humans through the ecological chain. The most studied persistent compounds are PCBs (polychlorinated biphenyls).

The third category of persistent organic compounds seen are *industrial waste emissions*. The most important are dioxins, brominated flame retardants like hexachlorbenzene (HCB), and polycyclic aromatic hydrocarbons (PAHs). Many of the compounds are not in use in the Western world today. For others, restrictions and improved waste management has lead to a reduced exposure.

5.5.1
Insecticides

Insecticides belong to a larger group, pesticides, that is any substance intended for preventing, destroying and repelling any pest. Pests can be insects, mice and other animals, unwanted plants (weeds), fungi, or microorganisms, like bacteria and viruses. The term pesticide applies to insecticides, herbicides, fungicides, and various other substances used to control pests.

Chemicals used to control insects have come into widespread use during the past century, with the development of a variety of synthetic insecticides. The principal classes of compounds that have been used as insecticides are organochlorine, organophosphorus, carbamate and pyrethroid compounds. Some of them, like DDT, have been banned in the Western world but are still in use in some developing countries. Insecticides are applied by aerial spraying and by various ground-based techniques. Occupational exposures occur in the preparation, spraying, and application of insecticides. Absorption resulting from dermal exposure is the most important route of uptake for exposed workers.

Several US studies have evaluated cancer risks in association with ecological measures of insecticide exposure (Pesatori et al. 1994). The risk for multiple myeloma tended to be greater for farmers residing in counties where insecticides were more heavily used, but that for leukaemia did not. In a study of a large cohort of grain millers in the USA, mill workers had excess risks for non-Hodgkin's lymphoma and pancreatic cancer but the risk for lung cancer was not increased (Alavania et al. 1990). Chronic lymphocytic leukaemia has been associated with use of DDT in Sweden (Flodin et al. 1988).

Overall, the strongest evidence that exposure to insecticides causes cancer in humans comes from the cohort studies of applicators. These findings were based on small numbers of cases and applicators in some of these studies had potential contact with arsenical insecticides. In an evaluation by the International Agency for Research on Cancer, it was concluded that there was *limited evidence* that occupational exposures in spraying and application of non-arsenical insecticides entail a carcinogenic risk (IARC 1991.

5.5.2
Industrial chemicals

PCBs accumulate in adipose tissue in the body and exposure to persistent organic compounds has been related to decrease in sperm counts, decreased male reproductive capacity, risk of leukaemia (Nordstrom et al. 2000), and malignant melanoma (Robinson et al. 1999). However, the endocrine-active behaviour of PCBs (and other organochlorines) has attracted the largest attention. Controversy exists over the role that environmental oestrogens, such as PCBs, might play as risk factors for breast cancer. Several laboratory studies have suggested that these chemicals function as weak oestrogens, binding to the oestrogen receptor and inducing various measures of oestrogen response, although results are conflicting.

Several epidemiological studies have linked PCB exposure, measured in plasma lipids or adipose tissue, with an increased risk of breast cancer (Woolcott et al. 2001; Demers et al. 2002). Most studies are small and lack statistical power and relationships are most often noticed in small subgroups of women. In 1993, five large US studies of women aiming at evaluating the association of levels of DDE and PCBs in blood plasma or serum with breast cancer risk were conducted (Laden et al. 2001). In a combined analysis of these results including 1,400 patients with breast cancer and 1,642 control subjects, no risk of breast cancer associated with PCBs was found (odds ratios = 0.94; 95 % CI = 0.73–1.21), or with DDE (odds ratio = 0.99; 95 % CI = 0.77–1.27).

Although in the original studies there were suggestions of elevated breast cancer risk associated with PCBs in certain groups of women stratified by parity and lactation, these observations were not evident in the pooled analysis. It was concluded that exposure to these compounds, as measured in adult women, is unlikely to explain the high rates of breast cancer experienced in the United States. Presently, no firm evidence exists of the carcinogenic effect of PCBs or its metabolites. Results are conflicting but larger pooled analyses seem to argue against a more profound effect on humans.

5.5.3
Industrial waste and emissions

The most important emissions are dioxins, brominated flame retardants like hexachlorbenzene (HCB), and polycyclic aromatic hydrocarbons (PAH). Dioxin is a general term that describes a group of hundreds of chemicals that are highly persistent in the environment. The most toxic compound is tetrachlorodibenzo-p-dioxin or TCDD. The toxicity of other dioxins and chemicals like PCBs that act like dioxin are measured in relation to TCDD.

Dioxin is formed as an unintentional by-product of many industrial processes involving chlorine. Burning chlorine-based chemical compounds is the major source (95 %) of dioxin in the environment. Dioxin pollution is seen in paper mills, which use chlorine bleaching in their process and with the production of polyvinyl chloride (PVC) plastics. It is believed that dioxin causes cancer, affects reproductive and cognitive functions, and that it can cause immune system damage and interfere with regulatory hormones. Since dioxin is fat-soluble, it accumulates in food and it is mainly (97.5 %) found in meat, dairy products, and fish.

Cancer mortality in relation to TCDD-contaminated products (trichloro-phenol or its derivatives) was evaluated in 3,538 male workers at eight US chemical plants (Steenland et al. 1999, 2001). Exposure scores were based on an estimated level of contact with TCDD, the degree of TCDD contamination of the product at each plant over time, and the fraction of a workday during which a worker was likely to be in contact with TCDD contaminated products.

The standardised mortality ratio (*SMR*) for all cancers combined was 1.13 (95 % CI = 1.02–1.25). Statistically significant positive linear trends in *SMR*s with increasing exposure for all cancers combined and for lung cancer were found. The *SMR* for all cancers combined for the highest exposure group was 1.60 (95 % CI = 1.15–1.82). No risk related to TCDD and heart disease or diabetes was found. The conclusion was that high TCDD exposure results in an excess of all cancers combined and that the excess was limited to the highest exposed workers, with exposures that were likely to have been 100–1,000 times higher than those experienced by the general population.

All-cause mortality and cancer mortality was evaluated in 6,745 inhabitants of the Seveso area, Italy (Bertazzi et al. 2001). Fifteen years after the accident, when large amounts of dioxin were released, an increased risk of dying from all cancers, rectal and lung cancer was found among men. The risk of haematological malignancies was elevated in both genders. Hodgkin's disease risk was elevated in the first 10-year observation period, whereas the highest increase for non-Hodgkin's lymphoma and myeloid leukaemia occurred 15 years after the accident. An overall increase in diabetes was reported, notably among women. Chronic circulatory and respiratory diseases were moderately increased, suggesting a link with accident-related stress factors and chemical exposure.

The International Agency for Research on Cancer (IARC) has evaluated the most potent dioxin, TCDD, and it is considered to be a class 1 human carcinogen. That is, a positive relationship has been observed between the exposure and cancer in epidemiological studies in which chance, bias and confounding could be ruled out with reasonable confidence (IARC 1997).

Polycyclic aromatic hydrocarbons (PAHs) are a group of over 100 different chemicals that are formed during the incomplete burning of coal, oil and gas, garbage, or other organic substances like tobacco or charbroiled meat. Some PAHs are manufactured and found in coal tar, crude oil, creosote, and roofing tar, but a few are used in medicines or to make dyes, plastics and pesticides. Exposure to PAHs usually occurs by breathing air contaminated through wild fires or coal tar, or by eating foods that have been grilled. The most important source of human exposure to PAHs is, however, tobacco smoke. Eleven PAHs were measured in 70 lung tissue samples (37 smokers and 33 non-smokers) from cancer-free autopsy donors

(Goldman et al. 2001). Smoking increased the concentration of five PAHs including benzopyrene, which increased approximately 2-fold. The sum of PAH concentrations was higher in smokers (P = 0.01), and there was a dose-response relationship for greater smoking (P < 0.01).

The risk of bladder cancer associated with occupational exposures to paint components, PAHs, diesel exhausts, and aromatic amines was evaluated in a population based study in the Netherlands of 532 cases and 1,630 controls (Zeegers et al. 2001). Men in the highest tertiles of occupational exposure had a non-significantly higher risk of bladder cancer. The associations between paint components and PAHs and risk of bladder cancer were most pronounced for current smokers. It was concluded that the study provided only marginal evidence for an association between occupational exposure to PAHs and bladder cancer.

A US cohort study including 4806 workers in the synthetic chemical industry showed an excess risk for cancer of the respiratory system (*SMR* = 1.5; 95 % CI = 1.1–2.0) (Waxweiler et al. 1981). Exposure of the patients with cancer of the respiratory system to vinyl acetate was below the mean exposure expected for the members of the cohort with the same year of birth and age at commencement of work in the plant. A subgroup of employees with undifferentiated non-small cell lung cancer had a slightly, but statistically non-significant cumulative exposure to vinyl acetate. Thus, the study of Waxweiler et al. (1981) does *not* give evidence for a carcinogenic effect of vinyl acetate in humans.

Nested case-control studies were performed in a cohort of 29,139 men employed in a chemical manufacturing industry (Ott et al. 1989). Risk of non-Hodgkin's lymphomas, multiple myeloma, nonlymphocytic leukemia and lymphocytic leukemia in relation to exposure to 21 specific chemicals were calculated. Exposure to vinyl acetate was associated with a decreased risk for non-lymphocytic leukemia (odds ratio: 0.5), but slightly increased odds ratios for non-Hodgkin's lymphoma (odds ratio 1.2) or multiple myeloma (odds ratio 1.6).

In conclusion, evaluation of epidemiological data on a possible carcinogenic effect of vinyl acetate is difficult, since most individuals in the existing epidemiological studies were exposed to several chemicals. Nevertheless, the existing data do not support a carcinogenic effect of vinyl acetate in humans.

5.5.4
Asbestos

Asbestos is a naturally occurring group of minerals and there are several types of different fibres. Asbestos has been added to a variety of products to strengthen them, provide heat insulation, and fire resistance. In most products, asbestos is combined with a binding material so that it is not readily released into the air. However, if asbestos is inhaled, it remains in the lungs with the potential to cause severe health problems.

More than 3,000 products in use today contain asbestos. Most of these are materials used in heat and acoustic insulation, fireproofing, and roofing and flooring. Some of the more common products that may contain asbestos include: pipe, duct, and building insulation, electrical wires, and cement.

Asbestos is defined as a type 1 carcinogen to humans by IARC (1977). From epidemiological studies it is known that asbestos causes four different diseases in humans: lung fibrosis (asbestosis), pleural fibrosis and plaques, pleural mesothelioma, and lung cancer. These disorders follow heavy exposure and, in industrialised countries, are mainly confined to occupational exposure.

All different asbestos fibre types seem to exert a similar effect on lung cancer risk and a multiplicative interaction with tobacco smoking has been suggested. The relationship between asbestos exposure and smoking indicates a synergistic, close to a multiplicative, effect of smoking with regard to lung cancer (Saracci 1977). Pleural mesothelioma is a malignant neoplasm specifically associated with asbestos exposure, the risk is linked with the cubic power of time since first exposure, after an induction and latency period of 10 years.

In contrast to lung cancer, the risk of mesothelioma depends on fibre type, as the risk is about three times higher for amphiboles as compared to chrysotile fibres. Not only occupational, but also environmental exposure to asbestos is associated with an increased risk of mesothelioma. In contrast to lung cancer, there seems to be little or no effect modification of smoking for pleural mesotheliomas.

Numerous reports describe pleural and peritoneal mesotheliomas in relation to occupational exposure to various types and mixtures of asbestos (Gardner et al. 1985; Paur et al. 1985). Not all asbestos exposure is environmental, and it has been estimated that a third of the mesotheliomas occurring in the US may be due to non-occupational exposure (Enterline 1983). In some of the studies asbestos fibres have been identified in the lung (Magee et al. 1986). There seems to be an indication of a dose-response relationship since an increasing risk of mesothelioma has been seen with increasing duration of exposure (Hauptmann et al. 2002).

It was recently concluded that exposure specific risk of mesothelioma from the three commercially provided asbestos types are 1 : 100 : 500 for chrysotile, amosite, and crocidolite (Hodgson and Darnton 2000). The same pattern was not seen for lung cancer. In contrast to mesothelioma, the risk ratio for lung cancer has usually been moderately increased (Beck and Schmidt 1985) although more recent findings has showed a 8-fold increased risk related to asbestos exposure (Yano et al. 2001). Risk ratios of about 2–5 have been reported in some studies, but did not exceed unity in another studies. Laryngeal cancer has been considered in two case-control studies, resulting in risk ratios of 2.4 and 2.3 that relate to shipyard work and unspecified exposure, respectively (Blot et al. 1980; Burch et al. 1981). Although most studies have been confined to occupational exposure, evaluation in population based cohorts has been undertaken. In a series of 924 non-selected lung cancers, without notification of occupational history, histological asbestosis was demonstrated in 56 (6 %) cases. These lung cancers were considered induced by asbestos (Mollo et al. 2002).

Among the numerous studies evaluating the risk of other cancers than respiratory tract cancers and mesothelioma after exposure to asbestos, the results are inconsistent. No significant increase related to asbestos has been seen for stomach (Sandén et al. 1985) or colo-rectal cancer (Liddell et al. 1984. In a review of 6 cohort and 16 case-control studies, published up to 1999, it was stated that a causal relationship between asbestos exposure and the subsequent development of lymphomas could not be found (Becker et al. 2001).

5.6
Summary

For atomic bomb survivors and in numerous studies of patients receiving radiotherapy it has been convincingly shown that cancer incidence and mortality are related to exposure to ionising radiation. We do not know if there are radiation doses below which there is no significant biological change, or below which the damage induced can be effectively dealt with by normal cellular processes. Analyses of Japanese data could not exclude a threshold of 60 mSv, i. e. below this dose limit no significant excess of cancer or leukaemia could be identified. The possibility that ecological studies of environmental exposure to ionising radiation can contribute to our knowledge of the effects of low-dose exposure is limited. The effects of natural background radiation are low and other risk factors will distort the results making data unreliable.

Advances in molecular biology and genetics will hopefully increase the likelihood of finding the "true" effect of ionising radiation at low doses. Research will focus on understanding cellular processes responsible for recognising and repairing normal oxidative damage and radiation-induced damage.

Malignant melanoma is found to be related to excessive UV-light exposure. However, there are still questions to be answered, e. g. the difference in effect of accumulated UV-light exposure as in contrast to sunburns and the effect of accumulated exposure. The effect of UV-light immuno-suppression is also not understood.

Mobile phones were introduced in the mid 1980s and, given the wide spread current use of mobile phones, any possible harmful effect is of considerable public health importance. In the largest study to date no increased risk of CNS tumours was noticed. It was concluded that no evidence of an increased risk was found for a person using a mobile phone for 60 or more minutes per day or regularly for five or more years. The potential carcinogenic effects of extremely low frequency electromagnetic fields have been evaluated (brain cancer or leukaemia) but no increased risk has been observed.

Tobacco is probably the most studied human carcinogen and also the largest preventable risk factor for morbidity and mortality in most countries. Active smoking causes a number of health problems including cancer at several sites besides lung cancer. Elevated risks of lung cancer after passive smoking have been identified. In a large pooled analysis an increased risk of lung cancer in non-smokers, living with a smoker, was seen. A significant dose-response relation was also seen and the risk increased with the number of cigarettes smoked per day and by duration of smoking.

The late health effects in humans exposed to persistent pesticides have been extensively studied. DDT has for many years been the main focus. Today there is no firm evidence that pesticides cause cancer.

Polychlorinated biphenyls accumulates in the adipose tissue and could theoretically influence the male reproductively, risk of leukaemia and breast cancer. A definite confirmation of the toxicity is still lacking.

Polycyclic aromatic hydrocarbons and the most potent dioxin, TCDD, have been considered potent carcinogens. That is, a positive relationship has been observed between the exposure and cancer in studies in which chance, bias and confounding could be ruled out with reasonable confidence.

There is convincing evidence that asbestos cause lung cancer and mesothelioma and that the carcinogenic effect is increased by tobacco smoke. No study has identified other tumours to be related to asbestos exposure.

5.7
Conclusion

A causal relationship between ionising radiation, cancer and leukaemia is well established. More problematic is the determination of the dose-response relationship at low doses because of the numerous factors other than ionising radiation that could influence the risk. Direct estimation of risks at low dose is possible only in a few sufficiently large populations but pooling of data could increase the statistical power. Advances in molecular biology and genetics could increase the likelihood of finding the "true" effect of ionising radiation at low doses.

Non-ionising radiation, such as UV-light, is related to the sharp increase in the number of diagnosed malignant melanomas. No study has convincingly shown that electromagnetic fields caused by mobile phones produce any harmful effect in humans.

Active smoking causes a number of health problems including cancer at several sites besides the lung. It could be considered an established fact that passive smoking causes lung cancer. There is convincing evidence that asbestos causes lung cancer and mesothelioma and that the carcinogenic effect is increased by tobacco smoke.

The late health effects in humans exposed to pesticides have been extensively studied but there is no firm evidence that pesticides cause cancer or any other adverse health effect in humans. Today there is no firm epidemiological evidence that polychlorinated biphenyls influence the risk of leukaemia or breast cancer. Polycyclic aromatic hydrocarbons and the most potent dioxin, TCDD, have been shown to cause cancer in humans.

5.8
References

Ahlbom A (1988) A review of the epidemiologic literature on magnetic fields and cancer. Scand J Work Environ Health 14: 337–343

Alavania M, Blair A, Masters MN (1990) Cancer mortality in the U.S. flour industry. J Natl Cancer Inst 82: 840–848

Allwright SPA, Colgan PA, McAulay IR, Mullins E (1983) Natural background radiation and cancer mortality in the Republic of Ireland. Int J Epidemiol 12: 414–418

Armstrong B, Kricker A (2001) The epidemiology of UV induced skin cancer. J Photochem Photobiol B 63: 8–18

Astakhova LN, Anspaugh LR, Beebe GW, Bouville A, Drozdovitch VV, Garber V, Gavrilin YI, Khrouch VT, Kuvshinnikov AV, Kuzmenkov YN, Minenko VP, Moschik, K V, Nalivko AS, Robbins J, Shemiakina EV, Shinkarev S, Tochitskaya SI, Waclawiw MA (1998) Chernobyl-related hyroid cancer in children of Belarus: a case-control study. Radiat Res 150: 349–356

Baverstock K, Egloff B, Pinchera A, Ruchti C, Williams D (1992) Thyroid cancer after Chernobyl. Nature 359: 21–22

Beck EG, Schmidt P (1985) Epidemiological investigations of deceased employees of the asbestos cement industry in the Federal Republic of Germany. Zbl Bakt Hyg, I. Abt Orig B 181: 207–215

Becker N, Berger J, Bolm-Audorff U (2001) Asbestos exposure and malignant lymphomas–a review of the epidemiological literature. Int Arch Occup Environ Health 74: 459–469

Bertazzi PA, Consonni D, Bachetti S, Rubagotti M, Baccarelli A, Zocchetti C, Pesatori AC (2001) Health effects of dioxin exposure: a 20-year mortality study. Am J Epidemiol 153: 1031–1044

Bithell JF, Stiller CA (1988) A new calculation of the carcinogenic risk of obstetric X-raying. Stat Med 7: 857–864

Blot WJ, Morris LE, Stroube R, Tagnon I, Fraumeni JF Jr (1980) Lung and laryngeal cancers in relation to shipyard employment in coastal Virginia. J Natl Cancer Inst 65

Boice JD Jr (1997) Leukaemia, Chernobyl and epidemiology. Invited editorial. J Radiol Prot 17: 129–133

Boice JD Jr, Holm LE (1997) Radiation risk estimates for leukemia and thyroid cancer among Russian emergency workers at Chernobyl. Letter. Radiat Environ Biophys 36: 213–214

Boice JD Jr, Preston D, Davis FG, Monson RR (1991) Frequent chest X-ray fluoroscopy and breast cancer incidence among tuberculosis patients in Massachusetts. Radiat Res 125: 214–222

Burch JD, Howe GR, Miller AB, Semenciw R (1981) Tobacco, alcohol, asbestos, and nickel in the etiology of cancer of the larynx: a case-control study. J Natl Cancer Inst 67: 1219–1224

Cardis E, Gilbert ES, Carpenter L, Howe G, Kato I, Armstrong BK, Beral V, Cowper G, Douglas A, Fix J, Fry SA, Kaldor J, Lavé C, Salmon L, Smith PG, Voelz GL, Wiggs LD (1995) Effects of low doses and low dose rates of external ionizing radiation: cancer mortality among nuclear industry workers in three countries. Radiat Res 142: 117–132

Cellular Telecommunication and Internet Association [CTIA] (2000) How many people use wireless phones? Cellular Telecommunication and Internet Association, Washington, DC

Chen D, Wei L (1991) Chromosome aberration, cancer mortality and hormetic phenomena among inhabitants in areas of high background radiation in China. J Radiat Res (Tokyo) 32, Suppl 2: 46–53

Committee on the Biological Effects of Ionizing Radiations [BEIR V] (1990) Health effects of exposure to low levels of ionizing radiation. Natl Acad Sci USA, Natl Res Council. National Academy Press, Washington, DC

Committee on the Biological Effects of Ionizing Radiations [BEIR VI] (1999) Health effects of exposure to radon: BEIR VI. Natl Acad Sci USA, Natl Res Council. National Academy Press, Washington, DC

Court Brown WM, Doll R, Bradford Hill A (1960) Geographical variation in leukaemia mortality in relation to background radiation and other factors. Br Med J 1: 1753–1759

Darby SC (1991) Contribution of natural ionizing radiation to cancer mortality in the United States. In: Brugge J (ed) Origins of human cancer: a comprehensive review. Cold Spring Harbor Laboratory Press, Plainview, NY, pp 183–190

De Gruijl F, van Kranen H, Mullenders LH (2001) UV-induced DNA damage, repair, mutations and oncogenic pathways in skin cancer. J Photochem Photobiol B 63: 19–27

Delongchamp RR, Mabuchi K, Yoshimoto Y, Preston DL (1997) Cancer mortality among atomic bomb survivors exposed in utero or as young children. Radiat Res 147: 385–395

Demers A, Ayotte P, Brisson J, Dodin S, Robert J, Dewailly E (2002) Plasma concentrations of polychlorinated biphenyls and the risk of breast cancer: a congener-specific analysis. Am J Epidemiol 155: 629–635

Dickman P, Holm LE, Lundell G, Boice J, Hall P (2003) Thyroid cancer risk after examinations with 131-I in relation to reason for referral and previous disorder. Int J Cancer 106(4): 580–587

Doll R, Peto R (1981) The causes of cancer: quantitative estimates of avoidable risks of cancer in the United States today. J Natl Cancer Inst 66: 1191–1308

Enterline PE (1983) Cancer produced by nonoccupational asbestos exposure in the United States. J Air Pollut Control Assoc 33: 318–322

Ewings PD, Bowie C, Phillips MJ, Johnson SA (1989) Incidence of leukaemia in young people in the vicinity of Hinkley Point nuclear power station, 1959–86. Br Med J 299: 289–293

Feychting M, Ahlbom A (1993) Magnetic fields and cancer in children residing near Swedish high-voltage power lines. Am J Epidemiol 138: 467–480

Feychting M, Forssen U, Floderus B (1990) Occupational and residential magnetic field exposure and leukemia and central nervous system tumors. Epidemiology 8: 384–389

Flodin U, Fredriksson M, Persson B, Axelson O (1988) Chronic lymphatic leukaemia and engine exhausts, fresh wood, and DDT: a case-referent study. Br J Ind Med 45: 33–38

Gardner MJ, Jones RD, Pippard EC, Saitoh N (1985) Mesothelioma of the peritoneum during 1967–82 in England and Wales. Br J Cancer 51

Goldman R, Enewold L, Pellizzari E, Beach JB, Bowman ED, Krishnan SS, Shields PG (2001) Smoking increases carcinogenic polycyclic aromatic hydrocarbons in human lung tissue. Cancer Res 61: 6367–6371

Greaves MF (1988) Speculations on the cause of childhood acute lymphoblastic leukemia. Leukemia 2: 120–125

Hackshaw AK, Law MR, Wald NJ (1997) The accumulated evidence on lung cancer and environmental tobacco smoke. Br Med J 315: 980–988

Hall P, Mattsson A, Boice JD Jr (1996) Thyroid cancer after diagnostic administration iodine-131. Radiat Res 145(1): 86–92

Hauptmann M, Pohlabeln H, Lubin JH, Jockel KH, Ahrens W, Bruske-Hohlfeld I, Wichmann HE (2002) The exposure-time-response relationship between occupational asbestos exposure and lung cancer in two German case-control studies. Am J Ind Med 41: 89–97

Heasman MA, Kemp IW, Urquhart JD, Black R (1986) Childhood leukaemia in northern Scotland. Letter. Lancet i: 266

Hodgson JT, Darnton A (2000) The quantitative risks of mesothelioma and lung cancer in relation to asbestos exposure. Ann Occup Hyg 44: 565–601

Inskip P, Tarone R, et al. (2001) Cellular-phone use and brain tumors. N Engl J Med 344: 79–86

International Agency for Research on Cancer [IARC] (1977) Asbestos. IARC Monogr Eval Carcinog Risk Chem Man 14: 1–106

International Agency for Research on Cancer [IARC] (1991) Occupational exposures in insecticide application, and some pesticides. IARC Working Group on the Evaluation of Carcinogenic Risks to Humans. IARC Monogr Eval Carcinog Risks Hum 53: 5–586

International Agency for Research on Cancer [IARC] (1997) Polychlorinated dibenzo-para-dioxins and polychlorinated dibenzofurans. IARC Monogr Eval Carcinog Risks Hum 69 1–631

International Commission on Radiological Protection [ICRP] (1990) ICRP Publication 60. Recommendations of the International Commission on Radiological Protection. Annals of the ICRP 21(1–3), Pergamon Press, Oxford, 1991

Ivanov V (1996) Health status and follow-up of liquidators in Russia. In: Karaoglou A, Desmet G G, Kelly N, Menzel HG (eds) The radiological consequences of the Chernobyl accident. Proceedings of the first international conference, Minsk, Belarus, 1996. Office for Official Publications of the European Communities, Luxemburg, EUR 16544. pp 861–870

Ivanov VK, Tsyb AF, Gorsky AI, Maksyutov MA, Rastopchin EM, Konogorov AP, Korelo AM, Biryukov AP, Matyash VA (1997) Leukaemia and thyroid cancer in emergency workers of the Chernobyl accident: estimation of radiation risks (1986–1995). Radiat Environ Biophys 36: 9–16

Jacob P, Goulko G, Heidenreich WF, Likhtarev I, Kairo I, Tronko ND, Bogdanova TI, Kenigsberg J, Buglova E, Drozdovitch V, Golovneva A, Demidchik EP, Balonov M, Zvonova I, Beral V (1998) Thyroid cancer risk to children calculated. Letter. Nature 392: 31–32

Johnson KC, Hu J; Canadian Cancer Registries Epidemiology Research Group (2001). Lifetime residential and workplace exposure to environmental tobacco smoke and lung cancer in never-smoking women, Canada 1994–97. Int J Cancer 93(6): 902–906

Kinlen L (1994) Leukaemia. In: Doll R, Frameni JF, Muir CS (eds) Trends in cancer incidence and mortality. Cold Spring Harbor Laboratory Press, Plainview, NY, pp 475–491

Kossenko MM, Degteva MO, Vyushkova OV, Preston DL, Mabuchi K, Kozheurov VP (1997) Issues in the comparison of risk estimates for the population in the Techa River region and atomic bomb survivors. Radiat Res 148: 54–63

Laden F, Collman G, Iwamoto K, Alberg AJ, Berkowitz GS, Freudenheim JL, Hankinson SE, Helzlsouer KJ, Holford TR, Huang HY, Moysich KB, Tessari JD, Wolff MS, Zheng T, Hunter DJ (2001) 1,1-Dichloro-2,2-bis(p-chlorophenyl)ethylene and polychlorinated biphenyls and breast cancer: combined analysis of five U.S. studies. J Natl Cancer Inst 93: 768–776

Liddell FDK, Thomas DC, Gibbs GW, McDonald JC (1984) Fibre exposure and mortality from pneumoconiosis, respiratory and abdominal malignancies in chrysotile production in Quebec, 1926–75. Ann Acad Med 13: 340–344

Little MP, Muirhead CR (2000) Derivation of low-dose extrapolation factors from analysis of curvature in the cancer incidence dose-response in Japanese atomic bomb survivors. Int J Radiat Biol 76: 939–953

Liu BQ, Peto R, Chen ZM, Boreham J, Wu YP, Li JY, Campbell TC, Chen JS (1998) Emerging tobacco hazards in China: 1. Retrospective proportional mortality study of one million deaths. Br Med J 317: 1411–1422

Lubin JH, Boice Jr. JD, Edling C, Hornung RW, Howe G, Kunz E, Kusiak RA, Morrison HI, Radford EP, Samet JM, Tirmarche M, Woodward A, Yao SX, Pierce DA (1994) Lung cancer and radon: a joint analysis of 11 underground miners studies. U.S. National Institutes of Health, Bethesda MD; Publication No. 94-3644

MacMahon B (1962) Prenatal X-ray exposure and childhood cancer. J Natl Cancer Inst 28: 1173–1191

Magee F, Wright JL, Chan N, Lawson L, Churg A (1986) Malignant mesothelioma caused by childhood exposure to long-fiber low aspect ratio tremolite. Am J Ind Med 9: 529–533

Mason TJ, Miller RW (1974) Cosmic radiation at high altitudes and U.S. cancer mortality, 1950–1969. Radiat Res 60: 302–306

McKinlay A (1997) Possible health effects related to the use of radiotelephones: recommendations of a European Commission Expert Group. Radiol Prot Bull 187: 9–16

Meinert R, Kaletsch U, Kaatsch P, Schuz J, Michaelis J (1999) Associations between childhood cancer and ionizing radiation: results of a population-based case-control study in Germany. Cancer Epidemiol Biomarkers Prev 8: 793–799

Miller AB, Howe GR, Sherman GJ, Lindsay JP, Yaffe MJ, Dinner PJ, Risch HA, Preston DL (1989) Mortality from breast cancer after irradiation during fluoroscopic examinations in patients being treated for tuberculosis. N Engl J Med 321: 1285–1289

Modan B (1991) Low-dose radiation epidemiological studies: an assessment of methodological problems. Review. Ann ICRP 22: 59–73

Mole RH (1990) Childhood cancer after prenatal exposure to diagnostic X-ray examinations in Britain. Br J Cancer 62: 152–168

Mollo F, Magnani C, Bo P, Burlo P, Cravello M (2002) The attribution of lung cancers to asbestos exposure: a pathologic study of 924 unselected cases. Am J Clin Pathol 117: 90–95

Muirhead CR, Butland BK, Green BM, Draper GJ (1991) Childhood leukaemia and natural radiation. Letter. Lancet 337: 503–504

Muirhead CR, Goodill AA, Haylock RG, Vokes J, Little MP, Jackson DA, O'Hagan JA, Thomas JM, Kendall GM, Silk TJ, Bingham D, Berridge GL (1999) Occupational radiation exposure and mortality: second analysis of the National Registry for Radiation Workers. J Radiol Prot 19: 3–26

Muirhead CR, Kneale GW (1989) Prenatal irradiation and childhood cancer. J Radiol Prot 9: 209–212

Naumburg E BR, Cnattingius S, Hall P, Boice JD Jr, Ekbom A (2001) Intrauterine exposure to diagnostic X rays and risk of childhood leukemia subtypes. Radiat Res 156: 718–723

Nordstrom M, Hardell L, Lindstrom G, Wingfors H, Hardell K, Linde A (2000) Concentrations of organochlorines related to titers to Epstein-Barr virus early antigen IgG as risk factors for hairy cell leukemia. Environ Health Perspect 108: 441–445

Oppenheim BE, Griem ML, Meier P (1975) The effects of diagnostic X-ray exposure on the human fetus: an examination of the evidence. Radiology 114: 529–534

Ott MG, Teta MJ, Greenberg HL (1989) Assessment of exposure to chemicals in a complex work environment. Am J Ind Med 16: 617–630

Parkin DM, Kramarova E, Draper GJ, Masuyer E, Michaelis J, Neglia J, Qureshi S, Stiller CA (1998) International incidence of childhood cancer, Vol. II. IARC Sci Publ 144: 1–391

Paur R, Woitowitz HJ, Rodelsperger K, John H (1985) [Pleural mesothelioma following asbestos exposure during brake repairs in the automobile trade: case report] (Ger.). Prax Klin Pneumol 39(10): 362–366

Pershagen G, Åkerblom G, Axelson O, Clavensjo B, Damber L, Desai G, Enflo A, Lagarde F, Mellander H, Svartengren M, Swedjemark GA (1994) Residential radon exposure and lung cancer in Sweden. N Engl J Med 330(3): 159–164

Pesatori A, Sontag J, Lubin JH, Consonni D, Blair A (1994) Cohort mortality and nested case-control study of lung cancer among structural pest control workers in Florida (United States). Cancer Causes Control 5(4): 310–318

Pierce DA, Preston DL (2000) Radiation-related cancer risks at low doses among atomic bomb survivors. Radiat Res 154: 178–186

Pierce DA, Shimizu Y, Preston DL, Vaeth M, Mabuchi K (1996) Studies of the mortality of atomic bomb survivors. Report 12, Part 1: 1950–1990. Radiat Res 146(1): 1–27

Poirier MC (1997) DNA adducts as exposure biomarkers and indicators of cancer risk. Environ Health Perspect 105 Suppl 4: 907–912

Poole C, Rothman KJ, Dreyer NA (1988) Leukaemia near Pilgrim nuclear power plant, Massachusetts. Lancet 2(8623): 1308

Richardson S, Monfort C, Green M, Draper G, Muirhead C (1995) Spatial variation of natural radiation and childhood leukaemia incidence in Great Britain. Stat Med 14(21–22): 2487–2501

Robinson CF, Petersen M, Palu S (1999) Mortality patterns among electrical workers employed in the U.S. construction industry, 1982–1987. Am J Ind Med 36(6): 630–637

Ron E, Lubin JH, Shore RE, Mabuchi K, Modan B, Pottern LM, Schneider AB, Tucker MA, Boice JD Jr (1995) Thyroid cancer after exposure to external radiation: a pooled analysis of seven studies. Radiat Res 141(3): 259–277

Rothman KJ, Chou CK, Morgan R, Balzano Q, Guy AW, Funch DP, Preston-Martin S, Mandel J, Steffens R, Carlo G (1996) Assessment of cellular telephone and other radio frequency exposure for epidemiologic research. Epidemiology 7(3): 291–298

Salonen T, Saxén L (1975) Risk indicators in childhood malignancies. Int J Cancer 15: 941–946

Sandén A, Näslund PE, Jarvholm B (1985) Mortality in lung and gastrointestinal cancer among shipyard workers. Int Arch Occup Environ Health 55(4): 277–283

Saracci R (1977) Asbestos and lung cancer: an analysis of the epidemiological evidence on the asbestos-smoking interaction. Int J Cancer 20: 323–331

Sharp L, Black RJ, Harkness EF, McKinney PA (1996) Incidence of childhood leukaemia and non-Hodgkin's lymphoma in the vicinity of nuclear sites in Scotland, 1968–93. Occup Environ Med 53(12): 823–831

Souchkevitch GN (1996) Main scientific results of the WHO International Programme on the Health Effects of the Chernobyl Accident (IPHECA). World Health Stat Q 49: 209–212

Steenland K, Deddens J, Piacitelli L. (2001) Risk assessment for 2,3,7,8-tetrachlorodibenzo-p-dioxin (TCDD) based on an epidemiologic study. Am J Epidemiol 154(5): 451–458

Steenland K, Piacitelli L, Deddens J, Fingerhut M, Chang LI (1999) Cancer, heart disease, and diabetes in workers exposed to 2,3,7,8-tetrachlorodibenzop-dioxin. J Natl Cancer Inst 91: 779–786

Stewart A, Webb J, Giles D, Hewitt D (1956) Preliminary communication: malignant disease in childhood and diagnostic irradiation in utero. Lancet 2: 447–448

Streffer C, Bücker J, Cansier A, Cansier D, Gethmann CF, Guderian R, Hanekamp G, Henschler D, Pöch G, Rehbinder E, Renn O, Slesina M, Wuttke K (2000) Umweltstandards. Kombinierte Expositionen und ihre Auswirkungen auf die Umwelt. Wissenschaftsethik und Technikfolgenbeurteilung, Bd. 5. Springer, Berlin

Tao Z, Wei L (1986) An epidemiological investigation of mutational diseases in the high background radiation area of Yangjiang, China. J Radiat Res (Tokyo) 27: 141–150

Thompson DE, Mabuchi K, Ron E, Soda M, Tokunaga M, Ochikubo S, Sugimoto S, Ikeda T, Terasaki M, Izumi S, Preston DL (1994) Cancer incidence in atomic bomb survivors. Part II: Solid tumors, 1958–1987. Radiat Res 137(2 Suppl): S17–S67

Tynes T, Haldorsen T (1997) Electromagnetic fields and cancer in children residing near Norwegian high-voltage power lines. Am J Epidemiol 145: 219–226

United Nations Scientific Committee on the Effects of Atomic Radiation [UNSCEAR] (1993) Report to the General Assembly, with 9 scientific annexes. United Nations, New York

United Nations Scientific Committee on the Effects of Atomic Radiation [UNSCEAR] (1994) Report to the General Assembly, with 2 scientific annexes. United Nations, New York

United Nations Scientific Committee on the Effects of Atomic Radiation [UNSCEAR] (2000) Report to the General Assembly, with 10 scientific annexes. United Nations, New York

Walter S, King W, Marrett LD (1999) Association of cutaneous malignant melanoma with intermittent exposure to ultraviolet radiation: results of a case-control study in Ontario, Canada. Int J Epidemiol 28(3): 418–427

Waxweiler RJ, Smith AH, Falk H, Tyroler HA (1981) Excess lung cancer risk in a synthetic chemicals plant. Environ Health Perspect 41: 159–165

Wei L, Wang J (1994) Estimate of cancer risk for a large population continuously exposed to higher background radiation in Yangjiang, China. Chinese Med J 107: 541–544

Wei L, Zha Y, Tao ZF, He WH, Chen DQ, Yuan YL (1990) Epidemiological investigation of radiological effects in high background radiation areas of Yangjiang, China. J Radiat Res (Tokyo) 31(1): 119–136

Weinberg CR, Brown KG (1987) Altitude, radiation, and mortality from cancer and heart disease. Radiat Res 112: 381–390

Westermeier T, Michaelis J (1995) Applicability of the Poisson distribution to model the data of the German Children's Cancer Registry. Radiat Environ Biophys 34: 7–11

Whiteman DC, Whiteman CA, Green AC (2001) Childhood sun exposure as a risk factor for melanoma: a systematic review of epidemiologic studies. Cancer Causes Control 12: 69–82

Wilkins JR, III, Hundley VD (1990) Paternal occupational exposure to electromagnetic fields and neuroblastoma in offspring. Am J Epidemiol 131: 995–1008

Woolcott CG, Aronson KJ, Hanna WM, SenGupta SK, McCready DR, Sterns EE, Miller AB (2001) Organochlorines and breast cancer risk by receptor status, tumor size, and grade (Canada). Cancer Causes Control 12(5): 395–404

Yamashita S, Shibata Y (1997) Chernobyl: a decade. Proceedings of the Fifth Chernobyl Sasakawa Medical Cooperation Symposium, Kiev, Ukraine, October 1996. Elsevier Science BV, Amsterdam

Yano E, Wang ZM, Wang XR, Wang MZ, Lan YJ (2001) Cancer mortality among workers exposed to amphibole-free chrysotile asbestos. Am J Epidemiol 154(6): 538–543

Yoshimoto Y, Delongchamp R, Mabuchi K (1994) In-utero exposed atomic bomb survivors: cancer risk update. Letter. Lancet 344(8918): 345–346

Zeegers MP, Swaen GM, Kant I, Goldbohm RA, van den Brandt PA (2001) Occupational risk factors for male bladder cancer: results from a population based case cohort study in the Netherlands. Occup Environ Med 58(9): 590–596

6 Mathematical Models of Carcinogenesis

6.1
Introduction

6.1.1
Models of carcinogenesis

As discussed in chapters 3 and 4, the process of carcinogenesis is very complex. It is too complex to be described in full detail by a mathematical model. Therefore, modelling carcinogenesis implies simplifications that try to identify the most important features of the process. Resulting predictions can be tested in laboratory experiments with carcinogens or by analyses of epidemiological cohorts. Compared to conventional risk models used in epidemiology, as, e. g, the excess relative risk model, mathematical models of carcinogenesis have the advantage of a straightforward description of complex exposure scenarios. After identification of the action of a carcinogen on the parameters of the model, no additional parameters are necessary to calculate the effects of exposures that may change many times over lifetime or differ for the various members of the study cohort.

Normal cells that have the potential to develop into a cancer cell are called here susceptible. Most stem cells are assumed to be susceptible for a stimulation of the carcinogenic process. Carcinogenesis is generally modelled by a process in which a susceptible normal cell undergoes several alterations to become a cancer cell. In most cases, such alterations are genetic or epigenetic changes. The number of necessary genetic alterations appears to vary from cancer to cancer. In the case of retinoblastoma, two mutations that inactivate both copies of the RB tumour-suppressor gene apparently suffice for tumour development (Knudson 1971, Bishop 1991; Weinberg 1991). In the case of colorectal cancer alterations in three tumour-suppressor genes (*APC, p53* and *DCC*) and one proto-oncogene (K-*ras*) have been observed during progressive stages of tumour development (Fearon and Vogelstein 1990). The presence of four or more mutations in the genesis of colorectal cancer suggests the existence of events that increase the rates of subsequent mutations, i. e. the induction of a genomic instability (Aaltonen et al. 1993; Peltomäki et al. 1993) or mechanisms that inactivate proteins of tumour-suppressor genes, e. g. a dominant negative action, as it has been observed for a mutant form of the p53 protein (Gannon 1990). Also, the increase of the division rate of intermediate cells in the carcinogenic process may lead to an increase of critical mutational events.

Phenotypically altered cells, that are believed to represent intermediate stages of carcinogenesis, may form clones as papillomas of the mouse skin, enzyme altered foci in the rodent liver, or adenomatous polyps in the human colon. Correspondingly, a large family of mathematical models of carcinogenesis contains a clonal expansion of the intermediate cells.

Early models of carcinogenesis (Nordling 1953; Armitage and Doll 1954) were proposed to describe age-specific cancer mortality data that for many cancer types could be approximated by an increase with a power of age. section 6.2 describes these early models, in which carcinogenesis is modelled by the occurrence of a number of cellular transformations. If all transformation rates are small, then a model with k cellular transformations leads to a proportionality of the mortality rate to age to the power of $(k - 1)$.

Section 6.3 describes a further development in modelling of carcinogenesis, namely the explicit implementation of cell proliferation kinetics, as it was introduced by Armitage and Doll (1957). The two-step version of the model was successfully applied to the genesis of retinoblastoma (Knudson 1971). The model was further developed by Moolgavkar and Venzon (1979) by taking account of cell death and differentiation. Two-step models of carcinogenesis with a clonal expansion of intermediate cells are among the simplest models that can describe age and exposure dependencies in many observational data. This is possibly because the complex transition from a normal cell to a malignant cell can be approximated at least in some cases by two rate-limiting events. Independent of this, there is an increasing effort to develop and apply multistep models that take into account cell proliferation kinetics (Moolgavkar and Luebeck 1992; Little 1995; Little et al. 2002; Luebeck and Moolgavkar 2002).

Models of carcinogenesis and effects of environmental agents have been reviewed by Kopp-Schneider (1997) and by Moolgavkar et al. (1999), and the present chapter focuses on basic concepts and on more recent work. The application of mathematical models of carcinogenesis to leukaemia has been explored only to a limited degree and is therefore not treated here. The interested reader may find an analysis of the lymphocytic leukaemia incidence in England and Wales with mathematical models of carcinogenesis in Little et al. (1996).

6.1.2
Hazard and excess risk

In epidemiology, new cancer cases or cancer deaths observed during a given time period among a collective of people are expressed by the cancer incidence rate or cancer mortality rate, respectively. These rates are defined by the number of cases per person-years of observation. In modelling, both rates are often called the hazard $H(a)$, where a denotes the age. The hazard is related to the probability $p(a - t)$ that a susceptible cell has become a malignant cell at age $(a - t)$, where t is the lag time that is necessary for the malignant cell to grow in a detectable tumour (incidence data) or to lead to death (mortality data). If there are n_s susceptible cells in an organ, then the probability $\Psi(a - t)$ that none of these cells has become malignant is

$$\Psi(a - t) = [1 - p(a - t)]^{n_s}. \qquad \text{(eq. 6.1)}$$

In the modelling literature, sometimes the function $\Psi(a)$ is called the survival function $S(a)$. We do not follow this terminology, because the term survival function

$S(a)$ is used here as in the epidemiological literature for all causes of death and not just for cancer (see below).

The hazard of a tumour at age a is defined by the quotient of the number of new cancer cases and the number of people without cancer observed in a time interval, which can be expressed by the probability - that at least one cell has become malignant in the time interval , where the dot denotes the derivative with respect to age a, divided by the probability that up to the time interval no cell has become malignant:

$$H(a) = \dot{\Psi}(a - t) / \Psi(a - t). \tag{eq. 6.2}$$

With eq. (6.1) one obtains

$$H(a) = n_s \, \dot{p}(a - t) / [1 - p(a - t)]. \tag{eq. 6.3}$$

Under the influence of a carcinogenic agent, the hazard is conventionally divided into a baseline hazard $H_0(a)$ and the hazard that is attributed to the agent. Fig. 6.1 shows as an example the baseline hazard $H_0(a)$ to get a solid tumour at age a, as it was derived by an adaptation of a model of carcinogenesis to the solid cancer incidence data for the male atomic bomb survivors.

The difference between the total hazard and the spontaneous hazard is called the excess absolute risk (see also chapter 5):

$$EAR(a) = H(a) - H_0(a). \tag{eq. 6. 4}$$

The ratio of the excess absolute risk and the spontaneous hazard is called the excess relative risk

$$ERR(a) = H(a)/H_0(a) - 1. \tag{eq. 6. 5}$$

In order to calculate cancer risks over given periods of time or lifetime risks, a survival function $S(a)$ is used, which is defined by the probability that a newborn

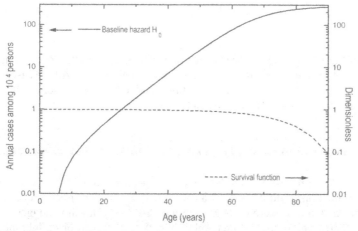

Fig. 6.1 Baseline hazard $H_0(a)$ to obtain a solid tumour at age a among male atomic bomb survivors with negligible radiation exposures, as derived with the TSCE model (see section 6.3.1), and the survival function $S(a)$ for German males in the period 1996–98 (Statistisches Bundesamt 2000).

person will survive until age a. In general, complete and reliable data on survival functions are not available. Therefore the survival function is approximated by products of age dependent death rates for a given calendar year (period) as they are easily available. Fig. 6.1 gives as an example a survival function that has been obtained from German death rates in the period 1996–98, according to which a newborn person has a chance of 9 % to survive until the age of 90.

The baseline lifetime hazard R_0 to get cancer or to die of cancer is defined by

$$R_0 = \int S(a)\, H_0(a)\, da. \tag{eq. 6.6}$$

For the example in Fig. 6.1 the baseline lifetime hazard R_0 (up to the age 90) has a value of 0.33, i. posure of a carcinogenic agent, the total lifetime hazard R is

$$R = \int S(a)\, H(a)\, da. \tag{eq. 6.7}$$

The excess relative lifetime risk ERR time due to the exposure is defined by

$$ERR = R/R_0 - 1. \tag{eq. 6.8}$$

Note that this is a slightly simplified concept, continuative references are Vaeth and Pierce (1990) and Kellerer et al. (2001).

Hazard functions and related excess relative risks have been derived by applying models of carcinogenesis to epidemiological cohorts and animal experimental data sets with exposures to ionising radiation (section 6.4), chemical carcinogens (section 6.5) and combined exposures (section 6.6). In addition, the sections give some more theoretical derivations as considerations of threshold values or properties of the hazard function for combined exposures with low doses.

6.2
Models without clonal expansion

For most cancer sites, carcinogenesis is considered as a complex process with several cellular transitions. In most cases these cellular transitions are mutations. However, they can be also epigenetic or other cellular events. For some of these transitions, it will not be of importance, which happens first. For other transitions, the sequence might be of importance. In the modelling, mainly two extreme cases, total independence on the sequence of the transitions, and definite sequence of all transitions (Armitage-Doll model) have been analysed.

6.2.1
A model in which the sequence of cellular transitions is inconsequential

Nordling (1953) proposed a model of carcinogenesis in which a cell has to undergo k transitions until it becomes a cancer cell. The order of occurrence of the transitions was not considered to be important. Although Nordling suggested that due to symmetric cell divisions the number of cells with transitions might increase leading to an increasing probability that a further mutation occurs in one of the cells, this effect was not taken into account in the mathematical formulation of the model.

In the following λ_κ denotes the rate of the κth cellular transition, where κ can have the values $1,\ldots, k$. Then the number of cells without the κth transition will change according to

$$\dot{n}(a) = -\lambda_\kappa(a)\, n(a). \tag{eq. 6.9}$$

If the rate λ_κ is constant over lifetime, then the number of cells without the κth transition is:

$$n(a) = n(0)\, \exp(-\lambda_\kappa a). \tag{eq. 6.10}$$

So the probability that the κth transition has occurred in a susceptible cell at age a will be

$$p_\kappa(a) = 1 - \exp(-\lambda_\kappa a). \tag{eq. 6.11}$$

According to the model the transitions are independent. If all k transitions have occurred in a cell, then the cell is called a tumour cell. The probability that a tumour cell has arisen

$$p(a) = \prod_{\kappa=1}^{k}[1 - \exp(-\lambda_\kappa a)]. \tag{eq. 6.12}$$

The hazard $H(a)$ at age a is calculated according to eq. (6.3):

$$H(a) = \frac{n_s\, p(a-t)}{1 - p(a-t)} \sum_{\kappa=1}^{k} \frac{\lambda_\kappa \exp[-\lambda_\kappa(a-t)]}{\{1 - \exp[-\lambda_\kappa(a-t)]\}}. \tag{eq. 6.13}$$

If all transition rates are equal, $\lambda_\kappa = \lambda$, then the hazard is given by

$$H(a) = \frac{n_s\, k\, \lambda \exp[-\lambda\,(a-t)]\, \{1 - \exp[-\lambda\,(a-t)]\}^{k-1}}{1 - \{1 - \exp[-\lambda\,(a-t)]\}^{k}}. \tag{eq. 6.14}$$

For a given number of susceptible cells and a constant lag time, the hazard decreases with decreasing transition rates and with an increasing number of necessary transitions (Fig. 6.2.). Under the assumption of $n_s = 10^9$ susceptible cells, in the two-step model rates of about 3×10^{-5} a^{-1} produce a cancer hazard that is comparable with observed cancer incidence rates. If more transitions are assumed to be necessary then the transition rates have to be considerably larger. This could be the case for epigenetic events, or if early mutations lead to an increase of later transition rates, e. g. by an induced genomic instability.

If for all transitions $\lambda_\kappa(a-t) \ll 1$, then the leading term in eq. (6.13) for $H(a)$ becomes

$$H(a) \approx n_s\, k\, (a-t)^{k-1} \prod_{\kappa=1}^{k} \lambda_\kappa. \tag{eq. 6.15}$$

Nordling (1953) derived for male adults a proportionality of all cancer deaths to the sixth power of age, indicating in the frame of a simple multistage model that seven transitions would be necessary to create a cancer cell.

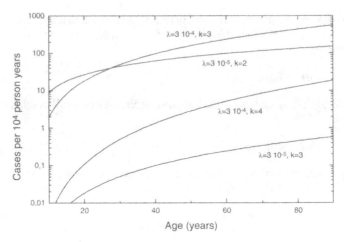

Fig. 6.2 Hazard as a function of age according to a multistep model of carcinogenis in which the sequence of cellular transitions is inconsequential (eq. 6.2.6); two transitions, each with λ = 3×10^{-5} yr^{-1}; three transitions, each with $\lambda = 3 \times 10^{-5}$ yr^{-1} or $\lambda = 3 \times 10^{-4}$ yr^{-1}; and four transitions, with $\lambda = 3 \times 10^{4}$ yr^{-1}. The number of susceptible cells n_s has been assumed to be 10^{9}, and the lag time to be 5 years.

6.2.2
The Armitage-Doll model

Armitage and Doll (1954) proposed a model of carcinogenesis in which a susceptible cell has to undergo k events to become a cancer cell when the mutations occur in the right order (Fig. 6.3). Let $p_0(a)$ be the probability that a susceptible cell has not undergone the first mutation at age a, and $p_k(a)$ is the probability that a malignant cell develops at age a from a susceptible cell. According to the Armitage-Doll model, the probabilities $p_\kappa(a)$; $\kappa = 1,...,k$ that the cell has undergone the first κ mutations are described by the following system of differential equations:

$$\dot{p}_0(a) = -\lambda_1 p_0(a)$$
$$\dot{p}_1(a) = \lambda_1 p_0(a) - \lambda_2 p_1(a)$$
$$\dot{p}_{k-1}(a) = \lambda_{k-1} p_{k-2}(a) - \lambda_k p_{k-1}(a)$$
$$\dot{p}_k(a) = \lambda_k p_{k-1}(a),$$

(eq. 6.16)

with the initial conditions $p_0(0) = 1$ and $p_\kappa(0) = 0$ for $\kappa = 1,..,k$. If all transition rates λ_κ are different and do not change over life time, then the solution of the system for

Fig. 6.3 The Armitage-Doll model. In order to become a cancer cell, susceptible cells have to undergo k transitions in the right order of occurrence. In addition, it is assumed here that the cancer cell needs a lag time t to develop to a detectable tumour.

$p_k(a)$ is a sum of exponentials of $\lambda_\kappa a$ with coefficients depending on λ_κ (see Mool-gavkar 1978). It is straightforward to calculate the hazard according to eq. (6.3). If $\lambda_\kappa(a - t) \ll 1$ for all $\kappa = 1,...,k$, then the leading term for $H(a)$ becomes

$$H(a) \approx n_s (a\text{-}t)^{k-1} (\prod_{\kappa=1}^{k} \lambda_\kappa) / (k\text{-}1)!, \tag{eq. 6.17}$$

which is apart from a factor k! the same as in the model in which the sequence of the mutations does not play any role (compare with eq. 6.15).

Armitage and Doll (1954) applied their model to the cancer mortality in different organs in England and Wales in 1950 and 1951. The mortality rates due to stomach, colon, rectum and pancreas cancer were proportional to the age to a power of 5 to 6.5. According to the model that would indicate that 6 to 7 or 8 cellular transitions would be necessary for a development of a healthy cell to a malignant cell. A second group of cancers – composed of cancer of lung, prostate, female breast, ovary and cervix and corpus uteri – did not follow the power law in eq. 6.17. Armitage and Doll attributed this to changes of endocrine secretions over lifetime which are considered to be important in these organs.

6.3
Models with clonal expansion

Clones of phenotypically changed cells as papillomas of the mouse skin, foci with altered enzyme activity in the rodent liver, or adenomatous polyps in the human colon are considered to be an intermediate state in the carcinogenic process. This indicates that intermediate cells have the ability to proliferate which has been introduced in models of carcinogenesis and is often called *clonal expansion*.

Multistage models with clonal expansion in general describe four main phases: (i) *Initiation* characterised by an accumulation of genetic changes leading to partial abrogation of growth control. (ii) *Promotion*, or clonal growth of the number of initiated or intermediate cells. (iii) *Malignant conversion* or *transformation*, which is an acquisition of further cellular changes leading to one (or more) malignant cells. (iv) *Tumour growth*, i. e. growth of the malignant cell to a detectable tumour or leading to the death of the organism. In general, carcinogens are considered to increase the transition rates of one or several of these phases.

6.3.1
The two-step clonal expansion model

One of the best analysed models of carcinogenesis is the stochastic two-step clonal expansion (TSCE) model (Moolgavkar and Venzon 1979). In this model, the promotion of intermediate cells is the net result of the stochastic processes of symmetrical cell division with rate α, death or differentiation with rate β, and asymmetric transformation with rate λ_2 (Fig. 6.4). In a first approximation, the promotion rate is equal to $\alpha - \beta$. The ratio β/α is the asymptotic probability of extinction of a clone (Luebeck et al. 2000). For $\beta > \alpha$ this probability is 1.

In the calculation of the hazard, clones of intermediate cells are followed until they have either died or until one of its cells has converted into a malignant cell. If

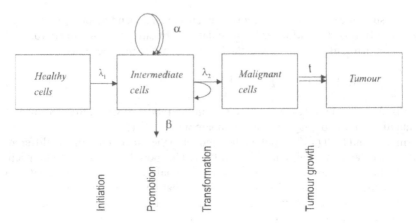

Fig. 6.4 Schematic presentation of the TSCE model with an initiation rate λ_1, and a division rate α, a death or differentiation rate β and a transformation rate λ_2 of intermediate cells[1]. During a lag time t a malignant cell is assumed to grow to a detectable tumour (if the model is adapted to incidence data) or to death due to the tumour (if the model is adapted to mortality data).

– for simplicity – the lag time is assumed to be a constant value t, then the hazard at an age a is proportional to the number of intermediate cells at the age $a - t$. Thus the assumption of a constant lag time facilitates the interpretation of model results. Analyses of the solid cancer incidence data among the atomic bomb survivors of Hiroshima and Nagasaki with constant lag times between zero and ten years gave similar results (Kai et al. 1997), indicating that the value of the lag-time or its distribution does not strongly influence the results. Also, Pierce and Mendelsohn (1999) obtained in simulation calculations for a multistep model of carcinogenesis, that there was little change of their results for the atomic bomb survivors if the standard deviation of the lag time is less or about 2–3 years. For simplicity, in this section a constant lag time t of five years is assumed.

6.3.1.1
The baseline hazard $H_0(a)$

In general, the solution of the stochastic two-step clonal expansion model needs complex mathematical tools (Moolgavkar et al. 1988; Moolgavkar and Luebeck 1990).[2] For model parameters that do not change over life time the model can be solved analytically (Kopp-Schneider et al. 1994; Zheng 1994). Such a model may be considered as the most simple approximation for spontaneous carcinogenesis. In this case the hazard $H_0(a)$ is completely determined by three combinations of the biological parameters (Heidenreich et al. 1996):

$$H_0(a) = \frac{X \{\exp[(\gamma + 2q)(a - t)] - 1\}}{q \{\exp[(\gamma + 2q)(a - t)] + 1\} + \gamma}, \qquad \text{(eq. 6.18)}$$

---

[1] In the literature, λ_1 or ($\lambda_1 \, n_s$) are often denoted by v, and λ_2 by μ.
[2] In mathematical terms, the model can be described by a Markovian process in which the probability generating function satisfies the Kolmogorov forward differential equation.

with

$$X = n_s \lambda_1 \lambda_2$$

$$\gamma = \alpha - \beta - \lambda_2 \qquad\qquad (eq. 6.19)$$

$$q = (\sqrt{\gamma^2 + 4\alpha\lambda_2} - \gamma)/2 .$$

Such parameter combinations that determine the hazard are called identifiable parameters because in principle they can be determined from fits of the model to epidemiological or experimental data. In the parameter set in eq. 6.19, the parameter γ is sometimes called the effective clonal expansion rate[3], because for intermediate ages the number of initiated cells grows exponentially with the product of age (minus lag time) and the rate $\gamma + 2q$ (see below, eq. 6.21), and in most applications $2q$ is considerably smaller than γ.

Equation 6.18 was found to fit the age dependence of the baseline risk in radioepidemiological data well resulting in estimates of the parameters X, γ and q. Fig. 6.5 shows an example with the values $X = 10^{-6}$ a^{-2}, $\gamma = 0.13$ a^{-1} and $q = 3.5 \times 10^{-5}$ a^{-1} as they have been obtained for all solid cancers among the atomic bomb survivors from Hiroshima and Nagasaki with negligible radiation exposures (Jacob and Prokić 2003).

If two values of or relations between biological parameters are known or assumed, then estimates of the other biological parameters can be obtained. Assuming n_s equal to 4×10^8 and $\lambda_1 = \lambda_2$, one would obtain $\lambda_1 = \lambda_2 = 5 \times 10^{-8}$ a^{-1} and $\alpha \approx \beta \approx$ $^{-1}$. Another possibility is to assume n_s equal to 10^8 and $\alpha = 9$ a^{-1}, resulting

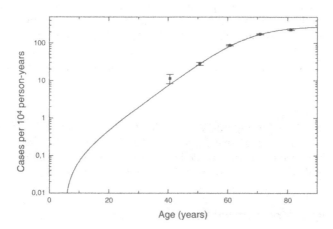

Fig. 6.5 Baseline hazard $H_0(a)$ for solid tumour incidence at age a among the male atomic bomb survivors with negligible radiation exposures, data points for the observation period 1958–1987 (black squares; Thompson et al. 1994), and as derived with the TSCE model (solid line; Jacob and Prokić 2003).

---

[3] A better approximation for the effective clonal expansion rate is, however, $\alpha - \beta + \mu \cdot (\alpha + \beta)/(\alpha - \beta)$.

in $\lambda_1 \approx 2 \times 10^{-8}\,a^{-1}$, $\lambda_2 \approx 5 \times 10^{-7}\,a^{-1}$ and $\beta = 8.87\,a^{-1}$. In this case, the mutation rate λ_2 of intermediate cells to malignant cells would be by about a factor of 25 larger than the mutation rate λ_1 of healthy cells to intermediate cells. In both cases, the proliferation rate α and the death or differentiation rate β of intermediate cells are similar, but their small difference is enough to cause a considerable exponential growth of the hazard over a large period of life. These numbers serve only as an illustration to what degree biological parameters can be determined. To obtain more meaningful conclusions, cancer-type specific estimates have to be made.

For young ages, the spontaneous hazard in eq 6.17 increases linearly with age:

$$H_0(a) \approx X\,(a - t); \qquad \text{for } a - t \ll (\gamma + 2q)^{-1}, \tag{eq. 6.20}$$

as it would also result from the approximate solution of the Armitage-Doll model (eq. 6.17). In the numerical example given above, $(\gamma + 2q)^{-1}$ is about 8 years and the linear growth applies to an age period of 5 to about 9 years. In this phase the hazard growth with age is dominated by the initiation of susceptible cells to intermediate cells.

For intermediate ages, and as long as $\gamma/q \gg \exp[(\gamma + 2q)\,(a - t)]$, $H_0(a)$ grows exponentially

$$H_0(a) = X \exp[(\gamma + 2q)(a - t)]/\gamma \qquad \text{for } a - t \gg (\gamma + 2q)^{-1}$$

$$\text{and } a - t \ll \ln(\gamma/q)/(\gamma + 2q). \tag{eq. 6.21}$$

In this phase, the hazard growth is dominated by the promotion of intermediate cells. The number of intermediate cells being produced in this age by initiating events becomes unimportant. In the example, this phase starts at an age of about 20 years. The value of $\ln(\gamma/q)/(\gamma + 2q)$ is about 60 years and the baseline hazard starts to deviate from the exponential growth at an age of about 40 years. In other examples, the period of the exponential growth of the hazard exceeds normal human lifetime.

For older ages, the exponential growth flattens and the hazard approaches an asymptotic value:

$$H_0(a) \Rightarrow X/q \qquad \text{for } a - t \gg \ln(\gamma/q)/(\gamma + 2q). \tag{eq. 6.22}$$

The existence of an asymptotic value is due to the fact that after a certain age clones have become so large that the sum of the probabilities for one cell within them being converted to a malignant cell and for an extinction of a clone is equal to the probability that a new clone is created by an initiating mutation.

6.3.1.2
Induced changes of parameters

Carcinogenic agents are assumed to change the parameters of the model for the time period of exposure, or changes may even persist after the time of exposure. An obvious example of a persisting effect is genomic instability. Another would be if parameters are changed during the period in which damage due to the exposure has to be repaired, either in the damaged cells or during the replacement of inactivated cells.

The effects of most exposure scenarios can be approximated by model parameters that are piecewise constant. Heidenreich et al. (1997b) showed that in this case the model can be solved by a set of iterative equations and discussed various properties of the hazard after constant exposures over finite periods of time.

As shown in the example in Fig. 6.5, an exposure to an initiating agent during the age period of 40 to 50 years leads to a linear increase of the hazard in the age period of 45 to 55 years. Subsequently, the excess risk continues to increase due to the spontaneous promotion of the initiated cells.

In the case of a high exposure ($ERR = 2$) to a promoting agent, the hazard increases strongly during the first years after the age of 45. Subsequently, the increase becomes smaller. After the age of 55, the hazard decreases and even falls below the baseline hazard. These effects may be explained by the induced rapid growth of the number of intermediate cells in the clones. This increases the probability that a cell in the clone converts to a malignant cell. Since the corresponding organism is then considered to develop a cancer after a certain lag time, at later ages the whole clone is not contributing to the hazard anymore.

For lower exposures ($ERR = 0.2$) the same effects can be found, but are much less expressed. Here and in the following, exposures leading to an ERR of 0.2 are sometimes used to describe effects of low doses. However, it should be kept in mind, that an ERR of 0.2 for the incidence of solid tumours corresponds to an acute exposure of external ionising radiation with a dose of about 0.3 Sv (Thompson et al. 1994). This is a considerable dose and doses due to environmental contaminations will in general lead to smaller effects.

As noted in the previous subsection, for asymptotically large ages the spontaneous hazard approaches a constant level because an equilibrium is reached between the gain of new clones and the loss of clones in which a cell has become malignant or which are extinguished. Correspondingly, if the biological parameters return after an exposure to a carcinogen to the spontaneous levels, also the hazard will approach for longer times after the exposure to the spontaneous level (not shown in Fig. 6.5 because the depicted age intervals are too short). However, lifetime risk (eq. 6.5) will be always higher in the exposed group.

6.3.2
Multistep models with clonal expansion

Tan (1991) and Moolgavkar and Luebeck (1992) were the first to generalise the two-step clonal expansion model to multistep models of carcinogenesis with a clonal growth of intermediate cells. Little (1995) examined the behaviour of the excess risk for small instantaneous perturbations of the parameters in a number of generalisations of the TSCE model. More recent applications of multistep models with clonal expansion are described below and in section 6.4.

Colorectal cancer is a well studied example of multistep carcinogenesis (Fearon and Vogelstein 1990; Aaltonen et al. 1993; Peltomäki et al. 1993). Colorectal cancer is a main area of application of multistep models with clonal expansion to population-based data on cancer mortality (Herrero-Jimenez et al. 1998, 2000) and incidence (Moolgavkar and Luebeck 1992; Luebeck and Moolgavkar 2002). This section will deal exemplarily with the analysis of incidence data for colorectal cancer.

Luebeck and Moolgavkar (2002) developed a mathematical model for colorectal carcinogenesis based on the assumption that several genetic changes, e. g. the loss

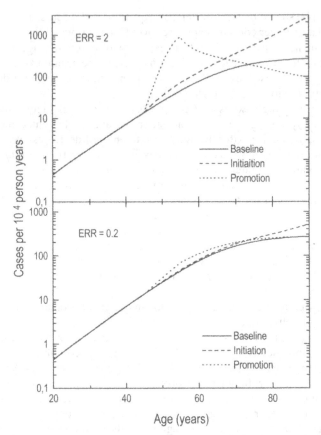

Fig. 6.6 Baseline hazard $H_0(a)$ to receive a solid tumor as obtained by an adaptation of the TSCE model to the solid tumour incidence data for the male atomic bomb survivors with negligible radiation exposures and total hazard after a constant exposure to an initiating and a promoting agent in the age period of 40 to 50 years. The exposures result in excess relative lifetime risks of 2 resp. 0.2 (Jacob and Prokić 2003).

or mutation of the tumour suppressing *APC* gene, precede the occurrence of clonally expanding adenomatous polyps in the colon. Correspondingly, they developed a series of multistep models (from the two-step (TSCE) to a five-step model), in which only cells that have undergone all but one changes in the carcinogenic process expand clonally by symmetric divisions (Fig. 6.6). They call these cells initiated, and those with less changes pre-initiated. Note, that for the pre-initiated cells division and death are not modelled explicitly. In the terminology used for the TSCE model, the initiated cells are called intermediate (Fig. 6.4).

For the application of the model to colorectal cancer incidence, it was assumed that there are 10^8 normal susceptible stem cells which is approximately the number of crypts in the colon. Incidence data for the period 1973–1996 and for nine geographic areas in the US Surveillance, Epidemiology and End Results (SEER) registry (Ries et al. 2001) were used. The data included about 256,000 colorectal cancer cases among the white population and 21,000 cases among the black population. Poisson regression was used to fit the models to the number of observed cases in the different age and calendar-year groups (see Appendix).

The four-step model was found to fit the data best. The two pre-initiating steps were interpreted as losses (or mutations) of the two alleles of the *APC* gene with transition rates of about 10^{-6} yr^{-1} (it was assumed that the two rates are equal).

The transition rate λ_3 of the initiating event was found to be of the same order of magnitude as the division rate α of initiated cells which is in the order of 10 yr^{-1} (Herrero-Jimenez et al. 1998), i. e. according to this model application the third transition is much more frequent than the pre-initiating events. Luebeck and Moolgavkar (2002) suggest the event to be an asymmetric division of the pre-initiated stem cell generating a progeny that moves from the stem cell zone into the proliferative zone of the crypt. Here, cell growth is not constrained by the microenvironment at the bottom of the crypt and the initiated cells can expand clonally via occasional symmetric cell divisions. Alternatively, the initiating step could be an epigenetic event.

The clones of initiated cells were interpreted to constitute adenomas and a growth rate α - β of the number of initiated cells of about 0.15 yr^{-1} was found. This corresponds to a doubling time of about 5 years. Based on the model results, Luebeck and Moolgavkar (2002) calculated the (age-dependent) number of these adenomatous clones for the healthy population and for individuals with familial adenomatous polyposis (FAP) who inherited a loss or mutation of one allele of the *APC* gene. Based on measurement data they estimated the number of aberrant crypt foci that are considered as the earliest stages of the formation of adenomatous polyps. The model data were by one order of magnitude lower, however, the model and the measurement-based estimation agreed well on the ratio of results for normal and FAP subjects.

As for the two pre-initiating events, the transition rate λ_4 for the initiated cells to convert to a malignant cell was found to be a rare event. The authors claim that the transition rate λ_4 is in the order of measured locus-specific mutation rates, indicating that there is no genomic instability in the carcinogenic process before the occurrence of this last rate-limiting step. They suggest that this rare event could be identified with mutation of one copy of the *p53* gene assuming a dominant negative action (Fearon and Vogelstein 1990; see also section 6.1.1). Alternatively, the occurrence of the cancer may be due to a combination of the rare mutation of one copy of the *p53* gene with the onset of genomic instability.

Multistep models of carcinogenesis should – at least in principle – allow a more realistic description of the carcinogenic process than the TSCE model. Taking into account the continuous growth of molecular genetic knowledge, it can be anticipated that multistep models will gain an increasing importance in the future.

6.4
Ionising radiation

Ionising radiation may be densely ionising, like α-particles and neutrons, or sparsely ionising, like X- and γ-rays. Densely ionising radiation has a considerably higher biological effectiveness per deposited energy, because it produces more complex DNA lesions per unit dose which have a higher probability of misrepair (see chapter 3). The pathway of densely ionising radiation that is most relevant to the exposure of the population, is the inhalation of radon progeny (see chapter 5). It is also the pathway best analysed with models of carcinogenesis (section 6.4.1).

External exposure to sparsely ionising radiation is the most relevant pathway for many occupational exposures. The cohort of the atomic bomb survivors remains to be the most important information source for acute external exposures and has also been analysed with models of carcinogenesis (section 6.4.2). There is less information on effects of external exposures that occur with low dose rates over longer times.

At low doses of ionising radiation, a number of non-linear radiobiological effects have been observed. The model development to integrate such effects in models of carcinogenesis in order to analyse cancer risks at low dose is still in an early phase (section 6.4.3).

6.4.1
Effects of inhaled radon progeny

Large experiments have been performed in which rats were exposed to radon progeny in air. A main aim of the studies is an improvement of the understanding of time-, dose-, and dose-rate dependencies of the carcinogenic effect of radon exposures. An advantage of the animal experiments is a high degree of consistency in the control of exposures and in diagnosis of lung tumours. A disadvantage is that the results cannot be transferred without caution to humans, because in contrast to humans, rats can live with some frequent types of tumours for a long time.

In the mid of the 20th century, uranium miners were exposed to high concentrations of radon resulting in a significant increase of the lung cancer rate among them. Numerical estimates for lung cancer hazard were derived from corresponding cohort data. A common aim of the two approaches (animal experiments and epidemiological studies of miners) is to contribute to a better quantification of the radiation risk due to residential radon.

6.4.1.1
Animal experiments

Two main series of experiments were performed with radon exposures of rats, one at the Pacific Northwest National Laboratory (PNNL), Richland, WA (Cross et al. 1993) and one at the Commissariat à l'Energie Atomique (CEA), France (Moncheaux et al. 1999).

PNNL rats. In the experiments at PNNL, male SPF Wistar rats were exposed to constant radon levels for 18 h d^{-1} for 5 d wk^{-1}. The exposure lasted between 2 days and about 100 weeks (about 90 % of the average lifetime of the rat). At the beginning of exposure, the rats were 75 to 110 days old. In order to simulate the situation in mines, uranium ore dust was always administered with radon exposure.

In a first analysis of the PNNL data with a model of carcinogenesis, all lung tumours were treated as incidental, i. e. it was assumed that the tumours have not caused the death of the animal (Moolgavkar et al. 1990). In later analyses tumours were differentiated of being incidental or fatal (Luebeck et al. 1996; Heidenreich et al. 1999). In all three studies the data were analysed with the TSCE model.

Heidenreich et al. (1999) used a data set for 4,276 rats. Of these 3,726 lived out their life span, 418 were sacrificed according to a planned schedule, 127 were euthanised for ethical reasons, and 5 were killed accidentally. A total of 487 rats developed at least one lung tumour. Based on the constant parameters for the TSCE model in eq. 6.18, the following dependencies on the average exposure rate d (accumulated exposure divided by time interval from beginning of first fraction of exposure until end of last fraction of exposure) were introduced:

$$X(d) = n_s \, \lambda_1(d) \, \lambda_2(0)$$

$$\gamma(d) = \alpha(d) - \beta(d) - \lambda_2(d) \qquad\qquad \text{(eq. 6.23)}$$

$$m(d) = \lambda_2(d)/\lambda_2(0)$$

$$q(d) = q(0).$$

$\lambda_1(d)$, $\gamma(d)$ and $m(d)$ were assumed to increase linearly with radon exposure rate, where additionally the initiating mutation rate $\lambda_1(d)$ had an exponential cell killing term for high doses, and the clonal expansion rate $\gamma(d)$ levelled off to a constant term at higher doses. Individual likelihood techniques (see Appendix) were used to fit the model to the data. Age and exposure patterns of the hazard for incidental and fatal lung tumour in rats were found to be quite different.

For incidental lung tumours in rats there is, according to the model fit, a large spontaneous initiation rate for intermediate cells. The initiation rate increases linearly with exposure rate. The main radiation effect, however, is to increase the size of clones of intermediate cells by an increased proliferation rate. The exposure-rate dependence of this promotion effect was found to be nearly a step function, i. e. small exposure rates have an effect that is comparable to high exposure rates. This might be due to the presence of uranium dust in the air for all radon exposures. The effect of the radiation on the second mutation rate was not significant.

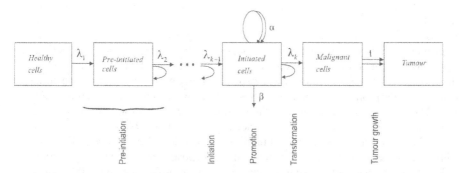

Fig. 6.7 k-step models of the carcinogenic process in which (k - 1) cellular changes are necessary to produce initiated cells that can undergo clonal proliferation (after Luebeck and Moolgavkar 2002)[2]

---

[4]   Luebeck and Moolgavkar (2002) denote λ_κ by $\mu_{\kappa\text{-}1}$.

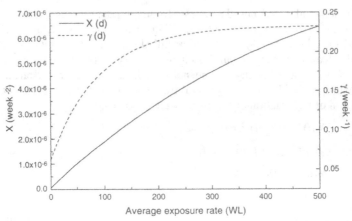

Fig. 6.8 Exposure-rate dependence of TSCE model parameters X (proportional to the initiation rate) and γ (effective promotion rate) for fatal lung cancer among radon exposed rats (after Heidenreich et al. 1999)

Concerning fatal lung tumours, the spontaneous initiation rate for intermediate cells was found to be small, the spontaneous promotion rate to be quite large. Radiation effects on both, initiation and promotion rate, turned out to be relevant for fatal tumours. For both rates, low radon concentrations had an higher effect per unit exposure than higher concentrations (Fig. 6.7). As for incidental tumours, the effect of the radiation on the second mutation rate was not significant.

Heidenreich et al. (1999) studied the excess relative risk per unit radon exposure (ERR WLM^{-1})[3] for fatal lung tumours at an average lifetime for rats exposed from an age of 10 weeks with different exposure rates until the requested exposure was applied (Fig. 6.8). According to the model fit, there is an inverse exposure-rate effect for high exposure rates which is especially expressed for high exposures. There are two reasons for the inverse dose rate effect. Firstly, for high exposure rates the radiation induced promotion is of large importance. Lower exposure rates last longer so that in this exposure rate range the radiation induced promotion acts in average on a larger number of spontaneously initiated and promoted cells (high dose rate protraction effect). Secondly, due to the curvature of the exposure rate dependence of the promotion, lower exposure rates with longer exposure times cause a higher promotion than high exposure rates with short exposure times (Fig. 6.7). The inverse dose rate effect for high exposures was also found by an analysis of the data with an excess relative risk model (Gilbert et al. 1996).

For low exposure rates, the model predicts a direct dose-rate effect (see Fig. 6.8). This is because for lower exposure rates in this exposure rate range, the radiation-induced intermediate cells have in average less time to proliferate due to the spontaneous promotion (low dose rate protraction effect). This effect is also found in situations in which radiation acts dominantly on initiation (Moolgavkar 1997). It should, however, be noted that the low dose rate protraction effect (spontaneous promotion

---

[5]   For a definition of the unit WLM see Section 3.5.1 or the glossary.

of radiation-induced intermediate cells) predicts an inverse dose rate effect for protracted exposures that end at the same time (not start at the same time).

Two points are worth noting concerning the protraction effects. Firstly, both effects occur in the discussed exposure scenarios, because the low dose rate exposures extend to a later period of life. Secondly, the dose rate effect (low dose rate protraction effect) is due to the age-at-exposure dependence in TSCE models with a non-negligible radiation effect on the initiation rate (see also section 6.4.3). A dose-rate dependence of DNA damage repair has not been taken into account in these versions of the TSCE model.

Besides the dose rate effects, two aspects of extrapolating from high doses to low doses can be observed for the model fit and the exposure scenarios. For high exposure rates, the excess relative risk per unit exposure is larger for high exposures than for low exposures (Fig. 6.8). This is again due to the high dose rate protraction effect. The radiation-induced promotion is important in this exposure rate range. Higher exposures last until higher ages, and therefore, in average the radiation-induced promotion acts on more spontaneously initiated and promoted cells. For low exposure rates the opposite trend is found due to the low dose rate protraction effect.

CEA rats. At CEA, male Sprague-Dawley rats were exposed to radon and/or various chemicals (Monchaux et al. 1994). Three groups of modellers analysed commonly a CEA data set of lung cancer among 3,503 rats that were exposed to radon and its progeny only and among 1,525 that served as controls (Heidenreich et al. 2000). For 3,342 of the rats the exposures started at an age of 3 months, exposure durations varied between 0.5 months and 16.4months, the cumulative exposures between 25 and 10,000 WLM. For 161 rats, the start of the exposure was in the age period 9 to 15 months. Of the rats 406 had at least one lung tumour. All tumours were treated as incidental. Individual likelihood techniques were used to fit models of carcinogenesis to the data. Using a simple test model, the authors showed that their mathematical approaches and fit routines gave comparable results. Subsequently, they analysed the data set with their own models of carcinogenesis. An overview over the models preferred by the three groups is given in Tab. 6.1.

The first group of modellers favoured a TSCE model with an action on promotion and did not find an effect on the second mutation rate, as for the application to the PNNL rats. Only identifiable parameters were used and a value for the lag time t was fitted. The model was optimised by assuming a constant value of the radiation inducing the promotion rate above a threshold of 1 WL[4] and below an exposure rate value of 250 WL. It has not been explained why the promoting action should disappear for higher exposure rates.

The second group favoured a TSCE model in which the two spontaneous mutation rates were set equal. They assumed linear exposure rate dependences of the mutation rates with exponential cell killing terms that also apply to the spontaneous mutation rate. The stochastics of the birth-death process were not found to be relevant in their approach, therefore only one parameter for the exponential growth of the number of intermediate cells was assumed. It was not tested whether the radiation had an effect on the promotion of intermediate cells. A number of 5×10^5 susceptible normal cells for lung tumours in rats was assumed.

---

[6] For a definition of the unit WL see Section 3.5.1 or the glossary.

Tab. 6.1 Parameters in models of carcinogenesis that were used by three groups of modellers in analysing the CEA data on lung cancer among radon exposed rats, p_i are the fit parameters (after Heidenreich et al. 2000).

TSCE model of first group	TSCE model of second group
$X = p_1 (1 + p_2 d)$	$\lambda_1 = (p_1 + p_2 d) \exp(-p_3 d)$
$\gamma = p_3 + p_4 \Theta(1 \text{ WL}; 250 \text{ WL})^a$	$\lambda_2 = (p_1 + p_4 d) \exp(-p_5 d)$
$q = p_5$	$\alpha - \beta = p_6$
$t = p_6$	$t = p_7$

TSCE model of third group	Three-stage model of third group
$\lambda_1 = p_1 + p_2 \cdot d$	$\lambda_1 = p_1 + p_2 d$
$\lambda_2 = p_1$	$\lambda_2 = p_1$
$\alpha - \beta = p_3 + p_4 (1 - \exp(-p_5 d))$	$\lambda_3 = p_1 + p_3 d \exp(-p_4 d)$
$T = p_6^b$	$T = p_5^b$
$n_{s+} = p_7^b$	$n_{s+} = p_6^b$

[a] $\Theta(1 \text{ WL}; 250 \text{ WL})$ is 1 for the exposure rate range of 1 to 250 WL and zero otherwise
[b]   It is assumed that after age T the number n_s (5×10^5) of susceptible healthy cells is changed to n_{s+}

The third group favoured two models, a TSCE model and a three-step model. In both models, the number of susceptible normal cells was allowed to change from 5×10^5 to another constant value at an age that is also a fit parameter. The spontaneous mutation rates were assumed to be equal. A linear exposure rate dependence of the first mutation rate was assumed. The death or differentiation rate of intermediate cells was set to zero. In the TSCE model, a linear increase of the proliferation rate with exposure rate was assumed, that levels off exponentially to a constant value. In the preferred three-step model clonal expansion was not taken into account at all, and the third mutation rate was assumed to increase linearly with exposure rate with an exponential cell killing term.

All four models were considered to describe the data well. In all four models the exposure rate dependence of the initiation rate for intermediate cells turned out to be similar. The ratio $\lambda_1(d)/\lambda_1(0)$ increases linearly (in one case with a slight downward bending due to the cell killing term) up to a value of 650–850 for exposure rates of 500 WL.

The radiation effect on the first mutation rate alone was not found to be sufficient for a satisfactory fit. The agreement with the data could be improved by assuming a radiation effect on the promotion rate or on the last mutation step. No further improvement was observed when both were increased. The three groups obtained comparably good fits by modelling radiation effects on promotion and last mutation rates quite differently.

The spontaneous hazard for lung cancer among radon exposed rats is modelled similarly by the four models for the whole life span with two exceptions (Fig. 6.9). The first group obtains lower spontaneous hazards at older ages, deviating by a fac-

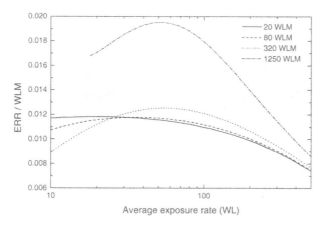

Fig. 6.9 Excess relative risk per unit radon exposure (ERR WLM^{-1}) for fatal lung tumours for rats with an age of 110 weeks having been exposed since an age of 10 weeks. Results are given according to an TSCE model application to the PNNL data (after Heidenreich et al. 1999).

tor of 2 from the other models at an age of 150 weeks. The three-step model gives considerably lower hazards than the TSCE models for ages younger than 30 weeks. It should be noted that the number of cases in the data set outside the age range of 30 to 150 weeks is small.

For the example of an exposure during an age period of 13 to 33 weeks with an exposure rate of 50 WL, the four models agree on the hazard within a factor of 3 only in the age period of 57 to 100 weeks. For older ages, the first group obtains considerably lower hazards. For younger ages, the TSCE models of the first two groups agree quite well, and so do the TSCE and three-step model of the third group.

In summary, the exercise of the three modeller groups has clearly demonstrated that a variety of models is equally well compatible with the data. Besides the quality of fit, plausibility and radiobiological knowledge are further criteria to support or rule out possible models. Independent of this, the application of a variety of plausible models is a possibility to explore model uncertainties and identify results that are obtained with all of the models.

6.4.1.2
Uranium miners

Moolgavkar et al. (1993) used the TSCE model to analyse the lung cancer mortality among miners who worked in the Colorado Plateau uranium mines between 1950 and 1964. In a more recent analysis (Luebeck et al. 1999), data with an extended follow-up (up to the end of 1990) were used. Detailed patterns of exposure to radon and cigarette smoke were simulated, and exposures from prior hard rock mining were taken into account. Data on 3,347 white male miners were analysed. For each of the miners times where given when their cumulative exposure exceeded 60, 120, 360, 600, 840, 1,800 and 3,720 WLM. For cigarette smoking, up to five times were

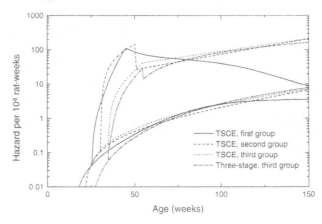

Fig. 6.10 Spontaneous and total hazard among male Sprague-Dawley rats exposed at CEA to radon in an age period of 13 to 33 weeks with an exposure rate of 50 WL. Results are given according to three TSCE model versions and a three-step model (after Heidenreich et al. 2000).

given when smoking levels (in packs d^{-1}) changed. For each miner birth year, age at entry into the study (first medical examination), attained age, and vital status were given. If the miner died before the end of the study, the listed ICD code indicated his cause of death and 354 of the miners died because of lung cancer with the ICD code 162 (malignant neoplasms of trachea, bronchus and lung).

Luebeck et al. (1999) used the TSCE model version defined in eq. 6.23 to analyse the data. They assumed that the first mutation rate increases linearly with birth year and mean radon exposure, and with smoking rate according to exp[1 - const· (smoking rate squared)], see Fig. 6.10. Although detailed data on the smoking behaviour were taken into account, an unexpected strong dependence of the initiation rate on the birth year was found. This may be related to problems with the smoking data. According to the data set, for instance, miners below age 70 had fewer lung cancer cases in the highest smoking group compared to miners belonging to the second highest smoking group.

In the model application, the clonal expansion rate was assumed to increase logarithmically with mean radon exposure rate and was found to be only slightly larger for smokers than for non-smokers. In the preferred model, the second mutation rate is independent of smoking and mean radon exposure rate. The lag time t was estimated to be about 9 years. All parameter estimates (except for the spontaneous promotion rate) have a large uncertainty and no information is given on correlations. Therefore, all conclusions on age, time and exposure dependences have also large uncertainties.

For exposures centred around the age 40, the excess absolute lifetime risk (up to age 70, other causes of death neglected) per unit exposure is found to first increase with duration of exposure, reach a maximum and then decline (Fig. 6.12). Thus an inverse dose rate effect is found for short exposure times, and a direct dose rate effect for long exposure times. The lower the total exposure, the weaker the inverse dose rate effect and the sooner the direct dose rate effect sets on.

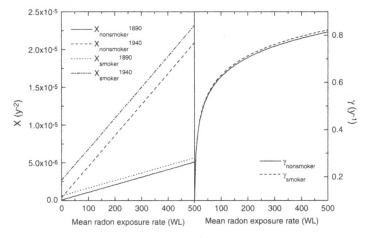

Fig. 6.11 Parameters of the TSCE model as assessed by Luebeck et al. (1999) for the Colorado Plateau miners. For smokers a smoking rate of 10 cigarettes d^{-1} has been assumed, the birth year is indicated in the legend by the parameter X. The estimates have large uncertainties.

As for the exposure pattern discussed for the data for the PNNL rats (section 6.4.1.1), the inverse dose rate effect is found for high exposure rates and the direct dose rate effect for low exposure rates (Tab. 6.2). In the scenario for the miners, the inverse dose rate effect is mainly caused by the downward curvature of $\gamma(d)$. The direct dose rate effect is due to the strong promotion of the initiated cells in the intermediate time interval which is not fully compensated by a weaker promotion over a longer time interval. Radiation induced initiation plays only a negligible role for the parameters and ages considered here.

For exposures with high exposure rates centred around the age 40, the excess absolute lifetime risk is higher for low exposures than that for high exposures (Fig.

Tab. 6.2 Dose and dose-rate effects for lung cancer risks due to protracted exposures to radon.

Exposure rate	Excess relative risk for rats at age of 110 days for exposures starting in young adult age (after Heidenreich et al. 1999)	Excess lifetime absolute risk up to age 70 for humans with exposures centred around age 40 (after Luebeck et al. 1999)
High	Excess risk per unit dose decreases with decreasing dose; *Inverse* dose rate effect	Excess risk per unit dose increases with decreasing dose; *Inverse* dose rate effect
Low	Excess risk per unit dose increases with decreasing dose; *Direct* dose rate effect	Excess risk per unit dose decreases with decreasing dose[a]; *Direct* dose rate effect

[a] For exposures smaller than 200 WLM

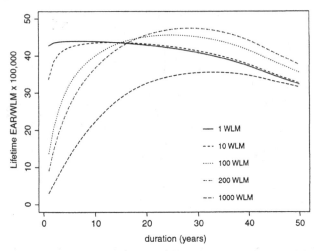

Fig. 6.12 Excess absolute lifetime risk WLM^{-1} (up to age 70, competing risks are neglected) after exposures to radon that are centred around age 40, as derived with the TSCE model for smokers among the Colorado Plateau miners (from Luebeck et al. 1999). A smoking rate of 10 cigarettes d^{-1} starting at age 15 has been assumed.

6.11). For low exposure rates, the inverse is found (Tab. 6.2). These trends are opposite to what has been observed in the exposure scenario for the PNNL rats (Fig. 6.8).

It may be noted that in contrast to the exposure scenarios shown in Tab. 6.2, for protracted exposures with low exposure rates that end in the same age, in most cases an inverse dose rate effect is predicted. It may be concluded that for protracted exposures the excess risk per unit dose depends in a complex manner on dose, dose rate, exposure pattern and action mechanisms of the radiation on the carcinogenic process. Correspondingly, the discussed applications of models of carcinogenesis do not contribute further evidence to the assumption that the LNT extrapolation does not lead to an underestimation of the risk at low doses and dose rates.

6.4.2
Effects of external exposures

6.4.2.1
Atomic bomb survivors

The data on solid cancer incidence among the atomic bomb survivors from Hiroshima and Nagasaki (Thompson et al. 1994) are the radioepidemiological data most frequently analysed with models of carcinogenesis. Soon after the data became available, work was published on two analyses of the data with the TSCE model assuming that radiation acts only in the first mutation rate (Heidenreich et al. 1997a; Kai et al. 1997). Later, the data were analysed with multistage models (Pierce and Mendelsohn 1999), which was then compared with results obtained by the TSCE model (Heidenreich et al. 2002). The data were also used to explore possible effects of low-dose hypersensitivity on lung cancer risk estimates (section 6.4.3).

The data cover the observation period 1958-1987. They are stratified in age-at-exposure categories (five years intervals up to 60 and older than 60), attained age (five year intervals), and colon dose equivalents D (cut points 0.005, 0.01, 0.1, 0.2, 0.5, 1.0, 2.0, 3.0 and 4.0 Sv).

Most of the analyses were performed with TSCE models in which the γ-radiation acts only as an initiator and causes instantaneously an increase of the initiation rate that is proportional to the dose:

$$\lambda_1(a, D) = \lambda_1(0) + c_1 \, D \, \delta(a - a_e), \tag{eq. 6.24}$$

where a_e is the age at exposure and $\delta(a)$ the Dirac δ-distribution. Alterations of the second mutation rate were not considered because resulting cancer cases would have been probably observed before 1958 and therefore not have been included in the data. Under these conditions, the hazard $H(a, D)$ can be calculated analytically (Heidenreich et al. 1997a):

$$H(a, D) = H_0(a) + \frac{c_1 \, \mu_2 \, n_s \, D \, (\gamma + 2q)^2 \, \exp[(\gamma + 2q) \, (a - a_e - t)]}{\{\gamma + q + q \, \exp[(\gamma + 2q) \, (a - a_e - t)]\}^2}, \tag{eq. 6.25}$$

where $H_0(a)$ is the spontaneous hazard (eq. 6.18).

Kai et al. (1997) analysed the data for solid cancer in lung, stomach and colon. Poisson regression techniques were applied to fit the models to the data. A variation of the lag time t in the range of 0 to 10 years made little difference in the results. Kai et al. tested various dependences of the model parameters on birth year which is in the case of an acute exposure directly related to the age at exposure. In the preferred models of Kai et al. (1997), the spontaneous initiation rate depends linearly on birth year (only for stomach cancer among males no significant birth year effect was found). The spontaneous lung and colon cancer incidence at a given age were found to increase with birth year. On the contrary, stomach cancer incidence was found to decrease with birth year (only for females). These trends are in agreement with data from the tumour registry in Hiroshima.

Tab. 6.3 Ratios of the radiation induced initiation rate $\Delta\lambda_1(D)$ to the spontaneous initiation rate $\lambda_1(0)$ for organ equivalent doses of 1 Sv, as derived by Kai et al. (1997) with the TSCE model from the cancer incidence data among the atomic bomb survivors with different ages at exposure (a_e).

Organ	Sex	$\Delta\lambda_1(1 \text{ Sv})/\lambda_1(0)$		
		$a_e = 5$	$a_e = 30$	$a_e = 45$
Lung	Male	11	15	19
	Female	25	208	804
Stomach	Male	8	8	8
	Female	53	28	22
Colon	Male	39	66	115
	Female	41	66	103

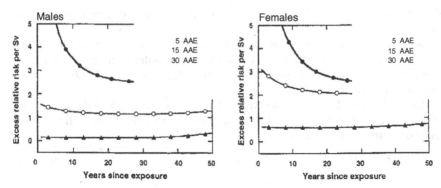

Fig. 6.13 Excess relative risk per unit dose to obtain colon cancer after an acute exposure to external radiation, as assessed with a TSCE model from the atomic bomb survivors data (from Kai et al. 1997).

In general, Kai et al. (1997) found that the radiation induced initiation is independent of birth year (only for lung cancer among females a birth-year dependence was found). Correspondingly, the ratio of the radiation induced initiation rate $\Delta\lambda_1(D)$ to the spontaneous initiation rate $\lambda_1(0)$ increased with age at exposure, with the exception of stomach cancer (Tab. 6.3). The authors argue that these ratios are broadly consistent with ratios that can be estimated from experimental radiobiological data for the *Hprt* gene and for the *GPA* gene. For lung cancer among females that were adult at the time of exposure, however, the values for the initiation ratio in Tab. 6.3 are exceptionally high, which might be due to a correlation of smoking and dose among female survivors in Hiroshima (Preston 1999).

Kai et al. (1997) found that the *ERR* decreases with age at exposure (Fig. 6.13). This is because the number of radiation induced intermediate cells are assumed to be independent of age and the number of spontaneously initiated intermediate cells increases with age. For most cancer sites a decrease of the *ERR* with age at exposure was also found by analyses of the data with conventional constant excess relative risk models (Thompson et al. 1994). However, in some cases the trend is inversed leading to contradictions of the two models. For lung cancer among males, e. g. Kai et al. (1997) found the excess relative lifetime risk for lung cancer (up to an age of 70) to decrease with age at exposure from 5 to 45 by a factor of 280, whereas the authors found for the ICRP-1990 model a decrease by a factor of 4 and for the BEIR V model an increase by a factor of 4.

For young ages at exposure, the *ERR* decreases according to the analysis of Kai et al. (1997) for all three cancer types analysed within the first decade after exposure, subsequently approaching a constant value. Again this is because the number of spontaneously initiated cells increases with age which is important for young age and only a minor relative increase in larger age.

For older ages at exposure, the *ERR* is essentially constant up to 40 years after exposure, because in this age range the growth of the number of intermediate cells is determined by the proliferation of intermediate cells. For even longer periods after exposure, the *ERR* is predicted to increase because the flattening of the exponential increase of the hazard at high ages comes later for the clones of intermediate cells

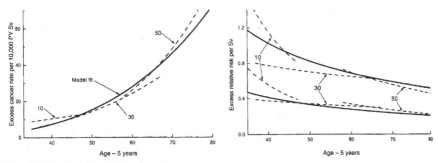

Fig. 6.14 Excess absolute risk (left panel, both sexes pooled) and excess relative risk (right panel, upper curves for females and lower curves for males) for solid cancer (excluding thyroid cancer and sex-specific cancers) among the atomic bomb survivors. The solid lines represent a fitted multistage model, the dashed lines empirical descriptions of the data. The numbers at the curves indicate the age at exposure (from Pierce and Mendelsohn 1999).

that were initiated by radiation (Fig. 6.6). It may be possible to test this prediction by the upcoming data for a longer follow-up. It may be noted that the increase of the *ERR* at long times after the exposure is a characteristic of TSCE models, in which the radiation acts dominantly on the initiation rate. It is less expressed or even vanishes if the radiation acts dominantly on the promotion rate (Fig. 6.6).

Age-time patterns of excess risks for solid cancer incidence in organs that are not strongly influenced by hormones, i. e. excluding thyroid cancer and all sex-specific cancers, have been analysed by Pierce and Mendelsohn (1999) with Armitage-Doll type multistep models (Fig. 6.3). The main aim of the study was to identify models of carcinogenesis explaining that the excess absolute risk mainly depends on age attained. The authors could explain main age-time characteristics of the excess risk without assuming additional dependences on age at exposure or time since exposure. They show that the approximate a^{k-1}-dependence of the spontaneous hazard in the k-step model also holds, if there are some restrictions on the order of the mutations, and if mutations can substantially alter the rate of subsequent ones.

Pierce and Mendelsohn (1999) assumed that radiation can act on any of the mutation rates and that the radiation action on the mutation rate does not depend on the age a. They claim that then for k being not too large, the excess hazard has an approximate a^{k-2} dependence. This implies that the excess relative risk decreases with age proportionally to a^{-1}. The fit of the model to the solid tumour incidence results in a value of k of 5.3. In general, the model fits well the age dependence of the excess hazard and of the relative risk of solid cancer among the atomic bomb survivors (Fig. 6.14). However, the agreement is not that good for cohort members who were exposed during childhood.

Stimulated by the work of Pierce and Mendelsohn (1999), Heidenreich et al. (2002) published a comparative analysis of the incidence data among the atomic bomb survivors with the TSCE model and with multistage models of carcinogenesis. In the case of the TSCE model, it was assumed that the radiation acts only on the initiation rate, and in the multistage model on any one of the k mutations. Heidenreich et al. (2002) used exact solutions of the multistage model in which the sequence of cellular transitions is inconsequential and of the Armitage-Doll model (section 6.2).

Numerically, all models fitted the data comparably well. So, none of the different models could be ruled out due to statistical reasons. However, in contrast to the approximate solutions used by Pierce and Mendelsohn (1999), the exact solutions of the multistep models gave implausible results for the number k of steps in the model. For the selection of solid tumours analysed by Pierce and Mendelsohn (1999), e. g. the estimated number of steps is for males (about 15) by a factor of 3 larger than for females (5–6). Similar discrepancies were found for single cancer sites analysed (lung, colon and stomach). Heidenreich et al. (2002) argue that the reason for this discrepancy is that the exact solutions of the multistep model without clonal expansion of intermediate cells exhibit an unrealistic form of levelling-off of the hazard for older ages and suggest that models with clonal expansion should be used instead.

Mortality data for the atomic bomb survivors have also been analysed with multistage models and the interested reader is referred to Little (1996).

6.4.2.2
Threshold considerations

Prokić and Jacob (unpublished) applied TSCE models with and without a threshold to the solid cancer incidence (Thompson et al. 1994) and mortality (Pierce et al. 1996) data among the atomic bomb survivors from Hiroshima and Nagasaki. The radiation was assumed to act instantaneously on the first mutation rate and eq. 6.24 was replaced by

$$\lambda_1(D) = \lambda_1(0) + c_1 \, (D - D_t) \, \Theta(D_t) \, \delta(a - a_e), \qquad \text{(eq. 6.26)}$$

where D_t is the threshold dose and $\Theta(D_t)$ is zero for $D < D_t$ and 1.0 otherwise. Calculations were performed for the following values for D_c: 0 (no threshold), 10, 20, 50, 100, 150, 200, 300, 400 and 500 mSv.

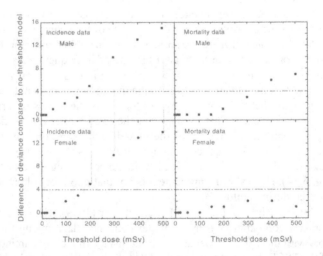

Fig. 6.15 Differences of deviance when fitting TSCE models to the cancer incidence and mortality data for the atomic bomb survivors from Hiroshima and Nagasaki. Models with a linear action of the radiation on the first mutation rate with and without a thresholds have been used (from Prokić and Jacob 2002).

Poisson regression was performed to fit the model to the data. The deviance was used as a measure of the quality of the fit, a lower deviance indicating a better fit. The deviance for the model without threshold is subtracted from the deviance for the models with threshold (Fig. 6.15). Negative values would indicate that the threshold model fits the data better. Positive values exceeding 4 were considered to indicate that the model is not supported by the data (see Appendix).

The deviance for the threshold models was found in no case to be smaller than the deviance of the no-threshold model. So, none of the considered threshold models fits the data better than the no-threshold model. For the cancer incidence among males and among females, the difference in deviance exceeds the value of 4 for threshold doses larger than 200 solid tumour incidence among the atomic bomb survivors cannot exclude the existence of a threshold in the dose range of 0 to 200 mSv. In the case of mortality data, fits for males can not exclude the existence of a threshold below 400 mSv; data on females are not at all conclusive on the existence of a threshold.

6.4.3
Low-dose radiobiological effects and low-dose risk estimation

Radiobiological experiments have revealed various effects at low doses which exhibit a non-linear dose dependence, e. g. adaptive response, low-dose hypersensitivity, bystander effects and genomic instability (see chapter 3). Up to now, there is only very limited work on exploring possible low-dose implications of these effects with the help of models of carcinogenesis. Some work has been performed on the application of a TSCE model incorporating low-dose hypersensitivity for lung cancer incidence among the atomic bomb survivors from Hiroshima and Nagasaki (Jacob and Prokić 2002; Prokić et al. 2002). Earlier applications of the TSCE model (section 6.4.2.1) had problems with the age-at-exposure dependence of the *ERR*.

Low-dose hypersensitivity. A number of radiobiological studies have indicated the presence of a hypersensitive region in the dose dependence of the survival response of many cell lines after exposure to low-LET radiation (Joiner et al. 1996, 2001). For most cell lines the number of inactivated cells per unit dose is higher for low doses (up to a few hundred mGy) than for high doses where an increase in radioresistance (IRR) is observed. Low-dose hypersensitivity in cell inactivation has also been observed *in vivo* for rhabdomyosarcoma R1H in rats (Beck-Bornholdt et al. 1989) and for normal human skin cells that were irradiated off the centre of the beam during radiotherapy of prostate cancer patients (Joiner et al. 2001).

The dose dependence of cell inactivation after exposures to low-LET radiation is modelled by survival curves that account for increased radioresistance (IRR)

$$S(D) = \exp[-a_r(1+(a_s/a_r-1)\exp(-D/D_c))D - bD^2], \qquad \text{(eq. 6.27)}$$

where D is dose, a_r and a_s are the coefficients of the linear dose term in the high-dose (IRR) and low-dose regime, respectively, b is the coefficient of the quadratic dose term and D_c is the dose characteristic for the transition from the low-dose to the high-dose regime. An increased radioresistance at higher doses is expressed by $a_r < a_s$. For normal human lung epithelial cells L132 Singh et al. (1994) obtained the values $a_r = 0.15$ Gy^{-1}, $a_s = 1.19$ Gy^{-1}, $D_c = 0.58$ Gy and $b = 0.07$ Gy^{-2} by fitting the function in eq. 6.27 to their experimental cell survival data.

Promoting effect of cell inactivation. It has been proposed that cell inactivation by carcinogens may cause a promotion, i. e. an increase of the number of intermediate cells (Trosko et al. 1983; UNSCEAR 2000). The suggested reason is that the inactivation of normal cells reduces the suppression of the proliferation of intermediate cells. Another possible reason is that intermediate cells have a higher potential to replace inactivated normal cells (Heidenreich et al. 2001).

A strong effect of promotion on the development of lung cancer among the Colorado miners has been reported (section 6.4.1). From these data a quantitative estimate of the promotion of intermediate cells due to a radiation exposure of lungs was derived (Heidenreich et al. 2001). The ratio $\Delta\alpha/\Delta\beta$, where $\Delta\alpha$ is the radiation induced increase of the proliferation rate and $\Delta\beta$ the radiation induced increase of the death or differentiation rate of the intermediate cells, was assessed to be 2.0, if basal cells are the sensitive cells for the induction of lung cancer, and 1.2, if secretory cells are the sensitive cells.

TSCE model and low-dose hypersensitivity. The radiation induced division rate of intermediate cells was assumed by Jacob and Prokić (2002) to be proportional to the cell inactivation of healthy L132 lung epithelial cells

$$\Delta\alpha = c_\alpha\,(1 - S(D))/\Delta t_p, \qquad\qquad \text{(eq. 6.28)}$$

and their inactivation rate to be the same as for the L132 cells

$$\Delta\beta = (1 - S(D))/\Delta t_p, \qquad\qquad \text{(eq. 6.29)}$$

where Δt_p is one week, the time span between exposure and measurement of inactivation. Two model variants were studied, one with $c_\alpha = 2.0$ and one with $c_\alpha = 1.2$. In addition, calculations have been performed with the conventional approach to assume that radiation acts only on the initiation rate.

Application to lung cancer incidence among atomic bomb survivors. Whereas there is evidence for the promotional action of protracted radon exposures, it is much less clear whether acute exposures to γ-radiation also cause a promotional effect. Jacob

Tab. 6.4 Parameters of three TSCE models adapted to the lung cancer incidence among male atomic bomb survivors: the spontaneous promotion rate $\gamma(0)$, its change $\Delta\gamma$ for a week due to the promotional effect of the exposure (for two different doses), and the product of the number $c_1\,n_s$ of intermediate cells created per unit dose and the spontaneous transformation rate λ_2 (after Jacob and Prokić 2002).

Radiation acts on:	γ (yr^{-1})	$\Delta\gamma$ (yr^{-1})	$c_1\,n_s\,\lambda_2$ (yr^{-1} Sv^{-1})
Initiation	0.18 0 for 1.0 Sv	0 for 0.1 Sv	4.3×10^{-6}
Initiation and promotion (secretory cells)	0.18	1.2 for 0.1 Sv 4.2 for 1.0 Sv	3.8×10^{-6}
Initiation and promotion (basal cells)	0.18	4.6 for 0.1.0 Sv	2.5×10^{-6}

and Prokić (2002) explored the possible consequences by applying models with promotional effects to the lung cancer incidence data for the atomic bomb survivors. The above discussed model with a radiation action on the promotion of intermediate cells by inactivating basal cells fitted the data slightly better than other models. The fit value of $\gamma(0)$, the spontaneous promotion rate of intermediate cells, is the same in all three models (Tab. 6.4 for males, similar results have been obtained for females). For the period of the promotional effect, the radiation induced promotion exceeds the spontaneous promotion rate considerably, even for doses as low as 100 mGy. The radiation induced initiation of intermediate cells decreases with an increasing effect of the radiation on the promotion.

The excess relative lung cancer risk varies with age at exposure and with sex (Tab. 6.5). According to the constant excess relative risk model, for males the *ERR* was smaller in the age-at-exposure group of 20 to 39 years than in the age-at-exposure group older than 40 years. In contrast, the TSCE model, assuming an exclusive radiation action on initiation, resulted in an risk estimate that is by a factor of 7.5 larger for the younger age-at-exposure group than for the older age-at-exposure group. For the TSCE model, with an assumed radiation action on initiation and promotion, this factor has the value of 2, which is closer to the ratio of the estimates performed with the constant relative risk model. Thus, the implementation of the low-dose hypersensitivity in the TSCE model improved the description of the age-at-exposure dependence of the observed data.

Implications for estimates of excess lifetime risk. In the TSCE model variant with a radiation action on the promotion, the dose dependence of the excess relative risk depends on age at exposure (Fig. 6.16). For low and medium ages at exposure, the dependence of the *ERR* on dose does not differ significantly from linearity. For large ages of exposure, however, the radiation induced promotion of intermediate cells has a large effect because intermediate cells are relatively frequent at higher age. The assumed low-dose hypersensitivity leads to an estimate of the excess risks per unit dose that is larger at low doses than at high doses. At low doses (10 mSv)

Tab. 6.5 Excess relative risk (*ERR*) for lung cancer incidence among the atomic bomb survivors as estimated by a model of constant *ERR* for two age-at-exposure groups and by two versions of the TSCE model for the cohort as a whole (after Jacob and Prokić 2002). Median values and 95 % confidence intervals are given.

Model	Age-at-exposure: 20–39 years		Age-at-exposure: ≥ 40 years	
	Male	Female	Male	Female
Constant *ERR*	0.24 (-0.2; 0.9)	2.0 (1.0; 3.3)	0.7 (0.2; 1.3)	1.9 (0.9; 3.3)
TSCE (radiation acts on initiation only)	0.6 (0.2; 2.0)	2.9 (1.7; 5.2)	0.08 (0.03; 0.3)	0.6 (0.3; 1.1)
TSCE (radiation acts on initiation and promotion[a])	1.2 (0.9; 2.4)	2.8 (1.9; 4.6)	0.5 (0.4; 0.7)	1.0 (0.7; 1.4)

[a] Basal cells are assumed to be the sensitive ones for lung cancer induction.

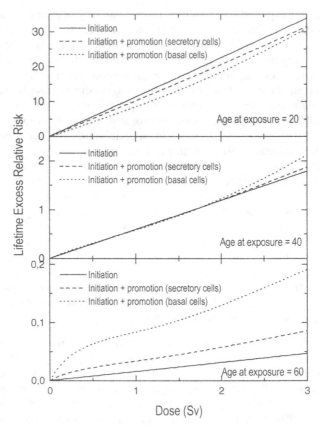

Fig. 6.16 Excess relative lifetime risk for lung cancer incidence among males as derived with three variants of the TSCE model for the cohort of the atomic bomb survivors from Hiroshima and Nagasaki (from Jacob and Prokić 2002). In the first variant radiation acts only on initiation, in the other two variants it acts also on the promotion of intermediate cells, where susceptible cells are either secretory or basal.

the risk per unit dose derived by the TSCE model variant with radiation action on the promotion of intermediate cells (if basal cells are the susceptible cells for lung cancer) happens to be close to the results obtained with the constant relative risk model.

In summary, there is a considerable uncertainty in age, time and dose dependences of organ-specific cancer risk estimates due to a lack of statistical power in radioepidemiological data. The evaluation of these data with a variety of models will elucidate what is known and with which uncertainty. Models of carcinogenesis should be further developed to contain main radiobiological and molecular genetic facts in order to contribute to this task.

6.5
Carcinogenic substances

A main field of modelling the influence of chemical carcinogens on the carcinogenic process is the evaluation of data on the development and size distribution of intermediate lesions. The experimental set-ups allow a sophisticated description of the early processes of carcinogenesis. The application of stochastic models of carcinogenesis on experimental data on the effects of carcinogenic substances has been reviewed by Moolgavkar et al. (1999). Here the state of the art was described exemplarily by reviewing more recent work of modelling of hepatocarcinogenesis in rats, one on the role of cell replication and apoptosis in tumour initiation and one on combined effects of an initiating and a promoting agent. Carcinogenesis in rat liver is of interest for an understanding of carcinogenesis because intermediate stages can be determined experimentally, e. g. by measuring changes of the expression of enzymes (see also section 4.5.1). Placental glutathione S-transferase (GST-P) is a useful marker for most (pre)neoplastic lesions (Sato 1989). GST-P positive (G^+) cells appear in liver preneoplasias after administration of various genotoxic hepatocarcinogens, but not after administration of non-genotoxic agents (Moore et al. 1987).

The number of applications of mathematical models of carcinogenesis to epidemiological studies on chemical carcinogens is limited. In the last part of this section, an analysis of lung cancer due to coke oven emissions is summarised.

6.5.1
Replication and apoptosis of intermediate cells in hepatocarcinogenesis

Cell replication and apoptotic activity have been observed in rats to increase from normal liver cells to preneoplasias and neoplasias (Grasl-Kraupp et al. 1997). At all stages of the carcinogenic process, cell replication rates were found to be higher than apoptosis rates.

Recently, first stages of hepatocarcinogenesis were studied by applying N-nitrosomorpholine (NNM) to rats (Grasl-Kraupp et al. 2000). NNM is one of the nitrosamines occurring in tobacco smoke and in a variety of foods and alcoholic beverages, and is likely to contribute to the development of human cancer. NNM was administered as a single dose of 250 mg NNM 10 ml^{-1} solution kg^{-1} body weight by gavage to male SPF Wistar rats at the age of 6–8 weeks. Measurements of the cellular dynamics were performed during a period up to 107.5 days after exposure. The exposure caused a high apoptotic effect in the liver. Regeneration started 48 h after NNM application. The total hepatic DNA content and absolute liver weights were back to their original level 24 days after the exposure. Rates of replication and apoptosis were still elevated after 31.5 days.

After NNM treatment, the number of G^+ single cells and small clones per unit area of evaluated tissue sections were scored (Fig. 6.17 A). The number of 3D clones per liver were estimated by a stereological method that took into account the non-spherical shape of the clones (Fig. 6.17 B). Both aspects indicated three phases after NNM application that may reflect alterations in the concentrations of growth factors in the liver. The modelling was performed with the first two stages of the TSCE model. The division rates α and death (or disappearance of G^+ phenotype)

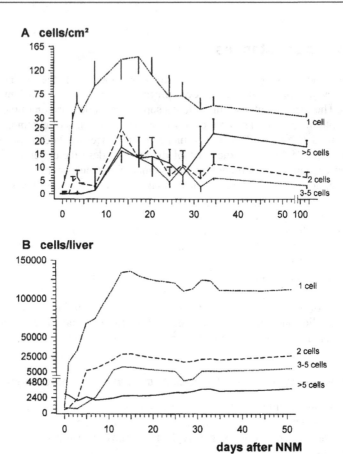

Fig. 6.17 Kinetics of appearance of G^+ single cells and G^+ foci. Experimental data are expressed in cm^{-2} evaluated tissue sections (A), derived results for 3D clones as mean values per liver (B) (from Grasl-Kraupp et al. 2000). Note that the majority of cells that appear as G^+ single cells in the tissue sections belong in the 3D reality to G^+ foci.

rates for G^+ single cells and for G^+ cells in clones with more than 1 cell were allowed to be different.

According to the model fit, about 5,000 G^+ clones were present in the rat liver at the time of NNM administration. Phase I is characterised by a reduction of body weight of the rats with a subsequent recovery. It lasted until day 14 after NNM application and reflects the continuous appearance of G^+ single cells and their development into multicellular foci (Fig. 6.17 B). According to the model, about 170,000 new clones, mainly G^+ single cells appeared. Regeneration signals, released in response to severe damage to the liver, are probably responsible for the growth. Both, measurements and model indicate that the G^+ cell division rate is larger in clones than for single cells (Tab. 6.6).

Phase 2, lasting until 28 days after NNM, was the first two weeks after recovery. The phase is characterised by a loss of G^+ cell clones (about 40,000 after the model

Tab. 6.6 G^+ cell division and disappearance rates in different phases after NNM administration to rats (after Grasl-Kraupp et al. 2000). Experimental results for disappearance rates are apoptotic rates, model results include in addition the loss rate of G^+ phenotype. Mean values (MV) and 95 %-confidence intervals (CI) are given for the model results.

Time after NNM administration (d)	G^+ cell division rates α (% d^{-1})		G^+ cell disappearance rate β (% d^{-1})	
	total (mainly single G^+ cells)	in clones > 1 cell	total (mainly single G^+ cells)	in clones > 1 cell
0–14	2.8	14	–	2
Model: MV and CI	4.2 (3.1; 5.2)	23 (19; 27)	0.0	0 (0; 10)
14–28	1.1	6	–	3
Model: MV and CI	0.5 (0.0; 2.5)	13 (4; 22)	6.2 (4.6; 8.6)	19 (8; 30)
28–107.5	0.02	2.5	–	1
Model: MV and CI	0.5 (0.0; 2.5)	13 (4; 22)	0.2 (0.0; 5.5)	0 (0–90)

fit) and a loss of constituent cells. The division rates of G^+ cells has gone down by a factor of about three, the disappearance rates are significantly increased. Experimentally, the disappearance rate is the apoptotic rate. The model calculations relate to a decrease of the number of G^+ cells, i. e. here no differentiation is made between apoptosis and loss of G^+ phenotype. Both, the division and disappearance rates, are in the clones higher than for single G^+ cells.

In the subsequent phase, the number of G^+ single cells and of multicellular foci appeared to stabilise and an increase in larger foci was observed. According to the model, in this phase about 11,000 new clones appeared until day 51. Compared to phase 2, the gain is mainly due to a significant decrease of the death rates of G^+ cells.

In conclusion, the consecutive development of G^+ single cells into foci with an increasing number of G^+ cells supports the concept that liver preneoplasias are of monoclonal origin and that expression of the G^+ phenotype is heritable by daughter cells. G^+ cells in clones with more than one G^+ cell exhibit an accelerated turnover. Consequently, a large number of clones disappears due to apoptosis or loss of G^+ phenotype, especially during the first weeks after the recovery of the animals from the NNM treatment (as expressed by their body weight).

6.5.2
Combined effects of an initiating and a promoting agent on hepatocarcinogenesis

2,3,7,8-Tetrachlorodibenzo-p-dioxin (TCDD) is known for its tumour-promoting property in rat liver (Dragan et al. 1992), lung, and skin of hairless mice (see also section 4.5.10). Moolgavkar et al. (1996) and Portier et al. (1996) analysed data on TCDD applications to rats with the TSCE model and found that TCDD has as initiating capacity as well. At the same time, Stinchombe et al. (1995) performed continuative experiments on G^+ cells in the liver of rats after the application of on initiating agent and of TCDD.

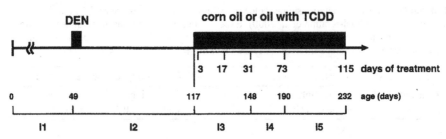

Fig. 6.18 Experimental protocol and intervals (I1,…, I5) for modelling the age dependence of initiation, division and disappearance rates of G^+ liver cells in rats that have been exposed to the initiating agent diethylnitrosamine (DEN) and subsequently to 2,3,7,8-tetra-chlorodibenzo-p-dioxin (TCDD) (from Luebeck et al. 2000).

In the experiments, 7-week-old female Wistar rats were exposed for 10 days to diethylnitrosamine (DEN), an initiating agent, with a dose rate of 10 mg kg^{-1} body weight d^{-1}. After a recovery period of 8 weeks, the animals received biweekly injections with TCDD corresponding to a chronic dose rate of 100 ng kg^{-1} body weight d^{-1} (Fig 6.18). The experiment showed a clear evidence that TCDD suppresses apoptosis in G^+ clones after ten weeks of application. Before this time, G^+ clones are smaller in the TCDD treated animals than in the controls, at later times they are larger.

Luebeck et al. (2000) analysed the data with the TSCE model assuming that the values of the parameters λ_1, α, and β are constant in five time intervals that are defined by the exposure patterns and long term behaviour of the G^+ cells. The number and sizes of G^+ clones in the liver were derived from results on G^+ positive foci in two-dimensional liver transections. The division rate α of G^+ cells was assumed to be independent of age and exposure, as it was indicated by the measurements. A value of 0.06 d^{-1} ml^{-1} of liver was obtained. Three main conclusions can be derived:

- Firstly, the acute treatment with DEN leads to a protracted appearance of initiated G^+ cells, with a massive increase in a period of about 100 to 150 days after the application. This may be related to long-lived DNA adducts that are converted into mutations as cells undergo division.
- Secondly, TCDD treatment appears to accelerate the formation of new clones possibly due to an accelerated conversion of DEN-induced DNA damage into fixed mutations required for an overexpression of the enzyme marker (Fig. 6.19). After the first 30 days of TCDD treatment, the initiation rate does not seem to exceed background any more. In total, less new clones are created than without TCDD exposure, possibly because the cytotoxicity of TCDD removes some of the cells with DEN-induced damage. The accelerating effect of TCDD may have shamed an initiating effect as it was reported in earlier analyses (Moolgavkar et al. 1996; Portier et al. 1996).
- Thirdly, the dynamics of the clonal growth of G^+ cells is complex, however, after longer periods of TCDD treatment (more than about 70 days), TCDD down-regulates the apoptosis of G^+ cells to about 60 % of the control value. This leads to a faster growth of the clones compared to the control implying a larger hazard for the incidence of later stages of the carcinogenic process.

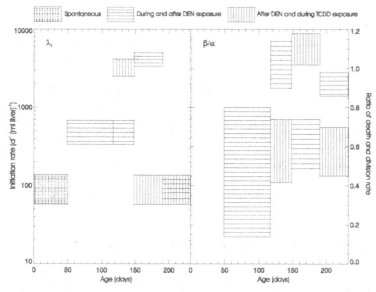

Fig. 6.19 Estimated 95 % confidence ranges of the initiation rate λ_1 and of the ratio β/α of the death and division rates for G^+ cells in clones in female Wistar rats that were exposed to DEN and TCDD (after Luebeck et al. 2000).

6.5.3
Lung cancer due to coke oven emissions

Coke oven emissions contain a complex mixture of polycyclic aromatic hydrocarbons. Moolgavkar et al. (1998) used the TSCE model to analyse the lung cancer mortality among a cohort of coke oven workers and not exposed controls from the same steel plants in the US. All cohort members started to work in the plants in the early 1950s. Detailed work histories were obtained for each worker up to 1982 and also the mortality follow-up extended up to the year 1982. Based on a survey taken in the late 1960s, average concentrations of coal tar pitch volatiles (CTPV) of 3.13 mg m^{-3} were assumed for working places on the top of the ovens and of 0.88 mg m^{-3} on the side of the ovens. Mathematical models of carcinogenesis are particularly suited for analysing such a data set, because complex exposure scenarios can be taken into account without further assumptions or parameters. Information on individual smoking behaviour was not available. The model, however, contained a term for a birth-year dependence in the promotion rate that can be due to a general change of the smoking behaviour within in the cohort.

The cohort was divided in four subcohorts, according to whether the workers were White or non-White, and whether they worked inside or outside of the Allegheny County. For all subcohorts except for Allegheny County Whites, a statistically significant difference was found in the lung cancer mortality among the unexposed and exposed subgroups. The authors concluded that this was possibly a substantial misclassification of exposures for the Allegheny County Whites, because the lung cancer mortality was highest in the second quartile of exposures.

Another possible reason for this pattern might be differences in the individual smoking behaviour about which no information was available. Similarly, for the non-Allegheny County Whites the lung cancer mortality was highest in the third quartile of exposure and there was little indication of an exposure-response relationship.

The main conclusions of the study were based on non-White cohort members who did not work in the Allegheny County. A significant effect of the coke oven emissions on both, the initiation and the promotion rate was found. The preferred model had exponential dependencies of the model parameters on the exposure rate (concentration of CTPV). No effect on the transformation rate was found. The optimal value of the lag time was found to be 3.5 years. Although these finding are remarkably similar to what has been found for uranium miners (see section 6.4.1.2), one should bare in mind that the results could be derived only for one of the four subcohorts.

6.6
Ionising radiation and carcinogenic substances

Although there is quite an extensive literature on effects of combined exposures to ionising radiation and carcinogenic substances (for a review see e. g. Streffer et al. 2000), there is only limited experience on modelling the carcinogenic process after such combined exposures. Work that has been performed on Colorado uranium miners and on Chinese tin miners is summarised in section 6.6.1. Some theoretical work based on TSCE model calculations for combined exposures are outlined in section 6.6.2 in order to demonstrate exemplarily low-dose risk estimates of the model for combined exposures.

6.6.1
Radon, smoking and arsenic

Luebeck et al. (1999) analysed the interaction of radon and smoking in the lung cancer mortality among the Colorado Plateau miners (section 6.4.2). It was assumed that radon and smoking affect the initiation and promotion rates independently. No effect on the transformation rate was found. The increase of the initiation and the promotion rates relative to their spontaneous values consists in both cases of two summands, one for smoking and one for radon exposure (Fig. 6.11). For the exposure scenarios shown in Fig. 6.20, the resulting relative risk is nearly multiplicative in the time period where both exposures apply (plus lag time). This is because smoking and radon are assumed to act additively on the promotion rate, and during the considered age range the hazard increases approximately exponentially with the product of the promotion rate and age (eq. 6.21). At the end of the assumed exposure scenario (plus lag time), the hazard is a bit submultiplicative due to the clones that were initiated by smoking or radon. Subsequently, the hazard becomes considerably smaller than the multiplicative risk and decreases with age (for high ages even below an additive risk model).

Hazelton et al. (2001) used the TSCE model in analysing lung cancer mortality data for a cohort of Yunan (China) tin miners who were exposed to arsenic and radon. The analysis was performed for 12,011 male miners with complete records

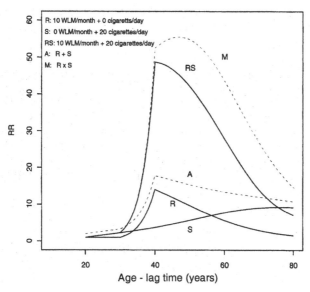

Fig. 6.20 Relative risk of lung cancer mortality at age (a-lag time) according to a TSCE model fit to the Colorado Plateau miners data for single exposures [smoking (S): 20 cigarettes d^{-1} starting age 20; radon (R): 10 WLM month^{-1} between ages 30 and 40], and for combined exposure (RS). Dashed lines (A and M) show strictly additive and multiplicative models, respectively (from Luebeck et al. 1999).

of arsenic, radon, and cigarette and pipe smoke exposures. In the follow-up period 1976 to 1987, 842 lung cancer deaths were reported that amounted to 48 % of all deaths. Additional risk factors like workplace exposures to silica or other carcinogens, residential exposure to smoke from poorly ventilated indoor cooking and heating, or others may have contributed to this high lung cancer mortality.

In the modeling, the complex information on the different exposure patterns was translated in piecewise constant parameters for several periods from birth to censoring date (end of follow-up, date of death or loss to follow-up). The authors optimised exposure rate dependencies of the parameters by performing calculations for a large number of possible models. Allowance was made for a birth year effect on the initiation rate. Finally they eliminated all parameters that contributed insignificantly to the likelihood, and allowed the lag time to vary according to a γ-distribution. This "final model" contained 14 free parameters.

Although tobacco smoke has been taken into account, a strong birth year dependence of the initiation rate was found. The analysis indicates that exposure to arsenic, radon and tobacco smoke increases division, death and transformation rates α, β, and λ_2 of intermediate cells. The fitted lag-time distribution has a mean value of 4.1 years with a variance of 2.9 years. The results are not compatible with spontaneous transformation rates $\lambda_2(0)$ exceeding a value of 8.6 10^{-12} yr^{-1}. This suggests that at least two sequential mutations are required after initiation for malignant conversion. However, taking into account the large number of fitting parameters, one should be careful in generalising these results.

6.6.2
Low-dose risk after exposure to an initiating and a promoting agent

As demonstrated by the example in section 6.5.2, interactions of initiating and promoting carcinogens may be very complex. However, it is illustrative to study the effects of combined exposures to initiating and promoting agents with a model under the assumption that the effect of the promoting agent depends only on the number and sizes of the clones of initiated cells, and not on other biochemical changes that were induced by the initiating agent (Jacob and Prokić 2003).

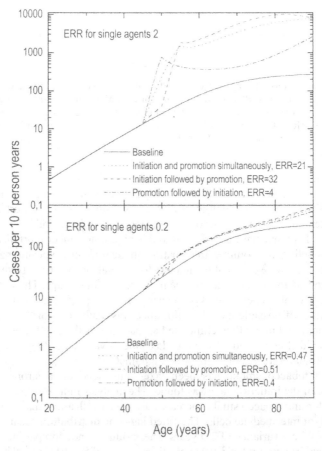

Fig. 6.21 Hazards for solid tumours as calculated with the TSCE model for simultaneous and consecutive exposures to initiating and promoting agents in the age period of 40 to 50 years. The exposures to the single agents result in an excess relative lifetime risk of 2 and 0.2, respectively (Jacob and Prokić 2003).

The TSCE model hazard due to single exposure during the age period of 40 to 50 years has been discussed in section 6.3.1.2. For a combination of the exposures to the initiating and promoting agent, there is a strong synergistic effect for high exposures (ERR = 2 for the single agents), if the agents act simultaneously or if the exposure to the promoting agent follows exposure to the initiating agent (Fig. 6.21). In the first case the synergistic effect amounts to a factor of 5, in the second case to a factor of 8. The reason for the strong synergistic effect is that in these two cases the promotion acts on the clones that are created by the initiating agent. This is not the case, if the exposure to the initiating agent follows the exposure to the promoting agent. The effects on the hazard are additive.

For low exposures (ERR = 0.2) the synergistic effect becomes negligible. Compared to the high exposure it decreases from a factor of 5 to 1.2, resp. from a factor of 8 to 1.3. The reason is that for low exposures the change of the distribution of spontaneously created clones of intermediate cells due to the initiating agent is minor. Therefore, even if the promoting agent can act on the clones created by the initiating agent, the corresponding contribution is small compared to the effect of the promoting agent on the spontaneously created intermediate cells. Having in mind, that an exposure leading to an excess relative risk of 0.2 is in general large compared to exposures resulting from environmental contaminations, the synergistic effect is negligible for environmental contaminations with initiating and promoting agents, if the effect of the promoting agent depends only on the number and sizes of the clones of initiated cells, and not on other biochemical changes that were induced by the initiating agent. These features of the TSCE model are in full accordance with what has been observed experimentally and with other mechanistic models (see Streffer et al. 2000).

6.7
Conclusions

Mathematical models of carcinogenesis have been applied to data of radioepidemiological studies and of animal experiments with exposures to ionising radiation and to carcinogenic substances. Compared to heuristic epidemiological models, as the excess relative risk model, mathematical models of carcinogenesis have in general a smaller number of free parameters, fit the data equally well, and have the advantage of a straightforward treatment of complex exposure scenarios.

The large potential of mathematical models of carcinogenesis in the application to epidemiological data is in the assessment of risks in the low-dose range by taking into account epidemiological, radiobiological and molecular genetic data. Further, the models relate numbers and sizes of intermediate stages as adenomas or liver foci to cancer incidence or mortality rates in the same population.

In the application to an evaluation of experimental studies the potential of mathematical models of carcinogenesis lies in an improvement of the understanding of the dynamics of early stages of the process and in the generation of hypotheses that can be tested experimentally. This has been proven, e. g. for studies on the behaviour of placental glutathione S-transferase positive (G^+) cells which have the capacity to form preneoplasias in rat liver.

Comparative studies with lung cancer data for rats that were exposed to radon progeny and with cancer incidence data for the atomic bomb survivors have shown that the statistical power of these data sets is not large enough to decide which is the most appropriate model. Simplicity, plausibility, and radiobiological and molecular genetic knowledge are additional aspects to identify appropriate models. According to the current state of art, a variety of models is possible.

The two-step clonal expansion (TSCE) model has been studied quite intensively. It has been shown to predict direct and inverse dose rate effects, depending on the exposure scenario and on the exposure rate dependence of the initiation and promotion rates of intermediate cells. At the same time, in most applications to epidemiological data, parameter estimates had a large uncertainty. Therefore, conclusions on age, time and exposure dependences for single cancer types remain to be uncertain. The application of a variety of models (including phenomenological epidemiological models) offers, however, the possibility to identify common predications and their uncertainties including the model uncertainty.

Interactions of initiating and promoting carcinogens have been studied with the TSCE model. As to be expected, the degree of synergistic effects depends on the exposure scenarios, e. g. which carcinogen is applied first. However, according to the model, synergistic effects tend to become negligibly small if the doses of both carcinogens are small.

In the past one or two decades, several low-dose radiobiological effects have been discovered. The biological response to low doses of ionising radiation may be quite different from that at medium and high doses. These include low-dose hypersensitivity, bystander effects, gene expressions and genomic instability (see chapter 3). Models of carcinogenesis offer the possibility to integrate such effects, which is not so easily done with phenomenological models. A first example of such work is the integration of low-dose hypersensitivity concerning cell inactivation in the TSCE model and its application to lung cancer incidence among the atomic bombs survivors. The example demonstrates possible consequences for low-dose risk estimates. More work in this direction is to be expected in the future.

For a large class of models, the stochastic multistage models with clonal expansion, the exploration of their features is still in an early stage. A main weakness of the currently existing approaches in this field is the vast number of possible processes and free parameters. The growing experimental experience on processes of carcinogenesis and genetic pathways, however, may offer the possibility of a fruitful application of such models in the future. A first impressing step has been made with applying a four-step model to population data on colorectal cancer.

6.8
References

Aaltonen LA, Peltomäki P, Leach FS, Sistonen P, Pylkkänen L, Mecklin JP, Järvinen H, Powell SM, Jen J, Hamilton SR, Peterson GM, Kinzler KW, Vogelstein B, de la Chapelle A (1993) Clue to the pathogenesis of familiar colorectal cancer. Science 260: 812–816

Armitage P, Doll R (1954) The age distribution of cancer and multistage theory of carcinogenesis. Br J Cancer 8: 1–12

Armitage P, Doll R (1957) The two-stage theory of carcinogenesis in relation to the age distribution of human cancers. Br J Cancer 11: 161–169

Beck-Bornholdt HP, Maurer T, Becker S, Vogler H, Würschmidt F (1989) Radiotherapy of the rhadomyosarcoma R1H of the rat: hyperfractionation – 126 fractions applied within 6 weeks. Int J Radiat Oncol 16: 701–705

Bishop JM (1991) Molecular themes in oncogenesis. Cell 64: 235–248

Cross FT, Buschbom RL, Dagle E, Gideon KM, Gries RA (1993) Radon hazards in homes. In: Annual Report for 1992 to the DOE Office of Energy Research. Pacific Northwest Laboratory, Richland, WA, pp 31–37

Dragan YP, Xu X, Goldsworthy TL, Campbell HA, Maronpot RR, Pilot HC (1002) Characterization of promotion of altered hepatic foci by 2,3,7,8-tetrachloodibenzo-p-dioxin in the female rat. Carcinogenesis 13: 1389–1395

Fearon ER, Vogelstein B (1990) A genetic model for colorectal tumorigenesis. Cell 61: 759–767

Gannon JV, Greaves R, Iggo R, Lane DP (1990) Activating mutations in p53 produce a common conformational effect: a monoclonal antibody specific for the mutant form. EMBO J. 9: 1595–1602

Grasl-Kraupp B, Luebeck G, Wagner A, Löw-Baselli A, de Gunst M, Waldhör T, Moolgavkar S, Schulte-Hermann R (2000) Quantitative analysis of tum0r initiation in rat liver: role of cell replication and cell death (apoptosis). Carcinogenesis 21: 1411–1421

Grasl-Kraupp B, Ruttkay-Nedecky B, Müllauer L, Taper H, Huber W, Bursch W, Schulte-Hermann R (1997) Inherent increase of apoptosis in liver tumors: implications for carcinogenesis and tumor regression. Hepatology 25: 906–911

Hazelton WD, Luebeck EG, Heidenreich WF, Moolgavkar SH (2001) Analysis of a historical cohort of Chinese tin miners with arsenic, radon, cigarette smoke, and pipe smoke exposure using biologically based two-stage clonal expansion model. Radiat Res 156: 78–94

Heidenreich WF (1996) On the parameters of the clonal expansion model. Radiat Environ Biophys 35: 127–129

Heidenreich WF, Atkinson M, Paretzke HG (2001) Radiation induced cell inactivation can increase the cancer risk. Radiat Res 155: 870–872

Heidenreich WF, Brugmans MJP, Little MP, Leenhouts HP, Paretzke HG, Morin M, Lafuma J (2000) Analysis of lung tumour risk in radon-exposed rats: an intercomparison of multistep modelling. Radiat Environ Biophys 39: 253–264

Heidenreich WF, Jacob P, Paretzke HG (1997a) Exact solutions of the clonal expansion model and their application to the incidence of solid tumors of the atomic bomb survivors. Radiat Environ Biophys 36: 45–58

Heidenreich WF, Jacob P, Paretzke HG, Cross FT, Dagle GE (1999) Two-step model for the risk of fatal and incidental lung tumors in rats exposed to radon. Radiat Res 151: 209–217

Heidenreich WF, Luebeck EG, Hazelton WD, Paretzke HG, Moolgavkar SH (2002) Multistage models and the incidence of cancer in the cohort of A-bomb survivors. Radiat Res 158: 607–614

Heidenreich WF, Luebeck EG, Moolgavkar SH (1997b) Some properties of the hazard function of the two-mutation clonal expansion model. Risk Anal 17: 391–399

Herrero-Jimenez P, Thilly G, Southam PJ, Tomita-Mitchel A, Morgenthaler S, Furth EE, Thilly WG (1998) Mutation, cell kinetics, and subpopulations at risk for colon cancer in the United States. Mut Res 400: 553–578

Herrero-Jimenez P, Tomita-Mitchel A, Furth EE, Morgenthaler S, Thilly WG (2000) Population risk and physiological rate parameters for colon cancer. The union of an explicit model for carcinogenesis withthe public health records of the United States. Mut Res 447: 73–116

Jacob P, Prokić V (2002) Increased radioresistance, modelling of carcinogenesis and low-dose risk estimation. J Radiol Prot 22: A51–A55

Jacob P, Prokić V (2003) Zur Bewertung radiologischer und chemisch-toxischer Wirkungen von Umweltkontaminationen. In: Bundesministerium für Umwelt, Naturschutz und Gesundheit (ed) Aktuelle und neue Aufgaben in der Radioökologie. Urban & Fischer, München

James F (1994) Minuit. Function minimization and error analysis. Reference Manual, Version 94.1. CERN Program Library Long Writeup D506. CERN, Geneva

Joiner MC, Lambin P, Malaise EP, Robson T Arrand JE, Skov KA, Marples B (1996) Hypersensitivity to very low single radiation doses: Its relationship to the adaptive response and induced radioresistance. Mut Res 358: 171–183

Joiner MC, Marples B, Lambin P, Short SC, Turesson I (2001) Low-dose hypersensitivity: current status and possible mechanisms. Int J Radiat Oncol 49: 379–389

Kai M, Luebeck EG, Moolgavkar S.H. (1997) Analysis of the incidence of solid cancer among atomic bomb survivors using a two-stage model of carcinogenesis. Radiat Res 148: 348–358

Kellerer AM, Nekolla EA, Walsh L (2001) On the conversion of solid cancer excess relative risk into lifetime attributable risk. Radiat Environ Biophys 40: 249–257

Knudson AG (1971) Mutation and cancer: Statistical study of retinoblastoma. Proc Natl Acad Sci 68: 820–823

Kopp-Schneider A (1997) Carcinogenesis models for risk assessment. Stat Meth Med Res 6: 317–340

Kopp-Schneider A, Portier CJ, Sherman CD (1994) The exact formula for tumor incidence in the two-stage model. Risk Anal 14: 1079–1080

Little MP (1995) Are two mutations sufficient to cause cancer? Some generalizations of the two-mutation model of carcinogenesis of Moolgavkar, Venzon, and Kundson, and of the multistage model of Armitage and Doll. Biometrics 51: 1278–1291

Little MP (1996) Generalisations of the two-mutation and classical multistage models of carcinogenesis fitted to the Japanese atomic bomb survivor data. J Radiol Prot 16: 7–24

Little MP, Haylock RGE, Muirhead CR (2002) Modelling lung tumour risk in radon-exposed uranium miners using generalizations of the two-mutation model of Moolgavkar, Venzon and Knudson. Int J Radiat Biol 78: 49–68

Little MP, Muirhead CR, Stiller CA (1996) Modelling lymphocytic leukemia incidence in England and Wales using generalizations of the two-mutation model of carcinogenesis of Moolgavkar, Venzon and Knudson. Stat Med 15: 1003–1022

Luebeck EG, Buchmann A, Stinchcombe S, Moolgavkar SH, Schwarz M (2000) Effects of 2,3,7,8-tetrachlorodibenzo-p-dioxin on initiation and promotion of GST-P-positive foci in rat liver: A quantitative analysis of experimental data using a stochastic model. Toxicol Appl Pharmacol 167: 63–73

Luebeck EG, Curtis SB, Cross FT, Moolgavkar SH (1996) Two-stage model of radon-induced malignant lung tumors in rats: Effects of cell killing. Radiat Res 145: 163–173

Luebeck EG, Heidenreich WF, Hazelton WD, Paretzke HG, Moolgavkar SH (1999) Biologically based analysis of the data for the Colorado uranium miners cohort: age, dose and dose-rate effect. Radiat Res 152: 339–351

Luebeck EG, Moolgavkar SH (2002) Multistage carcinogenesis and the incidence of colorectal cancer. PNAS 99: 15095–15100

Monchaux G, Morlier J, Morin M, Chameaud J, Lafuma J, Masse R (1994) Carcinogenic and cocarcinogenic effects of radon and radon daughters in rats. Environ Health Perspect 102: 64–73

Monchaux G, Morlier JP, Altmeyer S, Debroche M, Morin M (1999) Influence of exposure rate on lung cancer induction in rats exposed to radon progeny. Radiat Res 152: S137–S140

Moolgavkar S, Krewski D, Schwarz M (1999) Mechanisms of carcinogenesis and biologically based models for estimation and prediction of risk. IARC Sci Publ 131: 179–237

Moolgavkar SH (1978) The multistage theory of carcinogenesis and the age distribution of cancer in man. J Natl Cancer Inst 61: 49–52

Moolgavkar SH (1997) Stochastic cancer models: Application to analyses of solid cancer incidence in the cohort of A-bomb survivors. Nucl Energy 36: 447–451

Moolgavkar SH, Cross FT, Luebeck G, Dagle GE (1990) A two-mutation model for radon-induced lung tumors in rats. Radiat Res 121: 28–37

Moolgavkar SH, Dewanji A, Venzon DJ (1988) A stochastic two-stage model for cancer risk assessment. I. The hazard function and the probability of tumor. Risk Anal 8: 383–392

Moolgavkar SH, Luebeck EG, Anderson EL (1998) Estimation of unit risk for coke oven emissions. Risk Anal 18: 813–825

Moolgavkar SH, Luebeck EG, Buchmann A, Bock KW (1996) Quantitative analysis of enzyme-altered liver foci in rats initiated with diethylnitrosamine and promoted with 2,3,7,8-tetrachlorodibenzo-p-dioxin or 1,2,3,43,6,7-heptachlorodibenzo-p-dioxin. Toxicol Appl Pharmacol 138: 31–42

Moolgavkar SH, Luebeck EG, Krewski D, Zielinski JM (1993) Radon, cigarette smoke and lung cancer: A re-analysis of the Colorado Plateau uranium miners' data. Epidemiology 4: 204–217

Moolgavkar SH, Luebeck G (1990) Two-event model for carcinogenesis: biological, mathematical and statistical considerations. Risk Anal 10: 323–341

Moolgavkar SH, Luebeck G (1992) Multistage carcinogenesis: population-based model for colon cancer. J Natl Cancer Inst 84: 610–618

Moolgavkar SH, Venzon DJ (1979) Two-events models for carcinogenesis: Incidence curves for childhood and adult tumors. Math Biosci 47: 55–77

Moore MA, Nakagawa K, Satoh K, Ishikawa T, Sato K (1987) Single GST-P positive liver cells – putative initiated hepatocytes. Carcinogenesis 8: 483–486

Nordling CO (1953) A new theory on the cancer-inducing mechanism. Brit J Cancer 7: 68–52

Peltomäki P, Aaltonen LA, Sistonen P, Pylkkänen L, Mecklin JP, Järvinen H, Green JS, Jass JR, Weber JL, Leach FS, Petersen GM, Hamilton SR, de la Chapelle A, Vogelstein B (1993) Genetic mapping of a locus predisposing to human colorectal cancer. Science 260: 810–812

Pierce DA, Mendelsohn ML (1999) A model for radiation-related cancer suggested by atomic bomb survivor data. Radiat Res 152: 642–654

Pierce DA, Shimizu Y, Preston DL, Vaeth M, Mabuchi K (1996) Studies of the mortality of atomic bomb survivors. Report 12, Part 1: 1950–1990. Radiat Res 146(1): 1–27

Portier CJ, Sherman CD, Kohn MC, Edler L, Kopp-Schneider A, Maronpot RR, Lucier G (1996) Modeling the number and size of hepatic focal lesions following exposure to 2,3,7,8-TCDD. Tocicol. Appl Pharmacol 138: 20–30

Preston D (1999) Cigarette smoking and radiation dose in the Life Span Study. RERF Update 10: 9

Preston D (2002) Private communication. Radiation Effects Research Foundation. Hiroshima, Japan

Prokić V, Jacob P (2002) Cancer incidence and mortality among atomic bomb survivors. Threshold calculations with the TSCE model. Europäische Akademie zur Erforschung wissenschaftlich-technischer Entwicklungen. Bad-Neuenahr Ahrweiler, Germany (unpublished)

Prokić V, Jacob P, Heidenreich W (2002) Possible implications of non-linear radiobiological effects for the estimation of radiation risk at low doses. Radiat Prot Dosim 99: 279–281

Ries LAG, Eisner MP, Kosary CL, Hankey BF, Miller BA, Clegg L, Edwards BK (eds) (2001) SEER Cancer Statistics Review, 1973–1998. National Cancer Institute, Bethesda, MD

Sato K (1989) Glutathione transferases as markers for preneoplasia and neoplasia. Adv Cancer Res 52: 205–255

Singh B, Arrand JE, Joiner MC (1994) Hypersensitive response of normal human lung epithelial cells at low radiation doses, Int J Radiat Biol 65: 457–464

Statistisches Bundesamt (2000) Statistisches Jahrbuch 2000 für die Bundesrepublik Deutschland. Metzler-Poeschel, Stuttgart

Stinchcombe S, Buchmann A, Bock KW, Schwarz M (1995) Inhibition of apoptosis during 2,3,7,8-tetrachlorodibenzo-p-dioxin-mediated tumour promotion in rat liver. Carcinogenesis 16: 1271–1275

Streffer C, Bücker J, Cansier A, Cansier D, Gethmann CF, Guderian R, Hanekamp G, Henschler D, Pöch G, Rehbinder E, Renn O, Slesina M, Wuttke K (2000) Umweltstandards: kombinierte Expositionen und ihre Auswirkungen auf die Umwelt. Wissenschaftsethik und Technikfolgenbeurteilung, Bd. 5. Springer, Berlin

Tan WY (1991) Stochastic models of carcinogenesis. Marcel Dekker, New York

Thompson DE, Mabuchi K, Ron E, Soda M, Tokunaga M, Ochikubo S, Sugimoto S, Ikeda T, Terasaki M, Izumi S, Preston DL (1994) Cancer incidence in atomic bomb survivors. Part II: Solid tumors, 1958–1987. Radiat Res 137(2 Suppl): S17–S67

Trosko JE, Chang CC, Medcalf A (1983) Mechanisms of tumor promotion: Potential role of intercellular communication. Cancer Invest. 1: 511–526

United Nations Scientific Committee on the Effects of Atomic Radiation [UNSCEAR] (2000) Report to the General Assembly, Volume II, Annex H. United Nations, New York

Vaeth M, Pierce D (1990) Calculating excess lifetime risk in relative risk models. Environ Health Perspect 87: 83–94

Weinberg RA (1991) Tumor suppressor genes. Science 254: 1138–1146

Zheng Q (1994) On the exact hazard and survival functions of the MVK stochastic carcinogenesis model. Risk Anal 14: 1081–1084

6.9
Appendix

6.9.1
Likelihood, model fits and deviance

In general, in a cohort study for each cohort member i, the age of entry a_{1i} and the censoring age a_{2i} are known. In a model that describes the probability $\Psi_i(a\text{-}t)$ that for the exposure history i none of the susceptible cells has become malignant at age $a\text{-}t$, the probability or likelihood that the cohort member did not receive a cancer in the period (a_1, a_2) is

$$L_{\text{nocase},i} = \Psi_i(a_{2i}\text{-}t) \,/\, \Psi_i(a_{1i}\text{-}t). \qquad (\text{eq. A.1})$$

The likelihood, that he died of a cancer (mortality data) or received a cancer (incidence data) is

$$L_{\text{case},i} = \dot{\Psi}_i(a_{2i} - t) \,/\, \Psi(a_{1i}\text{-}t). \qquad (\text{eq. A.2})$$

In order to fit a model to the data, the product L of the likelihoods $L_{\text{nocase},i}$ for those who did not die because of (receive a) cancer and $L_{\text{case},i}$ for those who died because of (received a) cancer is maximised by varying the model parameters. This is called an individual likelihood technique.

In some studies, e. g. studies the atomic bomb survivors with publicly available data, instead of individual data, only the numbers of cancer cases N_j in subcohorts (strata) j are known. The expectation value E_j for the number of cases in a stratum is calculated by multiplying the person-years of observation in the stratum with the model hazard $H_j(a_j)$ for the age, time and exposure values for the stratum. The likelihood to have the number of observed cases N_j in the given stratum is calculated with a Poisson distribution with the expectation value E_j. The product L of the likelihoods over all strata j is then maximised to varying the model parameters. This is referred to a Poisson regression technique.

For most cases discussed here, there are no rigorous procedures to compare the quality of fit of different models. However, the deviance, which is twice the difference of the natural logarithm of their likelihoods, is often used as an approximate measure for the fit quality. In a family of models, compared to a model with a small number of parameters, a model with one additional parameter is considered to be an improvement on the 95 % confidence level, if its deviance is smaller by a value of 4, a model with two additional parameters, if its deviance is smaller by a value of 6, and a model with three additional parameters, if its deviance is smaller by a value of 8 (James 1994).

7 Comparing Summary of Action of Ionising Radiation and Chemicals in Carcinogenesis

Toxic agents cause a wide variety of adverse health effects. In the low-dose range the risk of stochastic effects and predominantly the induction of cancer is of dominating concern. Some other toxic effects may be discussed after low-dose exposures like genetic effects, disturbances of the immune system and developmental effects especially of the central nervous system. However, in the debate about protection of humans and the environment against industrial toxic releases and other anthropogenic burdens these later damages have only minor positions. Therefore, this report has focussed to discuss cancer risk.

Exposures to low doses of a carcinogenic agent have to be considered especially carefully when the risk of a certain agent can be described by a dose-response curve without a threshold as by the linear-no-threshold (LNT) model. In chapters 3 to 6 it has been elaborated that the dose-response at low doses and dose-rates may be quite complex. As long as no more definite information is available, the LNT model is considered as most appropriate for the assessment of cancer risk by ionising radiation and certain chemicals. Under these conditions a complete elimination of the potential risk does not take place even in the very low-dose range. However, the remaining risk is decreasing with decreasing doses and becomes extremely small. Thus, since it cannot be accurately quantified due to the reasons that have been pointed out before (chapters 3 and 4). However, while the LNT model is generally used for risk evaluation of all types of ionising radiation (chapters 3, 5 and 6), large differences exist between toxic chemicals with respect to the shape of the dose-response curve (chapter 4). In the present chapter the carcinogenic effects and mechanisms of ionising radiation and of the various types of chemicals are compared with respect to consequences for risk evaluation in the low-dose range.

The complex mechanisms that finally lead to cancer require years or even decades after exposures. It is generally accepted that multi-step processes including several genetic alterations are involved. Several mutation events and proliferative processes have to occur before a cancer with several hundred million to billions of cells becomes clinically manifest (chapters 3, 4 and 6). Concerning genotoxic carcinogens, DNA is apparently the target molecule where energy absorption of ionising radiation or reactions with chemicals have to occur. These primary events result in DNA damage which can be repaired or lead to mutation and cell transformation. Usually, the dose-response of such genotoxic carcinogens can be described by an LNT model. However, exceptions such as formaldehyde or vinyl acetate may exist, where a LNT model is not adequate.

Ionising radiation can interact directly with DNA (direct action) or highly reactive radicals are primarily formed by interaction of radiation with the cellular water. Subsequently, these radicals may interact with DNA (indirect action). The types of

DNA damage which have been described in chapter 3 originate from both types of action. Genotoxic and carcinogenic chemicals interact with DNA and damage often occurs in principle in a similar way as with ionising radiation. However, a difference occurs with respect to localisation of DNA damage: ionising radiation frequently lead to clusters of DNA damage. In contrast, DNA damage caused by chemicals usually occurs relatively equally distributed over the genome, although actively transcribed genes are often more affected than silent genes. From these differences it follows that even in the low-dose range, multiple damaged sites after radiation exposures occur whereas the damaging events originating from chemicals are more isolated. Another important difference between carcinogenic chemicals and ionising radiation occurs with respect to metabolism. Metabolic processes are necessary for many chemicals in order to generate an active genotoxic carcinogen that can interact with DNA (Fig. 4.1). Thus, metabolism is a further key parameter influencing the carcinogenic potential of a chemicals. In contrast, no metabolism is necessary for the genotoxic action of ionising radiation besides biokinetic processes in some cases which are necessary for the distribution of incorporated radioactive substances within an organism. Many chemicals are metabolised through multiple metabolic pathways for activation, but also metabolic inactivation of a carcinogen can take place. This adds considerable uncertainty to the process of risk evaluation, due to the complex interactions between metabolic activation, inactivation and toxicokinetics as well as the insufficient knowledge about these processes in humans.

The extrapolation between species e. g from rodents to humans additionally complicates the situation, since metabolism may differ considerably between species. Therefore, determination of exposure, which is a necessary prerequisite for risk evaluation, is much more complicated and uncertain for chemicals than for ionising radiation. In the case of ionising radiation with most radiation qualities, the radiation dose (absorbed energy per mass) can be determined or calculated in a tissue or cells with corresponding models even in the low-dose range with an acceptable accuracy. This is especially the case for external exposures with penetrating radiation but it is also possible for internal exposures after incorporation of radioactive substances, although in these latter cases with some higher degree of uncertainty. Nevertheless, it must be seen that differences exist between low-LET (β-, γ- and X-rays) and high-LET radiation (α-particles and neutrons) with respect to biological effectiveness. With high-LET radiation the dose distribution is much more heterogeneous in the low-dose range than exposures to low-LET radiation (chapter 3). Exposures to high-LET causes more and larger clusters of DNA damage. Repair of these multiple damaged sites is slower and less efficient. Frequently misrepair occurs.

Concerning chemicals, metabolism differs between individual organs and tissues even in the same species. In addition, large differences between individuals of the same species have been described. This leads to a considerable organ specificity of chemicals. Usually a carcinogenic chemical induces tumours only in few organs or tissues or even only in one tissue of a mammalian species (chapter 4). These additional differences in metabolic activities for activation of procarcinogens to genotoxic carcinogens or for deactivation of genotoxic carcinogens tremendously increases the degree of variability for individual susceptibility to a certain carcinogenic chemical.

After induction of DNA damage by a genotoxic agent (possibly after a metabolic activation of chemicals), which represents the initial step in carcinogenesis, a number of modifying mechanisms exist like DNA repair, adaptive response, apoptosis, immune response, cell cycle arrest and other mechanisms of regulatory processes for cell proliferation. Most of these mechanisms reduce the degree of malignant cell transformation. These processes are very similar for cancer induction by ionising radiation and chemical carcinogens. Differences only occur in some details. Thus, the specific enzymatic processes for repair of DNA damage after exposure to ionising radiation may be different from those after exposure to genotoxic chemicals. However, individual differences have been observed with respect to the repair capacity in different organs and between individuals. Especially the latter effect is caused by genetic predispositions. This circumstance leads to an appreciable variability in the individual susceptibility for cancer induction. Such effects have been observed for all carcinogenic principles.

The later steps in cancer development, which are connected to the processes of promotion and progression, are very similar or even identical in causation of cancer by ionising radiation and genotoxic chemicals. The first transformed, malignant cells are stimulated for cell proliferation and the development of malignancies in the tissues and organ systems occur during these phases. These processes can be influenced by endogenous factors like hormonal activities, by a number of chemicals with similar activities but also by ionising radiation(chapters 3 to 6). Again, genetic predispositions, for example with mutations in tumour suppressor genes (mutation in the *p53* gene, Li Fraumeni syndrome), can increase the susceptibility in the same way for ionising radiation as well as for genotoxic carcinogens.

Cancers whether induced by ionising radiation or by chemicals apparently show no differences in their clinical appearance, in any specific molecular or in cellular features. No such differences have been seen until now. Exposures to ionising radiation or carcinogenic chemicals may lead to specific patterns of tumour types. However, to our knowledge all tumour types induced by ionising radiation or chemicals, concerning histological as well as cellular appearance and oncogene expression patterns or other molecular features, can also occur spontaneously due to endogenous processes. In the same way, the cancers induced by the discussed exogenous factors cannot be distinguished from "spontaneous" cancers which are seen in the population without any apparent toxic exposures. As this background of cancers shows a considerable spread which is influenced by gender, age, lifestyle and other factors, it is generally not possible to observe a significant increase of the cancer rate at certain exposure doses and below this exposure limit. This holds true for the effects of ionising radiation as well as for genotoxic chemicals.

Therefore, the quantitative cancer risk in the low-dose range can only be obtained by extrapolation from exposures that lead to significant increases of cancer rates in the intermediate and high-dose range. The impossibilities of distinguishing induced cancers from the background cancers are probably the main reasons why the LNT model cannot be proven by epidemiological data. Experimental data with animals, cellular and molecular systems have revealed linear and also non-linear dose relationships at low doses. Such data can support the verification of the dose-response model and the extrapolation procedure. Therefore, risk evaluation decisions have to be made under uncertainty. These uncertainties and the scientific data which support the LNT model justify the introduction of the precautionary principle for the protection of humans and of the environment.

While the LNT model can be applied for ionising radiation in almost all cases, this is not possible for all chemical carcinogens. Based on the case studies presented in chapter 4, a differentiation may be made between groups of compounds, as far as low-dose extrapolation of cancer risk is concerned. For these reasons the following classification of carcinogens is proposed:

Group 1

Genotoxic carcinogens without a threshold, where a LNT model should be applied.

1A: Linearity of the dose-response and absence of a threshold is supported by scientific data, such as from biomarker studies in dose ranges below the sensitivity of experimental cancer bioassays. This is the case for vinyl chloride. For ionising radiation linear and non-linear dependencies have been observed for various endpoints. However, the LNT model is still the most appropriate model at low doses. Also other carcinogens, such as dimethylnitrosamine and the tobacco smoke-specific 4-[*N*-methyl-*N*-nitrosoamino]-1-[3-pyridyl]-1-butanone (NNK) are considered to belong to this subgroup.

1B: There is uncertainty about the dose-response at low exposure levels which leads to introduction of the precautionary principle. No scientifically plausible evidence demonstrates existence of a threshold. There may be differences in dose-response between different tissues, uncertainties in extrapolation between species, and gender specificities which contribute to these uncertainties. Examples are 2-acetylaminofluorene and 4-aminobiphenyl.

Group2

Genotoxic substances where sufficient scientific data, concerning the underlying mechanisms, suggest existence of a practical threshold. This is the case for non-DNA reactive genotoxins and for compounds which need repetitive local tissue damage and associated cell proliferation to exert a carcinogenic effect. For these compounds, standards may be derived from threshold assumptions, starting from an established No-Observed-Adverse-Effect-Level (NOAEL) (see UBA 1998). The proof of existence of a threshold effect is the prerequisite for compounds of this group. Examples are formaldehyde, vinyl acetate and motor protein inhibitors.

Group 3

Non-genotoxic carcinogens. These compounds display a clear threshold of carcinogenicity, and health-based exposure limits may be established starting from NOAEL and introduction of a safety/uncertainty factor, if applicable. Examples are tumour promoters, such as 2,3,7,8-tetrachlorodibenzo-*p*-dioxin (TCDD) and hormones.

Quantitative risk factors for cancer induction after exposures to ionising radiation have been obtained from a number of epidemiological studies with exposed human populations chapters 5 and 6). Investigations with radiation-induced cancers in animals and experiments with cellular as well as molecular systems have contributed to the knowledge about the dose-response and have led to the LNT model (chapter 3). It has already been pointed out that in most cases the dose-response for the carcinogenic and hereditary risk by ionising radiation can be best described by the LNT model. Nevertheless it has to be realised that in the low-dose ranges, due to the exposures in the environment from natural and anthropogenic sources, health effects have usually not been observed besides after high exposures to radon and its radioactive daughter products or high exposures after severe accidents (Chernobyl). Therefore the risk after low radiation exposures can only be obtained by extrapolation under uncertainty.

With chemical carcinogens extensive epidemiological studies are scarce. Animal data with indications from human studies including metabolic investigations are the main sources for risk evaluations of chemical carcinogens (chapter 4). With carcinogenic chemicals the dose-response depends on the involved mechanisms. In the case of genotoxic chemicals, frequently the absence of a threshold is the best description for the dose-response but also practical thresholds are possible whereas the scientific data show a clear threshold in the dose-response of non-genotoxic chemicals. As with ionising radiation the risk in the low exposure ranges which are usually experienced in the environment can only be obtained by extrapolation with the corresponding uncertainties.

With increasing numbers of epidemiological studies after exposures to ionising radiation it can be concluded that radiation-induced cancers can principally be induced in almost all human organs and tissues by ionising radiation although the cancer risk per unit dose is different in the various organs and tissues. Thus, cancer risk is high for example in bone marrow, female breast, thyroid and low in brain, rectum, head and neck (chapter 5). Risk factors of 10×10^{-2} Sv^{-1} (10 % per Sv) for exposures to high-LET radiation as well as acute exposures to low-LET radiation and of 5×10^{-2} Sv^{-1} (5 % per Sv) for chronic exposures to low-LET radiation for stochastic effects (hereditary effects and cancer mortality) in a total population have been derived from the epidemiological studies. Mainly the studies with the atomic bomb survivors in Japan have been used for the cancer estimates (ICRP 1991).

In order to take into consideration the differences of radio-sensitivity of different organs and tissues the effective dose (E) has been introduced, which represents the total stochastic radiation risk including genetic effects. It is defined as the sum of the products of the equivalent doses in a specific tissue or organ (H_T) multiplied with the tissue weighting factor (w_T), which is derived with respect to the gonads for the genetic effects and with respect to the various tissues and organs for the carcinogenic risks (focussed on cancer mortality). It results the expression:

$$E = \Sigma \, w_T \, H_T. \qquad \text{(eq. 7.1)}$$

The effective dose should only be used in the low-dose range, presupposes a linear dose-response (LNT) and allows to consider the stochastic risk not only after a homogeneous total body exposure, but also after an exposure where the radiation dose varies considerably between different tissues and organs. This concept has certain limitations. It should only be used for prognostic estimates of expected doses or dose limits in radiological protection and the connected values for total stochastic risk. It does not represent retrospective risk values of a defined individual person for a defined radiation exposure. Under these conditions individual factors have to be taken into account. In this context it should also be considered that the individual radiosensitivity may vary, although the number of hypersensitive individuals is small (chapter 3).

The comparison of the mechanisms and principles for risk evaluation of ionising radiation and chemicals as well as the classification of the carcinogenic chemicals (as outlined above) shows that a number of similarities exist between ionising radiation and genotoxic chemicals. A unifying system of risk evaluation is possible to a certain degree for these carcinogenic agents. However, the diversity within the total group of carcinogenic chemicals is so large that different principles have to be applied for regulatory purposes.

On the basis of the risk factor mentioned before for stochastic effects by ionising radiation (5×10^{-2} Sv^{-1}) the International Commission on Radiological Protection recommended an "acceptable" radiation dose (effective dose E) of 1 mSv yr^{-1} for individuals of a general population (ICRP 1990). This includes exposures to ionising radiation besides exposures from natural sources (average E: 2.4 mSv yr^{-1}) and from medicine. In an analogous way "acceptable" levels of risk and of exposures have been estimated for carcinogenic chemicals.

As an example for different levels of risks that have been considered as „acceptable" one can take the EU directive on drinking water (European Union 1998). The lifetime cancer risk for some selected carcinogens varies from 0.44 cases for benzene when considering the lowest estimation from EPA (EPA-IRIS 2003) and 500

Tab. 7.1 Cancer risk of carcinogens in drinking water when observing the maximum values indicated in the EU-Drinking water directive (European Union 1998). The risk units are taken from EPA-IRIS (2003).

Substance	Parameter (European Union 1998) $\mu g\ l^{-1}$	EPA Risk Unit = cases per $1\ \mu g^{-1}\ l^{-1}$	Lifetime cancer risk at maximum contamination level cases per 10^6
Acrylamide	0.1	1.30×10^{-4}	13
Arsenic	10	5.00×10^{-5}	500
Vinyl chloride, lifetime exposure from birth	0.5	4.20×10^{-5}	21
Vinyl chloride, exposure during adulthood	0.5	2.10×10^{-5}	10.5
1,2-Dichlorethane	3	2.30×10^{-6}	6.9
Benzene, lower estimation	1	4.40×10^{-7}	0.44
Benzene, upper estimation	1	1.60×10^{-6}	1.6
Benzo (a) pyrene	0.01	2.10×10^{-4}	2.1

cases per 1 million people for arsenic (Tab. 7.1). These examples clearly show that the standard setting is far from being consistent and that a lot of other aspects like technical feasibility, economic considerations, political and sociological arguments will be included.

7.1
References

European Union (1998) Council Directive 98/83/EC of 3 November 1998 on the quality of water intended for human consumption. OJ L 330, 05/12/1998: 32–54

International Commission on Radiological Protection [ICRP] (1990) ICRP Publication 60. Recommendations of the International Commission on Radiological Protection. Annals of the ICRP 21(1–3), Pergamon Press, Oxford, 1991

Umweltbundesamt (ed) [UBA] (1998) Aktionsprogramm Umwelt und Gesundheit. UBA Berichte 1/98. Erich Schmidt, Berlin, pp 571–594

US Environmental Protection Agency [EPA-IRIS] (2003) U.S. Environmental Protection Agency Integrated Risk Information System (IRIS) www.epa.gov/iriswebp/iris/index.html

8 Hypothesis Testing and the Choice of the Dose-Response Model

8.1
Introduction

Scientists have been debating the shape of dose-response curves for ionising radiation since exposures were found to be harmful to health. The initial assumption was that there was a threshold to the appearance of a health detriment, including cancer. But developments in the knowledge and aetiology of cancer led toxicologists to first suggest a non-threshold model during the early 1960s (Roderick 1992). Since that time a number of models have been suggested, and present work in health physics, toxicology and epidemiology is concerned with questions about both the shape of the dose-response curve and whether or not a threshold exists.

Much of the controversy over the choice of the "best" dose-response model is concerned with the quality of such models according to scientific standards. Criticisms have been directed towards the use of the LNT model or hypothesis as being "unscientific", "irrational" or simply wrong (Filyushkin 1991; Mossman et al. 1995; Becker 1997; Patterson 1997; Tubiana 1998; Jaworowski 1999). Extrapolation from cases where one has strong, consistent empirical data to situations of uncertainty is central to the debate. In order to examine this problem it can be helpful to consider what exactly we mean by scientific, what a scientific fact may or may not represent, and what criteria are used to judge the scientific credibility of a particular theory or hypothesis. As much of the work on what good science "is" has occurred within the philosophy of science, this section aims to illustrate some of the challenges facing radiation protection by reference to the philosophical discussion on the theory of scientific method.

After a brief summary of the main sources of uncertainties, the choice of dose-response model is evaluated against some contemporary theories of scientific method – hypothetico-deductive, Popperian falsification, standards of scientific proof, and Kuhnian Social Construction – and implications for the selection and use of such models in practice is considered.

8.2
Uncertainties in dose-response modelling

In examining the question of uncertainties one needs to consider two main issues: what these uncertainties are and why they exist. One of the main sources of uncertainty in the choice of dose-response model at low doses is the need to extrapolate from a situation with strong empirical data (i. e. consistent, statistically significant observations) to a situation of uncertainty. When choosing a model, scientists need to consider both the shape of the curve and the existence of a threshold. For radiation protection purposes, disagreements over the existence or not of a threshold are arguably more controversial than debates over the shape of the curve. While the shape of the dose-response curve determines the risk of detriment associated with exposure to ionising radiation and can be the source of disputes as to the size of that risk, the possible existence of a threshold raises the possibility that society is investing resources to reduce risks that are in fact non-existent.

As stated in section 3.6, for exposure to ionising radiation, no significant difference in human cancer incidence over background rates can be observed in the general population below about 100 mSv. Although, there have been reports of significant increases for other biological endpoints (e. g. chromosome aberrations, development damage to fetus) for doses of a few 10s mSv. Chapter 4 showed that similar problems exist for effects from exposure to chemical carcinogens.

8.2.1
Extrapolation

Situations where it is considered that we have relatively consistent empirical data are obviously those where the dose is large enough to allow observation of statistically significant effects, but the problem of extrapolation is actually more complex than the question of how one extrapolates from "high dose" to low dose (see chapter 1, Fig. 1.1). The underlying assumptions can depend both, on the choice of effect and the method of dose calculation, and these give rise to a number of different issues for extrapolation.

8.2.1.1
Extrapolation from data obtained from one subpopulation or species (experimental or epidemiological) to an extended population or different species

The most familiar situation in toxicology is extrapolation from animal studies to humans, where problems arise from, *inter alia*, differences in metabolism, biokinetics, toxic mechanism and physiology between species. Even for studies on humans — or indeed any single biological species a number of other variables between the specific (test population) and the general (the whole population) can influence observations and predictions. Population variations can include factors such as age, life-stage, gender, genetics, tissue and organ sensitivity, environmental and/or laboratory conditions and, for humans, ethnicity and lifestyle factors. Many of these are acknowledged in the problems identified within toxicology, such as the use of safety factors when deriving human risk factors from animal studies (see discussions in chapters 4 and 10) and epidemiology, for example, confounding factors, bias, etc. (chapter 5). The influence of metabolic and kinetic issues has been extensively discussed in chapters 3 and 7, and the important point has been made that this is an area where chemical toxins can differ from radionuclides. However, a number of factors may also affect the biokinetics of radionuclides, including radionuclide speciation and interactions with other pollutants.

8.2.1.2
Extrapolation from an observed biological endpoint to the specific disease or effect of interest

As shown in previous chapters, an observed frequency of "lower level" effects, such as observed chromosome aberrations, frequency of DNA adducts, genetic mutations, genomic instability or bystander effect need not be correlated, or not linearly correlated, with an increased risk of cancer incidence or other health detriments, such as reproductive, hereditary or immune system effects. The issue is of particular relevance if the response is extended from the biological endpoints commonly addressed for humans (and studied in test mammals) to other possible consequences in flora and fauna, which may warrant consideration of different biological endpoints (e. g. fecundity, colour mutations, ecological stability). For both radioactivity and chemical toxins, dose-response curves can differ considerably depending on the effect or endpoint in question (see the various graphs in chapters 3 and 4).

8.2.1.3
Extrapolation from knowledge on mechanisms to specific health risks

The plausibility, simplicity and experimental support for suggested mechanisms (e.g. for induction and promotion of cell mutations, metabolic factors, gene expression or tumour development) can vary for toxins, species and biological endpoint. Effects of ionising radiation in the low-dose range are most commonly linked to cancer or mutation, but even within those categories there are different types of cancer (i. e. leukaemia and a variety of solid tumours) and a multitude of mutations. Exposure to ionising radiation and chemical carcinogens can induce various biological processes, some of which may increase the risk of disease (e. g. DNA mutations, chromosome aberrations, genomic instability), and some which might reduce the risk (e. g. cell cycle arrest, stimulation of DNA repair mechanisms). These mechanisms can interact in complex ways (moderating, synergistic, additive) that vary according to both, the organism and the disease in question, and the dominant mechanism can clearly be different at high/acute and low/chronic doses. A better understanding of the underlying mechanism(s) can provide a stronger basis for extrapolation between different effects and exposure characteristics (see chapter 6), but a full awareness of the possible confounding variables may also raise serious questions on the logic of attempting to carry out such extrapolations in the first place.

8.2.1.4
Extrapolation from studies having differences in exposure characteristics

Variability in the exposure includes the radiation type (α-, β-, γ-radiation or neutrons) or secondary particles (auger electrons), energy deposition (high and low-LET), acute/chronic, internal/external (averaging of deposition over organ vs. cell) and combined exposures (Streffer et al. 2000). These differences have implications both for the calculation of dose and the types and levels of biological effects induced. As already pointed out, a number of other factors can influence the metabolism of carcinogens, causing significant uncertainties in the resultant calculations of the "dose-to-target-organ". Although scientists have reasonable data on metabolism for most of the common radionuclides in humans, knowledge data gaps remain for the more "exotic" radionuclides, particularly long-term metabolism, and very little data is available on the internal distribution of radionuclides in non-mammalian species. With respect to extrapolation to the low-dose range, however, the relationship between acute and chronic exposures and the assessment of internal exposure from α-emitters are two areas attracting considerable debate (see chapter 6). Disagreements over dose-response relationships have caused some scientists to question the validity of the absorbed dose concept itself, suggesting that it is not the transfer of energy that is the prime quantifier of biological effectiveness, but rather the collision/interaction cross section or "fluence" (Simmons and Watt 1999).

8.2.1.5
Conversion from concentration measurements to dose calculation

Extrapolation problems contribute to the inherent increase in uncertainty in going from measurements of concentrations in food or environmental/biological media, ($Bq \ kg^{-1}$ or $mg \ kg^{-1}$) through modelling of biological transfer to calculations of dose (rate) exposure. A number of factors can influence the biological uptake and metabolism of pollutants, including both environmental and biological variables, and dose calculations will suffer if sufficient knowledge is not available on radionuclide speciation. The source term speciation can have a direct effect on the radionuclide metabolism and dose distribution within an organism (e. g. from "hot particles"), but can also influence both the environmental measurement itself and the prognosis of future doses. For example, the predominant source term for Chernobyl radionuclides, particularly close to the reactor, was uranium oxide fuel particles. Underestimation of the levels of Sr^{90} in soils contaminated by Chernobyl could be attributed to the standard hydrochloric acid extractant not dissolving the uranium oxide particles (Oughton et al. 1992). Following particle weathering, Sr^{90} mobility and biological uptake from soils to plants and animals increased, resulting in increases in doses to populations living close to the exclusion zone (where particle deposition was highest) with time after deposition (Kashparov et al. 1997). This was quite the opposite to predictions based on analysis and modelling from weapon's testing experience, where Sr^{90} was deposited primarily in an ionic form, and where bioavailability deceased over time. Finally, even in cases where there are good dosimetry measurements, such as for workers, these are not always completely reliable. Examples include instances of workers removing dosimeter badges so that they won't get "too high doses" and be moved to other duties; and cases where positioning of the dosimeter may not record doses correctly. The latter has been suggested for workers changing fuel at Windscale, where it was postulated that the relatively narrow beam of radiation emerging from the reactor would not have "hit" the dose badges (Reay/Hope v. BNFL 1993).

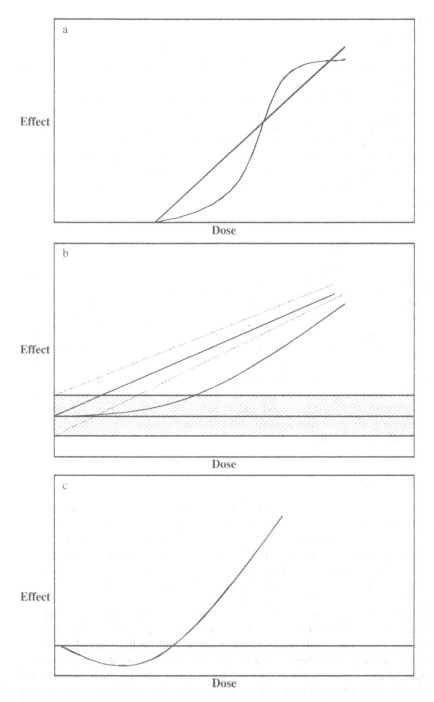

Fig. 8.1 Sketches of possible threshold mechanisms: a) absolute or perfect threshold; b) practical or statistical threshold; c) apparent or protective threshold

8.2.2
Thresholds

The LNT model includes hypotheses, both about the shape of the dose-response curve (linearity) and the existence of a threshold (no). Confounding factors that cause problems in determining the shape and extrapolation of dose-response curves also bear on the question of whether or not there is a threshold. This complexity arises partly because (like for linearity) there are a number of different (and inter-acting) effects and biological endpoints for which one might postulate a threshold, and partly because scientists have different ways of defining what exactly a thresh-old is. Different types of threshold have been described for both radiation and chemical toxins (chapters 3 and 4), and these can be roughly divided into three classes (Fig. 8.1):

8.2.2.1
Absolute threshold

An absolute (or perfect) threshold arises when there is a toxicant concentration or exposure below which no adverse effects will occur in the organism. This can be the case for a range of biological endpoints in the whole organism, and is a common assumption for "traditional" non-genotoxic and/or non-carcinogenic chemicals. The dose-response curve often shows a classic S-shape, reaching saturation at high doses (Fig. 8.1a). Absolute thresholds also apply to certain deterministic effects from radiation and chemical exposure (e. g. organ damage brought about by cell death) and carcinogens that inhibit DNA repair mechanisms or require some prior metabolic activation (e. g. motor protein inhibitors, vinyl acetate). Acceptance of an absolute threshold requires a good understanding of the underlying mechanisms, consistent experimental and epidemiological data, and, usually, relatively simple cause – effect relationships.

8.2.2.2
Practical threshold

A practical (or statistical) threshold has been used to describe situations where there is a concentration or level of exposure below which there is no *observable* level of damage above the expected "naturally occurring" or baseline incidence rate of effects. Such cases are common, but not exclusive, to stochastic effects, and are often classified as a NOEL – (no observed (adverse) effect level). In ecotoxicology NOELs are sometimes, and confusingly, also used to characterise exposure-response relationships that have absolute thresholds. The reasons for a practical threshold can be many, including the statistical uncertainty both in experimental observations and in the random fluctuation of baseline effects (Fig. 8.1b). In some cases, that fluctuation may reflect uncontrollable confounding variables between different populations and might be reduced by better compatibility between control and test groups. For low-dose exposures, specific mechanisms such as metabolite inactivation or DNA repair may act to reduce the incidence of effects at low as com-pared to high dose (sublinear curve). But only if these effects are 100 % successful, should the dose-effect response be classified as having an absolute threshold. If, however, exposure to the toxin *stimulates* certain repair or protective mechanisms, then an apparent or protective threshold may be seen.

8.2.2.3
Apparent threshold

An apparent (or "protective") threshold can apply in cases where exposure to the toxin stimulates or enhances the occurrence of a protective or adaptive biological response that modifies or "cancels-out" another detrimental effect – typically one caused by exposure to endogenous and environmental toxins other than the toxin in question. Enhanced rate of DNA repair, cell cycle progression, regenerative prolif-eration, and stimulation of immune response are thought to be the most important protective mechanisms (see chapters 3 and 4), and have been observed following exposure to both chemicals and radiation. It follows that exposure to the "toxin" is protective *up to* a threshold, above which the initiation of detrimental effects out-weighs any positive compensation (Joiner et al. 1999; Feinendegen and Pollycove 2001ab). These adaptive responses may also result in "practical protective thresh-olds", meaning that any beneficial effect would be not observable. In other cases, the protective effects may be experimentally observable and constitute evidence of hormesis (Fig 8.1c). Some scientists argue that J- or U-shaped dose-responses have been shown to exist for a wide variety of biological responses for both carcinogens and non-carcinogens and suggest that hormesis should be the preferred toxicologi-cal model for all stressors (Calabrese and Baldwin 2003). Others suggest that cer-tain mechanisms, together with experimental or epidemiological observations, pro-vide evidence for increased radiosensitivity at low doses (e. g. genomic instability, bystander effect, immune system suppression, see Nussbaum and Köhnlein 1994; Joiner et al. 1996; Barcellos-Hoff and Brooks 2001). A mechanism that may offer short term protection against one disease, e. g. reduced cell cycle proliferation to reduce the rate of a particular cancer incidence, may have little effect on other dis-eases, e. g. other types of cancer in different cells, or be outweighed in the long-term by other detrimental effects, e. g. tissue regeneration.

To conclude, the question of thresholds is one of immense complexity, and one that is confused by different experimental observations depending, *inter alia,* on the effect, the sensitivity of individual organisms, intra and inter population and species variability, the dose, dose rate and time after exposure. When discussing the prob-lem of thresholds it is important to be clear whether the hypothesis or model postu-lated refers to the existence of an absolute threshold, a practical threshold or an apparent threshold, and with regard to what specific biological effect. This might help to reduce some of the controversy surrounding low-dose effects – or at least make it clearer what the argument is about.

8.3
Theories of science

Since the start of "science" philosophers have tried to answer the question of what exactly makes something scientific or what is a measure of a good scientific theory. Many scientists would feel they know the answer to such questions, and have little time for philosophy of science. However, although at present there is no universal agreement on what exactly science is, among neither scientists nor philosophers, there is still a certain amount of consensus, particularly on what "bad" science might be. Philosophers of science have tried to look at the history of science and establish what might constitute a reliable scientific theory. Some of these evaluations have a descriptive aim, others fall into a more proscriptive role, – i. e. what *should* scientists do to achieve good science. This section looks at some theories of scientific method and evaluates the question of dose-response models against these theories.

8.3.1
The hypothetico-deductive method

In 1901, Henri Becquerel noticed a skin reddening next to the vial of radium he habitually carried in his waistcoat pocket. To test his conjecture that radiation was responsible for the observed effect, Becquerel carried out a simple experiment whereby he moved the vial to the other pocket and, as a control, Pierre Curie started to carry radium in his waistcoat pocket as well (O'Riordan 1996). Sure enough, the subsequent observation of skin reddening close to where the vials were carried supported the hypothesis, leading to the conclusion that exposure to radiation could cause erythema. This simple experiment carried out by Becquerel and Curie is illustrative of one of the main theories of science: the hypothetico-deductive method (Føllesdal et al. 1996).

The hypothetico-deductive method is characterised by the proposal of hypotheses, the deduction of logical consequences from those hypotheses, and experimental testing of those consequences against experience. From where these hypotheses may arise is a question that we do not have space to discuss here, although intuition, observation, and prior knowledge have all been proposed as possible criteria (Chalmers 1982). Whatever its source, a scientific hypothesis can be defined as a proposition that satisfies the two following criteria: i) we are not sure whether it is true; and ii) that a logical consequence can be derived from the hypothesis either to test the hypothesis or to predict an outcome.

Although the prediction of a hypothesis's outcome usually refers to something that has not yet been *observed*, it need not mean that the outcome has not yet *occurred*. Many of the hypotheses proposed in geology, for example, concern events occurring thousand to millions of years ago. Becquerel's early observations of erythema were soon followed by reports of more serious effects such as tumours and cancers (see Box 8.1), eventually leading to the hypothesis that increased rates of health detriments might also have occurred at lower doses. This stimulated an array of *(a posterori)* human epidemiological studies designed to test this hypothesis in populations previously exposed to radiation, as well as a variety of animal and biological laboratory experiments. The eventual outcome of these studies, of course, being the controversial linear-no-threshold hypothesis (see Box 8.1).

In many scientific studies, both the hypothesis and deduced consequences under investigation are clear, e. g. irradiation of cells will cause an increase in cell mutation rates; parents exposed to radiation will show increased incidence of genetic mutation in offspring; increased incidence of chromosome aberrations will result in an increased cancer incidence. In other cases the hypothesis is less clear, or may even be non-existent. The apparent lack of a clear hypothesis in many radioecological studies has been criticised, particularly those designed to monitor levels of radionuclides in biological or environmental materials (Hinton 2001). But the data produced by such investigations may be of vital importance in the derivation and testing of other hypotheses.

Although experiments may be designed to test one specific hypothesis, scientists make use of a whole host of auxiliary assumptions and subsidiary hypotheses in

---

Box 8.1: History and Relationship of the Scientific Arguments for the LNT Hypothesis to Practical Radiation Protection Policy

The first medical X-ray machines were in operation a little over a year after Röntgen's 1895 discovery; and less than ten years after its isolation by Marie and Pierre Curie, radium was introduced as a treatment for cancer (Pochin 1983). But it soon became clear that radiation could harm as well as cure. Many radiologists working without protection with primitive, high-dose X-ray machines died of radiation burns and cancer, the first tumour being reported as early as 1902. By the 1920's, doctors had discovered tumours in women painting clock dials with radium pigment and children who were irradiated for acne, bronchitis, ringworm and tonsillitis (Caldicott 1994). Knowledge of the harmful nature of radiation took an explosive turn in 1927 with Dr HJ Muller's discovery that irradiation with X-rays caused an increase in the number of mutations in the offspring of fruit flies (Muller 1927). Radiation could not only bring about harm to exposed individuals, but might also cause mutations capable of being passed down the genetic line.

Up to the early 1950s, radiation biologists assumed that there was a threshold level to almost all health detriments associated with radiation exposures (detriments caused by damage to cells in the developing embryo being an important exception). Indeed, for several decades after the discovery of radiation it was believed that small doses were beneficial to health, inspiring a fad for radium baths. In 1957, following studies of cancer rates among radiologists and Japanese atomic bomb survivors, EB Lewis suggested that cancer risks might exists at all doses greater than zero (Lewis 1957; Rodricks 1992). The acceptance of the no-threshold model by the radiation protection community meant that it was no longer possible to set a defined level over which a radiation dose is considered "dangerous" and under which the dose is considered "harmless". This resulted in a need to limit the risk of health detriments to an acceptable rather than a safe level, in other words, radiation protection became a question of values rather than a matter of fact (see also chapter 2). A number of writers have noted that this might lead to confusion in communication of risks:

> The layman, who wants an assurance that a situation is "safe", fails to understand the expert's aversion to the unconditional use of the word "safe". And the expert who tries to use a simplified language and call a low-risk situation "safe" may immediately be reproved by those who know that there will still be some risk. (Lindell and Malmfors 1994).

Such problems with communication, as well as difficulties in deciding what an acceptable level of risk might be, present ethical challenges for the management of any genotoxin. But within the past ten years, radiation protection policy has been subject to increasing pressure from scientists who reject the assumption of a linear-no-threshold (LNT) model at low doses. The resultant disagreement over dose-response models has fuelled controversy over the regulation of radiation risks, but despite the above distinction between "safe" and "acceptable", much of the debate has still revolved around the scientific strength of the LNT hypothesis.

their deduction and testing of consequences (Føllesdal et al. 1996). In dose-response experiments, for example, such assumptions might include that the dose calculation is correct, including measurement of radionuclide concentrations and energy deposition; that the doses are homogeneously distributed over space and time; or that the biological effectiveness of various types of radiation is known. Of course, the scientist will also appeal to such more widely accepted theories and laws such as the radioactive decay law, or theory of DNA and cell replication. Clarification of the underlying assumptions and premises is central to understanding both the relevance and the significance of the hypothesis in question.

8.3.2
Laws, hypotheses, theories and models: differences between biology and physics

Laws, theories, models and hypotheses are used in all branches of scientific study. As stated above, all *hypotheses* retain the suggestion that they may not be universally true. In philosophy of science a *law* is defined as a hypothesis having universal form that holds in all circumstances covered by the wording of the law. In common scientific use, law is often reserved for those universal hypotheses which, at least in approximate form, have also survived the rigor of scientific testing, and for which there is a large body of consistent experimental evidence. Examples include Boyle's law, the law of conservation of mass, Newton's law of gravitation. Universal laws apply in all circumstances, whereas statistical laws need to be formulated such that probabilities are part of the statement, e. g. "the probability of throwing a 6 with a die is 1/6". The difference between the two is not sharp, for example the law of radioactive decay, and whether or not the law is described as universal or statistical is often determined by the degree of specification one uses.

A description of nature that encompasses more than one law or hypothesis is sometimes called a *theory*. Theories are often described in the sense of being families of theoretical models – interpreted according to specific empirical circumstances – and include an explicit description of how the various laws and hypothesis hang together. Examples include Darwin's theory of evolution and Einstein's theory of relativity. A *model* is a representation of a system intended to capture the essential aspects of the system in a sufficiently simple form to enable the mathematics to be solved. Models are used extensively by scientists and come in different forms and guises, e. g. physical, theoretical, mathematical. All types involve some kind of analogy between the model and either reality or some other scientific claim, and in practice, require approximation techniques to be used. Clearly, there is a degree of overlap between the four concepts, and the way scientists use the concepts can vary between different disciplines. For example in radiation protection, one might talk about the linear dose-response model, the bystander effect theory, and the hypothesis that all doses carry a finite risk of cancer.

Physics and chemistry are said to be concerned with the derivation and testing of hypothesis concerned with universal laws; biology and medicine are often said to be statistical sciences, capable only of *explaining* observations. The complexity of biological systems tends to make the biological sciences less robust than the physical sciences at *predicting* absolute outcomes (Føllesdal et al. 1996). The discipline of

---

Box 8.2. The Sellafield Leukaemia cluster and court case to test the "Parental Preconception Irradiation Hypothesis"

The Sellafield leukaemia cluster first came to public attention in a television programme, "Windscale: The Nuclear Laundry", screened in November 1983, in which it was alleged that the childhood leukaemia rate in the nearby village of Seascale was 10 times the national average. The excess incidence was quickly confirmed by an official government enquiry (Black 1984) and subsequent epidemiological studies confirmed both the excess at Sellafield and identified clusters at other sites (Craft et al. 1993) [1]. In 1993, the dispute culminated in an unsuccessful, 8 month litigation case against British Nuclear Fuels (BNFL) in the London High Court, *Reay/Hope v. British Nuclear Fuels plc* (1993). The two plaintiffs were a woman suffering from non-Hodgkin's lymphoma (NHL) and the mother of a child who had died of leukaemia. The claim for damages was that of breach of statutory duty under the Atomic Energy Authority Act, 1954 (Day 1992). This act places BNFL under strict liability "to secure that no ionising radiations...cause injury to any person or damage to any property, whether he or it is on any such premises or elsewhere". Thus, only proof of causation of injury was required and negligence did not have to be established (Pugh and Day 1992). The plaintiffs' case was novel in that it was based on a hypothesis proposed by the epidemiologist Prof. Martin Gardner. A study published by Gardner in 1990 had indicated a correlation between childhood leukaemia risk in Seascale and the occupational radiation dose received by the child's father.[2] Gardner suggested that parental pre-conceptual irradiation (PPI) had predisposed the children to leukaemia (Gardner et al. 1990a,b). The hypothesis was controversial because it required a biological mechanism wherein radiation acted through the germ line rather than by a direct, stochastic effect on the individual. The plaintiffs built upon this theory and alleged that PPI had caused a mutation in the spermatogonia and increased the risk of leukaemia in offspring. The fathers of both children had been exposed to occupational doses of greater than 100 mSv, doses with which Gardner had identified a relative risk of 6–7.

The defendants claimed, successfully, that existing evidence on the effects of radiation made it causally impossible for the agreed levels of exposure (either environmental or occupational) to give rise to the excess leukaemia cases (Howarth 1992; Reay/Hope v. BNFL 1993). They supported this claim with evidence on the genetic effects of radiation and by questioning the validity of the Gardner study. They pointed out that a statistically significant increase in leukaemia incidence was seen neither in the children of men exposed to higher (but instantaneous) doses from the atomic bombs in Hiroshima and Nagasaki (Ishimara et al. 1981; Yoshimoto et al. 1990; Neel and Schull 1991), nor in other occupationally exposed workers (McLauglin et al. 1993); nor, most damaging, in children of occupationally exposed fathers born outside of Seascale (Draper et al. 1993; Parker et al. 1993; Doll et al. 1994). They concluded that another factor must be responsible for the cluster, although it was never explicitly determined which of the suggested alternative hypothesis was the most likely cause.

---

[1] The absolute risk of leukaemia for British children is about 1 in 4000. Between 1955–1983, 7 cases of leukaemia were found in Seascale residents under age 25, 5 of those were in children under 10 years (Black 1984). It is generally agreed that this represents a relative risk of about 10 (Gardner et al. 1987ab; Craft 1993).

[2] Occupational exposure of the fathers to radiation was associated with an increased of leukaemia in <u>all</u> children born in Seascale (relative risk 2), although, increases were also found for other occupations, e. g. farming. However, Gardner showed a correlation between parental pre-conceptual dose and leukaemia risk. This apparent "dose-response effect" lent considerable support to the assumption of a causal relationship. The correlation, although heavily disputed in court, has been confirmed by other studies (HSE 1993; Roman et al. 1993). A preconception dose over 6 months of more than 10 mSv, or a total dose of more than 100 mSv, increased the relative risk to about 6–7 compared to all Seascale children.

health physics might be said to lie at the frontier between the physical sciences and the life sciences, being concerned both with the quantification of the energetics of ionising radiation as well as the understanding of its interaction with biological material and prediction of subsequent effects. Nevertheless all sciences use hypotheses and models to explain observed data and predict future consequences, and all scientists are faced with choices between different models, theories and hypotheses.

8.3.3
Choosing between alternative hypotheses

There may be a number of different hypotheses that could explain observations. For example the observation of increased leukaemia rates around certain nuclear installations has been the subject of scientific debate for nearly two decades (e. g. Black 1984; Heasman et al. 1986; Gardner et al. 1987ab; Roman et al. 1987, 1993; COMARE 1988, 1989; Michaels et al. 1992). The largest (in terms of relative risk) and most discussed leukaemia cluster is that around the Sellafield reprocessing plant (previously called Windscale). Today, the *existence* of a childhood leukaemia cluster in the vicinity of Sellafield is accepted as scientific "fact". What is under dispute is the *cause* of these excess leukaemias. Many alternative hypotheses have been put forward to explain the cluster, including: environmental exposure of children to radionuclides; exposure of fathers working at Sellafield; a virus due to the mixing of populations (i. e. worker migration); occupational exposure to other carcinogens; or "chance" (Cook-Mozaffari et al. 1989; Doll 1989; Kinlen 1989; HSE 1993; Kinlen 1993; Roman et al. 1993; Doll et al. 1994). Interestingly, the cause of the cluster was the focus of a court case in England, leading to perhaps one of the most rigorous cases of "testing" of a scientific hypothesis on radiation effects in the past century (see Box 8.2).

Other examples where scientists need to choose between different hypotheses include explanations of observed non-linearity or thresholds in dose-response models. A number of different alternatives such as hormesis, DNA repair, etc. have already been presented above and in previous chapters. But what factors should one use to select the "best" hypothesis? What should one do faced with evidence contradictory to one's hypothesis? When should we reject a hypothesis or conclude it has been falsified?

8.3.4
Popperian falsification

One of the most famous proponents of the hypothetico-deductive model was Karl Popper, who suggested that the key to a good scientific hypothesis was its ability to be falsified (i. e. refuted by evidence). Scientists do not prove their hypothesis, they test them, try to falsify them, and do this in the most rigorous methods. Take the hypothesis that bats use radar to navigate. Blocking the bats' ears and observing whether or not they crash into obstacles when released into a closed room is a good example of an experiment which will falsify the hypothesis if it fails; blindfolding them and seeing whether or not they continue to avoid flying into the walls will simply confirm the hypothesis. Along similar lines, if Curie had simply stopped carrying the vial of radium and waited to see if the skin condition healed, this would have been a far inferior experiment to the one he devised. Had he *not* developed another erythema next to the other pocket this would have been strong grounds to falsify the hypothesis – much stronger than a healing when the radium source was removed.

If a hypothesis survives a number of tests then it may, eventually, be accepted as "true" by the scientific community. Most suggestions of what the dose-response model might look like at low doses are best described as hypotheses. The crucial issue here is whether or not they satisfy Popper's criterion of falsification. For example, one might argue that the hypothesis that annual doses of below 10 μSv (above existing average background doses) will lead to no observable increase in cancer rates in human populations was not possible to falsify, since the expected variation in cancer incidence would make a statistically significant case-control experiment impossible. However, this appeal to infalsifiability is based on a number of auxiliary assumptions, for example, that the presumed dose estimations and risk factors have been correctly calculated, or that the dose-response is linear.

Difficulties with Popper's theory include the problem that just as one can never be sure that one has proved a hypothesis, it is likewise not possible to prove that it has been falsified. Something that one scientist sees as a refutation (such as observations of non-linearity), another may see as an anomaly. Nevertheless it is generally agreed, and especially by experimental scientists, that sign of a *good* hypothesis is that it is capable of being falsified, and that a *good* scientific experiment is one that genuinely does try to falsify (rather than confirm) the hypothesis in question. In fact the correct, if somewhat outdated, definition of "proving" the hypothesis is "testing" the hypothesis; and the "proof" is the exception to the rule. In all cases, it is imperative that one is absolutely clear about what exactly the hypothesis is.

Many would argue that the LNT hypothesis has been falsified, albeit under specific circumstances, hence it is unreasonable to allege that such a hypothesis is unfalsifiable *per se*. But one needs to be specific as what exactly that LNT hypothesis refers to, which effect, from which exposure and to what. In this respect it helps to consider some other attributes of hypotheses besides their falsifiability, namely safety, strength and simplicity.

8.3.5
Safety, strength and simplicity

The philosopher Nelson Goodman has discussed three criteria that scientists might use when choosing between alternative different hypotheses: safety, strength and simplicity (Goodman 1961). Safety and strength have both been suggested as desirable criteria in a hypothesis, and Popper thought that the best scientific hypotheses were the "bold conjectures", since the strongest hypotheses should be the easiest to falsify. But Goodman suggests that safety and strength are insufficient and claims that another important criterion, simplicity, should also be taken into account. Take a hypothetical example modified after Goodman: say a number of experiments on different rodents had shown that the populations showed a significant increased rate of chromosome aberration in animals exposed to low-dose radiation (i. e. 10 mSv), but that experiments had not been carried out on all rodent strains. One might postulate the following hypotheses:

1) all rodents, except perhaps for chinese hamsters, will show increased chromosome aberrations if exposed to low-dose radiation, or
2) all rodents will show increased chromosome aberration rates when exposed to low-dose radiation.

Goodman suggests that a principle of maximum safety will quickly lead to absurdity as it would demand that one should always choose those hypotheses that do not go out over that which we already know. Caution would prompt scientists to pick the weakest hypothesis as this is less likely to let them down later. In the above example, the second hypothesis is both simpler and stronger than the first. We should choose 2) and look for another hypothesis if evidence concerning Chinese hamsters (or indeed any other rodent) should show that 2) was wrong. However, although simplicity is often taken to be synonymous with strength (and sometimes safety) this is not the case. Although 1) is clearly safer than 2), we could always construe another stronger hypothesis, for example:

3) all rodents will show increased chromosome aberration rates when exposed to low-dose radiation, and Chinese hamsters will show increased immune system deficiency.

Hypothesis 3) is even stronger than 2) but not as simple, and therefore less acceptable. Thus neither safety nor strength is synonymous with simplicity, and simplicity is more important than strength. The challenge for scientists is how to balance safety and strength, when faced with two, or more, equally simple hypotheses: for example, that dose-effect curves for all carcinogens will show hormesis (Calabrese and Baldwin 2003), or that all biological effects following exposure to genotoxic carcinogens will show a LNT dose-response curve.

Following Goodman's and Popper's criteria, both these simple hypotheses *would* seem impossible to falsify due to the enormous number of variables. There is experimental evidence for non-linearity for cancer and other biological endpoints for both radiation and other chemical toxins, including sublinear (reduced sensitivity at low doses), and supra-linear (increased sensitivity at low doses). Evidence for a threshold for cancer following exposure to some chemical carcinogens (e. g. vinyl acetate) can be supported by knowledge on plausible mechanisms and experimental observation. Hormesis (beneficial effects followed by detrimental effects) has been shown for some specific exposure scenarios and effects, but certainly not for all biological endpoints. The question one might ask is what is the simplest and strongest hypothesis on dose-effect curves that has *not* been falsified? To answer these questions one needs guidance for what to do when faced with contradictory experimental evidence on dose-response functions, especially in the low-dose region. Scientists must choose between revision of an auxiliary hypothesis, and adjustment or rejection of the hypothesis being tested.

8.3.6
Ad hoc hypothesis

Despite appeals to objectivity, scientists are notoriously reluctant to give up on their hypothesis. Although in many cases a slight adjustment to the hypothesis may be sufficient, for example from the hypothesis that the probability of contracting *all cancers* is linearly proportional to the ionising radiation dose, to the hypothesis that the probability of *solid tumours* in humans is linearly proportional to the external γ-dose. Alternatively one might revise a hypothesis by, for example, going from a hypothesis that the dose-response curve for *cancer* has no threshold, to a safer hypothesis that the dose-response curve for *cell mutation/DNA damage/etc.* has no threshold (see examples from chapters 3–6). Problems occur, however, when scientists resort to the use of *ad hoc* ("just so") hypothesis to explain findings that contradict their prior beliefs.

A possible example of appeal to an *ad hoc* hypothesis is a study on British atomic weapons test veterans. Comparisons of the medical records of about 22,300 veterans exposed under weapons tests with 23,300 servicemen chosen as controls indicated that exposure to fallout had increased the incidence of leukaemia. By 1984, 20 of the exposed group had died of leukaemia compared to 6 in the control group (Darby et al. 1988, 1993). However, the authors dismissed the increased risk as biased due to the *ad hoc* hypothesis that the control cohort had an extraordinarily low rate of leukaemia compared to the general population. The expected leukaemia incidence in the general population would have been 16. However, critics have pointed out that the control group was selected to be as close as possible to the exposed group, hence one can not exclude the possibility that the servicemen may have a lower leukaemia rate than the general population (Bland 1994; Edwards 1996). In other words, that medical selection and monitoring procedures for servicemen would lead to a "healthier" and thus non-representative group. Even though later studies found a gradual reduction of the difference in relative risk between the two groups, this may simply reflect an earlier onset of leukaemia, and thus loss in life expectancy in the exposed group.

Although scientists present risk identification and estimation as the objective side of risk evaluation, these disciplines are not quite as objective as some would imply. As the above case illustrates, scientists are often forced to make *value judgements* in their choice of methods as well as in their interpretation of data (Shrader-Frechette 1991, 1985). In scientific methodology, such value judgements refer to decisions about experimental set up, the choice of experimental subjects and variables and the interpretation of observations. For societal decision-making, additional value judgements arise, particularly those related to choices about the acceptability of actions. The types of ethical and social values that can legitimately influence choices about whether or not it is acceptable to expose people to risk (e. g. consent, control, equity) have been presented in chapter 2. In both, science and applied science, the theories and hypotheses that scientists select and defend, may reflect many factors of a non-scientific kind. In fact, due to the number of variables involved, as well as the auxillary hypothesis needed to support both theory and observations, it would seem preferable in most cases to talk of the LNT *model* rather than a LNT *hypothesis* when describing general effects of ionising radiation.

8.4
Scientific plausibility and standards of proof

Scientists have different ways of judging the strength of scientific evidence and different disciplines appeal to different rules and have different standards. The following two sections review some of those different rules and standards, first reviewing the rather general "degrees of evidence" used by toxicologists and epidemiologists to support various claims and extrapolations, and second considering the more specific statistical "standards of proof" used in hypothesis testing.

8.4.1
Degrees or weight of evidence

Scientists appeal to a variety of criteria when judging the plausibility of their experimental observations, models and theories. A number of these criteria or "degrees of evidence" have been mentioned in other chapters, and this section presents a brief summary of the main criteria used in the assessment of carcinogens and their possible effects on human health and the environment. The aim is to show that that there is not one specific rule which one can apply in order to demark scientific "facts from fiction", but rather a set of different criteria and that scientific plausibility is a question of scale, depending largely on how many of those criteria are met. Two aspects are considered: first the classification of toxic substances and prediction of health effects, including extrapolation from animal studies to human risk factors; and second, the relationship of scientific plausibility to the precautionary principle.

To *classify* a substance as hazardous, some minimum of empirical data, usually based on animal testing, is required; structure-activity relationships alone are not sufficient (section 10.3.1). To classify a substance as a class 1 human carcinogen requires observation of a positive relationship between the exposure and cancer in epidemiological studies in which chance, bias and confounding could be ruled out with reasonable confidence (see chapter 5). Epidemiologists have specific criteria to judge plausibility of the association between the postulated cause and effect (i. e. *evidence of causality)*, namely: i) biological plausibility; ii) strength of the effect; iii) dose-response correlation (effect increases with dose); iv) consistency among studies; and v) lack of explanation by confounding factors (see chapter 1 and section 5.1). Some scientists have questioned whether or not biological plausibility is indeed an epidemiological criterion, but it is undoubtedly a factor that will influence the degree of evidence for causality. A hierarchical distinction is made between the different types of epidemiological study wherein the weight of evidence is said to increase according to i) case report; ii) observation (cohort) study; iii) retrospective (case/control) study; and iv) intervention trial (preferably randomised and double blind). Proof of causality "beyond reasonable doubt", is said to be made on the basis of *two* independent trials of the last kind. An related area of study "*evidence-based medicine*" attempts to systematically rank the types of evidence from both medical and epidemiological studies, both for the purpose of aiding doctors' diagnosis of possible diseases and judgements of their likely cause and treatment (Katz 2001; McGovern et al. 2001). Also here it is acknowledged that judgements on weight of evidence require a multicriteria analysis. The ideal basis for *prediction* of risk factors for human effects associated with exposure to a classified hazardous substance would include: i) animal testing; ii) cellular and molecular studies; iii) clinical studies; and iv) epidemiological research (see section 10.3).

A lack of data from human epidemiological studies means that toxicologists often have to carry out *interspecies extrapolation* from animals to human. In this case greater weight is placed on the biological plausibility and knowledge on the toxic processes. A scientifically justified extrapolation for a specific compound would require: i) an adequate understanding of the metabolism and mechanism of toxicity of the compound; ii) data on its metabolism and toxicity in primary hepatocytes and other relevant cell types of humans as well as adequate experimental animal species; and iii) toxicity data including carcinogenity studies in these experimental animal species (see chapter 4). However, the inherent uncertainties in carrying out such extrapolations underlie the use of *safety* or *extrapolation factors* in going from laboratory bioassays on test animals to risk estimations of effects in humans. Safety factors and considerations of the weight of evidence are also widely used in ecological risk assessment, which faces similar difficulties with extrapolation from acute to chronic exposures as well as having to deal with a variety of effects on different species within complex ecosystems (EEC 1996; CCME 1997; USEPA 1998; RIVM 1999). Ecological risk assessors often include additional criteria to judge the weight of evidence, such as the number of independent observations of no observed effect concentrations (NOEC), the number of species, taxonomic groups and trophic levels for which evidence is available and the exposure/effect incubation time.

The question of scientific plausibility is a source of much debate about the precautionary principle. As defined in the Rio declaration, a precautionary approach should be adopted in cases where there is *lack of full scientific certainty* (UN 1992), and can apply to both human and environmental effects. Different types of evidence that could trigger precautionary measures have been identified (section 10.3.1):

i. a science-based suspicion of risk (empirical evidence on cause-effect relationships, but contradictory studies);
ii. a partly plausible suspicion of risk (empirical evidence on relationship between effect and exposure but no evidence about adversity);
iii. a hypothetical (i. e. merely theoretical) suspicion of risk; and
iv. a mere apprehension of risk (without any evidence).

With the possible exception of (i), the above criteria fall well short of those needed for classification of substances as hazardous, or the type of evidence required for positing causality. But remember that the precautionary approach does not apply under scientific *certainty*, although many suggest that there should be some degree of scientific *plausibility*. Indeed whether or not the above criteria are sufficient to claim scientific plausibility is the source of much controversy surrounding the use of the precautionary principle. Other examples where scientific plausibility is doubtful are conclusions: i) from empirically determined acute toxicity to chronic toxicity; or ii) from known persistence and bio-accumulation to ecotoxicity (see section 10.3.1). Characterisation of the types of risks for which a precautionary approach may be needed include those where there is an "ignorance of mechanism", meaning that our knowledge of the physical processes that determine the likelihood and magnitude of the risk is poor (section 9.2). In many cases there is a "low subjective probability" associated with the feared outcome, however, there is often no consensus about the probability of occurrence, due to the rarity or one-of-a-kind nature of the risk, with little or no acturial history upon which to base estimates (chapter 2).

Scientific plausibility is just one factor that should be considered in decisions on the adoption of the precautionary principle in policy. Other important characteristics of risks include that:

– there is a "potential for catastrophic loss": the harm to the affected individuals and to society as a whole can be very great if the activity or technology entails such a risk; or
– there is a "relatively modest benefit" associated with the activity or technology, especially when compared to the potential harm (section 9.2).

These factors, however, have more to do with the potential consequences of the actions than the degree of scientific evidence. The example of possible fetal effects of topoisamerase presented in chapter 4 is a good illustration of a precautionary approach being adopted under conditions of doubtful scientific plausibility (extrapolation from acute to chronic, few criteria of scientific evidence). However, issuing advice that pregnant women may wish to avoid the food is somewhat different to decisions to halt an industry or ban discharges. Here, one might argue that it is more the consequences than the degree of evidence that directs the policy.

To conclude, there is no one thing that scientists can apply to determine scientific plausibility. There is, however, a set of criteria and the more of these criteria are satisfied, the stronger the evidence and the more scientifically plausible the claim (Tab. 8.1). For toxicology, and particularly at low doses, both the complexity of many of the biological mechanisms and the difficulty with statistical significance would be expected to lead to situations with contradictory scientific evidence. Based on the above analysis, there are four main criteria (or sources of evidence) used to judge the plausibility of toxicological studies: laboratory experimental data, epidemiological studies, and knowledge on the underlying mechanisms and the availability of models.

Tab. 8.1 Criteria and degrees of evidence used to judge scientific plausibility and social factors influencing societal decision-making

Criteria for judging scientific plausibility	Social aspects influencing decision-making on the acceptability of toxic substances[a]
Laboratory experimental data	The size of the potential harm
Epidemiological studies	The benefits from the activity bringing about exposure to the toxic substance
Knowledge on the underlying toxic mechanisms	Ethical aspects (personal consent and control, distribution of risks and benefits, etc.)
Availability of models	

[a] – see chapter 2 for a full discussion on these issues.

8.4.2
Standards of proof

Another issue of importance under hypothesis testing is the question of statistical significance. Scientists are reluctant to accept as scientific fact theories that have not passed the test of falsification and use rigorous statistical tests to eliminate those hypotheses that do not meet this criterion. Standards of proof reflect the level of *uncertainty* that we can tolerate before accepting or rejecting a hypothesis (Crane 1987). Such hypotheses can refer to statements about scientific facts, about the safety of an industry, about the results of a medical test, or about the guilt or innocence of a person in the eyes of the law. Science (and to a certain extent law) employ more or less fixed standards of proof. In order to test hypotheses, statisticians employ a number of different types of tests, all of which are based on the same approach:

The standard of proof for confirmation of a hypothesis, H:

- X did cause or give rise to A,

is determined by the maximum degree of uncertainty permitted in the *null hypothesis*, H_0,[3]:

- X did not cause or give rise to A.

A measurement of the uncertainties associated with the hypotheses H/H_0 can involve the estimation of probabilities for two interrelated types of statistical error (Tab. 8.2):

- the probability of making a wrong decision in rejecting the null hypothesis, H_0, i. e. the probability of a false positive (*type I error*); or
- the probability of making an error in not rejecting the H_0, i. e. the probability of a false negative (*type II error*).

Scientific practice involves testing hypotheses and experimental results or effects, i. e. of whether X did/did not give rise to effect, A.

Scientists would like to minimise the probability of both false negatives and false positives. But small sample sizes combined with the mathematical relationship between the two types of errors means that reducing the chance of a type-I error (the acceptance of H and the proposed cause-effect where none exists) increases the

Tab. 8.2 False positive and false negatives in hypothesis testing

	H_0 is true	H_0 is not true
H_0 is accepted *(H is rejected)*	No error	Type-II error False negative
H_0 is rejected *(H is accepted)*	Type-I error False positive	No error

---

[3] In statistical denomination, H_0 is used to represent the null hypothesis, and this term should not be confused with the use of H_0 as hazard in modelling carcinogenisis, nor H as used to denote dose equivalent in radiation protection.

chance of making a type-II error (the rejection of H when the cause-effect is real). In terms of the progress of scientific knowledge, the consequences of being wrong in accepting something as "fact" when it is actually not "correct" is deemed more damaging to science and the scientific community than rejecting a "fact" that later turns out to be "true". So scientists tend to focus primarily on the avoidance of type-I errors, and set higher standards for type-I errors as compared to type-II errors. Conventionally, the probability of a false positive (i. e. the probability that chance gave rise to A, when in fact H_0 is true) must be less than 0.05 before one can reject H_0 and accept H.[4] This conditional probability, α, is often expressed as a *significance level* or *acceptance level*. Thus, a scientific standard of proof for H requires a significance level of α ee of certainty" (meaning that 95 % of our rejections of null hypothesis are correct). Note that this confidence level is *not* the same as saying that there is a 95 % probability that our hypothesis true, a mistake that is commonly made by both scientists and non-scientists. Furthermore, the decision as to what a level of α might be tolerable is somewhat subjective, and in many fields (e. g. physics) $\alpha = 0.05$ would be unacceptably high (Cranor and Nutting 1990).

The use of α can be illustrated with a simple example. If a scientist wished to confirm the hypothesis that radiation doses of 5 mSv increased the rate of chromosome aberration in mice, he would use statistics to determine the *significance* of any difference between an exposed and the control group. If there was a less than 5 % probability that the observed rate of chromosome aberrations in exposed mice had arisen by chance then, and only then, could he maintain an effect, reject the null hypothesis and retain the hypothesis that radiation exposure increases chromosome aberration. The sensitivity of the experiment will depend on the number of mice, the definition of the "effect", the statistical variation in aberration rates in the control group (i. e. baseline rates) and the relative risk of irradiation (i. e. the size of effect and number of events). With small sample sizes, one can confirm only relatively large increases in the aberration rate (i. e. large relative risks); with large sample sizes one could identify smaller increases in aberration rates (see also Tab. 3.7). This reflects the statistical power of an experiment $(1 - \alpha)$ and determines the conditional probability of making a false negative, β.

8.4.3
The standard of proof in criminal law

Criminal law also tries to avoid type-I errors. In determining whether the accused, X, did/did not commit (cause) crime, A, the jury is instructed to be *beyond reasonable doubt* before passing a guilty verdict. Criminal law regards a type-I error (convicting an innocent person) to be the greater miscarriage of justice as compared to a type-II error (acquitting a guilty person). The conservative and stringent standard of being beyond reasonable doubt has been likened to the scientific standard of proof (Shrader-Frechette 1991; Meeran 1992): there need only be a relatively small probability that H_0 obtains before one is forced to reject H. It follows that the burden of proof is on the prosecution to show that the defendant is guilty, not on the defendant to prove his innocence, in Anglo-Saxon jurisdiction. In civil cases, however, both burdens and

---

[4]   The conditional probability of false positives, $p[(-H_0)/(H_0 \text{ is true})]$ is designated α and the conditional probability of false negatives, $p[(H_0)/(H_0 \text{ is false})]$ is designated β.

standards of proof may vary and it is interesting that, like for science and criminal law, this can also depend on the consequences of our being wrong (see Box 8.3).

8.4.4
Burden of proof and consequences of errors

Science and law operate with more or less fixed standards of proof and rigid ideas of where the burden of proof lies. In both cases the burden lies with those positing a causal effect: respectively, with the scientist (against chance or accepted knowledge) and the prosecutor and plaintiff (against the defendant). If authorities or risk assessors also take the "scientific" stance that those asserting an effect should bear the burden of proof, then this burden will often lie with the public (or their representatives) against a potentially harmful practice. It is important to realise that the *burden* of proof determines who has "to make the showing" or who has the benefit of the doubt under uncertainty; the *standard* of proof determines how much uncertainty we can tolerate.

The standard of proof reflects what probability is *acceptable* either in erroneously rejecting H_0 or in erroneously rejecting H. In determining what is acceptable, the *consequences* as well as the rationality of erroneously rejecting H_0 or H must have significance (Jeffrey 1956). In pure science, epistemology carries the most weight. If scientists did not exert rigorous tests to determine what can be accepted as scientific fact, the credibility of science would suffer. The consequences of making a false positive are considered more damaging to the accumulation of scientific knowledge than a false negative. In criminal law, moral considerations are taken into account (Shrader-Frechette 1991): it is considered far worse to convict an innocent man than to set a guilty man free (Thomson 1986). In cases of environmental risk, however, the potential consequences of making either a false positive or a false negative can vary considerably and both benefits and costs, and their distribution within society, need to be taken into account (Tab. 8.3).

If authorities commit a false positive in deciding that a practice or industry is too risky to allow, this may cause society to lose important economic or heath benefits. Critics of the LNT model in radiation protection are essentially accusing authorities of making a false positive in assuming that low-dose radiation causes cancer and thus burdening society with unnecessary costs. On the other hand a false negative could result in society erroneously assuming that environmental or health risks from a particular action are negligible, resulting in large costs to society when the eventual damage is realised. Historically, it is the false negatives (DDT, thalidomide, BSE) that society remembers best and that cause the most serious damage to the reputation of science. The consequences of being wrong in assuming that low doses are safe, when they are in fact harmful, is arguably much worse for science and society than the opposite. But this, of course, depends on the actual health and economic consequences at stake, and how those benefits and risks are distributed. An industry profiting by exposing non-consenting members of the public to its "not-proven-to-be harmful" discharges is a more serious ethical breach than, for example, consenting patients agreeing to participate in trials of a new medical treatment, with possible negative side-effects, but for their own potential health benefit. To sum up, a *fixed* standard of proof will rarely be sufficient grounds upon which to judge the strength of hypotheses concerned with health and environmental risks.

Box 8.3: Proof of Causation in a Civil Claim

Science and criminal law tend to regard type-I errors as more serious than type-II errors, but civil law often shows weaker or no preferences. For example, in an English civil claim the standard of proof is that the plaintiff (the victim) must prove that it was *more probable than not* that: the defendant, X, caused harm to the plaintiff, P.

Hence, success depends on the plaintiff establishing a probability of causation (H) significantly greater than 50 %, i. e. a probability of H_0 less than 50 %. It follows that the standard of proof required from each side in a civil claim is equal, however, the burden of proof still falls on the plaintiff. If neither party can prove a significantly greater than or less than 50 % probability of causation, then the benefit of the doubt goes to the defendant.[5]

Other legal systems would not change the standard of proof (from "beyond reasonable doubt") but rather alleviate the onus of proof or even shift it onto the defendant when certain conditions are met. For, example, under the German Environmental Liability Act of 1990, where the plaintiff proves that a potentially dangerous facility was *liable* to cause the harm in question, considering the operation of the facility, its equipment, the kind and concentration of substances used or emitted, the meteorological conditions, kind and time of occurrence of the harm, the pattern of harm and any other relevant circumstances, there is a presumption of causation. This presumption would have to be refuted by the defendant. However, the presumption of causation is not applicable when the facility is operated in compliance with the conditions of the permit except when there is a disturbance or accident.

The question of causation under tort law (a civil claim for damages) is nicely illustrated by the Sellafield case described above. In English tort law, a plaintiff must establish that her injury was *more likely than not* caused by the defendant, meaning that the plaintiffs (Reay/Hope) needed to prove that it was more than 50 % probable that their blood cancers were caused by parental preconception irradiation (PPI). Similarly, any person pursuing a case of radiation induced cancer would need to show that exposure to that radiation source had at least doubled their risk of leukaemia (statistically equivalent to a relative risk of at least two). Epidemiological studies show a relative risk of about 10 for children born in Seascale, meaning that for every case of leukaemia that could be attributed to "natural causes" there would be 9 excess cases. Although this implies that for each leukaemia case in Seascale there is a 9 in 10 chance, or 90 % probability, that the leukaemia was not due to "natural causes", it does not follow that there is a 90 % chance of each child's leukaemia being caused by radiation. All the other hypotheses as to the cause of the excess leukaemia (worker migration, site specific factors) undermine the radiation hypothesis, and thus the probability of radiation being the cause.

Criminal law instructs us to accept H_0 and find the accused innocent if H has not been shown beyond reasonable doubt. But in science not being allowed to accept H does not force one to accept H_0 as scientific fact. Especially if rejecting H means that one must *act* on the assumption that H_0 is true, standards of proof based on statistical power (i. e. the probability of a type-II error) should apply before one accepts H_0 as scientific fact. Because today we have no way of statistically confirming the hypothesis "annual radiation doses of below 50 mSv increase the risk of cancer" does not mean that we are thereby forced to act on the null hypothesis "annual radiation doses below 50 mSv do not cause cancer". In the case of the Seascale leukaemias, one might contend that existing scientific evidence provides sufficient grounds to reject

---

[5] One exception in English Civil Law being libel claims, wherein the defendant bears the burden of proof to defend their statement or publication.

Tab. 8.3. The societal consequences of false positives and false negatives in environmental science. Practices refer to activities bringing about exposure to substances having the potential to harm human health or the environment; interventions refer to actions taken to reduce or avoid exposure to such substances. B – benefits: C – costs. These can be direct monetary costs as well as indirect or "side-effect" costs such as environmental damage or health detriments.

PRACTICES	H_0 is true	H_0 is not true
	The practice will not cause harm	The practice will cause harm
Accept H_0	*No error*	*False negative*
Assume that the practice will not cause harm	Development benefits (B)	The damage costs (C) are less than any realised development benefits.
Reject H_0	*False positive*	*No error*
Assume that the practice will cause harm	Lost development and employment benefits (-B)	No benefits or costs

INTERVENTIONS	H_0 is true	H_0 is not true
	The countermeasure will not cause harm	The countermeasure will cause harm
Accept H_0	*No error*	*False negative*
Assume that the counter-measure will not cause harm	Benefits from prevention of harm (B)	The damage costs (C) are less than any realised prevention benefits.
Reject H_0	*False positive*	*No error*
Assume that the counter-measure will cause harm	Lost benefits of harm prevention (-B)	No benefits or costs

the following causal hypothesis: "the estimated external radiation doses from γ-emitters to environmentally exposed children from Sellafield cause a 10-fold increase in leukaemia incidence". But rejecting *this* hypothesis may still permit the retention of alternative hypotheses on the increased leukaemias being caused by chronic exposure to internal α-emitters or synergistic effects.

8.5
Choice of hypothesis in science and choice of hypothesis for policy

In deciding which hypothesis, evidence or facts to accept, it can be worth pointing to the difference "what it is rational to believe" and "what it is rational to do". For example, Kristin Shrader-Frechette makes a distinction between what she terms *epistemological objectivity and rationality* – concerned with what we believe to be true – and *ethical objectivity and rationality* – concerned with how we act (Shrader-

Frechette 1994). A scientist can guard her epistemological objectivity by being open and honest about both the probabilities and possible errors arising from research results. She might believe there is a low probability that a particular toxin will cause cancer or that dose-response curves are linear with no threshold. But she does not sacrifice epistemological objectivity by stating that, even though the probabilities are low, the consequences are such that one should not take the gamble of exposing humans to the toxin or that the LNT hypothesis is the best option for *management* of radiation risks. The latter statements represent opinions of how we should act and can, therefore, appeal to ethical norms and values such as described in chapter 2.

There is a great deal of work on the philosophical problems posed by rational belief and its relationship to probabilities and standards of proof (e. g. Jeffery 1976, 1992; Nozick 1993). A number of paradoxes have been identified, including the lottery paradox (Kyburg 1961, after Peirce): We know that one ticket will win the lottery, hence it is rational to hold that belief, but for each individual ticket the probability is so low that it would be irrational to believe for any one of the tickets that it would win the lottery. Similar problems arise when we have to make a judgement on a piece of irrefutable evidence (e. g. the Sellafield leukaemia cluster), but when none of a number of conflicting hypotheses passes our criterion for belief. If no hypothesis attains the scientific standard of proof we believe that one of the many proposed hypotheses would explain the observed effect leukaemia cluster, but for each hypothesis we do not believe it to be the cause.

If the preference for either type-I or type-II error depends on the size of the potential benefits and harms of one's choice, then scientific standards of hypothesis testing cannot be applicable to decisions about how to *act* in the real world. Imagine a situation where one is not sure whether or not a needle is free from contamination with the AIDS virus. Faced with the decision of injecting a child with this needle we would be unlikely to simply appeal to the scientific standard of proof in order to test our hypothesis that the needle is contaminated. Furthermore, the scientific criterion for accepting or rejecting hypotheses could not capture the *difference* between a situation where we must inject a child and one where we would inject a monkey. Nor could it do justice to the problem of how desperately the child needed the injection, what alternatives were available (e. g. was this the only needle on hand), or whether or not the child's guardian was present to voice an opinion. In short, the ethical principles derived in chapter 2, as well as sociological factors influencing risk perception and legal issues about their management, would all have a role to play in our evaluation of how to act in the situation, which hypothesis would be best for policy making and risk management. In these choices, both science and social values influence the way one acts and the choices one makes.

8.5.1
Paradigm shifts and societal influences

While proponents of the hypothetico-deductive method of science focused on the derivation of normative or prescriptive theories in order to demarcate good science from poor or pseudo-science, other philosophers, such as Kuhn or Lakatos, have been more concerned with the history of scientific development (Kuhn 1970;

Lakatos 1970). By analysing the major accepted scientific theories and their acknowledgement by the scientific community, Kuhn suggested that acceptance or rejection of a scientific theory has as much to do with the social, psychological and political aspects of the society within which it was developed, as the "scientific" quality of the theory itself. Kuhn's view is seen as radical and disquieting by much of the scientific community as it addresses serious questions as to the rationality and logic of accepted scientific "facts". If science is simply a social construction, what makes it "better" than religion or myth?

Outside of philosophy, Kuhn is perhaps most well known for introducing the notion of *paradigms* to describe the framework of concepts, results and procedures within which normal science progresses. He suggests that the paradigm is only upset in periods of revolutionary science, or "paradigm shifts", when a series of anomalies and stresses cannot be absorbed within the existing frame. What makes Kuhn's position controversial is the central claim that there can be no strictly logical reason for the change of a paradigm. Kuhn's use of paradigms has been widely criticised by philosophers as being too vague to be of any use in questioning the objectivity of science (Scheffler 1982; Ruse 1993). And even Kuhn himself questions the more relativistic implications of his thesis in later editions of *The Structure of Scientific Revolutions*.

Interestingly, there have been few, if any, paradigm shifts in the Kuhnian sense within the science of health physics. The science of radiobiology has grown steadily since the discovery of radioactivity one hundred years ago. A few "shocks" have arisen connected to the harmful nature of radiation, for example, in 1927 with Dr H.J. Muller's discovery that irradiation with X-rays caused an increase in the number of mutations in the offspring of fruit flies (Muller 1927). But these were not irreconcilable or incommensurable with established science. Genomic instability took a while to be taken seriously by the scientific community, from scepticism when the first paper was published in the late 80's (Pampfer and Streffer 1989) to a more widespread acceptance in the late 90's.

In fact, one might argue, that the main paradigm shifts in radiation protection have less to do with science than with social influences on the use of this science in policy making. Both Kuhn and Lakatos focus on the historical context in which theories are developed and particularly the way that society directs and influences scientists' choice of research subjects and selection of "problem solving" (e. g. by decisions related to funding). As examples of the historical context of radiation protection, one might mention the nuclear weapon's industry, the change from primarily medical applications and regulation to management of industrial sources, shifting attitudes towards the use of radiation in medicine and the role and influence of the ICRP. All these factors have had an influence on the public's perception of both radioactivity and radiation protection itself.

Both science and societal developments have certainly affected the way radiation protection has been implemented. The transition from "safe" to "acceptable" mentioned earlier being one example, and the more recent increase in the pressure for radiation protection to consider effects on non-human species another. The claim that "*if man is adequately protected then other living things are also likely to be sufficiently protected*" (ICRP 1977) is one particular hypothesis that has attracted attention over the past decade. Although ICRP recently added a caveat that "*individ-*

ual members of non-human species might be harmed but not to the extent of endangering whole species or creating imbalance between species" (ICRP 1990), a number of problems remain with the existing state of affairs. The statements are not sufficiently supported by concrete scientific evidence; they are potentially invalid in certain situations (such as deep-sea disposal), and out of line with the current thinking on environmental protection (Strand and Oughton 2002; Oughton 2003). Following an increasing amount of criticism over the past ten years or so (Pentreath 1998; IUR 2000, 2002; Strand and Larsson 2001) the issue is now being addressed by almost all organisations dealing with radiation protection (IAEA 2001, 2002; European Commission 2002, ICRP 2002; OECD 2003). Note, however, that the change in opinion has arisen not due to new scientific evidence on the potential detriments of radioactivity to non-human species, but more due to ethical, legal and social demands.

8.6
Conclusions

There is certainly an influence of a diversity of values (politics, economics, aesthetics, etc.) and circumstances on the types of choices made in radiation protection. Many of these decisions (such as what level of risk is acceptable) cannot be made on scientific grounds alone. Thus an understanding of the societal influences on the choice and application of dose-response models requires a more thorough examination of the influence of ethical, legal and social factors on decision-making (chapters 2, 9 and 10). All policy needs to be informed by both, science and values, including radiation protection and toxicology, and the decisions made about the management of low-dose exposures in practice are often better described as moral judgements than scientific decisions. Yet it is vital that the best available science informs these decisions by providing relevant facts and information, and this includes transparency in the uncertainties and assumptions associated with that scientific knowledge. There can be valid and rational reasons for choosing different dose-response models in "science" and "policy making". Nevertheless, it is important to remember that the public are not the only ones to be influenced by social and ethical issues; scientists can also have difficulty separating facts and values. Finally, the greater the uncertainty in prognosis of the consequences of exposure to radiation, the more important ethical values will become in the judgements about risk management.

8.7
References

Barcellos-Hoff MH, Brooks AL (2001) Extracellular signaling through the microenvironment: a hypothesis relating carcinogenesis; bystander effects, and genomic instability. Radiat Res 156: 618–627

Becker K (1997) On the low dose problem in radiation protection. Swedish Radiation Protection Institute, SSI News 1: 7

Black D (1984) Investigation of the possible increased incidence of cancer in Cumbria. HMSO, London

Bland JM (1994) Cancer in nuclear test veterans – statistical analysis inappropriate. Brit Med J 308: 339

Calabrese EJ, Baldwin LA (2003) Toxicology rethinks its central belief. Nature 421: 691–692

Caldicott H (1994) Nuclear Madness (revised edition, 1st edition 1978). WW Norton, New York

Chalmers AF (1982) What is this thing called Science? (2nd edition) Open University Press, Milton Keynes

Canadian Council of Ministers of the Environment [CCME] (1997) Protocol for the derivation of Canadian tissue residue guidelines for the protection of wildlife that consume aquatic biota. Canadian Council of Ministers of the Environment, Water Quality Guidelines Task Group, Winnipeg, also http://www.ec.gc.ca/ceqg-rcqe/English/Html/tissue_protocol.cfm

Committee on the Medical Aspects of Radioactivity in the Environment [COMARE] (1988) Investigation of the possible increased incidence of leukaemia in young people near Dounreay Nuclear Establishment, Caithness, Scotland. HMSO, London

Committee on the Medical Aspects of Radioactivity in the Environment [COMARE] (1989) Report on the incidence of childhood cancer in the West Berkshire and North Hampshire area, in which are situated the Atomic Weapons Research Establishment, Aldermaston and the Royal Ordnance Factory, Burghfield. HMSO, London

Cook-Mozaffari P, Darby S, Doll R (1989) Cancer near potential sites of nuclear installations. Lancet 2(8672): 1145–1147

Craft AW, Parker L, Openshaw S, Charlton M, Newell J, Birch JM, Blair V (1993) Cancer in young people in the north of England, 1968–85: analysis by census wards. J Epidemiol Community Health 47(2): 109–115

Crane JA (1987) Risk assessment as a social research. In: Durbin PT (ed) Technology and responsibility. Dordrecht Reidel, Boston, pp 279–308

Cranor C, Nutting K (1990) Scientific and legal standards of statistical evidence in toxic tort and discrimination suits. Law Philos 9: 115–156

Darby SC, Kendall GM, Fell GP, O'Hagan JA, Muirhead CR, Ennis JR, Ball AM, Dennis JA, Doll R (1988) A summary of mortality and incidence of cancer in men from the United Kingdom who participated in the United Kingdom's atmospheric nuclear weapons test and experimental programmes. Brit Med J 296:332–338

Darby SC, Kendall GM, Fell TP, Doll R, Goodill AA, Conquest AJ, Jackson DA, Haylock RGE (1993) Further follow-up of mortality and incidence of cancer in men from the United Kingdom who participated in the United Kingdom atmospheric nuclear weapon tests and experimental programs. Brit Med J 307: 1530–1535

Day M (1992) The Sellafield leukaemia claims. Student Law Review, Spring 1992: 27–28

Day M (1990) Suing the goliath. Legal Action, May: 7

Doll R (1989) The epidemiology of childhood leukaemia. JR Statist Soc Ser A 152: 342–351

Doll R, Evans HJ, Darby SC (1994) Parental exposure not to blame. Nature 367: 678–680

Draper GJ, Stiller RA, Cartwright RA, Craft AW, Vincent TJ (1993) Cancer in Cumbria and in the vicinity of the Sellafield nuclear installation 1963–1990. Brit Med J 306: 89–94

Edwards R 1996. Written out of history. New Scientist, 18th May, 2030: 14

European Commission (2002) Stakeholder Conference on Protection of the Environment, 2–3. December, 2002, Luxembourg

European Commission (1996) Technical Guidance document in support of Commission Directive 93/67/EEC on risk assessment for new notified substances and Commission Regulation no 1488/94 on risk assessment for existing substances. Office for Official Publications of the European Communities, Luxemburg

Filyushkin IV (1991) Concept of a «lifetime dose» of 350 mSv. Health Phys 61: 401–407

Føllesdal D, Walløe L, Elster J (1996) Argumentasjonsteori, Språk og Vitenskapsfilosofi (6th edition) (in Norwegian). Universitetsforlaget, Oslo

Feinendegen LE, Pollycove M (2001) Biologic response to low doses of ionizing radiation: Detriment versus hormesis. Part 1: Dose responses of cells and tissues. J Nucl Med 42: 17N–25N

Gardner MJ, Snee MP, Hall AJ, Powell CA, Downes S, Terrell J (1990a) Results of case-control study of leukaemia and lymphoma among young people near Sellafield nuclear planet in West Cumbria. Brit Med J 300: 423–429

Gardner MJ, Hall AJ, Snee MP, Downes S, Powell CA, Terrell J (1990b) Methods and basic data of case control study of leukaemia and lymphoma among young people near Sellafield nuclear plant in West Cumbria. Brit Med J 300: 429–434

Goodman N (1961) Safety, Strength, Simplicity. Philos Sci 28: 150–151

Health and Safety Executive [HSE] (1993) Health and safety executive investigation of leukaemia and other cancers in the children of male workers at Sellafield. Health and Safety Executive, London

Heasman MA, Kemp IW, Urquhart JD, Black R (1986) Childhood leukaemia in Northern Scotland. Lancet 1(8475): 266

Hinton T (2000) Strong inference, science fairs and radioecology. J Environ Radioactiv 51: 277–279

Howarth S (1992) High court claims "totally unfounded". BNFL News, August 1992, British Nuclear Fuels plc, Manchester, p 13

International Atomic Energy Agency [IAEA] (2001) Summary report of the specialists meeting on environmental protection from the effects of ionizing radiation: international perspectives, held from 26 to 29 November 2001, Vienna. Ref.: 723-J9-SP-1114.3. IAEA, Department of Nuclear Safety, Division of Radiation and Waste Safety, Vienna

International Atomic Energy Agency [IAEA] (2002) Ethical considerations in protection of the environment from the effects of ionizing radiation. IAEA-TECDOC-1270, IAEA, Vienna

International Commission on Radiological Protection [ICRP] (1977) ICRP Publication 26. Recommendations of the International Commission on Radiological Protection. Annals of the ICRP 1(3), Pergamon Press, Oxford, reprinted with additions 1987

International Commission on Radiological Protection [ICRP] (1990) ICRP Publication 60. Recommendations of the International Commission on Radiological Protection. Annals of the ICRP 21(1-3), Pergamon Press, Oxford, 1991

International Commission on Radiological Protection [ICRP] (2002) ICRP Publication 91. A Framework for Assessing the Impact of Ionising Radioation on Non-human Species. Pergamon Press, Oxford 2003

International Union of Radioecology [IUR] (2000) Doses and effects in non-human systems. Work of the IUR Environmental Transfer Action Group 1997–1999. IUR, Østeras

International Union of Radioecology [IUR] (2002) Protection of the environment from radiation: present status and future work. IUR, Østeras

Ishimaru T, Ichimaru M, Mikami M (1981) Leukemia incidence among individuals exposed in utero, children of atomic bomb survivors, and their controls, Hiroshima and Nagasaki, 1945-79. Radiation Effects Research Foundation (RERF), Technical Report No. 11-81

Jaworowski Z (1999) Radiation risk and ethics. Phys Today 52(9): 24–29

Jeffrey R (1956) Valuation and acceptance of scientific hypothesis. Philos Sci 23: 237–246 (Paper reprinted in Jeffrey 1992)

Jeffrey R (1992) Probability and the Art of Judgement. Cambridge University Press, Cambridge

Joiner MC, Larnbin P, Malaise EP, Arrand JE, Skov KA, Marples B (1996) Hypersensitivity to very low single radiation doses: its relationship to the adaptive response and induced radioresistance. Mut Res 358: 171–183

Joiner MC, Lambin P, Marples B (1999) Adaptive response and induced resistance. CR Acad Sci, Ser. III, Life Sciences 322: 167–175

Kashparov VA, Oughton DH, Protsak VP, Zvarisch SI, Protsak VP, Levchuk SE (1999) Kinetics of fuel particle weathering and ^{90}Sr mobility in the Chernobyl 30 km exclusion zone. Health Phys 76: 251–259

Katz DL (2001) Clinical epidemiology and evidence based medicine: fundamental principles of clinical reasoning and research. Sage Publishing, Thousand Oaks, CA

Kinlen LJ (1989) The relevance of population mixing to the aetiology of childhood cancer. In: Crosbie WA, Gittus JH (eds) Medical response to effects of ionising radiation. Elsevier, London, pp 272–278

Kinlen LJ (1993) Can parental preconceptional irradiation account for the increase of leukaemia and non-Hodgkin's lymphoma in Seascale? Brit Med J 306: 1718–1721

Kuhn T (1970) The structure of scientific revolutions. University of Chicago Press, Chicago, IL

Kyburg H (1961) Probability and the logic of rational belief. Wesleyan University Press, Middletown, CT, pp 196–199

Lakatos I (1970) Falsification and the methodology of scientific research programmes. In: Lakatos I Musgrove A (eds) Criticism and the growth of knowledge. Cambridge University Press, Cambridge

LewisEB (1957) Leukaemia and ionizing radiation. Science 125: 2355

Lindell B, Malmfors T (1994) Comprehending radiation risks. In: Lindell B, Malmfors T, Lagerlöf E, Thedéen T, Walinder G (eds) Radiation and society: comprehending radiation risk. Proc Int Conf (Vol. 1) IAEA, Vienna, pp 7–18

McGovern D, Valori R, Levi M, Summerskill W (2001) Key topics in evidence-based medicine. BIOS Scientific Publishers, Milton Park

McLaughlin JR, King WD, Anderson TW, Clarke EA, Ashmore JP (1993) Parental exposure and leukaemia in offspring: the Ontario case control study. Brit Med J 307: 959–965

Meeran R (1992) Scientific and legal standards of proof in environmental personal injury cases. Lancet 339(8794): 671-672

Michaels J, Keller B, Haaf G, Kaatsch P (1992) Incidence of childhood malignancies in the vicinity of West German nuclear power plants. Cancer Cause Control 3: 255–264

Mossman KL, Goldman M, Masse F, Mills WA, Schiager KJ, Vetter RJ (1996) Health Physics Society position statement: radiation risk in perspective. Health Physics Society (HPS) Newsletter 14(3): 3

Muller HJ (1927) Artificial transmutation of the gene. Science 66: 84–87

Organisation for Economic Co-opration and Development [OECD] (2003) Radiological Protection of the Environment: The Path forward to a New Policy? Workshop Proceedings, Taormina, Sicily, Italy 12-14 February 2002. OECD Publications, Washington, pp 29–32

Neel JV, Schull WJ (1991) The children of the atomic bomb survivors: a genetic study. National Academy Press, Washington

Nozick R (1993) The nature of rationality. Princeton University Press. Princeton, NJ

Nussbaum RH (1998) The linear no-threshold dose-effect relation: is it relevant to radiation protection regulation? Med Phys 25: 291–299

Nussbaum RH, Köhnlein W (1994) Inconsistencies and open questions regarding low –dose health effects of ionising radiation. Environ Health Persp 102: 656–667

O'Riordan MC (ed) (1996) Becquerel's legacy: a century of radioactivity. Proceedings of a conference, London, February 29 & March 1 1996. Radiat Prot Dosim 68 (1/2), Nuclear Technology Publishers, Ashford

Oughton DH (2003) Protection of the environment against ionising radiation: ethical issues. J Environ Radioactiv 66(1–2): 3–18

Oughton DH, Salbu B, Brand TL, Day JP, Aarkrog A (1993) Under-determination of strontium-90 in soils containing particles of irradiated uranium oxide fuel. Analyst 118: 1101–1105

Pampfer S, Streffer C (1989) Increased chromosome aberration levels in cell form mouse fetuses after zygote x-irradiation. Int J Radiat Biol 55(1): 85–92

Parker L, Craft AW, Smith J, Dickinson H, Wakeford R, Binks K, McElveney D, Scott L, Slovak A (1993) The geographical distribution of preconceptual radiation doses of fathers employed at the Sellafield nuclear installation, West Cumbria. Brit Med J 307: 966–971

Parker L, Pearce MS, Dickinson HO, Aitkin M, Craft AW (1999) Stillbirths among offspring of male radiation workers at Sellafield nuclear reprocessing plant. Lancet 354: 1407–1414

Patterson HW (1997) Setting standards for radiation protection: the process appraised. Health Phys 72: 450–457

Pentreath RJ (1998) Radiological protection for the natural environment. Radiat Prot Dosim 75: 175–179

Pentreath RJ (1999) A Ssystem for radiological protection of the environment: some initial thoughts and lieas. J Radiol Prot 19 117–128

Pochin E (1983) Nuclear radiation: risks and benefits. Oxford Science Publications, Clarendon Press, Oxford

Pollycove M, Feinendegen LE (2001) Biologic response to low doses of ionizing radiation: detriment versus hormesis. Part 2: dose responses of organisms. J Nucl Med 42: 26N–32N

Popper K (1998) Science: conjectures and refutations. In: Klemke ED, Hollinger R, Wÿss Rudge D (eds) Introductory readings in the philosophy of science. Promtheus Books, Amherst pp 38–47

Pugh C, Day M (1992) Toxic torts. Cameron May, London

Reay/Hope v. British Nuclear Fuels plc. (1993) 1990 R No 860, 1989 H No 3689; Med L Rev 1

Rijksinstituut voor Volksgezondheid en Milieu [RIVM](1999): Environmental Risk Limits in the Netherlands. Part I Procedure. RIVM, National Institute of Public Health and the Environment. 601640001

Rodricks JV (1992) Calculated risks: the toxicity and human health risks of chemicals in our environment. Cambridge University Press, Cambridge

Roman E, Beral V, Carpenter L, Watson A, Barton C, Ryder H (1987) Childhood leukaemia in relation to nuclear establishments. Brit Med J 294: 597–602

Roman E, Watson A, Beral V, Buckle S, Bull D, Baker K, Ryder H, Barton C (1993) Case-control study of leukaemia and non-Hodgkin's lymphoma among children aged 0-4 years living in West Berkshire and North Hampshire health districts. Brit Med J 306: 615–621

Ruse M (1993) Is the theory of punctuated equilibrium a new paradigm? In: Ruse M (ed) The Darwinian paradigm, essays on its history, philosophy, and religious implications. Routledge, London, pp 118–145

Sheffler I (1982) Science and Subjectivity. 2nd edition. Hackett, Indianapolis, IN

Shrader-Frechette KS (1985) Risk analysis and scientific method. Reidel, Boston, MA

Shrader-Frechette KS (1991) Risk and rationality. University of California Press, Berkeley, CA

Shrader-Frechette KS (1994) Ethics of scientific research. Rowman & Littlefield, Lanham, MD

Simmonds J, Watt D (1999) Radiation protection dosimetry: a radical reappraisal. Medical Physics Publishing, Madison, WI

Strand P, Larsson CM (2001) Delivering a framework for the protection of the environment from ionizing radiation. In: Bréchignac F, Howard BJ (eds) Radioactive pollutants, impact on the environment, Collection IPSN, EDP Sciences, les Veix, pp 131-145

Strand P, Oughton DH (ed) (2002) Radiation Protection in the 21st Century: Consensus Conference on Protection of the Environment, NKS/IUR, Østerås

Streffer C, Bücker J, Cansier A, Cansier D, Gethmann CF, Guderian R, Hanekamp G, Henschler D, Pöch G, Rehbinder E, Renn O, Slesina M, Wuttke K (2000) Umweltstandards: kombinierte Expositionen und ihre Auswirkungen auf die Umwelt. Wissenschaftsethik und Technikfolgenbeurteilung, Bd. 5. Springer, Berlin

Thomson JJ (1986) Rights, restitution, and risk. Harvard University Press, Cambridge, MA

Tubiana M (1998) The report of the French Academy of Sciences: problems associated with the effects of low doses of ionising radiation. J Radiol Prot 18: 243–248 (see also letter in New Scientist, 2nd December 1995)

United Nations [UN] (1992) Rio Declaration on Environment and Development. A/CONF.151/26 (Vol. I) United Nations General Assembly, New York

United Nations [UN] (2002) World Summit on Sustainable Development, Johannesburg

US Environmental Protection Agency [EPA] (1998) Guidelines for Ecological Risk Assessment. EPA Report, EPA/630/R–95/002F. Risk Assessment Forum, Washington, DC

Yoshimoto Y, Neel JV, Schull WJ, Kato H, Soda M, Eto R, Mabuchi K (1990) Malignant tumours during the first two decades of life in the offspring of atomic bomb survivors. Am J Hum Genet 46: 1041–1052

9 Risk Evaluation and Communication

9.1
Introduction

Virtually all human activities involve some associated risk. In going about their daily life individuals continuously evaluate situations and make decisions on whether the risk associated to a particular action is justified. Such decisions are mostly made involving value judgments which normally cannot be explicitly expressed in terms of quantitative criteria. This is frequently the case when the risk is of a voluntary nature, i. e. it is taken as a free choice (e. g. smoking, downhill skiing). On the other hand when the individual cannot fully choose to avoid exposure to risk, it is termed involuntary (e. g. natural disasters) and the decision making process needs to be more explicit using quantitative data. Moreover, people are generally willing to expose themselves to quite different levels of risk depending on whether they feel it was their own decision or they feel that the exposure is beyond their control. Decisions involving involuntary risks are often dominated by emotional arguments, as has been amply demonstrated in the controversy about electricity producing technologies, and here especially about nuclear power (Chiosila 1996; Scholz 1996; Ansel 1997; Stoll 1997; Tanguy 1997). In most of these controversies, the potential negative consequences rather than the related low probabilities play the dominant role, as can be expected in a discussion related to risk. This applies not only to the nuclear issue but also to other areas such as the environment, public health, etc.

However, while concentrating on the potential negative aspects, normally there are also positive aspects, or benefits, to be considered. The components contributing positively to health and safety may greatly exceed the more publicised negative effects associated with achieving it. What is even more important, if public benefits could be developed into a discipline based on a scientific methodology, these positive effects might be further increased. Unfortunately exactly the reverse view is commonly held. The reason why this fallacy has become and has remained 'conventional wisdom' is that matters perceived by the public as a threat to their safety cannot be discussed in the way that matters of science and technology can usually be discussed. Emotional pressures, arising often from man's fear of the unknown, inhibit consideration of important aspects; expert and informed views are disregarded unless they conform with established conclusions. The well-publicised risk, however small, dominates discussion and decision making.

9.2
Risk analysis

The overall objective of risk analysis is to provide a notional formulation for the decision-making process concerning this risk by combining the results of the risk analysis with risk criteria. In order to analyse risks, several processes can be applied. In risk assessment, the aim is to determine as accurately as possible, the risk in question. The most desirable approach is to obtain quantitative, i. e. statistical data. However, in many cases it is necessary to resort to probabilistic estimates. The results derived from these assessments are used in risk characterisation, also termed risk evaluation, where an understanding is to be developed on how the risk can occur and also on how it can be controlled. These results should also be traceable, repeatable and verifiable. The insights gained are then used for risk management (Risiko-Kommission 2002).

In general, the level of detail of the analysis will depend on the purpose and the intended application in risk management policies. Three classes of methods can be distinguished:

– Rapid ranking methods: the result is used for strategic purposes, i. e. setting priorities. The objective of this approach is to get a broad but not detailed view of a great number of different activities/situations.
– Detailed analysis: the result is used to conclude whether a certain riskful activity is acceptable or not. Acceptability is related to the setting of criteria by politicians. The objective is to maintain a certain level of general safety and to focus on those situations that are really serious or of particular societal concern.
– Analysis of specific measures/alternatives: the result is used to decide on measures that can be taken or to decide between alternatives. The objective of this approach is to get a basis on which one can decide if a safety measure is appropriate in relation to the costs or to decide among less riskful alternatives. These methods are often used as the input for a cost-benefit analysis.

In particular, a number of techniques are available to establish the various aspects of the risk in question. Hazard identification refers to the determination of whether a particular agent is or is not causally linked to a particular health effect. Therefore, relevant information must be collected about all technical, environmental, organisational and human circumstances to the activity/problem being investigated, with special emphasis on health or safety implications (Gheorghe 1996).

The risk estimation process then deals with frequency analysis and consequence analysis. Frequency analysis estimates the likelihood of each undesired event identified at the hazard identification stage. The methods used are (i) the processing of relevant historical data, (ii) use of models, (iii) use of expert judgment techniques. Sometimes a combination of these methods is necessary. As an integral part of such a process, the consequence analysis estimates the likely impact of an undesired event (e. g. for health risk, the number of people who will suffer injury or illness could be calculated) and has to consider both immediate and delayed consequences.

There are many uncertainties associated with the estimation of risks. The uncertainty is, in essence, embedded in the state of knowledge relevant to the assessement of the problem at hand. Uncertainty analysis involves the determination of the variation of imprecision in the model results. Sensitivity analysis is a field closely related to uncertainty analysis and involves the determination of the change in response of a model to changes in individual model parameters.

9.2.1
Definitions of risk

Risk is defined as a quantity which expresses the hazard, danger or chance of a harmful consequence associated with an actual or potential exposure. Thus, the definition of risk (R) contains, as a minimum, 3 elements (Renn 1998): the consequence (C), which has an effect on what humans value, the probability (P) of occurrence and a formula to combine both elements $(R = P \times C)$.

More comprehensive procedures take additional factors into account: the frequency (in the past) or the probability (in the future) of an undesirable consequence (death or injury) per unit number of people per unit number of time. In addition, the magnitude of the uncertainties and weighting factors for very large consequences or long time periods or geographical areas can be considered. All these components are based on a statistical concept and can be referred to as "rational risks".

However, this approach is not fully accepted by the population. Here, personal factors, such as perceived ability to control the exposure, are included and therefore different weights are given to potentially harmful effects from an action or event. Thus the concept of "subjective risk" needs to be taken into consideration (Streffer et al. 2000). These authors also stress that "tension between rational and subjective risk is one of the main forces driving the public debate on health risks and their management".

Another definition which appears to be more appropriate for common understanding has been formulated in (Covello and Merkhofer 1994) and holds that risk is a characteristic of a situation or action wherein two or more outcomes are possible, the particular outcome that will occur is unknown, and at least one of the possibilities is undesired. They stress that people talk about risk when there is a possibility that something undesirable might happen, but this is of course not certain. Thus, in common thinking, the notion of risk includes a consequence, which is both uncertain and undesired. In general, people can visualise consequences but they have more difficulties with the concept of probabilities (see also chapter 2).

9.2.2
Risk assessment

Risk assessment is the scientific process of defining the adverse effects of a substance, activity, lifestyle or natural phenomenon in preferably quantitative terms and often forms the methodological basis for public policy. Whereas it is relatively straightforward to assess situations, such as driving a car, where accident statistics provide the data for reliable after-the-fact-assessments, it becomes more cumbersome in any area, where no reliable statistical data are available. When attention is directed at forecasting risks whose cause is rather uncertain, or that are only conceivable but not measured or measurable in human experience, the objectivity of the result of such a risk assessment might be put into question. Most familiar examples are the prediction of cancer risks from ionising radiation, chemicals, electromagnetic fields, smoking and other potential carcinogens (see chapters 3–5), where these predictions could be realistic but could also be either exaggerated or even hypothetical. Nevertheless, society has decided to take a preventive approach and is prepared to devote resources to exploring and regulating such risks.

The analytic tools of risk assessment, especially in the low-dose range, are designed to determine how much risk to human health might be caused by various exposures to chemicals and agents such as radiation. In this assessment, adverse effects cannot always be observed as an immediate effect of a causing substance. The link between exposure and effect is often difficult to draw, sometimes not even measurable. In such risk assessments causal relationships have to be explored and modelled explicitly. Modelling is a necessary step to isolate a causal agent from among several intervening variables. Based on toxicological (animal experiments), see chapter 3, or epidemiological studies (comparisons of a population exposed to a risk agent with a population not exposed), see chapter 5, researchers try to identify and quantify the relationship between the potential risk agent (e. g. ionising radiation) and physical harm observed in humans or other living organisms. These risk assessments can serve as early warning signals to inform society that a specific substance may cause harm to humans or the environment even if the effects are not obvious.

9.2.2.1
Quantitative risk assessment

As has been stressed above, the quantification of data to be used in risk assessment is necessary for achieving objective results. Even the simplest question of whether the net impact of a particular activity could be beneficial or dangerous to public health cannot be answered without at least a rough estimate of the components.

Quantified risk assessment and the use of results for decision-making have identified two basic types of risk, individual and societal:

- Individual risk is the frequency or probability per given time period in which an individual may be expected to sustain a given level of harm (mortality, morbidity) from the exposure to a specified hazard.
- Societal risk is the relationship between frequency or probability and the number of people suffering from a specified level of harm (mortality, morbidity) in a given population from the exposure to a specified hazard.

For individual risks, recommendations about quantified numerical criteria can be derived from statistical data on exposure and/or effects. However, criteria for societal risk are more difficult to derive. In general, the areas where interventions can be envisaged encompass: prohibition, control and ranking according to importance. Analysis of the potential for harm and the quantitative assessment of risk probabilities and consequences provide a very powerful tool for revealing the principal characteristics which can impair health and safety. It provides a knowledge base for all subsequent actions taken by governments or concerned decision-makers. The quantitative assessment also provides a crude measure of the relative scale of the consequences of a specific risk in the total spectrum of all the other risks that we must live with.

9.2.2.2
Probabilistic risk assessment

Estimating risk through Probabilistic Risk Assessment (PRA) is predominantly a physical engineering approach to assessing the reliability of a system. PRA has been designed to predict the probability of failures of complex technological systems even in the absence of sufficient data for the system as a whole (IAEA 1998a). Using fault tree analysis, where an undesired event is selected and is traced back to all possible causes, or event tree analysis, where a single occurrence is identified and the potential consequences are determined using inductive logic, the failure probabilities for each component of the system are systematically assessed and then linked to the system structure. All probabilities of such logical trees are then synthesised in order to model the overall failure rate of the system. A PRA provides an average estimate of how many undesirable events can be expected over time as a result of a human activity or a technological failure.

9.2.3
Risk characterisation

The objective of the characterisation is the description of the exact nature and also the magnitude of the risk as objectively as possible, with reference to the inherent uncertainties. In the majority of instances, such risk figures are based directly or indirectly on the statistical processing of empirical experience as to the frequency of accidents, injuries, morbidities and mortalities in the many situations encountered. Clearly, decisions with subjective elements play a role even in judging whether a certain statistic is appropriate to the specific situation of interest, while subjective decisions are unavoidable when the original data must be converted to health detriments or if the frequency of rare events must be synthesised with the help of basic statistical effects of e. g. low level exposure. Normative international agreements such as the recommendations made by the International Commission on Radiological Protection (ICRP) and the International Basic Safety Standards (IAEA 1996) directed towards the Regulatory Authorities and, in the case of intervention, the Intervening Organisations, in the field of ionising radiation reduce the need to resort to subjective interpretation of data.

From a practical point of view, the characterisation of risks should include:

- Identifying problems related to health, environment and safety, and approaches to their solution;
- facilitating appropriate decisions on the acceptability of risk (individual or societal);
- meeting regulatory requirements.

From a policy-making perspective, risks can also be categorised by the nature of their consequences:

- Public safety (health impact on public),
- individual safety (health impact on individuals),
- occupational safety (health impact on workers),
- environmental impact (air, water flora, fauna),
- economic impact,
- public opinion.

However, in the strict sense there are no universally valid risk figures. Any study will have to determine beforehand in which frame the assessments are to be performed. An organisation attempting the presentation of more generally applicable risk figures has the choice of two solutions. The range of risks quoted can include the full scope of individual risk values in each specific category, from the lowest to the highest values found. This will in many cases lead to results, which, because of their imprecision, can be of little use for policy decisions. On the other hand, risks can be determined for clearly specified "regions", i. e. groups within the population or geographical areas, which have some similarity in the relevant boundary conditions.

If one limits a given study to health risks, a whole list of risk categories must be considered. A wide range of injuries due to accidents and illnesses can occur, beginning with those of a trifling nature up to events leading to morbidity or premature mortality. It is obviously difficult to define unequivocal limits for classes of injuries and illnesses of different seriousness and even define a cut-off below which such health effects shall not be considered. Thus, statistics of accidents (of their effects, consequences) and of non-lethal injuries and illnesses are notoriously incomplete and unreliable. Therefore the analysis regarding health risks is generally limited to the determination of risk of premature death. The frequency or risk of mortality is usually quite good an indicator of where the major health risks occur. In addition to the possibility of immediate mortality, either due to an accident or an acute disease, ionising radiation and chemicals can also have effects leading to delayed mortality. Of particular significance here is the possible occurrence of cancer, either after a latency period of some decades following an acute exposure or after an extended period of chronic exposure. The incorporation of such delayed risks forms an important part of risk characterisation. The agents mentioned, i. e. radiation and pollution, can finally also lead to teratogenic and/or mutagenic or genetic damage. Whereas considerable information is available on these consequences in the case of exposure to ionising radiation, knowledge in the case of exposure to many chemical pollutants is rather limited.

A category of risks which should be set apart from those risks which can be determined on the basis of actuarial evidence and statistics of mortality are the risks of delayed harm or morbidity, particularly cancer, as a result of radiation and chemical exposure. While, as mentioned above, the fatalities determined in the former group are deaths which have actually occurred and which will, in the statistical sense, occur again in the future with the frequency ascertained. The second group comprises fatalities which occur either after a long latency period of some decades following an acute exposure or after an extended period of chronic exposure and have been calculated on the basis of a theoretical dose-effect relationship. Such a relationship extrapolates the extent of harm, which has been determined at high doses down to doses frequently many orders of magnitude lower. At such low doses it is problematic to demonstrate any harm experimentally or epidemiologically, because the harm in question – cancer – is unspecific, i. e. can have many causes, most of them unknown, which cannot be separated from the agent considered. In fact, in this case, the risk of harm is based on a hypothesis; it can be just as likely that there is no harm whatsoever, at least at very low doses.

In the case of ionising radiation there is a worldwide agreement on the assumption of a linear dose-effect relationship, without a threshold, down to zero dose. It is agreed that, in most cases, this hypothesis is conservative. Knowledge of the harmful effects of the many kinds of chemical agents is far more rudimentary. Epidemiological studies to determine the dose-effect relationships are even more difficult than in the case of radiation. Here, too, a no-threshold dose-effect relationship is assumed. This is again a conservative assumption, but one which seems all the more justified by the fact, that there is, in contrast to radiation, no natural background dose level for most of these substances.

On the basis of a no-threshold dose-effect relationship, the collective risk can become quite considerable even if doses are extremely low, if this integration is performed over a large population over a very long time period – a case of multiplying a very small value with a very large one. One would need to give the concept of a "*de minimis*" dose (*de minimis non curat lex* – the law does not bother with trifles) some consideration, i. e. the normative definition of a very low dose, below which the effects are taken to be so small that they could be disregarded.

In the next step of the analysis, the magnitude of very different potential health detriments must be determined one by one. The analyst is now confronted with the problem as to what extent these risk categories should be kept separate from each other in the final presentation of results and what possibilities for their aggregation into combined categories he can afford to employ. Generally, it is quite evident, that an analysis of the impact of a single substance is far more transparent than an analysis of combined substances. It is also far easier to update as new information becomes available, for instance a new study on health effects. In contrast, if there is a change in one of the substances of a combination, the effect on the aggregated result is not obvious unless they are presented and documented in sufficient detail (see also Streffer et al. 2000).

The results then will have to be explained not only to specialists in the field but also to laymen such as politicians and policy (decision) makers. However, laymen perceive the various categories of risk very differently. Immediate death as a result of an accident and a late death are felt to be of a quite different quality. Likewise, an accident at work is judged wholly different from an accident to an uninvolved bystander.

9.2.4
Uncertainty, the precautionary principle and governance

Some risks, with both, high and low probability, are well documented and understood, while other risks, again with both, high and low probability, are highly uncertain. Thus, uncertainty is not the same as probability. All events, including risks, are somewhat uncertain, because the future cannot be predicted with certainty. The public, however, often assumes that regulatory action is based on factual knowledge about what is good or bad, safe or unsafe and has a tendency to believe that the facts are quite certain. Governments accept this perception of scientific knowledge to justify policy decisions. Such belief in the certainty of observations underlying regulatory decisions can be misleading and the failure to communicate the underlying uncertainty invites demands for simple choices and assured protection that can, in reality, not be satisfied. If the public discovers that the facts, upon which regulatory action was taken, are disputed and the protection promised is thus not certain, trust in governmental/regulatory decisions will be at stake.

The precautionary principle (PP) has emerged as one of the main regulatory tools of the European Union environmental and health policy. It was developed and implemented originally in Germany and Sweden (see also chapters 2 and 10).

The European Commission (2000) states that the Precautionary Principle is supposed to cover:

> ...those specific circumstances, where scientific evidence is insufficient, inconclusive or uncertain and there are indications through preliminary objective scientific evaluation that there are reasonable grounds for concern that the potentially dangerous effects on the environment, human, animal or plant health may be inconsistent with the chosen level of protection.

This statement stresses that a scientific evaluation of potential adverse effects must be carried out but realises, that data may be limited or uncertain. It states that: "the appropriate response in a given situation is thus the result of a political decision, a function of the risk level that is "acceptable" to the society on which the risk is imposed". The conclusion is thus, that for situations, where potential harm can occur, remedial actions will rather be based on precaution than on scientific evidence which has a certain amount of uncertainty incorporated (Bodansky 1991).

In a paper by Page (1978) a characterisation of the types of technological, health and environmental risks is presented, for which the need for precautionary approaches is most likely to arise:

– there is an "ignorance of mechanism": our knowledge of the physical processes that determine the likelihood and magnitude of the risk is poor;
– there is a "potential for catastrophic loss": the harm to the affected individuals and to society as a whole can be very great if the activity or technology entails such a risk;
– there is a "relatively modest benefit" associated with the activity or technology, especially when compared to the potential harm;
– there is a "low subjective probability" associated with the feared outcome. However, there is often no consensus about the probability of occurrence, due to the rarity or one-of-a-kind nature of the risk, with little or no actuarial history upon which to base estimates.

Several types of scientific uncertainty can prevent consensus about the optimal way to deal with these potential risks. They encompass e. g. the difficulty of adequate experimental testing for possible hazards, conflicting results from seemingly 'identical' studies conducted with laboratory or epidemiological methods, statistical limitations in setting risk levels, uncertain extrapolations from laboratory results to human beings, lack of knowledge about mechanisms of interaction, imperfect models and uncertain parameters needed for mathematical models of exposure or risk. In addition, risk assessment paradigms vary with the type of exposure and presumed mechanisms of action and also differ among various government agencies and research institutions, among countries and can be influenced by the interests of industry, members of the public and politicians.

The use of the precautionary principle for regulatory purposes is highly controversial, especially between the countries of the European Union and the USA, as elaborated in (Wiener and Rogers 2002). They contend in their conclusion, that:

> ...each actor has been more precautionary than the other as to some risks and less precautionary as to others. The degree of precaution exhibited appears to depend less on some overarching national regulatory posture, and more on the context of the particular case: the risk, the technology, the location, the era, the politics, the public, the agency, the legal system.

The primary purpose of the precautionary principle is therefore to encourage actions where the evidence is inadequate to make a causal inference about some potential risk. Because the principle was developed to deal with insufficient scientific evidence, the precautionary principle purposely transcends scientific evidence and scientific methodology (Balzano and Sheppard 2002), which does not mean that the principle would be applied without a scientific basis.

There is a worldwide trend to be observed, where governance, i. e. political decision making, is getting increasingly aware of the existence and potential incorporation of risk-oriented concepts and tools, developed by theoretical and practical means, into the political process. This should eventually lead to a more effective role and also profile of science in governance, particularly in the areas of uncertainty and risk. The goal here can be described as: aiming at better and sounder policy decisions and their outreach, more accountability, transparency and wider participation. This approach could even assist in reducing unnecessary controversies.

9.2.5
Risk management

Risk management constitutes a system of decisions directed towards what needs to be done in order to reduce unacceptable risks. The problem of choice is in most cases based on economic analysis and every economic analysis is reduced to the question of the optimal utilisation of financial resources. Thus, risk management itself is reduced to the problem of selecting risk reduction/elimination measures to bring about maximum risk reduction with the available means. The approach selected can be either effect-oriented i. e. the necessity of risk-reducing measures is related to the (possible) effects of (riskful) sources to the population, regardless of the financial implications, or source-oriented, i. e. the application of a principle which permits risk reduction at the source to an "as low as reasonably achievable" (ALARA) level, with the utilisation of the best practical and technical means. However, this principle would entail that the reduction of a risk is carried to any point which is achievable, but this point might be lower than actually necessary. Therefore, the principle was somewhat modified into 'as low as reasonably practicable' (ALARP), where expenditures for risk reduction would only be expected to the point of tolerability and unjustifiable expenditures to accomplish marginal reductions would not be required (IAEA 1998b).

For instance, much of the UK legislative control of hazards and risks is based on the ALARP principle (UK Health and Safety Executive 1989). This legal decision process involves:

- quantification of likely risk with an understanding of the inherent uncertainties,
- reference to the benefits generated by the activity and the political and economic considerations associated with it,
- judgments as to "tolerability" and "acceptability" for groups directly or indirectly affected,
- sometimes, decisions as to further reductions in risk, taking costs (including effort) into account.

The difficulties in arriving at decisions, whenever health and safety are viewed by the public to be an important consideration, arise from several factors (IAEA 1995):

- No disciplined and systematic effort is made to find the appropriate balance, or compromise, between risks and benefits (including considerations of equity) because the established levels of safety lack uniformity of application. Safety levels are determined by mere expediency, which nevertheless acquires the force of law.
- The assessment of factors that contribute to the overall net benefit is generally ignored resulting in large sums being spent with the further consequence that real safety is compromised.

Therefore, a rational strategy for risk management is required because this provides a solid potential for improving the effectiveness of spending resources directed at health and safety concerns. This also includes a reasonable setting of priorities (Cross 1994).

As with most decision processes, risk management works better if clear criteria are established to guide the decision making process. Decision criteria can be expressed as 'tolerable' individual or societal considerations.

Once risk levels and criteria have been established, risk management is the process of determining whether risks should be controlled by standards or alternative approaches. Dose-effect investigations help risk managers to define standards in accordance with observed or modelled threshold values. If there is no threshold value as is the case of most carcinogens, risk assessments provide information about the probability of harm depending on the dose. Risks permitted by standards are called 'residual risks' because they represent the levels of risk that may persist after 100 % compliance with standards is achieved.

The legal and policy communities use various phrases to describe the regulatory status of a particular risk. If a risk is deemed 'significant' or 'unacceptable' risk managers are generally expected to take steps to reduce or eliminate it. A *de minimis* or "negligible" risk is one that can be ignored. If a risk is judged to be insignificant or acceptable, however, it is not necessarily de minimis or negligible. Cancer risks too small to exceed the 'significance' threshold but too large to fall below the *de minimis* threshold may have a discretionary status (see chapter 10). Regulations, especially when dealing with smaller risks, are not only determined by the extent of the exposure but also by other considerations such as weight of the scientific evidence, the feasibility and cost of control, the size of the population at risk and the cost-effectiveness of control (Sadowitz et al. 1995).

Under circumstances where the management of the risk is not under the control of the individual exposed to it, it takes a much higher degree of confidence to make the risk acceptable! For most involuntary exposures society has determined that the benefits of the activity involved are sufficiently large that the activity should not be foreclosed. In almost all existing cases the perceived benefits are so large that the option of abandoning the activity is not seriously considered and society´s efforts are focused on reducing the associated risks. When the benefits are clear, the issue is how much of society´s resources should be allocated to reduce a specific involuntary exposure to an acceptable level. The benefits of most activities are, however, not uniformly distributed and the involuntary risk exposure may be concentrated on a few.

Equity distribution ideally aims at restoring equity by suggesting ways to minimise risks and how risks should be spread over an area or a larger group of people or even a whole population. It is an attempt to make a fair distribution between the nature of good and harm, the roles involved in the distribution, its underlying values, rules applied to represent the values, measurement and decision making. This calls for a wide range of equity instruments. These instruments are not just technical but also economic and social. They are combinations of technical, participation, market and communication. They are designed to change misconceptions of certain hazards and are meant to upgrade knowledge and shift preferences to rationalise with societal goals. The equity instruments should be used in combined form, i. e. more than two:

– Technical: relies on statements and judgments by experts. However, disagreements among experts and opposition against certain technical issues often transforms into a political debate and thus increases general awareness. Trustworthy experts may alleviate the situation.
– Participation: a fully public-oriented decision-making procedure would theoretically be desirable but may not automatically lead to a just outcome because some groups are more influential than others.
– Market: here the question of fairness is addressed between the individual choice, which might perceive greater personal benefits and the collective choice, which could provide less personal benefits, but improved common gains.
– Communication: while the public is aware of the current and potential problems, issues often remain unsolved because the information is too superficial, broad based and not targeted to relevant groups, who have specific needs. Mass media often put up an agenda for open ended public debates instead of creating awareness and knowledge.

In an ideal world, any hazardous activity would not impose risks that were disproportionate to the benefits and any such risks would be equitably distributed amongst society in proportion to the benefits received. In practice, such distribution is not possible and the principles of distribution are applied in a more general way, involving tests to establish:

- whether a given risk is so great or the outcome so unacceptable that it must be refused altogether, or
- whether the risk is or has been made so small that no further precaution is necessary, or
- if a risk falls between these two states, that it has been reduced to the lowest level practicable, bearing in mind the benefits flowing from its acceptance, and taking into account the costs of further reduction.

These principles combine with other generally accepted approaches, namely that risk should never be imposed unnecessarily, and that no individual or community should bear an unfair proportion of any risk. The approach is applicable equally to the control of individual and societal risk. The legislator or decision maker is expected to propose figures for what should be judged acceptable, tolerable, intolerable, negligible, trivial, etc. The outputs of a quantitative risk assessment supply only one technical input into the arena of decision making in which value judgements of one kind or another are central. Such value judgements involve very complex social processes. Hazards and risks are viewed quite differently by experts and the public, depending on the origins of the hazards and the nature of the risks they present. In addition, the scientific approach to assessing and managing risks can arouse interest groups and public debates, as is stressed by Shrader-Frechette (1995):

> Too strong a distinction between risk assessment and risk management can lead to the hegemony of expertise in assessment areas where experts alone have no right to exercise complete control. Too strong a distinction can underemphasise the unavoidable value judgements in risk assessment, overemphasise the role of technical experts, thus disenfranchise the public who ought to have a voice in risk assessment as well as management.

9.2.6
Comparative aspects of interpreting risks

Comparison of risks is one approach for policy makers and legislative bodies to provide assistance in decision making. The intent is to use estimations of a risk in such a way as to present it in a 1:1 relation to other risks so that their relative significance can be better evaluated. Since the notion of hazards and risks has, inevitably, a negative aspect and can, when seen in isolation, arouse fear and opposition in the public, it is sometimes useful to put risks into perspective to demonstrate, that their consequences are within limits which have been determined as acceptable and/or controllable in the past (Fritzsche 1994).

The underlying philosophy for a well conducted comparative risk study comprises the following steps: First, boundaries need to be clearly identified that fit the objectives set for the analysis and the cases to be defined precisely. Identification of boundaries is fundamental to the whole analysis as inconsistencies in the treatment of different systems or agents can make the results of the study extremely misleading. The analysis then proceeds with a careful, sequential counting of burdens, impacts – where appropriate – and valuation. Quantification should be as complete as possible, accepting that some effects can be demonstrated to be negligible in the context of the question under investigation. The assumptions underlying the analysis and uncertainties that affect it need to be explicitly described and gaps in the analysis need to be clearly identified (IAEA 1999a).

The goal of comparing environmental and health impacts or risks is to inform the decision makers about the potential consequences of e. g. different forms of low level exposures. In performing the analysis one needs to pay attention to the different information needs of different decision makers (Davies 1996). This consideration determines the scope of the analysis, as well as the presentation of the results and their aggregation (over affected populations, over time or over impact categories):

The undertaking is very complex and in designing a study a considerable number of choices have to be made (IAEA 1992), such as:

- Specification of substances involved,
- Impacts to be included (public and occupational health),
- assessment boundaries (time and space),
- models and hypotheses for the analysis,
- economic parameters (value of life, cost of treatment),
- underlying philosophy (average case versus worst case).

These issues need to be well defined in order that the context of the analysis is properly understood. As has been stressed in the previous part, also here, results of comparative risk assessments cannot be considered robust unless specific attention is paid to the role of uncertainties. This means that particular impacts can have different uncertainties and the importance of the respective uncertainty depends on the final application of the findings. The existence of potentially very large uncertainties in the results of comparative risk assessments needs to be put into context. Technical and scientific uncertainty is usually expressed in terms of a threshold analysis. Policy/ ethical uncertainty is more in the domain of philosophical and conceptual understanding. Another important element, which contributes to the validity of an impact evaluation is expert opinion. It is woven into all assessments (predictions) and thus, a certain amount of subjectivity and uncertainty is unavoidable. The approach, which can be used to treat this uncertainty is to maintain awareness, rather than trying to control it, by keeping track of expert thinking and decisions. This is a form of quality assurance/quality control process.

It must be remembered that regulators and politicians are required to make decisions and even acceptance of the current state of affairs requires a decision to be made. The purpose of the comparative risk assessment must be to pass the best synthesis of information possible to decision makers and provide them with sufficient information to communicate their decision to the public (IAEA 1999b). Then it is their responsibility to respond to the data made available to them and to react to the declared uncertainties as they see fit. In turn, there is an obvious responsibility of analysts to adopt appropriate analytical techniques and to be clear in their explanation of where the principal uncertainties lie and of the effect of the adoption of alternative assumptions whenever these are possible.

Where it appears preferable to interpret the results of a comparative analysis with reference to standards as opposed to monetary values, it is vitally important to understand the basis and appropriateness of the standards being used. Ongoing progress in epidemiology, toxicology, cancerogenicity, etc. will eventually reduce the uncertainties in this process.

Considerable work has been devoted to determine the uncertainties between mathematical models in comparative analyses, but care must also be taken to ensure that the differences between deterministic and probabilistic approaches are understood when interpreting the results. Probabilistic results will generally be more appropriate for comparing risks of very unlikely events whilst deterministic results will provide a better indication of routine events.

The interpretation and application of results from a comparative analysis is also dependent on the time scale and spatial scale over which the assessments have been made. Time scales of over 25 years will result in issues of intergenerational equity, which can greatly affect the usability and implementation of the result.

After having identified, characterised and analysed risks and provided the decision maker with enough information to manage and also to put into perspective with already existing and even accepted risks, the next large step is to present the situation to the public in such a manner as to gain a maximum of acceptance and to prevent major misunderstandings.

9.3
Communication of information on risk

Effective communication of risk information is perceived as a possibility to achieve more democratic and participatory policy strategies. This approach has been identified as a mechanism to decrease the frequency and intensity of societal conflicts. Risk communication is considered as an indispensable component of interaction in society and effectively managing risks may not be possible without reasonable communication. A reasonable goal for health and regulatory officials is to create policies and engage in risk communication strategies that are appropriate and effective. Some characteristics which can be decisive in the communication process include: the importance of various aspects of the issue to be communicated, general prior beliefs about the risks in question which can be either natural or caused by human activities, the perceived fairness of the policy process and trust in agencies involved in risk management.

The circumstances of and objectives for risk communication can include: institutional educational campaigns to heighten public awareness of a hazard, obtaining informed consent when an activity or choice may be associated with an expected non-zero risk, reassuring individuals about the safety of a facility or process, obtaining community input regarding proposed risk management options, encouraging or promoting some behavioural change to mitigate the health effects of hazardous exposures. In all instances the information provided should be accurate, relevant and, if possible, delivered in a timely fashion.

Communicating to help people learn about something and communicating to persuade them to change their opinions/behaviors regarding an object are two very different things. A communicator needs to be clear about which goal is most important, as that choice will dictate a number of basic strategies. For example, individuals seem to prefer to use certain information channels to "learn about" a risk and different channels to help them decide whether to act in response to the presence of a risk. This means that a goal of "learning" may lead a communicator to use entirely different channels than would a goal of "persuading".

Making a distinction between learning and persuading also may lead a communicator to emphasise different types of message content. The former goal would call for more sophisticated and extensive use of explanatory strategies, for instance, than would the latter goal.

In addition, selecting a goal may force the communicator him/herself to be clear about intentions. Many information campaigns that profess to inform are really attempts to persuade, and that hidden agenda not only may frustrate members of the audience, who think they have detected it, but also may, quite simply, fail because the communicator has not employed effective communication strategies.

In general, it is considered inappropriate to try to persuade people as this might given them the impression that they are not expected to be able to make up their own mind. It is also not considered appropriate to use the word 'education' with respect to adults, especially if they are highly skilled (e. g. medical doctors) or have an important function in society (e. g. politicians or those occupying senior levels in administration), as it is too paternalistic.

As framework for the various aspects to be considered in a meaningful communication strategy the components proposed by Lasswell (1948) will be used. These components have served for many years as a useful guide to many thoughtful reflections on communication effects and can be presented in form of the question: "Who says to whom what in which channel with what effects?"

Each of these components will be addressed in the following with an emphasis on communication about risk.

9.3.1
Who (risk communicator)

Any communication about an issue needs somebody who wants or needs to relate this information to a concerned or interested audience. However, although the "who" for any risk communication exercise may seem obvious, some dimensions of the "who" get obscured routinely, to the detriment of the communication process. Two important dimensions need to be considered in this context to assure that the effort will have the desired or at least a desirable as opposed to undesirable effect, namely the credibility of the communicator and the importance of acquiring communication skills.

9.3.1.1
Credibility

There is considerable variability among the public in prior beliefs about the trustworthiness and competence of officials (Berman, Wandersman 1990). In addition, perceptions about the trustworthiness of various government agencies and beliefs about the risk management process are influenced not only by objective considerations but are also subject to the socio-cultural background of the recipients of the information provided. The fact, that communities and also individuals vary in social and risk histories, have different access to economic and other resources which might be relevant for dealing with a particular risk (e. g. cancer treatment) and could have diverging intuitive theories about e. g. toxicology and the relationship between exposure to substances and health outcomes, can have a considerable influence on the level of trust in those institutions and government agencies responsible for the protection of the environment and public health (Baker 1990). These socio-cultural aspects have been ignored at the inception of research on risk communication. Also the policy process has rather tended to focus on the quantified characteristics of a risk or hazard instead of taking the surrounding social and cultural dynamics into account, which do operate within a community (Vaughan 1995).

Trust includes a perception that the communication is technically competent, fair, honest, consistent and benevolent. Trust can be eroded by factors such as incompetence, poor performance, incomplete information, withholding of information, denial of obvious problems and denial of vested interests.

In order to enhance public confidence and trust it is essential to communicate both, negative and positive messages even if public attention is usually devoted more to messages with a negative content. In this context the studies of Siegrist and Cvetkovich (2001) are interesting to note. They investigated how the results of a scientific study influenced confidence in the study's validity and the resulting perceived risk. It could be demonstrated that prior positive or negative information has different effects on how subsequent information is interpreted. People appear to have more confidence in studies with negative outcomes than in studies showing no risks. This perception might be attributed to the fact that positive information usually reports on a normal course of activities, which is taken for granted anyway, whereas negative information mostly refers to single or multiple events which are a deviation from normal and thus not only noteworthy but also raising concerns and even fear (Slovic 1993). As practical implication this might indicate that communications about potential risks appear to be more influential than communications informing people not to worry about a hazard. In practice, this could mean that statements giving assurance that there is no harm to be expected might be counterproductive in the communication process.

Whether the communication is spontaneous or upon request, or whether the message to be communicated is positive or negative, it is essential that honest and trustworthy information is given. Sufficient openness and willingness to answer questions should complete the communication. This may be effective in preventing rumours and wrong or fuzzy information and will certainly contribute to a better perception of the message. The image of a communicator to be open about his own background, interests and convictions, and to relay the message in a context which is clearly not self-serving is essential.

A second important attribute placed by the audience onto the communicator and the institution or agency represented by the audience is the perception of the credibility of this information source. Credibility, traditionally defined as a combination of perceived "expertise" and "trustworthiness", is a crucial predictor of message success. Simply put, credible sources will be believed while non-credible sources will not. And once a source has been deemed "not credible", it may take a long time for any sense of trust to return regardless of the source's efforts.

Two issues, which are relevant to understand what underlies public credibility judgments, will be reflected on briefly: the futility of credibility and the possibility that the public can consider a source as credible only for a particular type of information but less credible with regard to other statements. Confidence and trust can easily be lost but are hard to gain. Once undermined, confidence is hard to restore. Building up trust may take quite a long time, however, one misunderstood message may ruin the efforts of several years. An explanation of the fragility and even futility of credibility judgments in risk perception comes from the sociologist Freudenburg (1988). He argues that, as the world becomes more complex, individuals can no longer make direct judgments about evidence. Instead, they are increasingly forced to trust the judgments of others, such as experts or spokespersons. In a world so dependent on trust, he says, individuals tend to adopt a very conservative standard and all it takes is one mistake for trust to evaporate. Thus, even a minor accident in a facility can result in a serious loss of credibility, if the accident is interpreted by the public as a violation of the trust placed in the management to ensure accident-free operation without disturbance. The operator should have prepared those most likely to be exposed with appropriate information about the likelihood of such a small accident and should also have given reassurance that all provisions to keep the consequences as low as possible have been implemented.

While credibility has been mostly viewed as an all-or-nothing judgment, there is some evidence that individuals are capable of making credibility distinctions about the same source but across different types of information. Many types of information may be available in the course of a risk discussion, from information about the regulatory environment to recommendations about appropriate risk-reduction behaviors for individuals, from information about the scientific background of a risk to suggestions about the possible legality of various options. Individuals are quite ready to evaluate a single source as not credible for some types of information but as credible for others (Jungermann, Pfister, Fischer 1996). For example, an agency dealing with nuclear power plants may be considered a less credible source of information about the benefits of nuclear power, since it is assumed to be supportive of the energy source, ("What else would they say?") but may be viewed as a credible source of regulatory information. This should be taken into account when selecting a particular communicator to disseminate or present a specific message.

9.3.1.2
Communication skills

The assumption that technical expertise is the primary prerequisite for good communication is still widely held. Most scientists are convinced that as long as the technical knowledge is there, dissemination of information to any audience should be adequately possible. Technical knowledge is indeed one of the important attributes for a good communicator, but this is not sufficient. Good communicators need to possess a number of additional qualities such as obtain information about the audience to be expected, possess narrative skills that can be tuned to that audience's needs, and have the ability to explain complex concepts and processes in everyday language. Failing to anticipate the needs and predispositions of the audience is one of the main problems, which an inexperienced communicator can encounter.

Audiences may bring a variety of "naïve theories", which are based on their understanding and experience of the world around them, to a discussion of a technical issue. These theories appear to work well in daily life but that can be actually in direct conflict with a more scientific understanding. If a communicator cannot anticipate and address those naïve theories in ways that will encourage the audience to set them aside, they may make it impossible for the audience to understand or accept the communicator's message (Rowan 1988) and in addition, the expert will have lost the trust which an audience is normally willing to grant at the beginning. For example, information about the movement of contaminants through ground water may fail to get across, simply because the audience is thinking of ground water as similar to an underground river whereas the communicator is referring to a slow filter-type process. Another example could be when an attempt is made to relate any information on radiation to an audience. The free associations which can be expected from a large part of this audience, when they hear the term 'radiation', will have negative connotations. The first ones coming to mind would be referring to their own well being, such as: bodily harm, sickness, cancer, burns, death. Subsequently, more general concerns would evolve, such as: dangerous, harmful, environmental hazard, bombs, fallout. To offset this negative cognitive setting of the audience, the communication would have to start first with some information on natural background radiation and medical benefits, such as X-rays and cancer treatment. Dispatching negative notions and replacing them with more positive "images" should help members of the audience to grasp the content of the message to be given.

9.3.2
To whom (Audience)

Since communication is a process which involves two parties, any communicator needs a receptor for the information to be given. This receptor is commonly referred to as the "audience" or more selectively as the "target audience". In risk communication it is often the "general public" that is considered to be the target audience of risk messages. However, experience has shown that the general public is much too heterogeneous to be meaningfully addressed by a single type of risk message. If risk messages are to be conveyed successfully it is more useful and promising to address specific segments of the public instead of the public in general, especially since the information and the message to be given will have to be geared towards the level of knowledge which could be anticipated. Some suggestions are also provided for specific channels to be used for addressing these audiences.

The specific segments of the public, which are quite clearly distinguishable from each other, could be divided and described in the following manner:

– Pupils and teenagers: Current public opinion on many risk related issues is, it seems, caused mainly by insufficient education. In particular, the Internet is a popular medium for young people nowadays, and special websites could address particular areas.
– Students: To ensure that the proposed education on the basic level will continue to the upper level, workshops on environmental issues and risk assessment techniques including the interpretation of results might be offered to various disciplines. As for the very young people, the Internet is also a very popular medium for students. Thus, appropriate information should be offered.

- Teachers: For the educational actions to be effective, it is necessary to first inform the teachers. This may be done by training, by distributing educational materials, such as documents written on different levels, videos and films, CD-ROMs, as well as guidance about websites.
- Academic world: The academic world has a specific role because these people are seen as having knowledge and they have credibility. Communication with them may be efficient by means of collaborations in conferences, visits and seminars, as well as in agreements for research and development. Particular attention should be given to the area of the social sciences, which may present a different view of technical issues.There could be more participation from the academic world using their expertise, offering membership in committees or organisations, and also inviting participation in the evaluation of technical developments.
- Medical professionals: Since medical professionals are considered as very credible by the majority of the population, and as it is expected that they have knowledge about radiation effects, it is advisable to particularly consider their input. Apart from the fact, that especially general practitioners are in direct contact with their patients, who relate their anxieties and fears to their doctors, it is the same doctors prescribing examinations and treatments involving radiation. Governmental organisations should offer educational courses, training and postgraduate education. It is also recommended to promote a re-evaluation of knowledge for those who are actually practicing nuclear medicine and to provide them with related materials, which are up to date.
- Politicians, decision-makers: It is crucial to provide especially politicians with understandable information since they are the ones answerable to the public.
- Media, journalists: Evidently the widest communication is achieved by mass media. However, risk mainly becomes an interesting topic for the media in case of an accident or disaster. But in journalism, stories not necessarily reflect truth and reality, just as people assimilate information in their personal way, which does not necessarily reflect the given facts. Thus, even though good relationships with journalist should be maintained, expectations to receive adequate coverage in the media might not be honoured. Scientific journalism plays a somewhat special role as these journalists are usually better informed and more objective. Their articles do not, however, normally reach a large readership.
- Associations: Various associations, in particular environmentally oriented ones, have an increasing influence in many countries. They have a clearly specified goal and are reluctant to accept information, which is not in line with or might even be contradictory to the principles they have subscribed to. Therefore, depending on their orientation, communication is in most instances difficult.
- General Public: An efficient way to approach the public in general is the Internet, providing not only information but also the opportunity for on-line debates. The distribution of local news magazines with information about possible environmental impacts might supplement the information on offer.
- Women appear to have a special sensitivity to what they perceive as risks, even if they originate from the low-dose range. When gender differences relating to the perception of risks were studied it could be demonstrated that women are generally more concerned, particularly with regard to nuclear technologies, pollution

and risk to health from man-made sources (Davidson and Freudenberg 1996). Thus, information particularly geared towards a female audience should be made available.

9.3.2.1
Information processing

A major distinction has to be made between systematic (or central) and heuristic (or peripheral) processing of information or how a person is prepared to deal with the facts presented. Systematic information processing refers to a thoughtful and careful consideration of an information. Here an individual is willing to really think about an issue, its meaning and implications. Information that is processed in this manner is likely to have a long lasting impact on the individual. Heuristic information processing, on the other hand, is done in a more superficial way, without devoting much thought, effort and time. Instead of considering the content of a message, the individual uses "external" cues (such as length of a message, the "image" of the communicator, etc.) to evaluate the message. As a consequence, the details of the information will not be remembered, the message as such might even be misinterpreted and the impact of information that has been processed heuristically is likely to be unstable and short term.

The key variables that decide whether information is processed systematically or heuristically appear to be the motivation to deal with the information and the ability to understand the information. (Chaiken 1980; Petty and Cacioppo 1986). Only if a person is both motivated and capable of understanding the information will he/she process the information in a systematic manner. Risk communication, no matter whether it is intended as persuasion or information, succeeds to the extent that it prompts systematic information processing. It follows that risk communication has to be tailored to the recipients and to their motivational and cognitive levels. This requires that, before risk communication efforts are made, the motivational and cognitive status of the target audience has to be investigated. Once this has been done, one can tailor the information in a way that fits the (probably various) cognitive levels of the recipients. More problematic are the motivational aspects. Although one will usually be able to find people who are very interested in a risk issue and thus highly motivated to receive information about it, these are typically persons who already have spent time and effort to collect additional information about the issue in question and are therefore often well informed about it. In contrast, those who have not developed an interest in the past will typically also have only a limited amount of knowledge about the risk. Since those are not motivated to listen or participate it would be difficult to capture their attention.

9.3.2.2
Intuitive risk perception

Dimensions of intuitive risk perception. While technical experts, in their assessment of risk, mainly focus on two parameters, namely probability of damage and severity of outcomes, the intuitive risk judgments of lay people are both, simpler and more complex. Intuitive risk perception is simple in that it usually does not rely on technical and/or quantitative aspects of risk, such as the result of a probability calculation. If lay people are required to give probability estimates, they will use a more simple way of thinking. People may judge the probability of an event according to the ease with which this event is accessible in their personal memory: the easier one can remember that, e. g. a certain accident has happened, the more probable the event is judged. But it has been demonstrated, that many people process information at the time when they are exposed to it, update their opinion or judgment accordingly and then tend to forget the detailed information and only retain some summary judgment (Lodge et al. 1995). This over-reliance on memory as opposed to objective facts has been termed 'availability heuristic' (Tversky and Kahneman 1974). While simplifying heuristics like availability, i. e. accessible in memory, will be useful most of the time in everyday life, they might lead to misperceptions under certain circumstances. Today most risks are not experienced directly but via the mass media. Mass media, however, often focus on spectacular events such as catastrophes or accidents that nevertheless may be very rare while common causes of death or injuries (e. g. coronary heart diseases or household accidents) are hardly reported. This mechanism is considered as a reason why the probability of rare but spectacular risks are often overestimated by lay people while the more common everyday risks are underestimated (Lichtenstein et al. 1978). One should note, however, that while lay people's probability estimates are often biased in their magnitude, they are nevertheless pretty capable of giving an appropriate rank order for the probabilities of events. That is, they may overestimate the absolute number of rare (but spectacular) events and underestimate the absolute number of more frequent (but ordinary) events, but they have a reasonably good sense of how these events are related to each other in terms of frequency (Boholm 1998).

A perhaps more important characteristic of intuitive risk judgments is that they are usually 'holistic'; which means that lay people do not evaluate aspects of risk separately (e. g. frequency, probability, severity) and then integrate their total evaluation into an overall risk judgment. Rather, individuals tend to make summary judgments that include a variety of risk dimensions or qualitative factors.

Specifically, this means that lay people seem to build more aspects of risk into their judgments than are contained in the technical concept of risk. Psychological research on risk perception, which started already some twenty years ago in the USA (Slovic et al. 1981), involved a variety of subgroups in society (Slovic 1987) and was later extended to research in Great Britain, the Netherlands, Austria and Germany (Streffer et al. 2000), has found that the following aspects might affect intuitive risk judgments:

– whether a risk is taken voluntarily or involuntarily,
– whether it is seen as personally controllable or uncontrollable,
– whether the risk can be perceived by human senses,
– the extent to which institutions held responsible for the risk are in control,
– the extent to which the risk is seen as understood by science,
– whether the information given about this risk is clear and understandable,
– whether one is familiar or unfamiliar with the risk,
– whether it is catastrophic or not,
– whether it threatens future generations or not,
– whether one feels personally affected by the risk or not,

- whether one had previous experiences with man-made and natural risks,
- whether the associated consequences (damages) are considered to be reversible or not,
- whether a risk is seen as particularly dreadful (e. g. a lingering death caused by lung cancer, as opposed to a sudden demise caused by an automobile accident),
- whether somebody can be blamed for the adverse consequences or not (which includes the often mentioned aspect whether a risk is considered to be natural or man-made),
- whether there are any benefits associated with the risky technology, activity, etc.,
- the likelihood that the risk could damage non-human organisms and the environment,
- if the source of the given information is considered trustworthy.

One should note, however, that while all these aspects may influence risk perception in general, it is hard to predict which will be relevant in a specific situation and/or for a particular person (or group of persons).

Nevertheless, the important lesson from this research is that what is considered relevant for a risk evaluation from a lay perspective is likely to differ from the criteria used by the expert. This implies that, before communicating about a risk, it will be important to understand how a particular target audience for risk communication actually perceives that risk, i. e. which aspects are considered to be relevant for evaluating the risk in this group or for this part of society.

Beliefs. Prior beliefs about a risk or control over an outcome are persistent and resistant to change. These a priori belief systems can have a significant influence over the risk communication process because novel information will be filtered and processed relative to pre-existing impressions and convictions.

In the media, conflicting opinions about the validity and meaning of scientific risk data are highlighted and disagreements about risk management strategies are featured. New information may have little effect on established judgments because it may be selectively evaluated relative to its consistency or inconsistency with previous beliefs.

In the previous section the multidimensional structure of intuitive risk perception was discussed. Another important issue to consider in the context of risk communication is that risk perception typically is embedded in a belief system that includes the subjective knowledge an individual has about a certain risk, how that risk is related to other relevant aspects, such as who is responsible for the risk or what can be done to mitigate it. This belief system may include some of the risk dimensions mentioned above but will also go beyond this list in important ways. Two aspects are of particular relevance here.

A person's beliefs about a particular risk may not be realistic from an expert's point of view, but those beliefs nevertheless puts the risks in question into a context. These beliefs may provide a mechanism to explain why a risk should be tolerated or rejected. Even though realistic perception should be expected for those types of risk with which people have had some personal experience, direct or indirect through friends or close family members, it has been demonstrated that people have a tendency to even underestimate exactly those risks (Weinstein 1989; Sjöberg 2000).

A useful way to think about individuals' beliefs about how unfamiliar risks operate is to regard them as naïve theories. These theories about a certain risk may be right in some regards and wrong in others. It is important to identify those elements of a naïve theory that are wrong and that lead the individual to false conclusions. Research has identified a number of false risk beliefs that seem to be held by many lay people (Kraus et al. 1992). For example, one common element of naïve theories about risk is that they seem oblivious to the level of exposure a person has to a noxious agent. While level of exposure plays a crucial role in scientific risk assessment, lay people may assume that contact with that agent may produce adverse health effects, regardless of level of exposure. Obviously this reasoning may lead to an overestimation of risk. Another area of public concern is e. g. the handling of radioactive waste and its deposition in underground depositories. While experts from the nuclear industry view the risks of such a site being predominately related to political instability, i. e. a changing political composition might jeopardise the safety regime of the depository and other socio-economic constraints, the sample from the public had their images concentrated on danger, toxicity, death and environmental damage (Flynn et al. 1993). Thus, individuals who are not aware of the technical solution available may be unnecessarily worried.

If the beliefs of a person or a group of citizens about a particular risk should be changed, the associated beliefs (i. e. the belief system) have to be taken into account and information that can be integrated into those associated beliefs needs to be provided. Beliefs are hard to change if they are closely linked to many other aspects in a belief system, for example, if they are a central and integral part of a person's belief system. Beliefs are much easier to change if they are rather isolated elements with only a distant relationship to other elements of a belief system.

Another problem that often makes it difficult to change beliefs is that what some might regard as a misconception may loom for others as a controversial topic, for example the problem of nuclear waste repositories. In these cases it is hard to convince an audience that their beliefs are wrong by referring to scientific evidence (as there is no unambiguous evidence). All that can be done is to explain the (general and case-specific) technical approaches. This, of course, requires the readiness (both cognitive and motivational) of the audience to deal with these rather complicated matters.

In principle, people are certainly capable of processing information and accessing information sources to make up for lack of factual knowledge, but there is little indication that large numbers of people use these strategies effectively, regularly and across a wide range of situations and issues (Althaus 1998). There will often be a relationship between the motivation and the ability of a person to search for and process information carefully and the 'quality' of his/her belief system. That is, those people who are willing and able to process carefully tend to be those who already have built up a rather detailed belief system about an issue.

Audience reliance on information channels. The dissemination of information, and particularly specific information such as messages on risk issues, is to a large extent influenced by the tendency of communicators to select those channels for the risk related messages, with which they are familiar or that have proved to be easy or efficient to use. This means that communicators of more scientific information put most of their messages in printed form; these articles, brochures, fact sheets then get disseminated to potentially interested parties.

There are a number of reservations regarding this strategy. One important issue, already discussed, is that information, which is received by a person who has no particular motivation to read it, is wasted information. Equally important is that information arriving via the wrong channel can often be ignored too. There is increasing evidence that individuals prefer to use different information channels to establish their knowledge of a risk versus their judgment of what to do about that risk. Specifically, individuals seem to prefer to use "mediated" channels, i. e. those that serve as intermediaries between source and audience, such as newspapers, television, magazines and radio, to learn about a risk but 'interpersonal' channels to develop their personal judgments about the risk.

This distinction between learning and doing may seem artificial, but it is actually a distinction between "others" and "self", whereby the former relates to the public in general, the latter, however, includes not only the person in question but also family members. Risk perception research has demonstrated that a clear distinction is made between "my personal likelihood of coming to harm" and "everyone else's likelihood of coming to harm" (Sjöberg 2000). As a consequence, people regard their own potential exposure to risk and the ensuing consequences much lower or less likely than what they perceive could happen to the public at large. Communication researchers have now discovered that this distinction between self and others influences channel choices, too. Individuals tend to interpret risk stories in mediated channels as offering information about a risk generally, but they resist seeing that information as relevant to them personally. Rather, they seem to require the intervention of another person to make them realise that the risk information might concern them personally.

9.3.3
What in which channel (risk message)

9.3.3.1
Types of risk information

Depending on the goal, such as persuasion or information, of a risk communication effort, a risk message will consist of different types of risk information. On a general level one may distinguish factual information about a risk from evaluative information.

Factual risk information is concerned with what is known about what harm can result from a technology, activity etc. and how likely that is to happen. Technically speaking, factual risk information is about possible adverse consequences and the associated probabilities. While it is usually not difficult to inform about the type and scope of consequences, it turns out that informing people about probabilities, particularly very small probabilities, is much harder. This is not surprising because, firstly, understanding the meaning of probabilities in a specific risk context often requires some knowledge of probability calculus to see how these numbers are derived, and secondly, the often very small probabilities presented in risk assessments are beyond the personal experience and imagination of human beings. Thus, research has shown that lay people often have difficulties in interpreting low probabilities (Camerer and Kunreuther 1989). Of course, an important variable here is the level of education about probabilities the target audience already possesses. In general, however, it is probably wise to assume less knowledge rather than more.

Although one can facilitate the understanding of probabilities by using graphics or analogies, little is known so far about which means are best applied under what conditions. Analogies especially carry the risk of upsetting people rather than reassuring them when applied inappropriately.

However, it bears repeating that even this factual risk information is not objective but also includes value judgments (see Otway and von Winterfeldt 1992). For instance, which consequences to consider within a risk assessment is not simply "given", but defined by experts, usually scientists or engineers. Should one consider only the (expected) mortalities or should one also include (expected) morbidities? In either case, which time span should be taken into account? And what about material or environmental damages? While these value judgments are not problematic *per se* (and cannot be avoided anyway), in the context of risk communication it is important to bear in mind that values play a role even at this factual level and that the target audience might have different values.

Often, comparative risk information is not limited to providing facts but also includes evaluative statements. This is, of course, particularly true for persuasive risk communication. For example, the information might state that the risks associated with system A are in the same order of magnitude as the risks of system B; and, since B (and, therefore, its risks) is already used in the society, A (and its risks) should also be accepted. While this may sound reasonable at least at first glance, there are at least two problems associated with this line of argumentation. First, as has been noted by Fischhoff (1983), tolerating the risks of a system is not the same as accepting them. A system may be a long standing component of a society, perhaps established even before society really became aware of the risks, and thus, may be difficult to remove. These risks have come to be tolerated. However, deciding about the implementation of a new, though equally risky, system is different in that one has the opportunity to reject the risk. That is, acceptance is an active act that requires a choice, while tolerance often is passive and without a choice dimension.

More important, however, is the notion that while two systems might be equal with regard to some technical concept of risk, they nevertheless may differ in other important ways. Here, it should suffice to note, that evaluative risk information is particularly prone to go astray if those value issues are not taken into account.

9.3.3.2
The importance of frames

A frame is basically an interpretative framework within which the information is placed. It can exist at the level of words and phrases, e. g. explaining a risk in terms of probability of dying rather than in terms of probability of surviving, or at the level of an entire narrative, e. g. the first few paragraphs of a newspaper story provide an explanation or frame to which all later facts ostensibly contribute.

The framing of a (risk) problem refers to how the issue to be addressed is conceptualised, defined or structured – a basic and primary step in the risk communication process. Frames represent the perspectives adopted and in controversies predispose the individual to see or emphasise certain aspects of the situation. The framing of a risk issue influences how risk communications are filtered or processed and the eventual choices or preferences expressed.

To communicate effectively, as a preliminary step, public concerns must be assessed to appropriately frame the risk issues. Differences in the framing of a risk issue can be a source of persistent disagreements and may preclude ever reaching a viable solution.

Decision or problem frames can result from the way in which risk information is presented but more often are derived from past experience, values and the socio-cultural context within which risk responses evolve. Framing differences can also be attributed to hidden agendas or self-interests.

If government officials and risk communicators frame an issue in a way that differs from community groups, they may be neither prepared for nor more able to adequately answer questions that citizen's view as being fundamental. For instance, the risk characterisation presented to lay populations usually reflects aggregate estimates derived from a consideration of risk across the entire exposed population. Traditional assessment may be perceived as ignoring or minimising the interests of vulnerable populations and their presentation viewed suspiciously by community members. Particularly when health and safety concerns are prevalent, questions about exposures often are geared towards especially vulnerable populations such as children.

As much as there is a frame in place with the receptor of the information, there is also a frame in existence from the side of the communicator. Most communicators "package" information in some way, understanding intuitively that 'raw' information, for instance elaborate statistical data, is meaningless to communicate. That packaging or framing can have a powerful effect on how the raw information will be interpreted by audiences.

Frames are potentially powerful because people seem to have an enduring tendency to seek explanations for things that occur in the world. The typical frame offers an explanation, thus permitting to facilitate some understanding of the facts that follow. Three points can be made here:

1. There is empirical evidence that frames do influence understanding. It has been demonstrated that, when confronted with novel information, we readily adopt the interpretative framework offered in concert with that information. For example, one experimenter asked individuals to read stories about poverty (Iyengar 1991). Some of the stories referred to poverty as stemming from individual limitations: Folks were poor, because they were ill, lazy or unlucky. Other stories presented the problem as a structural one: Individuals at the bottom of the social structure were rarely permitted to rise higher; they were thus doomed to a subsistence life regardless of their individual capabilities. When queried later, subjects' explanations for homelessness faithfully reflected the frames to which they had been exposed. This tacid acceptance of frames only works in the absence of strong beliefs. Individuals who already have relevant beliefs about an issue will interpret the information in ways that reinforce those beliefs.

2. When confronted with an issue, there is a tendency to merely accommodate one frame at a time, making the first available frame the powerful one. In other words, once an explanatory framework for an issue has been constructed, that framework prevails. This gives a strong interpretative boost to the frame that emerges before others and it explains the tendency of individuals to rush to express themselves regarding an issue. Their message may not be well consid-

ered and thought of, but, since it was the first, this explanatory framework also will most likely endure.

3. Once established, a frame is difficult to change. Cognitive psychology reiterates that it is human nature to cling to an established explanation, sometimes against all odds. Evidence that supports the frame is preferred and contradictory evidence is either ignored or reconfigured to fit.

Only if the catalyst is very strong, on the verge of being catastrophic, frames can be susceptible to change. One example may be the shift in some countries within the last 30 years from considering nuclear power as an acceptable technology to the process of phasing out this technology. In the beginning, journalists and the public faithfully followed the scientists' and engineers' description of the technology as efficient, affordable, and forward-looking in the 1950s and 1960s. But the emergence of counter frames that accompanied the environmental movement of the 1970s, the incident at Three Mile Island in 1979 and the dramatic accident at Chernobyl in 1986 effectively brought about a framing shift. The number of opponents of nuclear power increased, but mostly in those countries with high levels of radiation exposure. When analysing public responses in Europe (Renn 1990) the ratio of opponents to total respondents had increased in Finland, Yugoslavia and Greece by over 30 %, in Austria, West Germany and Italy by over 20% and in the UK, France, the Netherlands, Sweden and Spain by 12–18 %. A similar result was obtained in Japan after a criticality accident occurred at a uranium processing plant in Tokai Mura in 1999 (Tsuneda 2001). The author of this study also noted that there is a large group of respondents which can be described as neutral, i. e. not having a strong opinion. After the accident, however, neutrals were more likely to move to the opposition side than supporters of nuclear power were likely to become neutrals. But this effect would also work in the opposite way, where those, who are located in the neutral area, might, through a small shift already, generate a more positive overall public perception for a given issue.

Once frames have been developed, they generally are so enduring that individuals seem willing to fit prevailing frames to relatively new phenomena rather than construct new interpretations for those phenomena. This may explain, for example, why some newer, clearly beneficial uses of radiation – the diagnostic procedure called nuclear magnetic resonance imaging and the process of irradiating food to cut down on spoilage and decay – initially suffered from negative frames. The magnetic resonance imaging community responded by dropping 'nuclear' from its label and those promoting food irradiation are seeking to rename that process as well to create distance from the negative nuclear power frame. The success of this process will depend on two important aspects:

Time: In the United States and a number of other countries, it has taken literally decades to promote a shift in the frame of smoking as socially acceptable to one of smoking as negative.

An acceptable substitute frame: Frames are influential in part because they supply a needed cause-and-effect explanation for some process going on in the world. In order to eliminate one frame from cultural memory, it would have to be replaced with an equally powerful frame, but that may be rather difficult to do. For example, when citizens claim that a cluster of cancers must reflect some kind of underlying environmental cause, scientists are quite unsuccessful at replacing that frame with one that offers chance as the most likely explanation. Individuals with limited numerical skills may have reservations in accepting a probability or chance as an explanation, thus rendering the scientists' explanation ineffective and certainly not too convincing.

9.3.4
With what effects?

Consistent patterns of information effects are difficult to identify for several reasons:

1. Studies vary widely in quality and rarely try to replicate earlier findings. This makes comparability difficult. For example, studies that purport to look at the impact of messages on individuals' risk perceptions may actually be using very different dependent variables. Those that measure perception in terms of a person's estimated likelihood of coming to harm will get different responses than will studies that operationalise risk perception as a person's level of concern. Similarly, studies have often confused estimates of personal risk with general estimates of risk, two very different measures. In addition, follow up efforts can seriously be hampered by demographic constraints, i. e. same respondents are not available. For comparisons between studies, the socio-cultural element is an inhibitive factor, since this has a major influence on the perception of most issues in a population.
2. Many information communicators are interested in opinion or behavior change (these are the goals of the typical information campaign, for example). However, among the possible dependent variables – awareness, knowledge gain, opinion change, behavior change – the two *least* likely to be affected directly by information interventions are the last two. There is some suggestion that information acts directly on awareness and knowledge gain but indirectly on opinion and behavior changes. If so, then the typical study on effects will rarely demonstrate a direct link between message and either opinion or behavior change. Thus the impact of a message would rather be found in an increased awareness and/or knowledge of the issue in question.
3. Finally, the effects of information exposure, while strong, can be counterintuitive. Many communicators rather simplistically assume that provision of information will result in opinion or behavior change consistent with their goals. Some critics call this the "audience-as-blank-slate" approach, for it ignores the information brought to the message by audience members themselves. In fact, the most common effect of new information is not change but increased stability and/or intensity of existing beliefs: Information tends to reinforce existing patterns of beliefs, whatever those beliefs may be. For a communicator, who has been trying to influence a given set of beliefs by providing new information, this effect is quite frustrating, since instead of an observable impact he gets the impression that the communication had no effect at all.

Conceptualisations of the risk communication process have analysed factors influencing the form and effectiveness of communication, but have failed to include the social contexts within which individuals adapt to risk and information exchange. The more traditional approaches neglect cultural aspects, social experiences and motivations that could exacerbate conflict or facilitate communication. Risk communications should be conceptualised as being embedded and shaped within settings that vary in terms of salient cultural themes, values, norms and other socio-cultural features.

9.3.5
References

Althaus SL (1998) Information effects in collective preferences. Am Poli Sci Rev92: 545–558

Ansel P (1997) Democratie, scene mediatique et mesure des opinions. Le cas particulier du nucle-aire. In: Actes Colloque Atome et Societé. Arak Publications, Paris, pp 113–124

Baker F (1990) Risk communication about environmental hazards. J Public Health Pol 24: 341–359

Balzano Q, Sheppard AR (2002) The influence of the precautionary principle on science based decision making: questionable application to risks of radiofrequency fields. J Risk Res 5(4): 351–369

Berman SH, Wandersman A (1990) Fear of cancer and knowledge of cancer: a review and proposed relevance to hazardous waste sites. Soc Sci Med 24: 35–47

Bodansky D (1991) Scientific uncertainty and the precautionary principle. Environment 33(7): 4–5, 43 45

Boholm A (1998) Comparative studies of risk perception: a review of twenty years of research. J Risk Res 1(2): 135–163

Camerer CF, Kunreuther H (1989) Decision processes for low probability events: policy implications. J Policy Anal Manag 8: 565–592

Chaiken S (1980) Heuristic versus systematic information processing and the use of source versus message cues in persuasion. J Pers Soc Psychol 39: 752–766

Chiosila I (1996) Nuclear power in Romania, potential risks for environment and public health. Regional Center of Environment Protection for Central and Eastern Europe, Bucharest

Covello VT, Merkhofer MW (1994) Risk assessment methods: approaches for assessing health and environmental risks. Plenum Press, New York, London

Cross F (1994) The public role in risk control. Envir L 24: 887, 930–933

Davidson DJ, Freudenburg WR (1996) Gender and environmental risk concerns: a review and analysis of available research. Environ Behav 28: 302–339

Davies JC (ed) (1996) Comparing environmental risks: tools for setting government priorities. Resources for the Future, Washington, DC

European Commission (2000) Communication from the Commission on the precautionary principle. COM (2000) 1. Brussels

Fischhoff B (1983) Acceptable risk: the case of nuclear power. J Policy Anal Manag 2: 559–575

Flynn J, Slovic P, Mertz CK (1993) Decidedly different: expert and public views of risks from a radioactive waste repository. Risk Anal 13: 643–648

Freudenburg WR (1988) Perceived risk, real risk: social science and the art of probabilistic risk assessment. Science 242: 44–49

Fritzsche AF (1994) Strahlenrisiken im Vergleich mit anderen Risiken. Atomwirtschaft

Gheorghe AV (1996) The role of risk assessment in obtaining technical information for emergency preparedness and planning due to major industrial accidents: views from a UN international project. Int J Environ Pollut 6(4–6): 604–617

International Atomic Energy Agency [IAEA] (1992) Methods for comparative risk assessment of different energy sources. IAEA-TECDOC-671, Vienna

International Atomic Energy Agency [IAEA] (1995) Principles and recommendations for the integrated management of technological risks. IAEA Working Material, CT-2436 (limited distribution), Vienna

International Atomic Energy Agency [IAEA] (1998a) Data and comparisons of accident risks in different energy systems. IAEA Draft Working Material (limited distribution), Vienna

International Atomic Energy Agency [IAEA] (1998b) Guidelines for integrated risk assessment and management in large industrial areas. IAEA-TECDOC-994, Vienna

International Atomic Energy Agency [IAEA] (1999a) Estimating and comparing risks from very low levels of exposure resulting from emissions from energy systems. IAEA Working Material (limited Distribution), Vienna

International Atomic Energy Agency (1999b) Basic principles for communicating risk to the public: a focus on risk comparisons. Working Material (Limited Distribution), Vienna

International Atomic Energy Agency [IAEA], Food and Agriculture Organisation of the United Nations [FAO], International Labour Organisation [ILO], Organisation for Economic Co-operation and Development [OECD] – Nuclear Energy Agency, World Health Organisation [WHO], Pan American Health Organisation [PAHO] (1996) International basic safety standards for protection against ionising radiation and for the safety of radiation sources (BSS). Safety Series 115, IAEA, Vienna

Iyengar S (1991) Is anyone responsible? How television frames political issues. University of Chicago Press, Chicago, IL

Jungermann H, Pfister HR, Fischer K (1996) Credibility, information preferences, and information interests. Risk Anal 16: 251–261

Kraus N, Malmfors T, Slovic P (1992) Intuitive toxicology: expert and lay judgments of chemical risks. Risk Anal 12: 215–232

Lasswell H (1948) The structure and function of communication in society. In: Bryson L (ed) The communication of ideas: a series of addresses. Harper, New York, pp 37–51

Lichtenstein S, Slovic P, Fischhoff B, Layman M, Combs B (1978) Judged frequency of lethal events. J Exp Psychol – Hum L 4: 551–578

Lodge M, Steenbergen MR, Brau S (1995) The responsive voter: campaign information and the dynamics of candidate evaluation. Am Poli Sci Rev 89: 309–326

Otway HJ, Von Winterfeldt D (1992) Expert judgment in risk analysis and management: process, context and pitfalls. Risk Anal 12(1): 83–93

Page T (1978) A generic view of toxic chemicals and similar risks. Ecol Law Quart 7: 207–244

Petty RE, Cacioppo JT (1986) Communication and persuasion. Central and peripheral routes to attitude change. Springer, New York

Renn O (1990) Risk perception and risk management: a review. Risk Abstracts 7 (1): 1–9 (Part 1) and 7 (2): 1–9 (Part 2)

Renn O (1998) Three decades of risk research: accomplishments and new challenges, J Risk Res 1(1): 49–71

Risiko-Kommission (2002) Ad-Hoc-Kommission "Neuordnung der Verfahren und Strukturen von Risikobewertung und Standardsetzung im gesundheitlichen Umweltschutz in der Bundesrepublik Deutschland". Erster Bericht über die Arbeit der Risiko-Kommission. Bundesministerium für Umwelt, Reaktorsicherheit und Naturschutz, Bonn

Rowan K (1988) A contemporary theory of explanatory writing. Writ Commun 5(1): 23–56

Sadowitz M, Graham JD (1995) A survey of residual cancer risks permitted by health, safety and environmental policy. Risk: Health, Safety & Environment 6(1): 17–35

Scholz R (1996) Das Bund-Länder-Verhältnis am Beispiel Kernenergie. Energiewirtschaftliche Tagesfragen 46(6): 386–390

Shrader-Frechette KS (1995) Evaluating the expertise of experts. Risk: Health, Safety & Environment 6(2): 115

Siegrist M, Cvetkovich G (2001) Better negative than positive? Evidence of a bias for negative information about possible health dangers. Risk Anal 21(1): 199–206

Sjöberg L (2000) Factors in risk perception. Risk Anal 20: 1–11

Slovic P (1987) Perception of risk. Science 236: 280–285

Slovic P, Fischhoff B, Lichtenstein S (1981) Perceived risk: psychological factors and social implications. In: Warner F, Slater DH (eds) The assessment and perception of risk. Royal Society, London, pp 17–34

Stoll W (1997) Kernenergie in Deutschland. 25 Jahre Widerstand – die Folgen. Energiewirtschaftliche Tagesfragen 47(11): 666–671

Streffer C, Bücker J, Cansier A, Cansier D, Gethmann CF, Guderian R, Hanekamp G, Henschler D, Pöch G, Rehbinder E, Renn O, Slesina M, Wuttke K (2000) Umweltstandards: kombinierte Expositionen und ihre Auswirkungen auf die Umwelt. Wissenschaftsethik und Technikfolgenbeurteilung, Bd. 5. Springer, Berlin

Tanguy P (1997) Sureté nucléaire et débat public. In: Acte Colloque Atome et Societé. Arak Publications, Paris, pp125–129

Tsunoda K (2001) Public response to the Tokai nuclear accident. Risk Anal 21: 1039–1046

Tversky A, Kahneman D (1974) Judgement under uncertainty: heuristics and biases. Science 185: 1124–1131

UK Health and Safety Executive (1989) QRA: its input into decision making. HMSO, London

Vaughan E (1995) The significance of socioeconomic and ethnic diversity for the risk communication process. Risk Anal 15: 169–180

Weinstein ND (1989) Optimistic biases about personal risks. Science 246: 1232–1233

Wiener JB, Rogers MD (2002) Comparing precaution in the United States and Europe. J Risk Res 5(4): 317–349

10 Low Doses in Health-Related Environmental Law

10.1
Introduction: the legal framework

From a legal perspective the protection of human life and health against possible adverse effects of low doses raises two different preliminary questions: the first problem is whether the state or – in the case of the European Community – the Community is under an obligation to protect human life and health against low doses; secondly, one has to ask the question whether and to what extent the state or the Community is empowered to deal with low doses.

10.1.1
Protective obligations

Protective obligations of German authorities may originate with the Federal Constitution or statutory law or have their basis in European directives or regulations. Under the jurisprudence of the German Constitutional Court the fundamental right to physical integrity and health (article 2 II Federal Constitution) which primarily has a defensive function directed at encroachments by the state, also establishes an objective obligation of the state to actively protect human life and health. Moreover, article 20 a Federal Constitution sets forth that the state, in the interest of the present as well as future generations, is under an obligation to protect and maintain the natural bases of life. Insofar as risks to human life and health originate with discharges or releases into environmental media, the protective duty also encompasses the protection of human life and health, while article 2 II Federal Constitution is the only constitutional basis of protection regarding occupational health, indoor pollution, alimentary products, and medical application of radiation. Together, these two constitutional provisions mandate a comprehensive protection of human life and health against risks caused by exposure to toxic substances and radiation. However, the sweeping nature of the protective obligations is relativated by the recognition of a broad margin of political discretion. As regards the protection obligation arising from article 2 II Federal Constitution, the Federal Constitutional Court[1] has recognised a far-reaching political discretion of the legislature whose exercise, apart from the protection of human life against direct encroachments where the obligation is stricter, can only be reviewed by the court as regards its outer limits. The protective obligation is only violated where the state does not take any protective measures, the measures

---

[1] See, e. g. BVerfGE 49, 89 at 142; BVerfGE 56, 54 at 78; BVerfGE 77, 170 at 214.

taken are entirely unsuitable or inadequate to reach the required objectives or completely fall short of them.[2]

In the famous case on the constitutionality of German abortion law[3] the Federal Constitutional Court developed a more stringent requirement, namely to provide adequate protection ("prohibition of underprotection"). However, this requirement has since then been referred to only rarely, and this without drawing any practical consequences. It is indicative of the judicial restraint exercised by the Federal Constitutional Court in this respect that the Court has never found a violation of the duty to protect in matters relating to health and the environment.

Moreover – and this is particularly relevant for the problem of low doses – the scope of the duty to protect is far from clear. Under the Federal Constitution the state is obliged to protect individuals against impairments and endangerments of health. This includes anticipatory prevention when the risk of harm is substantial, and the necessary degree of probability is low when loss of life or serious harm to the health of a multitude of people could occur. However, apart from the latter case the prevailing opinion denies a protective duty when there is no sufficient probability of future harm and the question only is whether to take precautionary measures. Without going here into further details regarding the touchy delimitation between prevention of dangers and precaution, it is safe to say that even if one recognises a farther-going protective duty that also encompasses the sphere of precaution, the intensity of the relevant duty would be reduced and the margin of political discretion extended correspondingly. This is reflected by a recent decision of the Federal Constitutional Court[4] on protection from electromagnetic waves. While apparently extending the duty to protect to the outer limits of precaution, the Court denied a duty to protect against "purely hypothetical hazards to human health".

The general state obligation to protect the environment set forth by article 20 a Federal Constitution in principle also mandates precautionary protection and conservation of the environment and indirectly of human health, i. e. in cases where human health is affected by the discharge or release of substances or radiation into the environment. This may be the source of a somewhat higher standard of protection, although mitigated by the broad margin of discretion vested in the state by the legislature.

The Treaty on the Establishment of the European Community (EC Treaty) does not contain express duties of the Community organs to protect human life and health against impairments or endangerments from exposure to toxic substances or radiation. Article 2 EC Treaty establishes as one of the objectives of the Community achieving a high degree of environmental protection and improvement of quality of life and article 3 EC Treaty makes it clear that these objectives also comprise a Community policy in the field of the environment and a contribution for achieving a high level of protecting health. However, it is not possible to derive from this description of tasks of the European Community concrete obligations to protect individuals. If at all, such obligation can only be derived from article 174 EC Treaty

---

[2]   BVerfGE 56, 54 at 81; BVerfGE 77, 381 at 405; BVerfGE 79, 174 at 202; BVerfGE 92, 26 at 46; NJW 1996, 65; NJW 1997, 2509 ; NJW 1998, 2961.
[3]   BVerfGE 88, 203 at 251–255.
[4]   NJW 2002, 1638.

which is the principal basis for Community environmental policy and which expressly includes the protection of human health in the notion of environmental policy. This provision expressly sets forth that the environmental policy of the Community aims at a high level of protection and is based both on the preventative and precautionary principles. Moreover, in the framework of the harmonisation policy, article 95 III EC Treaty mandates the Community organs to aim for a high level of protection in the areas of health and environmental protection.

These central Treaty provisions in the first place confer competences and empowerments and regulate their exercise. It is doubtful whether they can be construed to the extent that they also confer positive obligations on the Community organs to actively protect individuals against impairments or endangerments of their life and health arising from toxic substances or radiation. Originally, the prevailing opinion even took the view that these mandates were only political guidelines. However, in a more recent decision, the European Court of Justice[5] in principle recognised the justiciability of article 174 II EC Treaty – high level of protection – as a standard for assessing the adequacy of Community measures taken for the protection of the ozone layer. The case dealt with alleged discrimination between substances that were subject to stringent obligations and other substances that were accorded privileges and only concerns the exercise of the empowerment set forth by article 174 II EC Treaty. It is but a first step towards recognising also protective obligations of the European Community. However, in the light of Court decisions in other areas it appears plausible to affirm the existence of a duty to protect (Winter 2002, 78; 2003, 139/140; Epiney 1999; more restrictive Frenz 1997, No. 53). Such a duty could also be derived from fundamental rights such as the right to physical integrity recognised by the EC Treaty (article 6 in conjunction with article 3 European Charter on Fundamental Rights) (Szczekalla 2002). However, in the absence of any pronouncement of the European courts in this direction this construction does not rest on safe doctrinal grounds.

As in the case of German constitutional law and even more clearly spelt out, the duty to protect is qualified by a broad margin of political discretion accorded to the Community organs. It is only in instances in which the hard core of article 174 EC Treaty or the right to physical integrity is not respected – e. g. inaction in case of unacceptable risk or evidently inadequate level of environmental protection taking into account countervailing interests (evident error of appreciation) – that the European Court could be expected to compel the Community organs to act (cf. Epiney 1999, 181; Schroeder 2002, 219).

10.1.2
Degrees of risk

At the level of statutory law, German environmental law does not expressly use the notion of low dose. Nevertheless, this notion is implicit in the key terms of danger and (simple) risk widely used in environmental laws. In most statutes, there is a general distinction between danger, simple risk below the threshold of danger and residual risk.

---

[5]   Case C-284/95 (Safety High-Tech), 1998 ECR I, 4301 at 4344; cf. case C-293/93 (Standley), 1999 ECR I, 2603 at 2647.

The terminology used does not correspond to that employed in toxicology, epidemiology and safety science as well as in foreign legislation. In the German legal terminology danger is a qualified form of risk which is unacceptable and therefore must be prevented. A danger is a course of (future) events which makes the expectation sufficiently probable that harm to a protected interest (for example harm to health) will occur in the future; the theoretical possibility of such harm is not sufficient. Risks below the threshold of danger are covered by the precautionary principle. They are in principle acceptable in the sense that they need not be prevented under any means. However, they are undesirable and must be reduced by precautionary measures considering the principle of proportionality. A precautionary situation exists where there is no sufficient probability of future harm; however, there is some reason to believe that such harm might occur. Thus the notion of simple risk as well as the precautionary principle address uncertainty. A mere residual risk which has to be borne by the citizens as a general burden of human civilisation exists in a variety of situations including situations of entire ignorance (possibly large but not ascertainable future harm) or of extreme improbability (possibly large but extremely improbable future harm).

The delimitation of danger, risk and residual risk is not easy to draw and to a certain extent also flexible insofar as in addition to the relevant degree of probability of harm also the magnitude and severity of potential harm, the value attributed to the affected interest by the Constitution and some other factors must be considered. In health-related environmental law in principle a relatively low degree of probability can already constitute a relevant danger and simple risk is extended into what one would consider as residual risk under other circumstances. The requirements as to the necessary degree of probability of possible harm are reduced and hence the scope of the notion of danger is greater depending on the extent and severity of possible harm. Especially where serious irreversible harm, e. g. by accidents or by substances or ionising radiation for which no threshold can be indicated, is possible, the relative notion of danger leads to a potential extension into an area which otherwise would be considered to be only covered by the precautionary principle (see also section 2.2.3). There also is a tendency towards extending the scope of the notion of danger where under a particular law precautionary measures are not provided. This is in particular true of parts of alimentary law and the law relating to the management of historic pollution (clean-up of contaminated buildings and land).

The distinction between danger, simple risk and residual risk is important since the duty to protect imposed upon the state, is more intensive in the area of classical prevention of dangers. Dangers must be prevented "unconditionally", although the principle of proportionality plays a certain role regarding the decision on the kind and extent of measures. With respect to simple risk below the threshold of danger the state is accorded discretion already as to the question whether to tackle them at all and, in application of the principle of proportionality, such risks must in any case be reduced by precautionary measures only subject to the availability of technical means and cost/benefit considerations. No action at all is required with respect to residual risk. Secondly, in the case of prevention of dangers the protection of vested interests of the owners of sources of potential harm against measures taken or ordered subsequent to the initial decision are weaker than in the case of precaution. Finally, judicial review is different since, under the prevailing view, only the duty to protect against dangers – in contrast to the duty to take precautionary measures – is deemed to protect individual interests and can serve as a basis for according the affected individual's standing to sue.

However, it must be noted that there are some environmental laws that do not follow the three-tier concept of risk but rather apply a simpler model that consists of two tiers, only distinguishing between unacceptable or undesirable risk and residual risk. This is in particular true of the law of nuclear energy (arguably also including the law of radiation protection) and biotechnology. As a response to the perceived higher risk potential of ionising radiation and works with, and releases of, genetically modified organisms, these laws employ a broader notion of precaution against risk which includes the classical field of danger, and they also extend the notion of risk to an area which under general environmental laws would be considered as residual risk. However, they partly make other distinctions, namely between mandatory and merely optimising precaution, which cause new problems of delimitation.

As German law, article 174 II EC Treaty distinguishes between prevention and precaution so that one could sustain that the criteria for delimiting danger and simple risk employed in Germany might also be applicable to EC law. However, the precautionary principle only entered into the EC Treaty by the Single European Act which went into force in 1987 and did not have a great practical relevance in EC environmental policy during the following decade. Therefore many EC environmental laws do not embody the precautionary principle or attach legal consequences to the notion of risk without making clear whether the relevant act contains precautionary elements. There are even environmental regulations, such as the Regulation on the Assessment and Control of Environmental Risks of Existing Chemical Substances of 1993, which do not provide for any precautionary measures. With the BSE crisis and the challenge of EC policy relating to the use of hormones in animal feeding under the WTO regime, the precautionary principle has emerged as the fundamental principle of EC environmental policy. This culminated in the adoption of the "Communication on the Applicability of the Precautionary Principle" (EC Commission 2000) which defines the precautionary principle as an approach to risk management that is applied in circumstances of scientific uncertainty, reflecting the need to take action in the face of a potentially serious risk without awaiting the results of scientific research. The European Court of First Instance, in a ruling on a prohibition of certain antibiotics in animal feeding stuff[6] strongly relied on this communication to uphold the measure which had been taken against the advice of the Scientific Committee on Animal Nutrition. Previously, the European Court of Justice[7], deciding on the validity of a Commission decision banning the exportation of beef from the United Kingdom with an eye to limiting the risk of transmission of BSE, had recognised the applicability of the precautionary principle to veterinary law (covered by article 95 EC Treaty, ex article 100 a EC Treaty):

> Where there is uncertainty as to the existence or extent of risks to human health, the Commission may take protective measures without having to wait until the reality and seriousness of those risks become apparent.

---

[6] Case T-13/99 (Pfizer Animal Health SA./. Commission), judgement of 11 Sept. 2002 (not yet published).

[7] Case C-180/96 (United Kingdom./. Commission), 1998 ECR I, 2265 at 2298; somewhat weaker Case C-157/96 (National Farmers Union), 1998 ECR I, 2211 at 2259.

Since 1996, precautionary clauses have been inserted into some existing environmental directives, such as the Directive on Integrated Pollution Prevention and Control (IPPC Directive 96/61), the Directive on the Release of Genetically Modified Organisms (2001/18) and the Regulation on General Principles and Requirements of Food Law (178/2002) (see Daemen 2003, 15–16). The Commission proposals on the reform of the EC chemicals policy (EC Commission 2001) also strongly reflect the new concern about the need to devise a precautionary environmental policy. However, it would seem that the focus is less on the delimitation of prevention of danger and precaution against simple risk; in this respect, apart from the IPPC Directive which has adopted the German approach, EC policy uses a unitary concept of significant risk. Rather, the focus is on identifying the limits of legitimate precaution vis-à-vis mere residual risk as well as proportionality of precautionary measures (Rengeling 2000, 1478–1480; Appel 2001, 398; Schroeder 2002, 218–221).

German environmental law does not expressly use the term of low dose. EC law only does so with respect to risk assessment. However, as stated, the notion of low dose is inherent in the three-tier notion of risk widely used in German law as well as the two-tier notion of risk employed in European law. In the light of the foregoing one may denote for legal purposes a dose of chemicals or radiation as low where the adverse effects potentially associated with the substance or the radiation as such or, as normally is the case, with a particular dose are entirely unknown or causation of an adverse effect is highly improbable. The lack of express rules using the term of low dose has the effect that in German law the central notions of risk assessment relating to the determination of low doses, namely "no observable adverse effect level" (NOAEL), "lowest observable adverse effect level" (LOAEL) or "benchmark dose" (i. e. effect in less than 10 %), do not form part of legal terminology unless introduced by European law. However, in administrative practice they are widely used.

The lack of conceptual focus on low doses may also explain that it has never been legally clarified whether the ultimate objective of risk assessment is to determine a level of exposure of individuals to substances or radiation where there is no adverse effect or whether one can be satisfied with stating the absence of a significant adverse effect. The EC Directive on Risk Assessment of New Substances speaks in this connection simply of "reason for concern" and the Regulation on Risk Assessment of Existing Substances is content with simply referring to "risk assessment". In any case, the problems of interpreting these notions are mitigated by the inherent imprecision of extrapolating from animal tests to effects on human health.

10.2
Risk assessment, low doses and environmental law

10.2.1
Risk assessment and risk management

The decision as to whether an agency should address risks presented by low doses, especially in application of the precautionary principle, in some cases even in application of the preventive principle, requires an objective assessment of the relevant risk identifying the degree of scientific uncertainty. In the international terminology, one distinguishes in this respect between risk assessment and risk management. A preceding step is the "preliminary procedure" which relates to the definition of the problem, the determination of the requisite level of protection (prevention or precaution), and the necessary kind of risk assessment. It is by and large determined by the applicable legislative framework (Risiko-Kommission 2002, 9, 38/39). Risk assessment concerns the determination of the probability, severity and extent of potential harm associated with exposure to a chemical substance or to radiation. Although it entails a variety of assumptions including value judgements, it is basically a science-based, i. e. cognitive-predictive process which aims to predict the occurrence of adverse effects to human health or the environment in the future. By contrast, risk management is a largely normative, interest-oriented decision-making process, in which the competent body, based on the results of the risk assessment, decides on the acceptability or tolerability of the predicted risk and on the taking of appropriate measures. This "separation model" is widely accepted, although not entirely uncontroversial and in practice not always observed (Risiko-Kommission 2002, 26/27).

In German law, apart from chemicals legislation implementing EC directives, there are normally no formal legal decision-making requirements that distinguish between risk assessment and risk management. However, this basic distinction has been developed by the US National Academy of Sciences (NRC 1983; 1996) and recognised by OECD (1991); it may be said to be inherent in any rational governmental decision in application of the precautionary principle. EC environmental law, especially the law relating to chemicals, expressly embodies the distinction between risk assessment and risk management. According to the relevant texts risk assessment is further divided into four components, namely hazard identification, hazard characterisation, appraisal of exposure and risk characterisation. The limits of scientific knowledge may affect each of these components. Hazard identification means identifying agents which due to their properties may have adverse health effects. Hazard characterisation consists of determining, in quantitative and/or qualitative terms, the nature and severity of the adverse health effects associated with the relevant agents. It includes the establishment of a dose-response (dose-effect) relationship. Appraisal of exposure aims at evaluating in quantitative or qualitative terms the probability, kind, scope and extent of exposure of individuals to the agent under study, which requires, at the outset, the determination of the release of the agent into the environment. Finally, risk characterisation is a quantitative and/or qualitative estimation of the probability of the frequency (incidence) and severity of the known or potential adverse health effects liable to occur in a given population. It may also include qualitative aspects of risk such as the nature of the adverse effect, the circumstances of its occurrence, the identity of the receptors (e. g. distributional aspects) and the nature of the consequences (NRC 1996). Risk characterisation is established on the basis of the three preceding components and considers the uncertainties inherent to the determinations made in the framework of these steps. Depending on the reliability of the data generated and assessed, the reliability of risk characterisation may differ from case to case. When the available data are inadequate, it may be necessary to rely on mere hypotheses. Risk characterisation is the least cognitive of all four tiers of risk assessment and already marks the transition towards risk management. The assessor must conclude whether there is a significant risk, i. e. a sufficiently serious potential harm with possibly irreversible effects for human health. Therefore, risk characterisation also embodies a judgement on the need for action from a purely health or environmental point of view. Even measures which enable the exposed users to protect themselves, such as classification according to hazardous properties, are recommended in this process.

Risk management can be divided into the decision on the acceptability or tolerability of the risk thus determined (risk evaluation) and the decision on appropriate measures. The decision on acceptability or tolerability will consider factors such as probability, extent and severity of harm, and at least with respect to precaution also previous decisions on comparable risks, equity including distribution of risk, costs and benefits of action. The selection of measures is primarily governed by the principles of equality and proportionality. Risk management embodies a normative decision which, however, is to a more or less large degree based on scientific data. This is even true when the decision-maker applies the precautionary principle. The application of the precautionary principle is not "unscientific"; rather, it only uses less or less corroborated scientific data.

Since risk characterisation being the last step of risk assessment and the decision on tolerability of risk which already belongs to risk management are closely related, one can also unite them under the heading of risk evaluation (Risiko-Kommission 2002, 9, 26 et seq., 36/37). However, this should not detract us from the need to uphold the analytical separation of the two steps.

There are a variety of normative assumptions that underlie risk assessment and risk management. This is in particular true of the concept of harm to human health and – in the front-line – of adverse effects, the persons or groups of persons to be protected (problem of relevant susceptibility), the criteria which enter into the determination of danger and simple risk (apart from probability and severity/magnitude of harm, for example latency, persistence, bio-accumulation, reversibility etc.) and economic and social balancing factors. These normative criteria do not only shape the decision on acceptability or tolerability of risk associated with a particular substance or radiation; rather, most of them are also relevant for the process of risk assessment.

10.2.2
The notion of harm to human health

The World Health Organisation in its declaration of 22 June 1946 defines health as "a state of complete physical, mental and social well-being and not merely the absence of disease or infirmity". Since health-related environmental policy deals with the protection of individuals against harm to their health originating with the environment rather than with a positive promotion of health, its core notion cannot be health as such. Rather, it must depart from the notion of impairment of health and hence disease. One can define disease as the presence of anormal physiological functions which under certain circumstances require a treatment or the taking of protective measures. Disease is not limited to physical well-being but also comprises psycho-somatic disturbances. From a medical point of view, disease comprises disturbances of the normal functions of organs and organic systems of the human body, the totality of subsequent anormal reactions of an organism or its parts to an adverse impulse, a permanent deviation from the normal limits of physiological balance and an anormal state of the body. A disease also exists where an agent influences physical systems which as such are not target organs or end points in the sense of a disease but contribute to an anormal functioning of other organs. According to the German court cases[8] an impairment of health is any disturbance of the physiological, morphological, mental or spiritual expressions of human life and any causation or increase of a state which adversely deviates from the normal physiological state, even if not associated with pain or fundamental change of human conditions. It includes burdens on physiological functions which lead to an anormal response only at a later (but foreseeable) time (e. g. a weakening of the immune system or significant increase of sensitivity to infections). However, it must be noted that there is a grey area between a healthy ("normal") reaction to impulses originating with the natural, work or indoor environment, including compensation of the effects of such impulses and a disease. Consequently a clear-cut delimitation is not possible. Rather, in the grey area the determination of whether a state is to be considered as healthy or unhealthy largely depends on the assessment of the decision-maker. The introduction of the notion of adversity (WHO 1994) and classification into the categories "undoubtedly adverse", "doubtful adversity and undesirable effects" and "undoubtedly irrelevant for human health" (Risiko-Kommission 2002, 51/52) which are meant to operate in the grey area reflect the problem but in essence only shift it to another level (Neus 1998, 60–62).

Another problem of defining harm to human health, particularly in the field of cancer, is the question as to whether in risk assessment and risk management total cancer risk or individual, including rare, tumours must also be considered. Since different types of cancer are associated with different mortality rates and impairments of quality of life, there are good arguments that militate for the proposition that in this field a stronger individualisation of the relevant disease is required.

Environment-related impairments of human health require that environmental factors cause or at least contribute to the origination or reinforcement of a disease. Causation is not excluded by the fact or possibility that besides influences on human health originating with the natural, work or indoor environment, the target person at the same time is exposed to other adverse influences such as food, tobacco, alcohol and drugs nor by the contribution of the social situation and other life factors which may also have an influence on the origination and the re-enforcement of a disease. It is the task of risk assessment to determine the significance of the causal relationship between exposure to a chemical substance or radiation and the occurrence. Whether the additional risk caused by exposure to the chemical substance or radiation is tolerable in view of prevailing concurrent causal factors must be decided in the framework of the risk management decision, applying the various causation concepts under relevant health-related laws.

10.2.3
Individual versus collective protection

There is scientific evidence that there are a number of inter-individual variations of toxicological response in the human population to exposure to hazardous agents.

---

[8] BVerfGE 56, 54 at 74/75; Administrative Court of Appeal Munich, NVwZ-RR 2000, 661 at 665; Münster, NVwZ 1990, 781.

These variations may relate to social groups but also be due to individual suscep-taibility such as genetic predisposition (see section 4.7). From a legal perspective the question arises as to what extent such variations may or even need to be consid-ered in regulating exposures to hazardous agents.

The general nature of regulation, reasons of practicality as well as gaps in scien-tific knowledge militate in favour of a protective concept that does not aim at the protection of every individual but rather concentrates on the public at large repre-sented by an average individual. However, the individualistic model of fundamen-tal rights, in particular the right to physical integrity and health which serves as a guideline for the state in performing its protective duties, serves as an argument for a protective concept that aims for preventing unacceptable risks and reducing undesirable risks for each individual. In line with an early decision of the Federal Constitutional Court[9], it cannot be considered permissible to rest the protection against dangers to health on the model of an average healthy individual because this amounts to a discrimination of susceptible persons. Taken at its face's value this would mean that an increased risk of every individual due to individual sus-ceptibility (eventually coupled with high exposure) would have to be addressed. However, it is now widely accepted that as a compromise for solving this conflict only typical differences of susceptibility need to be considered. This is embodied in the protective concept of "vulnerable groups". Groups of individuals such as children, aged people and sick people, but also more specific groups and their pro-tective needs have to be considered in regulating environmental health risks, at least as regards prevention of dangers to health. By contrast, a collective approach seems to be more acceptable in precautionary reduction of risk. On the whole, one is far from agreement on the limits of the group approach, and the practice of reg-ulation both in Germany and the European Community has not yet developed a systematic and consistent approach to protecting vulnerable groups (Wulfhorst 1994; Böhm 1996).

As regards the control of air pollution it has been sustained that the ambient quality standards for conventional air pollutants are sufficiently protective of the health of some risk groups such as elderly people, sick people, pregnant women and children.[10] However, these are only some typical vulnerable groups. It is not clear to what extent the regulating agencies have considered other aspects of susceptibility, for example allergic persons. The normal lack of transparency of the standard-set-ting process makes it difficult to ascertain whether the postulate of an effective pro-tection of vulnerable groups has really been complied with. There are some German court decisions[11] which go farther in that they seem to expressly consider typical diseases induced by air pollutants such as allergies, asthma and pseudo-croup as relevant for the setting of ambient quality standards.

The law of radiological protection contains uniform standards for the protection of the population at large. Particular susceptibility is not considered. The Regula-tion on Electromagnetic Fields contains special precautionary standards for the pro-tection of children and sick people.

---

9   BVerfGE 27, 220 at 230.
10 Administrative Court of Appeal Münster, NVwZ 1988, 173 at 176.
11 Administrative Court Berlin, GewA 1985, 228.

In alimentary law the orientation at an average consumer clearly prevails (Böhm 1996, 55 et seq.) and is also accepted by the courts.[12] In particular there are normally no requirements for specific information of vulnerable groups. A major exception are alimentary products consisting of or containing genetically modified organisms and dietary products where the specific protection and information needs of allergic persons or persons in need of a diet must be duly considered (e. g. by subjecting allergic effects of novel foods to permit and labelling requirements). A German criminal court decision[13] that takes a broader view in this respect specifically mentioning allergic persons and smokers as vulnerable groups has not had visible impacts on practice.

The law of occupational health does not unequivocally define the circle of persons to be protected (Böhm 1996, 83 et seq.). It is true that the German Chemicals Act, while generally denoting the protection of human health as one of its central objectives, addresses the specific protection of allergic persons in some of its provisions, especially in the definition of hazardous properties (sensitising effects). The Regulation on Hazardous Substances promulgated under the Act contains specific requirements for an additional protection against allergies even when the applicable standards are met. However, if one considers the two major groups of occupational health standards one cannot state that the law of occupational health generally and consistently addresses the problem of vulnerable groups. As regards the maximum exposure values at the work place, the starting point for establishing such values is an average healthy person of working age. Some values aim to protect vulnerable groups, especially allergic persons (contact allergies and to a certain extent allergies through respiratory passage). However, allergies are only considered with respect to some known allergenic substances. The problem of protecting pregnant female workers against potential hazards has been addressed by establishing risk classifications for some genotoxic substances (A: unequivocal proof, B: risk problem, C: safe, D: no sufficient information) without, however, establishing clear-cut standards. Moreover, due to a lack of scientific knowledge about embryotoxic and fetotoxic effects – animal tests are not deemed to be reliable – this classification is far from being complete. It appears that in the majority of cases the average healthy worker is the model on which standard setting rests. The second group of standards, namely biological tolerance values are expressly oriented at a healthy adult worker and do not rule out that compliance with them may lead to allergic reactions. Occupational radiological protection is more responsive to the protective needs of vulnerable groups. There are specific dose-limit values for occupational exposure of minors, women during pregnancy as well as women capable of giving birth.

EC health-related law has as yet not laid particular emphasis on the problem of susceptibility, major exceptions being the regulation of novel foods made of or containing genetically modified organisms, occupational radiation protection, and theoretically classification of hazardous substances. The German legal rules referred to in this respect implement EC directives.

As stated, although in protecting of human health the orientation at an average healthy person is not permissible because it does not take account of the protective

---

[12]   Federal Administrative Court, LRE 21, 174 at 179.
[13]   Federal Supreme Court, BGHSt 37, 106.

needs of susceptible individuals in the general population, some standardisation is possible and even mandated (Böhm 1996, 129, 147; Kunig 2000, Art. 2 No. 63). Environmental standards by their very nature require simplification and standardisation. The concentration on particularly susceptible (vulnerable) groups is permissible and on the other hand required as a minimum of protection. The susceptibility of individuals or of small groups need not be taken into account. Authors who sustain, that in view of the weight of the right to life and health in protecting human health a radical individualistic concept is required and therefore dangers to the health of individual, particularly susceptible persons must be prevented (Wulfhorst 1994, 101 et seq.), ignore the massive problems of practicality, costs and scientific knowledge which are presented by this approach (Böhm 1996, 135, 139). One may postulate, though, that particular susceptibilities of individuals can and should to a certain extent be considered by safety factors.

The definition of particular susceptibility is not really clear, given the fact that inter-species variation occurs at a continuous scale (Parkin and Balbus 2000). To overcome these difficulties, it is appropriate to admit, as one normally does, that the shaping of relevant vulnerable groups can in principle be oriented as typical risk groups. Such recognised groups are children, aged persons, pregnant women as well as allergic persons. Other risk groups such as small children, new-born children, women capable of giving birth or nursing women are used more seldom. It is doubtful whether one has to include in the concept of vulnerable groups also persons who are particularly susceptible because of their life style, such as smokers, alcoholics or drug addicts, but also members of diffuse groups, i. e. groups that cannot be easily delimited, for example chronically sick people (Bohl 2002, 10) or persons with genetic or immunal defects. In the first case the lack of reasonable self-protection may justify an exclusion. In the latter case it is not easy to strike a compromise between the postulates of human dignity and health and the requirements of practicality. It is true that the concern for minority groups is inherent in the protective concept of the Federal Constitution and possibly also European human rights, although the European concept of fundamental rights is less individualistic and more communitarian than the German one. However, this concern should not burden the risk assessment and management capacity of a society to a degree that the protection of the vast majority would suffer. Therefore it is submitted that the legislature or executive has a certain margin of discretion in this respect.

It is also unclear whether, beyond intrinsic susceptibility, high exposure must be considered, for example exposure above a 90 percentile (Parkin and Balbus 2000). The inclusion of external factors would mean that risk assessment could not be based on an approximation to average exposure; rather high-end exposure would have to be included. This is in principle justified. However, in parallel to intrinsic susceptibility, a worst case scenario need not be addressed. In essence, this is not a problem of susceptibility but rather of environmental justice which will be dealt with later (see section 10.4.3).

In keeping with the principal limitation of the constitutional duty to protect human health to (anticipatory) prevention of dangers (unacceptable risk), the legal requirements exposed so far are primarily applicable to situations of unacceptable risk. It is rather doubtful whether they can be generally transferred to the field of precaution. In this respect, a distinction between agents associated with a threshold and those for which a threshold cannot be determined appears appropriate.

As far as German law is concerned, under the traditional view precaution does not serve to protect individuals but rather to reduce the risk of the public at large in the public interest. The objective of precautionary measures is the reduction of the population risk rather than that of individual risk. Therefore one does not recognise legal standing to challenge the legality of precautionary measures before administrative courts. From this point of view there is no cogent need to grant members of vulnerable groups specific precaution against threshold agents, provided they are sufficiently protected individually against unacceptable risk to their health. It is yet another question whether the regulatory agency is empowered, rather than obliged, to consider vulnerable groups even in the field of precaution. This is not unreasonable from a point of view of public health and legitimate interests of (potential) polluters are not normally impaired.

Even if one accepts the distinction between an individual- or vulnerable group-oriented concept in the field of prevention of unacceptable risks and a collective concept in the field of precaution – which is disputed (Lübbe-Wolff 1986, 179 et seq.; Steinberg 1998, 310) – it is subject to serious objections when it is either impossible or in the light of the present state of scientific knowledge not justified to assume a threshold value such as regards carcinogenicity and mutagenicity. The no-threshold hypothesis makes a clear delimitation between unacceptable risk and merely undesirable and even residual risk impossible. If this is correct, the concept of protecting vulnerable groups (which in itself already sacrifices particularly susceptible individuals which do not belong to a recognised vulnerable group) must necessarily extend to the area of precaution. A justification for this position is served by the prevailing opinion in Germany that an affected person can base a court action on the alleged violation of the dose limit values as well as the discharge values for nuclear facilities although these values are (also) precautionary.[14] Even insufficiency of dose limit values can be asserted by a court action.[15] The same is true of risk-oriented emission standards for carcinogenic air pollutants which are set forth in concretisation of the minimisation requirement for such substances.[16] One can either associate these standards with prevention or speak of precaution that is protective of individuals (Köck 2001c, 203).[17]

Consequently it would seem that the concept of protecting vulnerable groups must, at least in these cases, be extended so as to reduce undesirable risks of cancer and mutations at least for members of recognised vulnerable groups. This has been recognised in the field of occupational radiation protection by providing precautionary dose limits for pregnant women as well as women capable of giving birth.

---

[14] Federal Administrative Court, BVerwGE 61, 256 at 261; BVerwGE 72, 300 at 315.
[15] Federal Administrative Court, NVwZ 1998, 631 at 632/633.
[16] Administrative Court of Appeal Münster, NVwZ 1991, 1200 at 1202.
[17] The Technical Guidelines for the Protection Against Air Pollution of 1986 classified emission standards for carcinogenic substances partly (known carcinogenics) as prevention of danger, partly (suspected substances) as precaution. The Guidelines of 2002 uniformly classify them as precaution. A similar ambiguity as that referred to in the text exists with respect to ambient quality standards for carcinogenic pollutants (below 4.7); see, e. g. Federal Administrative Court, NVwZ 2002, 726 at 727; Administrative Court of Appeal Munich, NVwZ-RR 2000, 661 at 665; Mannheim, NVwZ-RR 1999, 298 at 302; Köck 2001a, 206; Salzwedel 1992, 53/54, 64/65 (precaution); Breuer 1991, 171; Böhm 1997, 54 (prevention).

10.2.4
Criteria for the determination of risk

In the traditional German differentiation between danger (unacceptable risk), simple risk (undesirable risk) and residual risk but also in that between risk and residual risk, the probability of future occurrence of harm as well as the severity and extent of this harm play a central role. Depending on the product of these two criteria one can classify potential risks into these categories. However, it should be noted that probability is rarely used in the statistical sense. The notion of probability includes "probability of truth" (Letzel and Wartensleben 1989). Except for accidents and other disturbances of normal operations of a facility, probability is a qualitative term which reflects a subjective assessment not only as to the frequency (incidence) but also the scientific plausibility of future harm. Uncertainty plays a role at each tier of the three- or two-pronged notion of risk both regarding the probability or plausibility and the severity or extent of damage. Therefore it is appropriate to specifically include an assessment of the remaining uncertainty in the process of risk assessment.

In modern literature various proposals have been made to introduce some additional criteria for determining risk. These criteria relate either to the severity of possible harm or its geographical or temporal scope, namely ubiquity of agents and consequences, persistence of agents and consequences, bio-accumulative capacity of agents, the degree of reversibility of consequences and latency of occurrence of harm. The criteria are a useful supplement to the traditional risk analysis and deserve a higher degree of consideration in the process. However, they can merely be seen as refined elements of determining the severity and/or extent of harm which may be caused by exposure to a substance or radiation. EC toxic substances law implicitly recognises some of these criteria by considering them in the setting of extrapolation factors. Moreover, EC and national laws of radiation protection and biotechnology go clearly beyond traditional risk analysis in considering some of these criteria. Apart from that the criteria make it possible to define, between clearly unacceptable and merely undesirable risk, a grey area where risk reduction is the rule but, where this is not feasible, the risk may be tolerable based on a cost/benefit evaluation. To this extent problems of risk management rather than risk assessment are concerned.

10.3
Risk assessment: science and law

10.3.1
Requirements as to the scientific foundation of determining Risk

As stated earlier, the delimitation between unacceptable risk (danger) and undesirable risk (precaution) is determined by the criterion of "sufficient probability". The occurrence of future harm need not be beyond all reasonable doubt. Rather, based on a balance of evidence, it must be reasonably probable that such harm will occur. Since the determination of sufficient probability must also be in relation to the assumed severity and extent of harm, which includes an evaluation of the value attributed to the interest at stake, in case of harm to health a relatively low degree of probability is sufficient to find that a given situation presents an unacceptable risk. This is especially true in the framework of laws which do not empower the agency to act merely on grounds of precaution. While it is true that the notion of danger is based on a traditional concept of causation and scientific proof and hence aims to avoid the risk of error in the sense of false positives (Scherzberg 1993, 498; Wahl and Appel 1995, 93/94), this formula does not determine the methods for finding sufficient probability and allows to rely on evidence based on extrapolation from animal tests. This seems to be generally accepted although there are different opinions as to the necessary degree of empirical foundation of evidence (von Lersner 1990; Breuer 1991, 171, 174; Salzwedel 1992, 39, 48/49). Merely hypothetical risk or risk simply assumed on the basis of highly conservative, empirically unfounded estimates does not suffice to consider a particular course of events as an unacceptable risk. However this does not mean that conventional extrapolation is *per se* unsuited to establish sufficient probability. For example, the finding that the ratio between the predicted environmental concentration and the NOAEL or LOAEL is much higher than 1 may suggest a conclusion that there exists a danger to human health which has to be prevented (cf. Dieter and Konietzka 1995; contra Theuer 1996, 126 for ecological risk).

The definition of precaution and its delimitation from residual risk that does not warrant any corrective measures and has to be tolerated by citizens as part of the general burden of modern civilisation poses more serious problems. Apart from statistically established low probability, the precautionary principle mainly covers the field of uncertainty. However, a clear-cut delimitation between uncertainty and complete ignorance which is considered to be the realm of residual risk which does not justify any corrective measures is not easy to draw. Wiedemann et al. (2001) distinguish between 4 types of evidence which could trigger precautionary measures: a science-based suspicion of risk (empirical evidence on cause-effect relationships, but contradictory studies), a partly plausible suspicion of risk (empirical evidence on relationship between effect and exposure but no evidence about adversity), a hypothetical (i. e. merely theoretical) suspicion of risk and a mere apprehension of risk (without any evidence). The formulae used in health-related environmental law differ from this classification and are more vague.

The conventional opinion in Germany limited the precautionary principle to situations where there was a concrete suspicion of risk apparently requiring empirical evidence (Ossenbühl 1986; Feldhaus 1987; Kloepfer and Kröger 1990). Such a formula can be criticised on the grounds that it suggests a restrictive application of the precautionary principle and impairs its application to risks where it is most needed, namely risks whose magnitude or severity is unknown although there is some reason to believe that a risk exists. In a landmark judgement rendered in 1985 the German Federal Administrative Court[18] recognised that in nuclear energy law the competent agency must:

> ...also consider such possible harm which cannot be ruled out only for the reason that according to the present state of knowledge particular causal links can neither be affirmed nor denied and therefore there is not yet a danger but only a suspicion of danger or a 'potential for concern'.

---

[18] BVerwGE 72, 300 at 315; narrower still BVerwGE 69, 37 at 45.

The decision was subsequently transferred to other areas.[19] This liberalised understanding of the precautionary principle has been widely recognised in modern legal literature (Wahl and Appel 1995, 123 et seq; Rehbinder 1997b Nos. 50/51; Lübbe-Wolff 1998, 64/65), although, contrary to the demands of some authors (Callies 2001), it falls short of an outright reversal of the burden of proof. It leads to a justified attenuation of the requirements for applying the precautionary principle. It is sufficient that the competent agency determines only plausible reasons for concern. Formulae that describe this concept are "scientifically based suspicion", "reasonable grounds for concern", "credible scientific warning" or "scientifically serious assumptions" (EEA 2001, 184; very critical Bergkamp 2002, 27). This corresponds to the formula used by the European Commission communication on the precautionary principle (EC Commission 2000) which defines the triggering criterion for precautionary measures as "reasonable grounds for concern". Likewise, the European Court of First Instance, in a recent ruling[20] held that precautionary measures may not be justified by a merely hypothetical risk consideration based on assumptions not yet scientifically verified. Rather, there must be, on the basis of available scientific data, sufficient evidence as to the assumed risk. The criterion of "reasonable grounds" for concern primarily covers uncertainties in the field of dose/effect relationships of an agent that has already been determined as hazardous, such as ignorance about the agent's mechanism of action or conflicting or unreliable results from animal or epidemiological studies. Moreover, it relates to uncertainties as to the dispersion behaviour and exposure patterns of the agent (see 9.2.4). In some countries, the precautionary principle is also applied to uncertainties which relate to the hazardous properties of an agent (e. g. Section 17 IV German Chemicals Act).

The requirements for the necessary degree of plausibility of concern must be based on a value judgement which also considers the weight of the protected interest and the magnitude and severity of possible harm. This means that in case of possibly serious, irreversible adverse effects a weak reason for concern would be sufficient. Comprehensive empirical knowledge is not necessary. The reason for concern can also rest on theoretical scientific grounds, at least when there is some empirical foundation. An example is the linear no-threshold hypothesis in carcinogenicity which is suggested by empirical evidence as regards linearity of effects while with respect to the lack of a threshold one can only determine a dose below which a significant increase of risk cannot practically be proven or refuted (lowest observable cancer effect level – LOCEL) (Fischer 1996, 65/66). Moreover, there are some carcinogenic substances for which a threshold exists (see section 4.5). The extent as to which an empirical basis is necessary is unclear. Since the crucial test is scientific plausibility rather than empirical evidence, this question would seem to depend on the severity and magnitude of possible harm. Therefore in case of potential catastrophic or wide-spread and severe harm a merely hypothetical concern based on scientifically plausible models and calculations may be sufficient.

Merely speculative assumptions about adverse effects which are not based on grounds of scientific plausibility are not acceptable for justifying measures of pre-

---

[19] Federal Administrative Court, NVwZ 1997, 497 at 498; NVwZ 1992, 984 at 985.
[20] See supra note 6.

caution (Wahl and Appel 1995, 92/93, 112/113). This formula allows extending the precautionary principle from simple uncertainty also to cases of ignorance provided that the assumed severity and/or magnitude of the harm justify such extension and there are scientific grounds for suspicion. Examples where scientific plausibility is doubtful are conclusions from empirically determined acute toxicity to chronic toxicity or from known persistence and bio-accumulation to ecotoxicity (see also section 8.2.1). It should be noted, that the definition of what is scientific and what is merely speculative is somewhat fluid. It would be too narrow to only consider the knowledge generated by mainstream science and denote everything else, i. e. grounds for concern adduced by representatives of minority positions in science, as unscientific and merely speculative. However, there must also be safeguards against flawed science which is not rare with outsider scientists (Bergkamp 2002, 78–80). Therefore the determination of what constitutes a scientifically plausible reason for concern must be based on best professional judgement.[21] It is evident that the decision-maker disposes here of a relatively wide margin of appreciation, which has expressly been recognised by German courts.[22] This is one of the reasons why there are marked differences among various governmental agencies and among countries as to the application of the precautionary principle (Wiener and Rogers, 2002; Daemen 2003).

Although the existence and extent of exposure primarily are relevant in deciding on acceptability or tolerability of risk and measures for risk reduction (and therefore must be analysed in the process of risk assessment), they also play a certain role in determining the applicability of the precautionary principle. The reason is that, in using the relational formula of "reasonable concern" described above, the magnitude of possible harm also is a factor that shapes the requirements as to the necessary degree of evidence. In the absence of empirical knowledge about the number of persons affected, spatial and temporal indicators such as ubiquity, mobility and persistence can be relevant to determine the magnitude of possible harm. The possible source of error in multiplying two uncertainties – severity and magnitude of harm – must be considered in the risk assessment process.

10.3.2
Requirements as to the kind of scientific evidence: the legal status of scientific models in risk assessment

From a scientific point of view a combination of animal testing, cellular and molecular studies, clinical studies and epidemiological research presents an ideal basis on which to rest the prediction that a particular exposure harmful to human health either will or may occur or is practically excluded. The limits of animal experiments as well as cellular and molecular studies militate for supplementary clinical and especially epidemiological studies, which in case of a clear association of an adverse effect with a given exposure may appear even preferable (Risiko-Kommis-

---

[21] Federal Administrative Court, BVerwGE 72, 300 at 315/316; BVerwGE 92, 185 at 196; BVerwGE 106, 115 at 121.

[22] Federal Administrative Court, BVerwGE 72, 300 at 316/317; BVerwGE 78, 177 at 180; BVerwGE 92, 185 at 196; BVerwGE 106, 115 at 122; Federal Constitutional Court, NJW 2002, 1638.

sion 2002, 55/56). However, for obvious reasons, clinical studies are possible only to a rather limited extent (that is in case of pharmaceuticals). Epidemiological research is associated with its own problems due to inhomogeneous populations, the difficulty to discard confounding factors and low predictive power in the low-dose range. Therefore animal testing and in appropriate cases also cellular and molecular studies, and extrapolation from these studies play a central role in the process of risk assessment. A minimum of empirical data, especially based on animal testing is required, though. The European legislation on dangerous substances recognises the legitimacy of this approach as regards the classification of substances as hazardous. It does not allow to base this classification on simple scientific models without a strong empirical basis. For example, structure-activity relationships alone are not sufficient. However, once a substance has been classified as hazardous, and there are gaps in knowledge about dose-effect relationships and exposure patterns, theoretical considerations, analogies and scientific models may play a certain role.

Criticism of the conventional model of risk assessment is not infrequent among legal scholars. Although this criticism is at its face's value directed at risk assessment, the real target often is the decision on acceptability or tolerability of risk and hence risk management: because of underestimation, over-estimation or sluggish assessment of risk (in the latter case due to overambitious data requirements) allegedly wrong risk management decisions are taken (e. g. Winter 2000, 177–179). This criticism stems from two sides.

In the first place it is alleged that the unilateral reliance on the scientific risk assessment model keeps regulators from taking precautionary measures in the absence of sufficient scientific data which justify a scientifically sound risk assessment. This criticism primarily relates to the shortcomings of the assessment procedure established by the EC regulation for the risk assessment of existing substances for generating hazard and risk data on existing substances produced in high volumes in the European Community. It has in principle been accepted that this criticism is justified. As a response, a fundamental reform of the European regulation of toxic substances is underway. It aims for more pragmatism in risk assessment and a strengthening of the precautionary principle by introducing, in the absence of sufficient scientific data which enable a full risk assessment, stronger precautionary elements into the process of risk assessment and hence also risk management.

One approach, previously favoured by the German Council of Environmental Advisors (SRU 1999) and in principle preferable (Cranor 1995; Neus 1998, 91/92; Rehbinder 2002), consists of allowing a precautionary risk assessment. While insisting on the need for data on all steps of the normal risk assessment procedure – determination of hazardous properties, dose-effect relationship and exposure patterns -, it would allow for a preliminary risk assessment based on the available data. Accordingly, a risk management decision would be taken where this is justified for reasons of precaution. Other, farther-going demands would concentrate on a single component of the risk assessment process, especially certain inherent properties of the substance. Consequently, they would allow for the determination of a significant risk and hence taking of precautionary measures where the properties of the substance in question warrant this; for instance, the mere existence of persistent and bio-accumulative properties without determining any toxicity or ecotoxicity would enable the competent agency to ban or restrict the chemical (Müller-Herold and Scheringer 1999; WBGU 1999; Scheringer 2000). The assumption, that these properties can serve as "proxy" criteria, stands and falls with the inference of a causal nexus between them and potential harm to health and the environment. It would seem that from past experience with other substances having the same properties, only "pattern prediction" rather than a causal nexus can be derived, which may justify closer scrutiny but not regulatory action. As a compromise, one could set forth a generic risk assessment depending on a matrix of classifying harmful properties (very high concern, high concern, concern, no concern, no data) and the mere exposure potential (intermediary product and use in close circuits, industrial use, open but controlled use, consumer product) (The Netherlands 2001, 44; SRU 2002, No. 368).

The Commission proposals on the reform of the EC chemicals regulation (EC Commission 2001) reflect these radical demands only in an attenuated form, namely in that particular properties of existing substances – except for virtually safe uses in a closed circuit – may be the basis for subjecting them to closer scrutiny exercised in the framework of a new authorisation procedure. However, as an authorisation procedure entails, at least to a certain extent, a reversal of the burden of proof, this already is an important step towards stiffening precaution. In any case, a critical step would be taken if the Community, as the European Parliament has demanded in the ongoing debate on the reform of EC chemicals policy, developed a strong policy of phasing out chemicals merely based on certain properties (von Holleben and Schmidt 2002, 535–537; 2003, 25–30).

In the second place, there has been a more fundamental criticism mainly voiced in the United States (Latin 1987; Shere 1995; Wagner 1995) but partly also in Europe and in particular Germany (Ladeur 1994, 161 et seq.; Theuer 1996, 125/126; see also Risiko-Kommission 2002, 27, 50) that relates to the scientific status of the risk assessment process itself, especially the technique of extrapolation. In this connection some American authors speak of "science charade". They point to the great number of value judgements that are in their opinion inherent in the conventional process of risk assessment, especially extrapolating from the results of animal tests to man to the effect that the conventional separation of risk assessment and risk management is blurred. It is said that risk assessment contains a variety of premises which are not empirically founded but contain, in the form of "scientific conventions", hidden value judgements. This begins with the division of risk assessment into four segments which, due to the selection of data generated, may multiply errors. Moreover, the suitability of test animals in view of comparability of the metabolism of these animals and that of man and comparability of effects may raise doubts. It is questioned whether the transfer of findings on subchronic to chronic effects, the transfer of a determined LOAEL to a NOAEL when the latter cannot be determined, the transfer of the scope of variation from test animals to humans and the extrapolation from the results of high test doses to low doses of actual exposure when empirical data are not available, constitute purely scientific judgements. The multiplication of the various extrapolation factors may, in the opinion of critics, result in grossly overstating the actual risk when they are interdependent, although they do not avoid an underestimation where the animal test results are not meaningful. Moreover, the concept of extrapolation factors is said to concentrate risk assessment on existing knowledge about normal instances, while increasing the problem of ignorance by making it disappear. It is alleged that a distinction between a normative value judgement and a cognitive source of error in the extrapolation process is practically impossible.

Apart from extrapolation, it is sustained that the appraisal of exposure also only constitutes an approximation model but does not represent real exposure conditions.

Extrapolation aims to bridge the gap between the scientific data generated and the need to predict the occurrence or non-occurrence of future harm to human health (see section 8.2.1). There is no denying that the extrapolation process contains a number of estimates and assumptions which are subject to possible disagreement and may render quite different results (Neus 1998, 73 et seq.). This is evidenced by the fact that a uniform practice as to the kind of relationships to be considered, the magnitude of the relevant factors and their multiplication (Dieter 1995; Englert 1996, 78, 83/84; Kalberlah and Schneider 1998, 161 et seq.; Risiko-Kommission 2002, 50/51) has not yet fully developed. However, the relevant judgements are essentially science-based, i. e. cognitive-predictive in nature. The theoretical assumptions and methodology underlying extrapolation including conversion factors are more or less supported by empirical findings on well researched substances and there is agreement on the core elements of extrapolation (cf. Arnold, Gundert-Remy and Hertel 1998). Scientific conventions on extrapolation are means to ensure reliability, transparency and generalisation. It is true that extrapolation may contain a certain precautionary (or "prudential") element (Renn 2000), especially in making "conservative" assumptions, in the face of uncertainty, about conversion factors, variability and exposure patterns. Critical in this respect is the setting of safety factors relating to aspects of risk assessment as to which quantification cannot be justified, e. g. for lack of knowledge about mechanisms or in view of unquantifiable intraspecies differentiation. Moreover, schematic multiplication of exposure factors may distort the risk so that a correction by recourse to best professional judgement should be considered. However, these judgements do not relate to the acceptability or tolerability of risks (contra Bergkamp 2002, 74). On the whole, the criticism does not justify abolishing the fundamental distinction between risk assessment and risk management. Potential sources of error can be tackled by specifically addressing the remaining uncertainties and gaps of knowledge at each tier of the risk assessment process and making them transparent (Risiko-Kommission 2002, 54–55). Especially in the risk characterisation process the assessor must determine whether a conclusive assessment is possible or not and whether, in view of insufficient data, from a scientific point of view the taking of regulatory (precautionary) action or more in-depth testing and other scientific investigation should be recommended (Risiko-Kommission 2002, 58–60).

It would be possible for the legislature or competent agency to politically decide on standards for risk assessment such as "inference guidelines" that prescribe the kind of conclusions one can derive from existing scientific data, especially a set of fixed extrapolation factors, taking into account the degree to which already at this stage precautionary elements should be applied. There are demands in this respect in the more recent discussion (Risiko-Kommission 2002, 35, 51). It should be noted that in the risk assessment of chemicals the EC has already achieved a certain degree of harmonisation of procedures. The extrapolation factor of 1000 used in extrapolating from animal and organism testing results to adverse effects on ecosystems in the EC Regulation on the Risk Assessment of Existing Substances constitutes an example. Another pertinent example is the extrapolation factor of 10 prescribed, in principle, in the US Food Safety Act for protecting children.

The advantages of standardised risk assessment lie in greater political legitimacy, transparence, predictability, harmonisation and lower transaction costs (Neus 1998, 88). However, general rules of this kind have a doubtful scientific basis. They could only serve as a general guideline subject to qualification in the particular case (Risiko-Kommission 2002, 51). Therefore, the EC Technical Guidance Document on Risk Assessment (EC 1996), following an "opening clause" in the regulation itself, as well as the OECD Testing Guidelines (OECD 2002) provide for lower extrapolation factors for ecological effects (10 or 50 instead of 1000) when more meaningful data are available. The same is true for some risk assessment concepts in the field of human health. An alternative is deciding on the relevant extrapolation factors case by case on the basis of best professional judgement. This would seem to be preferable because questions of transferability of animal testing results to man depend on the extent of data availability and must consider a variety of factors that are not amenable to generalisation. This would not rule out generalisation as to proper elements of the process when justified by scientific knowledge, in particular as to the basic issues of risk assessment (see Neus 1998, 93–95). In the ultimate result, both methods – standardised risk assessment subject to relativation and individual risk assessment with standardised elements – would approximate one another (cf. Neus 1998, 84/85). However, as regards extrapolation factors, the potential for standardisation appears to be limited. In any case both the legislature and the competent agency dispose of a wide margin of appreciation in developing and applying appropriate assessment criteria. Whatever one thinks of the harmonisation needs and possibilities, in view of the increasing EC involvement in risk assessment and risk management harmonisation at national level is of limited relevance so that endeavours in this respect must primarily be directed at the EC level.

Somewhat different problems of the scientific status of extrapolation arise in relation to agents which are deemed not to be associated with a threshold, such as carcinogenic and mutagenic agents. Here mathematical extrapolation models are used which are based on the linear-no-threshold hypothesis (ICRP 1990; STOA 1998, 19 et seq., 47). Their empirical content is limited since it is not possible to generate meaningful data by exposing test animals to low doses of such substances. Moreover, the results of mathematical extrapolation depend on a variety of factors, so that the upper and lower bounds of estimates may diverge considerably (Neus 1998, 76 et seq.). Where epidemiological data based on human exposure exist, as in the case of ionising radiation, transferability may be problematic insofar as much as they relate to relatively short exposure to relatively high doses while the aim of risk assessment is to determine potential adverse effects of long exposure to low doses. Linear extrapolation normally overestimates the risk, although one cannot rule out, that in a particular case it may also underestimate it. Moreover, it appears that with respect to certain chemicals, for example vinyl acetate, formaldehyde or chloroform,[23] the hypothesis has been falsified (see sections 4.5, 8.2.2 and 8.6) so that the generality of the hypothesis must be reduced. However, it remains true that the no-threshold hypothesis is based on scientific models relating to the genesis of carcinogenic or mutagenic processes and has only been falsified in some particular cases. In such a situation it does not appear unreasonable from a legal point of view, considering both the obligation to protect and the constitutional rights of (potential) polluters, to maintain the no-threshold hypothesis for all substances and ionising radiation for which it has not been falsified. Then the solution of the problem can be sought in the decision-making process on acceptability or tolerability of the risk thus assumed where uncertainties have to be duly considered (Scherzberg 1993, 497/498).

These remarks are also relevant for the protection of vulnerable groups. With conventional hazardous substances (threshold substances) potentially susceptible parts of a testing animal population can be subjected to separate testing (targeted testing), although one may assume that at present this is done only to a limited extent. Of course, the transfer of these results to humans is fraught with the extrapolation problems that have been discussed. As regards carcinogenic substances, targeted testing is not meaningful in the low-dose area.

The variation of the (linear) dose-effect curve may be said to reflect differences in susceptibility as well. However, particularly susceptible persons remain unprotected or, more exactly, less protected if a threshold cannot be determined. This problem has to be referred to the risk management process in which, considering the situation of particularly susceptible groups, tolerable exposure concentration values can be set and, if needed, supplemented by a minimisation obligation.

10.4
The decision on the tolerability of exposure to low doses and measures to prevent or reduce risks

10.4.1
Health-related criteria

The first step in the process of risk management is a decision on whether a particular risk is to be considered as acceptable or not. This is the normative phase of risk evaluation which is based on scientific risk characterisation. Since, depending on the legal framework, this decision may also concern simple risk below the threshold of danger (precaution) and therefore require a delimitation also between undesirable risk on the one hand and residual risk on the other hand, a more appropriate expression would be the notion of "tolerable" risk. Generally speaking, one can say that the decision on tolerability of risk aims at ensuring that risks are prevented or reduced to a degree that they are practically excluded. However, this term is geared to problems that can be solved by technical means and where practical experience is available, such as accidents; it does not fully fit to the risk presented by permanent

---

[23] Chloroform is of little relevance in Germany. For a holding quashing a US chloroform standard based on the no-threshold hypothesis see Chlorine Chemistry Council v. EPA, 206 F. 3d 1286 (D.C. Cir. 2000).

exposure to hazardous substances and radiation where theoretical assumptions necessarily play a greater role (Ladeur 1994, 13). Therefore a more adequate formula is that it must be ensured that future harm to human health will most probably, i. e. with a probability close to certainty, not occur. One can also term this as prevention of significant risk. Since under the German and European constitutional order human health is an interest of high value, the scope of residual risk which has to be tolerated by every citizen as a part of human civilisation is reduced.

However, it should be noted that the objective of avoiding significant risks is liable to convey a degree of precision which in practice does not exist. The decision on tolerability of risk is a decision under more or less strong uncertainty about facts. This uncertainty may be due to the impossibility of precisely analysing cause-effect relationships or to their degree of complexity (Ladeur 1994, 12). In particular, one can distinguish between limited knowledge, uncertainty in the narrow sense and ignorance. Risk evaluation is a process in which the competent agency evaluates, on the basis of the science-based conclusions of the risk assessment, the existing knowledge with a view to decide whether the risk is tolerable or must be reduced. This means that the agency can also arrive at conclusions that differ from risk characterisation, even independent of balancing with countervailing concerns, just because it is either more or less risk averse.

In taking a decision on tolerability of risk not only the probability or plausibility of future harm but also its severity and extent play a crucial role. This "relational" concept of risk means that the application of health-related criteria already engenders some degree of implicit balancing. According to the severity of possible harm one can distinguish between fatalities, chronical diseases and disabilities, reversible harm to health, and mere functional and performance disturbances. The extent of exposure can be classified into three groups of high, medium and low exposure, considering the number of persons exposed and the geographical and temporal scope (Wiedemann et al. 2001, 70/71). In this connection criteria relating to spatial and temporal scope such as mobility, reversibility, persistence, bio-accumulation and latency must be considered, some of which are certainly more important with respect to ecological damage but are also relevant in the field of health, especially because occurrence of an agent in the environment may have indirect impacts on man through the food chain. Finally, the degree of remaining uncertainty is relevant. This concerns regulatory action or non-action or the carrying out of further tests but, when required by the applicable legal framework, also the delimitation of prevention and precaution (Risiko-Kommission 2002, 52–53).

With respect to substances not associated with a threshold level, such as carcinogenic or mutagenic substances, the decision on tolerability of risk has other dimensions. If the no-threshold hypothesis is correct, one cannot maintain that the possibility of future harm is practically excluded at a given low degree of exposure. Of course, one can make a subtle distinction between practical exclusion and theoretical non-exclusion of risk. This was advanced by the German Federal Administrative Court in a decision on ionising radiation caused by the operation of the Stade nuclear power plant.[24] A holding of the German Federal Constitutional Court[25] also

---

[24]   BVerwGE 61, 256 at 263, 266/267; to the same extent BVerwGE 72, 300 at 315; BVerwGE 106, 115 at 121.
[25]   BVerfGE 49, 89 at 143.

seems to go into the same direction. However, it relates to nuclear accidents so that it cannot be taken as unconditional support for the proposition that the risk originating with exposure to no-threshold substances must and can at all be practically excluded. In any case, reference to the existence of natural background exposure to carcinogenic substances does not support such a position; it could at best be adduced for falsifying the no-threshold hypothesis as such. On the other hand, in view of the practical impossibility of absolute protection, the merely hypothetical nature of risk[26] in the low-dose area and the countervailing needs of an industrial society, the acceptance of a certain risk associated with exposure to a particular substance or ionising radiation is normally reasonable unless we have to do with a highly potent carcinogen where any exposure may have to be practically excluded. This can best be circumscribed with the formula that only a "significant" risk of harm to human health caused by a no-threshold substance or ionising radiation must be avoided. Standards set forth in this field (tolerability standards) are essentially political standards whose scientific foundation is "attenuated".

It should be noted that EC legislation contains substantive criteria for deciding on tolerability of risk only in those cases where an authorisation procedure is prescribed, such as in the field of pesticides or biocides. With respect to human health an authorisation requires that the agent in question, according to the state of scientific knowledge when applied according to its destination and properly or as a result of such application, may not have adverse effects on human health. Beyond that requirement which in itself is rather vague and is concretised by risk assessment guidelines, substantive evaluation criteria are rarely provided. This is in particular true of the general regulation of dangerous substances where the only truly substantive criterion, applicable to environmental harm rather than harm to health, is that when the relation between the "predicted no-effect concentration" (PNEC) and the "predicted environmental concentration" (PEC) is higher than 1 : 10, risk reduction measures must be taken. This is tantamount to the assumption that under these circumstances the risk thus determined is not acceptable or at least not tolerable. Below this threshold the competent agency has a wide margin of appreciation.

In particular, the role of practical experience with exposure of people to (natural and anthropogenic) background pollution and the role of risk/risk comparison in decision-making on tolerability is unclear. In the following part of this section, these two themes will be addressed in greater detail.

When there is sufficient experience with the effects of background exposure to agents, especially based on the results of epidemiological studies, such as may be the case in the field of ionising radiation relating to natural background radiation loads, one arguably can sustain that exposure to radiation which remains within the scope of variation of natural background radiation (provided that the sum of local background radiation and additional anthropogenic radiation does not exceed this limit) is tolerable. When such experience does not exist one could consider the future occurrence of harm to human health as practically excluded when a sufficient safety margin between the variation of (natural or anthropogenic) background pollution existing elsewhere and the dose to which the recipient is exposed can be established. However, the typical situation is that background and additional pollution sum up. Here, German courts tend to take the view that the additional risk is tolerable when the overall risk to which the individual in question is already exposed is only insignificantly increased.[27] Such a de minimis approach appears practicable. However, it is subject to doubts if the extent of the pre-existing risk is not considered. This has been recognised by a recent Administrative Court of Appeal case[28] which in a dictum questions whether in view of significant pre-existing cancer risk a relatively small additional risk could be negligible.

In deciding on tolerability of risk, the comparison with other types of risks, for which a pertinent decision on their tolerability has already been taken, is in principle admissible in support of the decision in question because this is part of the political discretion accorded to the regulator. The EC Commission in its communication on the precautionary principle emphasises the need for consistency of the decision relating to acceptability or tolerability of risk (EC Commission 2000, 19). The requirement of consistency applies to precautionary measures that are comparable in nature and scope with measures already taken in equivalent or similar areas in which the scientific data were available. In particular, if a well known risk is not regulated at all, regulation of a less understood risk requires justification (Bergkamp 2002, 75). The recent German endeavours to achieve consistency in the criteria and procedures for environmental standard setting (Risiko-Kommission 2002) go in the same direction. However, risk/risk comparison is legally mandatory only to a very limited extent (Breuer 1978, 838/839; Wagner 1980, 671; Köck 2001b, 285). In German as well as European law there are no formal rules that would require a risk/risk comparison. From a constitutional point of view it is safe to say, that under German law (potential) polluters cannot invoke inconsistency of regulation to their benefit if a preceding decision on the tolerability of risk is not sufficiently protective of human health (Wahl and Appel 1995, 205–207). The question is whether this can be reversed to the detriment of potential victims. It is submitted that here the principle of equality (article 3 I Federal Constitution) which entails elements of horizontal proportionality may set restraints on the discretion of the regulator (Salzwedel 1992, 12 et seq.; Di Fabio 1997, 825; Köck 1999b, 92; 2001b, 293). It prohibits at least an arbitrarily unequal treatment.[29]

The European Court of Justice[30] seems to measure at least parallel risk reduction decisions directly according to the principle of proportionality rather than the principle of equality or non-discrimination. This suggests that one could proceed in the same way with respect to preceding risk reduction decisions, which would then supply a standard for risk/risk comparison. Independent of the legal bases of risk/risk comparison it should be noted that the decision concerns Community measures. As regards restrictive measures taken by the member states, the prohibition of discrimination set forth in articles 30 and 95 VI EC Treaty and inherent in article 28 EC Treaty provides more reliable grounds for requiring consistency of risk regulation.

---

[26] This has been emphasised by the Federal Administrative Court, NVwZ 1998, 1181 at 1182.
[27] Administrative Court of Appeal Mannheim, NVwZ 1996, 297; NVwZ-RR 1999, 298 at 302; 1995, 639 at 644; Administrative Court of Appeal Munich, NVwZ-RR 2000, 661 at 664; Administrative Court of Appeal Münster, NVwZ 1991, 1200 at 1202.
[28] Administrative Court of Appeal Munich, NVwZ-RR 2000, 661 at 665.
[29] See generally Federal Constitutional Court, BVerfGE 90, 145 at 196; BVerfGE 50, 142 at 166.
[30] Case C-284/95 (Safety High-Tech), 1998 ERC I, 4301 at 4349.

In any case, two caveats are necessary: Comparability of risks is difficult to establish. Even unvoluntary deaths do not necessarily lend themselves to comparison, and the same is true of cancer fatalities (Bergkamp 2002, 75/76). Therefore, risk/risk comparison is not a crucial element of the decision on tolerability of risk. Moreover, even if the relevant risks are deemed to be comparable, justification of unequal treatment must remain possible. One should in particular recognise the need not to rely blindly on a preceding risk management decision but to allow for its re-evaluation when a risk/risk comparison is at stake.

10.4.2
Balancing advantages and disadvantages

A crucial question in this context is whether in deciding on tolerability of a particular risk the decision-maker is empowered to balance possible advantages presented by reducing a particular risk to human health with countervailing concerns and, if this is permissible and/or mandated, to what extent economic analytical tools such as cost/benefit analyses can or even have to be applied. At the outset it should be noted that a distinction between the decision on tolerability of risk and the measures to be taken for risk reduction is not easy to draw because both decisions are intertwined. Countervailing factors, especially costs of regulation, can only be fully assessed in relation to concrete measures for risk reduction (Ginzky and Winter 1999, 95). In particular, balancing risks and benefits also is relevant for comparing different regulatory options.

Under German law one makes a categorical distinction between the prevention of dangers and precaution against simple risk (Winter 1995, 12 et seq., 25). The obligation to prevent dangers is said to be "absolute" in the sense that when the competent agency finds a sufficient probability of harm the risk is deemed to be unacceptable and the decision cannot be qualified by balancing the protection of human health against countervailing, especially economic and social factors. Such balancing only plays a role with respect to the selection of measures. From a constitutional perspective, one may doubt whether this assertion is correct. The constitutional duty to protect against dangers is not absolute; rather, the legislature is accorded a wide margin of discretion. The controlling standard is what harm and dangers can be imposed upon the citizen considering countervailing interests.[31] At the level of statutory law, authorisation procedures normally do not allow for any balancing with countervailing interests when dangers to human health are at stake. By contrast, powers to intervene ex post are discretionary. A pertinent example is Section 17 Chemicals Act which accords the executive political discretion as to the question whether and how to act. This also includes a margin of appreciation for the relevant facts (Ginsky and Winter 1999, 42). However, this does not rule out a distinction between prevention and precaution. At least in case of serious dangers to human health the discretion is reduced to the extent that categorical protection may have to be provided (Ginzky and Winter 1999, 44/45; see also Köck 1999a, 344/345).

---

[31] Federal Constitutional Court, BVerfGE 56, 54 at 80.

By contrast, in the field of precaution the competent agency disposes of a wide margin of discretion and can and even is obliged to decide that a particular degree of risk is tolerable because the costs to achieve a lower degree of risk are considered to be excessive. German courts have expressly held that measures for precautionary reduction of risks must be proportionate to the assumed risk[32] – a position which is shared by legal literature (e. g. Ossenbühl 1986, 167; Wahl and Appel 1995, 135 et seq.; Ginsky and Winter 1999, 67 et seq.). The principle of proportionality requires that the measure chosen be suitable, constitute the least burdensome alternative and be not disproportionate to the objective pursued and the expected benefits. The principle of proportionality is open to the consideration of costs of regulation and therefore in principle also economic efficiency (Führ 2001, 177 et seq.; Meßerschmidt 2001, 226 et seq.; Winter 2001, 109 et seq.). The least burdensome alternative may in the opinion of some authors (Di Fabio 1997, 833; Sachs 2002, Art. 20 No. 101) already require a certain lessening of the effectiveness of the instrument selected and thereby of the protective level when an alternative option is associated with considerably lower costs. In any case, in the framework of proportionality in the narrow sense (adequacy of regulation) the relationship between the level of protection and the costs for achieving it through concrete measures must be considered.

There is wide agreement on the factors that may play a role in the balancing decision (Ginzky and Winter 1999, 60 et seq.). Relevant are on the one hand the severity and magnitude of possible harm and the degree of risk, on the other hand, among others, possible losses of manufacturers on the market, the costs for a change of the production process, international competitiveness, effects on the labour market, effects on consumer prices, the availability of, and risks associated with, substitutes and in principle also the foregone benefit derived from the substance in question. Most of these aspects are covered by fundamental economic rights guaranteed by the German Constitution (articles 12, 14 and 2). Although these rights are not protected unconditionally, it is imperative to accord them an appropriate weight in the balancing process (Ladeur 1994, 11 et seq.). There is no categorical priority for precaution against merely undesirable risks to human health. The mere fact that the Chemicals Act empowers the executive to adopt prohibitions and restrictions also for reasons of precaution does not in itself justify the conclusion that countervailing interests are *per se* inferior and become only relevant when the costs of regulation are "exorbitant" (contra Winter 1995, 24; 2001, 121/122; Risiko-Kommission 2002, 69).

However, in contrast to American practice (Executive order 1229 and some court decisions constructing reasonableness clauses of environmental laws) (see Driesen 1997, 555–560; Murswiek 2003, 7–17;) a formal cost/benefit analysis or any other formal balancing procedure is not prescribed (Köck 2001b, 287, 288, 294). The legal balancing factors can be characterised as flexible, more or less soft in nature. The apparent rigour of the principle of proportionality is attenuated by the recognition of a broad margin of political discretion including the appreciation of the facts. Therefore, in principle, only evidently disproportionate decisions can be held by the courts to be illegal (Ginsky and Winter 1999, 67, 104/105; Führ 2001, 202/203;

---

[32] Federal Administrative Court, BVerwGE 69, 37 at 43/44; NVwZ 1997, 497 at 499, 500.

Meßerschmidt 2001, 239/240). This also solves the methodological problem that on the side of health protection a more or less uncertain and future risk has to be balanced against disadvantages that mainly are certain and present. In this situation one can be content with a determination that the expected disadvantages are not evidently disproportionate to the assumed improvement of health protection, considering the severity and extent of possible harm and the degree of plausibility of the assumed risk (Rehbinder 1997, Nos. 50/51; see also Petersen 1993, 288 set seq.; Wahl and Appel 1995, 138/139). This restriction of judicial review accommodates for existing uncertainties of knowledge and prediction, and reflects fundamental principles of the Constitution, in particular the democratic principle and the division of competences between the different state organs. Instead of a comprehensive scientific programme of material rationality the legal model relies to a large extent on procedural rationality.

Article 174 III EC Treaty prescribes that in developing precautionary environmental policy the costs and benefits of action and non-action must be considered and this can also be extended to article 95 EC Treaty. Article 5 EC Treaty contains a general pronouncement on the principle of proportionality which therefore must be respected by the Community organs in implementing precautionary policies (Bergkamp 2002, 74). Quantification is not required. The Community organs dispose of a wide margin of political discretion (Ginzky and Winter 1999, 100, 104/105). These standards also apply when member state measures are scrutinised under the reasonableness test of article 28 EC Treaty.[33]

However, European practice takes a position which contains strong elements of neoclassical economic theory and is closer to the American example. The Commission communication on the precautionary principle (EC Commission 2000) postulates that its application must include a cost/benefit assessment (advantages/disadvantages) with an eye to reducing the risks to a level that is acceptable to all the stakeholders. There must be a real net benefit to society in reducing the risk to an acceptable level. If the elimination of a risk is very costly to society in social-economic terms or it involves alternatives resulting in a higher risk or in shifting the risk to another population group, the decision-making agency, according to the Commission paper, must consider to take no measures at all. In a dictum it is expressly stated that the model of cost/benefit assessment is not restricted to the application of the precautionary principle but applies to all risk management decisions, which could mean to include preventing dangers. More generally, a recent Commission Paper (Action plan "Simplifying and improving the regulatory environment") (EC Commission 2002) proposes to subject all EC regulation to cost/benefit assessment. Likewise, the German Advisory Council on Global Environmental Change (WBGU 1999) has expressly taken the view that balancing also has a (limited) role in the field of preventing dangers. However, it should be noted that there is as yet no statutory basis for formal cost/benefit analysis in health-related environmental legislation of the EC. Rather it is based on the discretionary elements of rule-making powers of the EU organs.

The decision on tolerability of risk may also include a risk/benefit analysis which compares the risks presented by the agent with the benefits derived from its

---

[33]   European Court of Justice, Case C-47/98 (Toolex./.Alpha), 2000 ECR I, 5681 at 5714–5717.

use. In contrast to ionising radiation the need for or necessity of the agent as such is not normally questioned (but see Winter 1995, 58 et seq.). Rather, the need or necessity is only assessed in relation to the risk associated with the agent (Ladeur 1995, 13/14; Theuer 1996, 129; Ginzky and Winter 1999, 63 et seq.). In essence this amounts to a search for, and selection of, substitutes that serve the same purpose but present a lower overall risk. The idea underlying risk/benefit analysis is that when there is a considerably less risky alternative the acceptance of a higher risk associated with the agent in question cannot be justified. Therefore, risk/benefit evaluation is close to minimisation of risk; it is not applicable when the risk is so low that the level of residual risk is reached (Köck 1999b, 92). Risk/benefit analysis is required with respect to health in the regulation of occupational exposures, biocides, pharmaceuticals and under German law in an attenuated form with respect to chemicals in general (Section 17 II Chemicals Act). Pesticides law limits risk/benefit analysis to adverse effects on nature.[34]

However attractive risk/benefit analysis may appear, it is fraught with a number of methodological problems. It is controversial whether risk/benefit analysis is at all a separate kind of cost-related analysis (Wiegand 1997) or whether it is but a part of general cost/benefit analysis because substitutes may be integrated into the balance of benefits (avoided harm due to lower risk presented by the substitute) and costs (costs of regulation due to higher costs of, and lower benefits from, the substitute) (Hansjürgens 1999, 293). However, the real problems lie in more fundamental questions.

First of all, the meaning of benefit must be clarified. It could be confined to effectiveness of an agent for a particular purpose. It could also mean the effect or result achieved by using the agent such as the reduction of losses of life and diseases or the lowering of treatment costs. Since risk/benefit analysis forms a part of the balancing process which includes disadvantages incurred by society as a result of the use of an agent, it would make sense to opt for the wider perspective. On the other hand, the decision-making process may become over-complex. Apart from this fundamental question the scope of alternatives to be included in the comparison may be doubtful. The question is whether in addition to substances and other agents also non-material alternatives, such as processes, shall be considered. Furthermore, the proof (including burden of proof) of a lower risk potential of an alternative raises problems when the substitute has not yet been fully assessed. The comparison of a substance with incomplete data and a substitute whose risk potential is fully understood constitutes an ideal case which will be rare in reality. Moreover, in contrast to agents with limited use such as pesticides, biocides and pharmaceuticals, in regulating new chemicals for general use, a variety of potential uses not yet known to the producer could theoretically be considered. This would unduly burden both, the producer who has to generate data and the assessing authority. Finally it has to be decided as to what extent trade-offs are admitted (comparison of agents with high risk potential and high utility and agents with low risk potential and low utility and any combinations of these properties). Although this problem can be built into,

---

[34] See Federal Administrative Court, BVerwGE 81, 12 at 17 (pesticides). However, it is doubtful whether the EC Pesticides Directive (91/414) which refers to the acceptability of adverse effects on nature allows at all for a risk/benefit assessment.

and accommodated for, in the balancing process, decision-making may become over-complex. A possible response to these challenges is a pragmatic limitation of risk/benefit comparison to benefits constituted by effectiveness/utility rather than general advantages associated with the agents, to a limited kind of substitutes with a risk potential already assessed, to concrete uses of new substances and to alternatives that combine a clearly lower risk potential with a high effectiveness/utility (cf. Ginzky and Winter 1999, 66/67; Köck 1999b, 88/89). It has also been proposed to confine risk/benefit analysis to priority groups of substances classified for such an analysis (Ladeur 1995, 14).

Finally in this context the question arises of whether and to what extent risk/risk tradeoffs may or even need to be considered in the balancing process. This concerns ancillary risks that are associated directly or indirectly with "target" risk reduction. Inclusion should not be doubtful in case of ancillary risks directly presented by target risk reduction measures (e. g. adverse side effects of the use of pharmaceuticals and pesticides) or by substitutes selected to tackle the target risk. However, the consideration of general health/health tradeoffs which is often advocated in the United States (Sunstein 1996 a; Rascoff and Revesz 2002, 1778–1790) poses more serious conceptual and methodological problems. Health/health comparison rests on the assumption that expenditures for reducing target risks to health may, as they have negative income effects, reduce public and individual welfare and due to lower health care expenses ultimately contribute to increased health risks in other areas ("richer is healthier" paradigm). However, this concept is fraught with insurmountable data collection, assessment and attribution problems and would open the Pandora's box of an infinite regression. Under the auspices of increasing the rationality of public decision-making it would render the decision-making process entirely indiscriminate and speculative (Mc Garity 1998, 46–49; Heinzerling and Ackermann 2002, 650 et seq.; Murswiek 2003, 29–31). Moreover, this concept is unilateral because it is blind to ancillary benefits that may also be associated with measures to reduce target risks (Rascoff and Revesz 2002, 17901813). A certain reduction of social complexity is needed if one does not want to overburden the problem-solving capacity of a decision-making system.

10.4.3
The role of "soft" balancing factors

It is doubtful to what extent "soft" factors such as distributional justice and risk perception are relevant in the balancing process (see also sections 2.3, 2.3.8, 2.3.12 and 9.3). A uniform answer is not possible, and much depends on the legal framework that governs the decision-making.

There is wide-spread agreement that distributional issues have to be addressed in the balancing process concerning the determination of tolerability of risk (Britz 1997, 197 et seq.; Meßerschmidt 2001, 231 et seq.). This is mandated by the principle of proportionality. It follows that insofar as efficiency is a factor in the decision-making process it must be balanced with distributional concerns; thus the principle of proportionality may result in setting limits to considering costs. The distributional issue concerns both the protection of vulnerable groups and the reduction of unequal exposures.

As regards the protection of vulnerable groups, the regulator must decide on the shaping of the relevant group which primarily concerns the question as to what extent diffuse groups shall be included. As stated in section 10.2.3, the regulator disposes here of a margin of discretion. The severity of the expected disease, the size and kind of the groups, the quality of risk (imposed or voluntary) and the possibilities of self-protection have to be balanced against the overall costs for society incurred by specific regulation in favour of vulnerable groups. One can also draw different conclusions from the problem of susceptibility regarding the measures to be taken, which has a bearing both on societal costs and equity. One can orient the relevant standards at the susceptibility of the relevant vulnerable group thereby providing for a considerable and costly degree of overprotection for the public at large. When the members of the group can be separated, base standards for protecting the public at large accompanied by provisions that allow for such separation, for example by a provision that prohibits the employment of pregnant women in facilities where they would be exposed to relevant substances, would seem to be possible. However, this would result in a certain degree of discrimination which is in principle undesirable. The consideration of vulnerable groups may also depend on exposure patterns (e. g. exposure of children to benzene from car exhausts or to hazardous soil contaminants).[35] Clear-cut answers are not available.

In the second place, distributional issues raised by an unequal exposure to carcinogenic agents have to be addressed. It has already been stated that persons with high exposure may form a separate vulnerable group, which implies that the particular risk of this group must be reduced above all. However, the real problem goes beyond the protection of high exposure groups. The question is whether and to what extent, under the perspective of environmental justice, an unequal distribution of risk as well as an unequal distribution of risks and benefits ensuing from a risky activity can or even have to be addressed in the balancing process. Problems of environmental justice have as yet not played a particular role in German environmental policy and law (Kloepfer 2000; Maschewsky 2001, 145 et seq.; Rowe 2001). The Federal Land Use Planning Act aims for "equivalent living conditions" at the level of state-wide and regional planning. However, the law of local planning as well as the law relating to noise control follow a concept of local differentiation of environmental quality. There are no environmental justice clauses in health-related environmental laws. Regulatory practice has not focussed on environmental justice. A major exception is presented by the system of guide values for carcinogenic air pollutants established by the State Commission on Air Pollution and Noise Control (LAI 1992). In view of a strong divergence of the population risk of cancer fatalities due to air pollutants between urban and rural areas (1 : 5000 vs. 1 : 1000) the State Commission based the ambient guide values for six carcinogenic air pollutants on the objective of reducing the population risk in the whole territory of Germany to at least 1 : 2500 in order to ensure equivalent living conditions. Issues relating to the fair distribution of risks and benefits have played a certain role in informal local agreements on the siting of polluting facilities rather than in the development of generic policies for risk reduction.

---

[35] Administrative Court of Appeal Münster, NVwZ 1991, 1200 at 1202.

Environmental justice is normally no relevant issue with respect to prevention of unacceptable risk. Since all individuals must be protected against unacceptable risk categorically, the distribution of such risks at most plays a role when one has to decide on risk reduction measures. Moreover, one can also state that ensuring a minimum of safe living conditions for the whole population rather than environmental justice is the primary concern of health-related environmental policy and law. However, there should be no doubt that issues of environmental justice can be addressed in the balancing process with respect to precaution against undesirable risk; their inclusion is covered by the broad discretion the regulator is accorded here. Distribution of risk is part of the balancing process and has to be weighted against the cost for society incurred by taking precautionary measures for the benefit of environmentally disadvantaged groups. Likewise, distribution of risks and benefits can be considered to some extent. However, this is limited by the generic nature of environmental standard setting. While it may be possible to determine unequality in administrative decisions on individual cases, an attribution of risks and benefits to particular groups of the population is more complex with respect to standard setting; in particular the disadvantages of a higher risk in agglomerations are to a great extent compensated for by economic and infrastructural benefits (see sections 9.2.5, but also section 2.3.7).

Whether there even is an obligation to consider an unequal distribution of risk is open to doubt. The only "hard" legal standards for ensuring environmental justice can be derived from the constitutional guaranty of equality under article 3 I Federal Constitution (Salzwedel 1992, 12–14). The standards for determining an unconstitutional unequal treatment vary considerably, ranging from the prohibition of arbitrariness, i. e. the lack of any reasonable consideration which may support the regulation[36], to the requirement of justification, which means that the differences on which regulation is based must be of a kind and weight that they justify the unequal treatment.[37] The problem is that spatial differentiation of environmental quality as such can arguably be justified on various grounds, provided that a minimum of safe living conditions for the whole population is ensured.

Environmental injustice is normally not due to a targeted unequal treatment of particular persons or groups but frequently only the indirect result of area-wise differentiation of environmental standards. It may be associated with historic development of agglomerations and the market prices for housing and hence be part of the "general risks of life". This leads to the question of whether environmental injustice can be better addressed as a form of indirect discrimination, i. e. formally equal treatment on the basis of neutral criteria – spatial differentiation of environmental quality – but whose impact is, that in the vast majority of cases members of particular social groups are affected. Apart from discrimination based on gender (article 3 II) the German Constitution does not prohibit indirect discrimination.[38] The EC Directive 2000/43 which does, is limited to race or ethnic groups that, in contrast to income and social status, are not very relevant in our context. However, this does

---

[36] Federal Constitutional Court, BVerfGE 1, 4 at 52; BVerfGE 68, 237 et 250; BVerfGE 71, 39 at 53.

[37] Federal Constitutional Court, BVerfGE 63, 251 at 262; BVerfGE 75, 284 at 300; BVerfGE 81, 208 at 224; BVerfGE 82, 126 at 146.

[38] Federal Constitutional Court, BVerfGE 64, 135 at 156/157.

not mean that in applying the general equal protection tests relating to direct discrimination one must entirely disregard the factual impacts of spatial differentiation on people living in the affected area. Rather, space has to be considered as a composite notion that encompasses all its endowment factors. If this is correct, then one must also consider, in selecting the standard of review, whether and to what extent the regulation constitutes an encroachment on the protective scope of fundamental rights.[39] This suggests a stricter standard of review in the sense of a justification requirement – rather than the prohibition of arbitrariness. In case of carcinogenic agents, the weight of the right to human health and the practical impossibility of delimiting unacceptable risk (prevention of danger) and merely undesirable risk (precaution) militate in favour of the proposition that at least a certain equalisation of exposure is mandated (Salzwedel 1992, 13/14). Total equalisation of typical exposure to carcinogenic substances is clearly not required by the Constitution.

The remaining unequal distribution cannot be fully remedied under a legal perspective by granting compensation (see also 2.3.8). Under German public law compensation by the state is limited to environmental burdens that are equivalent to a taking or are otherwise intolerable. Although some authors advocate de lege ferenda a more extensive use of compensation with respect to siting conflicts, for promoting greater flexibility, securing environmental justice and improving acceptance (Suhr 1990; Voßkuhle 1999, 80 et seq.; Kloepfer 2000, 753/754), the prospect of thereby creating a "total compensation state" is a clear deterrent to legislative proposals in this direction. In private law compensation mechanisms are somewhat more developed. Significant emissions from neighbouring land that are customary in the neighbourhood and cannot be prevented or reduced by use of best available technology must in principle be tolerated by the victim. They may give rise to a claim for compensation. However, this only is the case when the victim's normal use of his or her land is impaired to an unreasonable degree. Once again, instead of a comprehensive compensation model, in order to enable dynamic development in the neighbourhood without privileging the status quo, the law takes recourse to balancing of interests and limits compensation to extreme cases of harm.

Subjective risk perception by affected citizens or the public at large and related factors such as divergences in societal evaluation of risk and conflict potential may not be a factor in deciding whether a particular risk constitutes a danger which must be prevented categorically. When the decision on the tolerability of the risk is subject to balancing, especially in the field of precaution, the subjective risk perception by citizens is in principle an element of decision-making and a prerequisite for securing acceptance of the decision by the citizens. In this respect both risk perception in the narrow sense and the perception of the fairness of risk/benefit distribution play a role. However, there are constitutional and there may be statutory limits to the consideration of risk perception. It would be contrary to the political responsibility of the regulator to consider evidently unreasonable risk perceptions such as phantom risks or extreme carelessness. Likewise, regulation implemented on the basis of a unilateral consideration of irrational risk perceptions would, insofar as it encroaches upon fundamental economic rights of polluters, violate the principle of

---

[39] Federal Constitutional Court, BVerfGE 74, 9 at 24; BVerfGE 88, 87 at 96; BVerfGE 91, 346 at 363.

proportionality and in appropriate cases also that of equality (SRU 1999, 80). Moreover, empowerments to set precautionary health standards often do not allow for the consideration of subjective risk perception. Risk perception may be relevant, though, for setting priorities in the standard setting process. All told, in a legal perspective, in contrast to ethical considerations (section 2.3.3) the role of risk perception is bound to be limited.

10.4.4
The role of cost/benefit analysis

It appears appropriate to contrast the juridical concept of decision-making on tolerability of risk with that of (neoclassical) economic theory, especially as regards formalised decision-making methods. Neoclassical economic theory already denies the reasonableness of "absolute" protection of human health and takes the view that questions of acceptability of risk must always be decided on the basis of balancing costs and benefits for society as a whole (Cansier 1994; Gawel 1999, 267/268, 299 et seq.; 2001, 259 et seq., Hansjürgens 2001, 74/75). However, practically, as will be exposed, we have to do in this respect with a divergence of concepts and methods rather than results. This is at least true of European thinking while in the United States more radical – and quite influential – modes of thought do exist which would valuate life and health on economic terms (including discounting future harm) and protect it only to the extent that their value exceeds the costs incurred; thereby one would subordinate individual life and health to societal welfare (e. g. Sunstein 1996b).

As has been stated, the empowerments and obligations to act in case of danger already embody a number of value judgements, although partly only in an implicit and highly differentiated form (Köck 2001b, 287). The attributed value of the interest to be protected and the need for protection are paramount but costs are not irrelevant. Therefore the accusation of "blindness to costs" or arbitrary reclamation of a "market-free space" is an overstatement. The conflict is on the degree of formalisation, especially monetarisation; it more or less reflects the general methodological differences as to decision-making on risks. Many European economists accept the division of labour between command-and-control regulation in the field of danger and (possible) market-based instruments in the area of precaution. Although this does not necessarily imply a renouncement to cost/benefit analysis it is conceded that in view of the value attributed to human life and health by the Constitution, the high degree of risk and – at the instrumental level – problems of effectiveness, cost/benefit analysis would normally not render different results and therefore a "rebuttable presumption of exclusion from the market" is acceptable (Gawel 1999, 299 et seq.; 2001, 261/262; Hansjürgens 1999, 288; 2001, 74/75). Then the more relevant divergence of opinion is that in contrast to the legal concept, economic theory favours the carrying out of formal cost/benefit analysis which would require a translation of all effects and foregone effects of a given course of action on hazardous agents into monetary terms. As an alternative, cost-effectiveness analysis is considered, that does not allow to question the objective – namely effective protection of human health – as such, but only aims to ensure that a given objective be achieved at least cost.

The divergence between legal concepts of decision-making about tolerability of risk and risk reduction measures and the respective economic concepts of formalised decision-making have recently been the subject of interdisciplinary debate in which a certain approximation of the respective positions can be ascertained. Cost-benefit analysis aims to weigh up, in a monetarised fashion, the advantages and costs of alternative action, e. g. in our context of decisions on the level of tolerable risk and in particular for risk reduction measures; that action shall be taken which yields the highest net benefit and thereby increases overall welfare. Cost-benefit analysis requires adequate valuation methods for converting those factors which are not by themselves valuated on the market, especially the benefits of alternative action, into monetary terms. Such methods include contingent valuation of the risk presented by the hazardous agent (willingness to pay method), valuation of evasion costs, valuation of damage restoration costs (treatment costs for curing diseases caused by exposure to the agent) and indirect valuation by reference to other diseases not related to exposure to the agent (Endres and Holm-Müller 1998, 33 et seq.; Hansjürgens 1999, 322 et seq.). Although one can point to considerable progress that has been achieved over the years in refining these methods, there are limitations to monetary valuation of risk to human health which according to the respective basic position can be seen as impediments to using cost-benefit analysis at all or as causes for a pragmatic relativation. These limitations have a wide range: ethical doubts as to overt monetary valuation of human life and health (implicit valuation in society cannot be denied); problems of adequately dealing with uncertainty and ignorance about risk and the resulting limitations to quantitative risk assessment (which are to a certain extent tackled by normative assumptions expressed in subjective decision rules for uncertainty); difficulties in extrapolating from empirical data gathered for persons exercising risky occupational activities to the population at large; the problem of valuation of benefits derived from presently reducing long-term, future risk of present and future generations (problem of discounting); neglect of societal costs associated with the omission of regulating risks; and criticism by modern ecological economics relating to fundamental assumptions of cost/benefit analysis (existence of a substitution potential and reversibility). Critics assert that, as a result quantification may be misleading (Driesen 1997, 583–600; Mc Garity 1998, 37 et seq.; Köck 1999b, 88/89; 2001b, 295; Revesz 1999, 987–1016; Heinzerling and Ackermann 2002, 648 et seq.). Moreover, cost/benefit analysis is cost intensive and time-consuming.

Cost estimates as well as semi-quantitative or even qualitative cost/benefit analysis may be a response to these difficulties (Hansjürgens 1999, 348, 366/367; 2001, 84). Although this would approximate cost/benefit analysis to the practiced legal concepts of decision-making (Führ 2001, 204/205) and run the risk of abandoning some of its methodological advantages, cost/benefit analysis would still provide for a high degree of transparency and ensure that risks and costs are considered simultaneously in a systematic fashion. Moreover, it is correctly emphasised that cost/benefit analysis only supplies an input into the decision-making process. It is a decision-making tool that does not pre-empt the ultimate political decision of the regulator on tolerability of risk and risk reduction measures. This decision must remain in the political sphere (Driesen 1997, 606–613; Hansjürgens 2001, 82 et seq.; Murswiek 2003, 44). It can consider the robustness of data and remaining uncertainties and also correct the unilateral reliance of cost/benefit analysis on efficiency by including, and even giving priority to, distributional aspects. There remains the problem of a possible distortion of the political process by the possibility that hard, quantitative data may gain an excessive weight over soft, unquantified risk data. This mandates adopting risk evaluation guidelines that emphasise the importance of qualitative data; it does not justify an outright rejection of cost/benefit analysis. Cost/benefit assessment is a useful tool for securing transparency as to the criteria of the decision-making process and to this extent may contribute to a more informed, rational decision (Führ 2001, 205). In view of its inherent limitations, even some economists tend to favour simple cost/effectiveness analysis which concentrates on costs of regulation and does not allow questioning the protective level as such whenever it appears that the health benefits derived from alternative risk reduction measures are about equal (Hansjürgens 1999, 348, 367/368). This of course is a minimal common denominator which is also acceptable from a legal point of view (Winter 2001, 122). However, it is only a partial solution since equivalence of benefits will not always exist.

10.4.5
The role of extrapolation and safety factors in the risk management process

Extrapolation is a means of translating, in the process of risk assessment, empirical results generated by animal tests to humans allowing for prediction of possible future harm. In spite of a certain discretionary element, extrapolation has an empirical and statistical basis and is essentially cognitive-predictive in nature. The empirical basis varies with the method of extrapolation. The general extrapolation factors used in practice only partly have an empirical content and thereby reflect to a certain degree the measure of desirable precaution the risk assessor assumes. Even if a modification depending on the degree of empirical data is allowed for or extrapolation is based on the dose-effect curve, there is no clear-cut borderline between the desire to avoid errors in the cognitive process of predicting future harm (reliability) and the desire to provide for an ample margin of safety which basically constitutes a precautionary measure (tolerability). This becomes particularly evident when one has to decide whether to base risk assessment on upper bound estimates or even a worst-case scenario or rather on best professional judgement. It would seem that the choice between these two approaches also depends on the weight of the interest to be protected, which may justify a conservative worst-case scenario where serious, irreversible harm to human health is at stake (Risiko-Kommission 2002, 52). However, this already is the domain of risk management. Similar considerations prevail regarding the no-threshold hypothesis with respect to carcinogenic and mutagenic substances. The no-threshold hypothesis serves as a criterion for predicting potential harm to human health in the low-dose range; on the other hand, it is a constituent element of the decision on the tolerability of the risk presented by such substances.

No true cognitive-predictive elements are contained in safety factors which are either used in the risk assessment process for considering factors that cannot be quantified at all (e. g. general uncertainty) or in risk management for setting envi-

ronment-related health standards. The most frequent example for the latter procedure is the setting of ambient quality standards at a certain percentage of the
NOAEL (e. g. at 50 or 25 %). Such safety factors are essentially of a normative
nature and therefore belong to the risk management process. This means that the
regulator, when the precautionary principle rather than the preventive principle can
be applied, disposes of a margin of discretion. However, this is not a licence for irrational decisions (cf. Risiko-Kommission 2002, 57). It is only when a reliable determination of the threshold of danger is not possible that safety factors are mandated
by the constitutional duty to protect (Winter 1986, 134/135; Salzwedel 1987, 277;
Bohl 2002, 5/6).

A pertinent example of problems of legitimacy in this respect is presented by the
translation of a decision taken on the acceptability of risk associated with exposure
to hazardous substances at the work place to exposure by the public at large. Here,
a conversion factor of 100 is frequently advocated. Although one could sustain that
such a safety factor reflects differences in temporal exposure and susceptibility,
arguably the major implication of this – disputed – concept is to provide the public
at large a higher margin of safety. Since there is no rational basis for selecting a
general factor of 100, the accepted higher vulnerability of the population at large
should be reflected by more complex quantitative and qualitative criteria, e. g. by
transposing the test results to a full day exposure and extrapolating to a whole population collective including relevant vulnerable groups using the normal extrapolation factors (Fischer 1996). Depending on individual circumstances, this may or
may not correspond to the composite factor of 100.

10.4.6
Problems of quantification

In some countries it is usual to delimit the field of unacceptable risk, undesirable
risk and residual risk by means of quantification. The most explicit programme of
this kind has been developed by the various environmental programmes of the Ministry for the Environment of the Netherlands. This position was first set forth in the
Indicative Environmental Programme of 1986–1990 and then further developed by
the report "Omgaan met Risico's" (Risk Management) which was published as an
appendix to the First National Environmental Plan of 1989. With respect to no-
threshold substances, the unacceptable individual risk is set at 10^{-6} and the residual
risk at 10^{-8} yr^{-1} (translated into life-long exposure 70–80 × 10^{-6} respectively, which
is close to 10^{-4} or 10^{-6} respectively). In the "grey area" between unacceptable and
residual risk, which in the German terminology would be the domain of precaution,
additional measures for reduction of risk, considering the principle of proportionality, must be carried out. If possible, the figure that denominates the beginning of the residual risk shall be attained. With regard to other substances, it is proposed to set forth a limit value considering a margin of safety whose exceedance
denotes the field of unacceptable risk. The residual risk is set at 1 % of the limit
value. In the "grey area" between unacceptable and residual risk additional measures for risk reduction, considering the principle of proportionality, shall be carried
out. Here too, the objective should be to reduce the risk until the figure which
denotes the threshold of residual risk is reached. Additional quantitative limits

apply to combined effects. After the publication of the report on risk management it was questioned whether the determination of a fixed quantitative risk value of 1 % of the value set forth for unacceptable risks was scientifically and economically reasonable. Especially the Dutch parliament took the view that the application of the minimisation principle (ALARA principle, article 8.11 Environmental Protection Act) without a rigid lower limit for risk reduction was sufficient. However, it appears that the Dutch government at least formally sticks to its position. In the Second Environmental Plan of 1994, the risk objectives were confirmed. Later plans do not mention the subject. In practice the simpler application of the minimisation requirement is apparently preferred.

A political concept of quantification also exists in the United States, although a formal distinction between danger and precaution is not accepted. Quantification concepts mainly relate to the determination of residual risks, in other words describe the borderline between risk that is unacceptable or otherwise must be reduced and residual risk that shall be tolerated. Generally speaking the individual risk presented by carcinogenic substances must be reduced up to a risk of 10^{-4} to 10^{-6} in case of life-long exposure. A risk of 10^{-6} is considered to be "virtually safe" (Overy and Richardson 1995). With respect to carcinogenic air pollutants, Section 112 (f) (2) of the Clean Air Act provides that a quantitative risk of no more than 10^{-6} shall be the basis of emission standards for such pollutants.

In Germany and the European Community similar quantification concepts are less frequent. However, a life-long individual risk of 10^{-6} also seems to be accepted in practice (BGA 1981). It has also been stated that 1/1,000 of the maximum exposure standard at the workplace is a value that describes a virtually safe level of exposure (Fischer 1996, 65). In the field of ionising radiation quantification of tolerable risk is widely practiced, although normally together with the ALARA principle and the principle of justification of exposure. Thus in the past, one considered a cancer fatality risk from all sources of 10^{-5} yr^{-1} and 10^{-3} at life-time exposure as tolerable if in addition justification and minimisation was required (Environmental Standards 1998, 154 et seq., 169/170). The new dose limits established by directive 96/29/Euratom accept an annual risk of members of the public at large of 7.3 in 100,000 (which is below 10^{-4}) and of workers in the range of 112 in 100,000 (which is about 10^{-3}). However, the relevant risk is a composite one ("detriment"), comprising besides the risk of cancer fatality other risks from exposure equivalent to fatalities such as non-fatal cancers and genetic disease which are weighted and transposed into additional fatalities (ICRP 1990; Clarke 1999). Recent German court decisions held that an additional cancer case stemming from carcinogenic substances of less than 10^{-6} or for members of vulnerable groups of 5×10^{-7} was not significant (in one case comparing it with the much higher risk of being killed by a strike of lightning).[40]

A major attempt at quantifying the tolerable risk in case of exposure to no-threshold chemicals has been undertaken by the Länderausschuß für Immissionsschutz (LAI – State Commission on Protection against Air Pollution and Noise) with

---

[40] Federal Administrative Court, NVwZ 1998, 1181 at 1182 (relating to cadmium emissions from a waste incinerator); Administrative Court of Appeal Munich, NVwZ-RR 2000, 661 at 664 (relating to carcinogenic substances in general).

respect to six major carcinogenic air pollutants (LAI 1992; see also Fischer 1996, 71 et seq.). The limitations of this concept are due to the fact that according to the Commission's own calculations the overall share of cancer caused by exposure to external air pollutants is about 2 % while other calculations set this share even much lower, namely at 0.2 % (Fischer 1992, 188; Franßen 1993, 37). The Commission report assumes that a population risk of roughly 1 : 2,500 (27 cancer cases in a population of 100,000 = 3×10^{-4}) is tolerable, the more so since it is deemed to largely correspond to the risk associated with the traditional 0.3 mSv concept in radiation protection. It is noteworthy that the Commission deliberately avoided a pronouncement on the question at which point an unacceptable risk (danger) to human health exists, probably because this would have required the setting of a standard for the individual risk. The risk factor of 1 : 2,500 has the main function of approximating the population risk in urban areas to that in rural areas (from 5 : 1 to 2 : 1) in order to create equivalent living conditions. On the basis of average frequency of occurrence (ambient load quota), carcinogenic potency and the assumption of dose additivity of all carcinogenic substances the State Commission set forth ambient quality guide values which are designed to ensure meeting the risk factor of 1 : 2,500 in the whole territory of Germany.

The fact that in Germany there is no general concept of translating the notions of danger, simple risk (precaution) and residual risk into numerical terms has two reasons: One seems to be that a qualitative determination immunises the decision-maker from political and legal challenges and better reflects the relative concept of risk recognised in Germany which, besides the degree of probability or plausibility, also considers the severity and extent of harm including the value attributed to the interest at stake by the Constitution. Secondly, the prevailing concept of precaution is to make use of technical means. This also implies that precaution is limited by availability of technical means. German environmental law focuses on the use of the state of the art or – limited to ultra-hazardous activities – minimisation in order to reduce the risk rather than establishing precautionary quality standards. Hence quantification is not seen as indispensable.

As regards the first rationale, the problem consists in choosing among two evils: possible arbitrariness of a merely qualitative determination of relevant risk which does not force the regulator to unveil the true reasons underlying the decision vs. possible pseudo-rationality of risk figures. It is submitted that a quantification concept is in principle useful because it allows for a more transparent and hence more rational discussion of acceptability and tolerability of a particular risk. However, the relevant risk figure can only be a guideline or target since the determination of the level of tolerability can only be done in a balancing process that considers all aspects of the individual case. The second rationale is acceptable if one accepts its underlying premises. This will be discussed later in the next section.

10.4.7
Health-related standards for low doses

Standard setting for the protection of human health has had a long tradition in Germany. As regards air pollution, the first standards were already set forth by the Prussian Guidelines for Industrial Facilities of 1895 without distinguishing between the population at large and workers' protection. However, these administrative rules did not yet contain numerical standards. In the second half of the 20th century, the regulation for the protection of workers and that of the population at large developed separately. Explicit ambient quality standards and technology-based standards for the protection of the population at large were adopted by the Technical Guidelines for the Protection against Air Pollution of 1964. With the enactment of the Federal Emissions Control Act in 1974 the regulatory activities were intensified, as evidenced by several new versions of the Technical Guidelines for the Protection against Air Pollution in 1974, 1983, 1986 and 2002. Chemicals received special attention with the enactment of the Chemicals Act in 1981, although particular uses of chemicals such as pesticides or chemicals in food products had been regulated much earlier. Since the 1970's (Regulation on Dangerous Substances at Workplace of 1971, based on the Act on Dangerous Substances at Workplace of 1939) workers have been protected explicitly by a series of consecutive regulations concerning risks presented by dangerous substances used at the workplace.

The first official standards for the protection against ionising radiation in Germany were set forth in 1960 (Regulation on Protection against Ionising Radiation) which then were modified several times (especially in 1976, 1989 and 2001). First regulations for radiological protection in medical uses of ionising radiation were adopted in 1987. Before that time, since the 1950's, the practice had followed the recommendations issued by private national and international bodies such as the US National Council on Radiation Protection and Measurements and the International Commission on Radiological Protection, which also deeply influenced official standard setting.

At European level, express powers to regulate adverse impacts of dangerous substances on human health and the environment were only introduced by the Single European Act of 1987 (article 130 r), powers for regulating public health as well as health and safety at work by the Maastricht Treaty of 1993 (articles 137, 138, 152). However, the European Community had used its harmonisation competences to regulate both in the fields of pollution control and of toxic substances and occupational health (e. g. Directive 67/548 on Dangerous Substances, Directive 80/778 on Drinking Water, Directive 80/779 on limit values for SO_2 and NO_x, and Directive 80/1107 on Workers' Protection). The extension of Community powers has led to an intensification of regulatory efforts, as evidenced by the adoption or amendment of numerous directives that set numerical standards in various fields of health and the environment. By contrast, the European Atomic Community from its very incipiency has had competences in the field of radiological safety both of workers and the public at large (articles 30, 38 EURATOM Treaty) and already in 1959 set forth the first minimum safety standards for radiological protection (amended several times, now Directive 96/29/EURATOM). Safety for the medical use of ionising radiation was first regulated in 1984.

The standard-setting concept regarding low doses differs considerably depending on whether the regulatory agency has to deal with a threshold or a no-threshold agent. For threshold agents normally ambient quality or recipient-oriented standards are set that are based on the results of the risk assessment, reduced by application of a safety factor. Prominent examples are low-dose exposure standards in the field of occupational health where the safety factor applied is relatively low, and ambient quality standards for air pollutants for protecting the health of the general population where a higher safety factor is employed. The relevant standards may attach to different steps along the pollution pathway. Emission standards limit the permissible concentration or volume of agents source by source. Ambient quality standards relate to the concentration of the agent over time (short-term and/or long-term exposure) in environmental media. Product standards limit the concentration of relevant agents in the product itself, e. g. alimentary products or drinking water. Recipient-oriented standards consider the occurrence of the agent in the object to be protected. Such standards sometimes consider all exposure paths (e. g. the acceptable-daily-intake concept in alimentary product standards), but more often isolate a particular exposure path. This may be justified when different modes of exposure (ingestion, inhalation or contact) lead to different adverse effects. However, it is not in conformity with the requirement of effective protection, at least not when the exposure paths are equivalent in the sense that the same end point is affected (cf. Bohl 2002, 11). In any case, the allocation of quota per exposure path causes some difficulties.

Numerical ambient quality or recipient standards for no-threshold agents, especially carcinogenic and mutagenic substances, are rare. The normal approach of German environmental law in this field has been a reduction of emissions or exposure according to the state of the art (best available technology – BAT) or – or even in addition to it – minimisation (ALARA – as low as reasonably achievable). This is in particular true of air pollution, water pollution and occupational health. Especially minimisation embodies rather stringent requirements for avoiding exposures. However, clearly disproportionate measures are not mandated.[41] Therefore, it would seem that the difference between ALARA and ALARP (as low as reasonably practicable) which is sometimes advocated as a less demanding option for radiation protection (see 2.2, 9.2.5) is not very great. This is at least true if one understands the latter principle as to mean that unjustifiably expensive measures to achieve marginal reductions are not required (IAEA 1998). Although the principle of proportionality is inherent even in minimisation in all its facets, so that the degree of assumed risk can be reflected in designing risk reduction measures, from an economic perspective one may argue that minimisation tends to either achieve excessive safety or is irrational (Gawel 2001, 264/265); for if there is no rational basis for setting a quality standard, there is none either for requiring minimisation. The preference for an emission-oriented concept (BAT or ALARA), which some authors have coined as the "taboo" of German environmental law (Franßen 1993, 28), has two interrelated reasons. One is that, as long as the no-threshold hypothesis is maintained, a delimitation of the different levels of risk is difficult, if not impossible. Hence standard setting would imply the deliberate acceptance of a certain risk of the individual. Politicians and administrators are reluctant to assume political responsibility for such an open political decision about tolerability of risk. Secondly, in German environmental law precautionary ambient quality or recipient standards are not widely used at all so that there is no pressure to use them, for reasons of consistency, with respect to carcinogenic or mutagenic substances.

---

[41] Federal Administrative Court, NVwZ 1997, 497 at 498/499; NVwZ 1991, 1187.

However, there are, to an ever increasing extent, instances where ambient quality or recipient standards for no-threshold agents do exist. The standards may be derived on the basis of a qualitative risk analysis, considering genotoxicity or proven or plausible human carcinogenicity. They may also be based on a quantitative risk assessment of carcinogenic potency, using the unit risk concept which represents the slope factor of the dose-effect curve (Mosbach-Schulz 1998). As the low quantitative reliability of unit risk estimates presents certain problems, a uniform practice has not yet been developed. The oldest standard approach of this kind has existed in protection against ionising radiation, although supplemented by emission standards and minimisation obligations (Regulation for Radiation Protection). There are tolerance standards for alimentary products and drinking water (Guide values of the former Federal Institute for Health-Related Consumer Protection and Veterinary Medicine, Regulations on Contaminants and Solvents in Food, Drinking Water Regulation). Moreover, there are deposition standards for the protection of the soil against carcinogenic air pollutants (Technical Guidelines for the Protection against Air Pollution, Federal Soil Protection Regulation) and ambient quality guide values for certain carcinogenic air pollutants (Regulation on Concentration Values for Traffic-Related Air Pollution relating to Diesel Soot and Benzene, guide values of the State Commission on Protection against Air Pollution and Noise relating to Arsenic, Asbestos, Cadmium, Diesel Soot, PAH, Dioxins and Furans). There are a number of protective soil standards (clean-up standards for contaminated sites) and also some precautionary soil standards for carcinogens relating to various exposure paths such as soil to humans, soil to cultivated plants and soil to ground water (Federal Soil Protection Regulation). Finally, as regards clean-up of buildings contaminated with PCB and asbestos, there are recommended federal protective and precautionary standards which have been introduced by the state authorities through administrative rules. By contrast, the emission standards for hazardous air pollutants based on a classification into three risk groups which include carcinogenic air pollutants (under the Technical Guidelines for the Protection against Air Pollution) and emission standards for carcinogenic solvents (under Directive 1999/13 and the two German regulations on solvents) constitute a concretisation of the principle of minimisation rather than a standard concept oriented at the concrete potential adverse effects associated with exposure. The same is true of the technical exposure limits at the work place established, after consultation with an advisory commission, by the competent Federal Minister (under the Dangerous Substances Regulation).

Precautionary ambient quality or recipient standards for carcinogenic substances are sometimes supplemented by a minimisation requirement such as in radiation protection as well as under the new Drinking Water Regulation. Thus the weakness of such standards, namely that they draw a somewhat artificial line between significant and residual risk, is to a certain extent compensated for.

In EC law, a change of paradigm in regulating carcinogenic substances is emerging, which already has had or will in the near future have a strong influence on German practice (possibly even beyond direct application or transposition of EC regulations and directives).

The ambient quality standards for benzene at the work place (Directive 97/42) and for the protection of the public at large (Directive 2000/69) deserve special mentioning; these are the first ambient air quality standards for a carcinogenic air pollutant that have ever been set by the EC. According to the program contained in Directive 96/62, in the future PAH, Cadmium and its compounds, Diesel Soot, certain Arsenic compounds and Nickel are candidates for the further setting of ambient air quality standards. Moreover, recently under the EC Contaminants Regulation (315/93) standards for carcinogenic contaminants in foodstuffs (Aflatoxines, Cadmium, Dioxins and Furans – Regulations 466/2001, 2375/2001 and 221/2002) have been established. The EC Drinking Water Directive (Directive 98/83) sets forth tolerance standards for Acrylamide, Arsenic, Benzene, 1,2-Dichlorethane, Nickel, PAH, Trichloroethane, Tetrachloroethane and Vinyl chloride.

An ambient quality standard concept has the advantage of tackling the "hot spot problem". Furthermore it allows for an explicit and transparent and hence more rational discussion of, and decision on, tolerability of risk, including the fact that a certain cancer risk must be borne as part of an industrial civilisation (Böhm 1996, 42/43; Lübbe-Wolff 2000, 23; Köck 2001c, 206). However, as can be demonstrated by the EC Drinking Water Directive, the mere concept of ambient or recipient quality standards in itself does not ensure uniformity of the degree of tolerable risk that is imposed on the population. For example, tolerable lifetime risk from exposure to carcinogens in drinking water varies between 0.44 fatalities for benzene (below 10^{-6}) and 500 fatalities per one million of exposed population for arsenic (5×10^{-4}) (see Tab 7.1).

This example shows that the standard setting is far from being consistent and that factors like technical feasibility, economic considerations, political, social and ethical concerns play an important role. As it is impossible to determine absolute maximum exposure limits for chemicals in parallel to ionising radiation, it appears indispensable to develop case by case a safety concept for individual substances considering the factors denoted above (sections 10.4.1–10.4.3). This almost invariably will lead to differences in tolerable risk. Therefore in spite of possible economic objections additional minimisation requirements should supplement the ambient quality standards in order to provide additional safety in the grey area between undesirable and residual risk. This is the risk reduction concept used in radiation protection and there are no cogent reasons not to transfer it to the regulation of carcinogenic substances as this has already been done in the regulation of carcinogenic substances in drinking water. However, the case for additional minimisation is much weaker when the regulator has to deal with a carcinogenic substance for which a threshold or practical threshold can be determined. Here ambient quality standards based on the threshold value and an additional ample margin of safety appear appropriate.

10.4.8
Measures for risk reduction

The selection of risk reduction measures is primarily governed by the principle of proportionality (see section 4.2). Equality may also play a role, especially in configurations of competition. As stated, the principle of proportionality applies to the decision on tolerability and to that on the selection of measures. Since both levels of the decision-making process cannot really be separated, it is sufficient to refer to the discussion in section 4.2 and add some remarks.

There are a variety of measures that can be taken for preventing or reducing risks presented by hazardous agents. Among these options are prohibitions or restrictions of production or marketing, substitution obligations, prohibitions or restrictions of use, reduction of emissions (e. g. through ambient quality/recipient or emission standards or minimisation obligations) (see 4.7), reduction of exposure by technical protection measures for persons at risk (e. g. closed circuit requirements or obligations to use protective equipment), labelling, information and other requirements for securing self-protection of users, liability rules etc. (cf. Theuer 1996, 126/127; Köck 1999b; EC Commission 2000, 18). If one wants to generalise, one can state that from a precautionary health perspective, prohibitions suggest themselves the closer the relevant risk is to unacceptable risk (Ladeur 1994, 13; Theuer 1996, 128); conversely in case of a weakly founded suspicion more limited measures appear appropriate. However, also the severity and magnitude of possible harm, including the geographical and temporal scope of risk and distributional issues are important decision factors. As countervailing concerns, all the economic and social factors described above (see sections 4.2 and 4.3) enter into the decision. Normally, the costs of regulation to be considered are the costs incurred by society as a whole rather than individual costs since regulation of hazardous agents is of a generic nature (Ginzky and Winter 1999, 67 et seq.). This does not rule out that, when a particular risk reduction measure adversely affects only a limited number of enterprises, the aggregation model normally underlying the proportionality test in generic policy-making may have to be relativated (Theuer 1996, 127/128; Winter 2001, 123).

10.5
Organisation of the decision-making process

10.5.1
The problem

From the previous discussion it can be derived that the risk assessment process, although largely science-based, contains considerable non-empirical elements and that decision-making on risk management, especially on tolerability of risk, involves value judgements and requires balancing of risks to be prevented and chances foregone, of benefits of risk regulation and regulatory costs for society as a whole. Therefore, the organisation of the process of risk assessment and risk management is of high importance for securing rationality and legitimacy of the relevant determinations and decisions. This question had been neglected for a long time so that unreflected and incoherent patterns of scientific advice and decision-making have developed in practice. However, in the more recent past the organisation of the relevant processes has emerged as a key issue of modern environmental policy in the interface between science and regulation.

10.5.2
Current practice and new proposals

In the German practice, risk assessment and risk management procedures are characterised by a low degree of coherence and transparency and a weak expression of public participation. Rather, a clear emphasis is laid on non-public consultation of advisory bodies or ad hoc participation of selected representatives of science and organised interests. The prevailing model is that of an integrated (pluralistic) advisory body in which scientific experts and representatives of affected interests cooperate in assessing risks and setting standards. Where only ad hoc participation exists, it is largely organised in the same way. This is in particular true of media-related regulation (air pollution, waste management, biotechnology), but also relates to general toxic substances control. However, there are also purely scientific advisory bodies, especially in parts of occupational health protection (threshold substances and classification of CMR substances), authorisation procedures for pesticides and biocides, and protection against ionising radiation. Finally, in food and drinking water regulation, risk assessment and recommendations on health standards are exclusively carried out within the administration.

In the European Community, in the field of general hazardous substances, pesticides and biocides regulation, occupational health and radiological protection risk assessment, including, where appropriate, classification and labelling, is organised as a predominantly internal process within the competent EC organs allowing only for organised consultation of scientific advisory bodies as well as member state representatives and experts nominated by them. This is by and large also true of pharmaceuticals and more recently food safety where risk assessment has been entrusted to the European Medicines Evaluation Agency and the European Bureau for Food Safety respectively. These agencies can be considered as bureaucratic scientific advisory bodies. However, while their powers are limited to risk assessment, the relevant procedural rules also grant them an important role in the authorisation and risk management process. Standard setting including the decision on tolerability of risk occurs in a legislative or quasi-legislative (regulatory committee) procedure under procedural rules that do not formally provide for transparency of the process nor require true public participation. However, a considerable degree of public participation is afforded in practice. Quite often (with the notable exception of general toxic substances regulation) a scientific advisory committee must be consulted whose opinion is not binding but has a strong de facto influence on the decision (Hankin 1997, 164–166). Moreover, a certain degree of lobbying, especially with the European Commission and the Parliament, takes place and scientific experts and representatives of affected interests, including environmental groups, may act as members of ad hoc working groups that prepare proposals or parts of them. In the "Institutional Declaration on Democracy, Transparency and Subsidiarity" of 1993 the Commission committed itself to consult representatives of affected interests in preparing proposals. This commitment has been confirmed by No. 9 of the Subsidiarity Protocol accompanying the Amsterdam Treaty (Sobotta 2001, 88 et seq.). However, with the exception of "visible" problems, the involvement of environmental interests is not important due to manpower restrictions (Viehbrock 1994, 19, 68).

Many national or European regulations rely on recommendations or guideline lists published by international permanent or ad hoc expert advisory bodies, established especially under the auspices of the International Commission on Radiological Protection (ICRP) or the World Health Organisation (WHO). Thus, organised science exercises an important influence on the standard setting process.

As a response to the existing or assumed shortcomings of the risk assessment and risk management procedure for hazardous agents, since the late 1980s various proposals have been made in Germany with a view to improve the transparency of the process, provide for a more extensive public participation and secure a balanced representation of scientific expertise and affected interests in the relevant bodies (see, e. g. von Lersner, 1990; Kloepfer et al. 1991, 94–100, 471–484; Viehbrock 1994, 201–218; Lamb 1995, 175–222, 237/238; SRU 1996, Nos. 865–887; WBGU 1999, 277–282). Some proposals also call for an independent scientific institution for risk assessment (Environmental Standards 1998, 374–378; UGB-Kommission 1998, 115/116, 475–480). The recent interim report of the German Commission on harmonising risk assessment (Risiko-Kommission 2002, 72–89) advocates the establishment of two different central coordination bodies for risk assessment and risk management (Risk Council and Regulatory Committee) whose task is to harmonise and steer the risk assessment and risk management process which, however, shall remain the primary responsibility of the competent area-specific agencies.

As regards the EC, the discussion has been less intensive probably because any proposal must fit in the complex multilevel decision-making system of the European Community. The emphasis of the discussion has been laid on standardisation by private institutions which is less important in the field of chemicals regulation and radiation protection. Apart from the discussion on European regulatory agencies one seems to advocate a strengthening of participatory elements of chemicals regulation by formally granting environmental associations access to preparatory groups within the regulatory committee procedure and/or providing for open notice and comment opportunities in controversial cases (Viehbrock 1994, 218–221; Falke and Winter 1996, 570/571). The Commission communication on the precautionary principle (EC Commission 2000, 20) emphasises the need to include all stakeholders in preparing the decision to study risk management options which, on the basis of the risk assessment, may be envisaged and that the procedure should be as transparent as possible. Participation in risk assessment has been neglected.

10.5.3
Key issues

Apart from the need to harmonise the criteria of risk assessment and the factors to be considered in risk management decisions, which is widely accepted, there are three strategic issues that have emerged in the public debate:

– centralisation of risk assessment for all areas of chemicals regulation (and even radio-protection) or area-wise organisation of the risk assessment process,
– separation or integration of cognitive-predictive and normative phases of the risk assessment and risk management process,
– composition of competent advisory bodies, especially with respect to pluralism of scientific views and societal interests.

The installation of a central risk assessment body for all fields of chemicals regulation and radiation protection which would also be entrusted, beyond risk assessment in the strict sense, the task of making recommendations regarding the tolerability of risk and standard setting is associated with obvious advantages. In particular, it is liable to overcome the present fragmentation and sectoralisation of the risk assessment process and development towards different "standard setting cultures".

However, harmonisation can also be achieved through common criteria and procedures, transparency, public participation and, if necessary, mere coordination bodies as proposed by the German Risk Commission. A central risk assessment body would be overburdened by the complexity of scientific issues and the ambiguity of impacts on societal interests which arise in each risk assessment and standard setting process anew and in a quite different manner. Moreover, there is a problem of democratic legitimacy created by the possible creeping development of the central commission towards a codecision-making body.

A crucial problem of organising the risk assessment and risk management process is the coordination of scientific expertise and interest representation. The case for integrated (pluralistic) models is, as formulated by Denninger (1990, 31) that "it is not possible to isolate consensual and cognitive processes as such or to combine them at will". Denninger (1990, 34) considers it as a "fundamental insight that a synthesis of voluntary (interest-oriented) and cognitive (result-oriented) components must be found in each phase of technological norm setting and concretisation if the intention to ensure the common good shall not be lost." In the opinion of Denninger's followers this requires the organisation of an interactive process between science and regulators by establishing pluralistic bodies for standard setting (Ladeur 1995, 136, 156–161; Lamb 1995, 35, 235).

As already stated a functional separation between the phases of the process in which scientific expertise is primarily relevant – risk assessment in the strict sense – and phases in which normative-political considerations prevail – framing of risk assessment (selection of relevant agents, protected interests, level of protection and relevant exposure patterns) and decision on tolerability of risks and standard setting – is possible and even advisable. The reason for this position is that there is a functional difference between cognitive-predictive uncertainty and normative-political evaluation (Mayntz 1990; SRU 1996, Nos. 854/855; Rehbinder 1999, 153–156; Risiko-Kommission 2002, 26/27, 85). In order to reduce the extent of uncertainty a scientific discourse must be organised. The ambiguity of normative-political evaluation in deciding on tolerability of risk and standard setting requires a high degree of social openness. This results in different institutional requirements for the shaping of procedures which would seem to militate for the separation model. However, the functional separation of cognitive-predictive knowledge ("truth") and normative-political evaluation ("effectiveness", "efficiency", and "fairness") does not mean that both processes need to be entirely isolated institutionally. Rather, a dialogue between science on the one hand and regulators and stakeholders on the other must be organised without negating the different logics of scientific cognition and prediction and normative-political decision-making. This can, among others, also be achieved by establishing integrated risk assessment and management bodies composed of separate branches with clearly defined tasks. A clear attribution of tasks together with transparency of the procedure would make it obvious when scientific knowledge ends and political-administrative responsibility begins; thereby, it would contribute to improving the rationality and accountability of the decision-making process even if one must concede that the process might be burdened exactly by this kind of transparency.

Finally, the composition of scientific advisory bodies or scientific branches in integrated advisory bodies is at issue. It is doubtful whether the requirement of a "balanced" representation (Lamb 1995, 249) is applicable to scientific advice as it obscures the analytical separation between interest pluralism and perspective pluralism (Führ 1994, 29; Ladeur 1995, 159). Interest pluralism is not appropriate when the role of scientific experts is, as suggested here, limited to the determination of the state of scientific knowledge and the representation of scientific positions in the discourse on tolerability of risk and standard setting. Rather, in this respect, there only is a need for ensuring quality of scientific work, i. e. competence, independence and impartiality, which, however, would not rule out the nomination of scientists from industry and environmental groups because of their competence and practical experience. What may be worth discussing is the demand for perspective pluralism in the composition of an advisory body that would reflect different positions in science. A cogent argument that militates for perspective pluralism is that minority positions have sometimes been the source of progress in scientific knowledge. On the other hand, the demand for perspective pluralism may be at odds with the requirements of competence, independence and impartiality (Environmental Standards 1998, 372–374; but see Ladeur 1995, 159; Lamb 1995, 249). There also is a problem of defining minority positions. Therefore, besides a formal representation of minority positions in an advisory body, there also is the option of simply obliging the relevant body to consider the whole state of scientific knowledge including minority positions and control compliance by ensuring a high degree of transparency.

Independent of these organisational issues it would seem that the whole process should be more open both to scientific discourse (risk assessment) and interest and value presentation (risk management). Hence, the participatory elements of the process should be strengthened. In this respect the proposals of the German Risk Commission (Risiko-Kommission 2002, 72/73) and others (WBGU 1999, 277–282) for innovative participation models such as expert workshops, Delphi procedures and consensus conferences with respect to risk assessment and notice and comment procedures, hearings, mediation procedures and consensus conferences with respect to risk management deserve further scrutiny.

10.6
References

Appel I (1996) Stufen der Risikoabwehr. Natur und Recht (NuR) 18: 227–235
Arnold D, Gundert-Remy U, Hertel RF (1998) Die Verwendung von Unsicherheitsfaktoren in der quantitativen Risikoeinschätzung – neue Wege. In: Umweltbundesamt (ed) Aktionsprogramm Umwelt und Gesundheit. UBA Berichte 1/98. Erich Schmidt, Berlin, S 547–556
Bergkamp L (2002) Understanding the precautionary principle. Env Liability 10: 18–30, 67–82
Bohl J (2002) Ableitung von Grenzwerten (Umweltstandards) – Juristische Gesichtspunkte. In: Wichmann HE, Schlipköter HW, Fülgraff G (eds) Handbuch Umweltmedizin. Kapitel III–1.3.12, Ecomed, Landshut
Böhm M (1996) Der Normmensch. Mohr Siebeck, Tübingen
Böhm M (1997) Abschied vom Vorsorgeprinzip im umweltbezogenen Gesundheitsschutz? In: Lange K (ed) Gesamtverantwortung statt Verantwortungsteilung im Umweltrecht. Nomos, Baden-Baden, pp43–56
Breuer R (1978) Gefahrenabwehr und Risikovorsorge im Atomrecht. Deutsches Verwaltungsblatt (DVBL) 93: 829–839
Breuer R (1991) Rechtliche Bewertung krebserzeugender Immissionen. In: Ministerium für Umwelt, Raumordnung und Landwirtschaft des Landes Nordrhein-Westfalen (ed) Neue Entwicklungen im Immissionsschutzrecht. Düsseldorf, pp158–181

Britz G (1997) Umweltrecht im Spannungsfeld von ökonomischer Effizienz und Verfassungsrecht. Die Verwaltung 30, 185–209

Bundesministerium für Umwelt, Naturschutz und Reaktorsicherheit [BMU] (ed) (1998) Umweltgesetzbuch (UGB-KomE). Entwurf der Unabhängigen Sachverständigenkommission zum Umweltgesetzbuch. Duncker & Humblot, Berlin

Callies C (2001) Vorsorgeprinzip und Beweislastverteilung im Verwaltungsrecht. Deutsches Verwaltungsblatt (DVBL) 116, 1725–1733

Cansier D (1994) Gefahrenabwehr und Risikovorsorge im Umweltschutz und der Spielraum für ökonomische Instrumente. Neue Zeitschrift für Verwaltungsrecht (NVwZ) 13: 642–647

Clarke R (1999) Control of low-level radiation exposure: time for a change? J Radiol Prot 19: 107–115

Cranor CF (1995) The social benefits of expedited risk assessment. Risk Anal 15: 352–358

Daemen T (2003) The European Community's evolving precautionary principle – comparison with the United States and ramifications for Doha Round trade negotiations. Eur Environ Law Rev 12: 6–19

Denninger E (1990) Verfassungsrechtliche Anforderungen an die Normsetzung im Umwelt- und Technikrecht. Nomos, Baden-Baden

Di Fabio U (1997) Voraussetzungen und Grenzen des umweltrechtlichen Vorsorgeprinzips. In: Kley MD, Sünner E, Willemsen A (eds) Steuerrecht Steuer- und Rechtspolitik Wirtschaftsrecht und Unternehmensverfassung Umweltrecht, Festschrift für Wolfgang Ritter. Dr. Otto Schmidt, Köln, pp807–838

Dieter HH (1995) Risikoquantifizierung: Abschätzungen, Unsicherheiten, Gefahrenbezug. Bundesgesundheitsblatt (BlB) 38: 250–257

Dieter HH, Konietzka R (1995) Which multiple of a safe body dose derived on the basis of safety factors would probably be unsafe? Regul Toxicol Pharm 22: 262–267

Driesen D (1997), The societal cost of administrative regulation: beyond administrative cost-benefit analysis. Ecol Law Quart 24: 545–617

Endres A, Holm-Müller K (1998) Die Bewertung von Umweltschäden. Theorie und Praxis sozioökonomischer Verfahren. Kohlhammer, Stuttgart

Englert N (1996) Ableitung von Grenzwerten für Stoffe in der Luft. In: Umweltbundesamt (ed) Transparenz und Akzeptanz von Grenzwerten am Beispiel des Trinkwassers. UBA Berichte 6/96. Erich Schmidt, Berlin, pp78–87

Epiney A (1999) Rationalitätsgebote im Recht – unter besonderer Berücksichtigung der Vorgaben im europäischen Gemeinschaftsrecht. In: Gawel E, Lübbe-Wolff G (eds) Rationale Umweltpolitik – rationales Umweltrecht. Nomos, Baden-Baden, pp167–191

European Commission (2000) Communication from the Commission on the precautionary principle. COM (2000) 1. Brussels

European Commission (2001) Strategy for a future chemical policy. COM (2001) 88 final. Brussels

European Commission (2002) Action plan "Simplifying and improving the regulatory environment". COM (2002) 278 final, Brussels

European Environmental Agency [EEA] (2001) Late lessons from early warnings: the precautionary principle 1896–2000. Office for Official Publication of the European Communities, Luxemburg

Falke J, Winter G (1996) Management and regulatory committees in executive rule-making. In: Winter G (ed) Sources and categories of European Union law. Nomos, Baden-Baden, pp 541–582

Fischer M (1992) Kanzerogene Luftschadstoffe und ihre Bedeutung für die Krebsmortalität. Bundesgesundheitsbl. 35: 184–189

Fischer M (1996) Vergleich und Bewertung von Krebsrisiken durch Luftverunreinigungen. In: Umweltbundesamt (ed) Transparenz und Akzeptanz von Grenzwerten am Beispiel des Trinkwassers. UBA Berichte 6/96. Erich Schmidt, Berlin, pp54–77

Franßen E (1993) Krebsrisiko und Luftverunreinigung – Risikoermittlung und rechtliche Bewertung. In: Gesellschaft für Umweltrecht (ed) Dokumentation zur 16. Wissenschaftlichen Fachtagung der Gesellschaft für Umweltrecht e.V. Berlin 1992. Erich Schmidt, Berlin

Führ M (1994) Wie souverän ist der Souverän? Technische Normen in demokratischer Gesellschaft. VAS Verlag für akademische Schriften, Frankfurt/Main

Führ M (2001) Ökonomische Effizienz und juristische Rationalität. Ein Beitrag zu den Grundlagen interdisziplinärer Verständigung. In: Gawel E (ed) Effizienz im Umweltrecht. Nomos, Baden-Baden, pp157–213

Gawel E (1999) Umweltordnungsrecht – ökonomisch irrational? In: Gawel E, Lübbe-Wolff G (eds) Rationale Umweltpolitik – rationales Umweltrecht. Nomos, Baden-Baden, pp237–322

Gawel E (2001) Rationale Gefahrenabwehr. Marktsteuerung und ökologische Risiken. In: Gawel E (ed) Effizienz im Umweltrecht. Nomos, Baden-Baden, pp349–370

Ginzky H (2001) Vermarktungsbeschränkungen von gefährlichen Chemikalien. Neue Zeitschrift für Verwaltungsrecht (NVwZ) 20: 536–538

Hahn RW (1996) Regulatory reform: what do the government's numbers tell us? In: Hahn RW (ed) Risks, costs and lives saved: getting better results for regulation. Oxford University Press, New York, pp 208–253

Hankin R (1997) The role of scientific advice in the elaboration and implementation of the Community's foodstuffs legislation. In: Joerges C, Ladeur KH, Vos E (eds) Integrating scientific expertise into regulatory decision-making. Nomos, Baden-Baden, pp 141–167

Hansjürgens B (1999) Ökonomische Bewertung der Regulierung von Gefahrstoffen. In: Winter G, Ginsky H, Hansjürgens B (eds) Die Abwägung von Risiken und Kosten in der europäischen Chemikalienregulierung. UBA Berichte 7/99, Erich Schmidt, Berlin, pp283–370

Hansjürgens B (2001) Mehr Effizienz im Umweltrecht durch Kosten-Nutzen-Analysen? Zu den Möglichkeiten aus ökonomischer Sicht. In: Gawel E (ed) Effizienz im Umweltrecht. Nomos, Baden-Baden, pp 63–95

Hanusch H (1994) Kosten-Nutzen-Analyse, 2. Aufl. Vahlen, München

Heinzerling L, Ackermann F (2002) The humbug of the AntiRegulation movement. Cornell Law Rev 87: 648

International Atomic Energy Agency [IAEA] (1992) Methods for comparative risk assessment of different energy sources. IAEA-TECDOC-671, Vienna

International Commission on Radiological Protection [ICRP] (1990) ICRP Publication 60. Recommendations of the International Commission on Radiological Protection. Annals of the ICRP 21(1–3), Pergamon Press, Oxford, 1991

International Programme on Chemical Safety, World Health Organisation [IPCS-WHO] (1994) Assessing health risks of chemicals. Derivation of guidance values for health-based exposure limits. IPCS Environmental Health Criteria 170, International Programme on Chemical Safety, World Health Organisation, Geneva

Kalberlah F, Schneider K (1998) Quantifizierung von Extrapolationsfaktoren. Schriftenreihe der Bundesanstalt für Arbeitsschutz und Arbeitsmedizin. Wirtschaftsverlag NW, Bremerhaven

Kloepfer M (2000) Environmental Justice and geographische Umweltgerechtigkeit. Deutsches Verwaltungsblatt (DVBL) 115: 750–754

Kloepfer M, Rehbinder E, Schmidt-Aßmann E (1991) Umweltgesetzbuch – Allgemeiner Teil. Erich Schmidt, Berlin

Köck W (1999a) Umweltordnungsrecht – ökonomisch irrational? In: Gawel E, Lübbe-Wolff G (eds) Rationale Umweltpolitik – rationales Umweltrecht. Nomos, Baden-Baden, pp324–359

Köck W (1999b) Risikobewertung und Risikomanagement im deutschen und europäischen Chemikalienrecht. In: Hansjürgens B (ed) Umweltrisikopolitik. Zeitschrift für angewandte Umweltforschung (ZAU) 12 (S10): 76–96

Köck W (2001a) Zur Diskussion um die Reform des Chemikalienrechts in Europa – Das Weißbuch der EG-Kommission zur zukünftigen Chemikalienpolitik. Zeitschrift für Umweltrecht (ZUR) 12: 303–307

Köck W (2001b) Rationale Risikosteuerung als Aufgabe des Rechts. In: Gawel E (ed) Effizienz im Umweltrecht. Nomos, Baden-Baden, pp 273–302

Köck W (2001c) Krebsrisiken durch Luftverunreinigungen – Rechtliche Anforderungen an genehmigungsbedürftige Anlagen nach dem BImSchG. Zeitschrift für Umweltrecht (ZUR) 12: 201–206

Kunig P (2000) Artikel 2. In: Von Münch I, Kunig P (eds) Grundgesetz-Kommentar, Bd 1, 5. Aufl. Kohlhammer, Stuttgart

Ladeur KH (1994). Berufsfreiheit und Eigentum als verfassungsrechtliche Grenze der staatlichen Kontrolle von Pflanzenschutzmitteln und Chemikalien. Natur und Recht (NuR) 16: 8–14

Ladeur KH (1995) Das Umweltrecht der Wissensgesellschaft. Duncker & Humblot, Berlin

Lamb I (1995) Kooperative Gesetzeskonkretisierung. Verfahren zur Erarbeitung von Umwelt- und Technikstandards. Nomos, Baden-Baden

Länderausschuß für Immissionsschutz [LAI] (1992) Beurteilungsmaßstäbe zur Begrenzung des Krebsrisikos durch Luftverunreinigung. In: Ministerium für Umwelt, Raumordnung und Landwirtschaft des Landes Nordrhein-Westfalen (ed) Krebsrisiko durch Luftverunreinigung, Teil III. Düsseldorf

Latin H (1988) Good science, bad regulation, and toxic risk assessment. Yale J Reg 5: 89–148

Letzel H, Wartensleben H (1989) "Begründeter Verdacht" und "Jeweils gesicherter Stand der wissenschaftlichen Erkenntnisse". Pharma Recht (PharmR) 1: 2–8

Lübbe-Wolff G (1986) Die rechtliche Kontrolle incremental summierter Gefahren am Beispiel des Immissionsschutzrechts. In: Dreier H, Hofmann J (eds) Parlamentarische Souveränität und technische Entwicklung. Duncker & Humblot, Berlin, pp167–188

Lübbe-Wolff G (1998) Präventiver Umweltschutz – Auftrag und Grenzen des Vorsorgeprinzips im deutschen und europäischen Recht. In: Bizer J, Koch HJ (eds) Sicherheit, Vielfalt, Solidarität – Ein neues Paradigma des deutschen Verfassungsrechts? Nomos, Baden-Baden, pp 47–74

Lübbe-Wolff G (2000) Sind die Grenzwerte der 17. BImSchV für krebserzeugende Stoffe drittschützend? Natur und Recht (NuR) 22: 19–24

Mayntz R (1990) Entscheidungsprozesse bei der Entwicklung von Umweltstandards. Die Verwaltung 23: 137–151

McGarity TO (1998) A Cost-Benefit State? Admin Law Rev 50: 7–79

Meßerschmidt K (2001) Ökonomische Effizienz und juristische Verhältnismäßigkeit – Gemeinsames und Trennendes. In: Gawel E (ed) Effizienz im Umweltrecht. Nomos, Baden-Baden, pp 216–247

Mosbach-Schulz O (1998) Probabilistische Modellierung in der Prioritäten- und Standardsetzung. In: Umweltbundesamt (ed) Aktionsprogramm Umwelt und Gesundheit. UBA Berichte 1/98. Erich Schmidt, Berlin, pp 571–594

Müller-Herold U, Scheringer M (1999) Zur Umweltgefährdungsbewertung von Schadstoffen und Schadstoffkombinationen durch Reichweiten- und Persistenzanalyse. Graue Reihe 18, Europäische Akademie zur Erforschung von Folgen wissenschaftlich-technischer Entwicklungen. Bad Neuenahr-Ahrweiler

Murswiek D (2003) Umweltrisiken im amerikanischen Recht: Höhere Rationalität der Standardsetzung durch Kosten-Nutzen-Analyse? Jahrbuch des Umwelt- und Technikrechts, UTR 71 (in press)

National Research Council [NRC] (1983) Risk assessment in the federal government: managing the process. National Academy Press, Washington DC

National Research Council [NRC] (1996) Understanding risk: informing decisions in a democratic society. National Academy Press, Washington DC

Neus H (1998) Ziele und Rahmensetzungen für eine Harmonisierung der Standardsetzung. In: Umweltbundesamt (ed) Aktionsprogramm Umwelt und Gesundheit. UBA Berichte 1/98. Erich Schmidt, Berlin, pp 1–116

Office for Scientific and Technical Options Assessment [STOA] (1998) Survey and evaluation of criticism of basic safety standards for the protection of workers and the public against ionising radiation. Final Workshop Study (revised version), PE 167.161/Fin.Wksp.St./Rev. European Parliament, Luxemburg

Organisation for Economic Co-operation and Development [OECD] (1983) OECD Guidelines for testing of chemicals. Section 4: Health effects. OECD, Paris

Ossenbühl F (1986) Vorsorge als Rechtsprinzip im Gesundheits-, Arbeits- und Umweltschutz. Neue Zeitschrift für Verwaltungsrecht (NVwZ) 5: 161–171

Overy D, Richardson A (1995) Regulation of radiological and chemical carcinogens: current steps toward risk harmonization. Environmental Law Reporter (ELR): News and Analysis 25: 10657–10670

Parkin R, Balbus J (2000) Variations on concepts of "susceptibility" in risk assessment. Risk Anal 20: 603–611

Petersen F (1993) Schutz und Vorsorge. Duncker & Humblot, Berlin

Pinkau K, Renn O (eds) (1998) Environmental standards. Scientific foundations and rational procedures of regulation with emphasis on radiological risk management. Kluwer, London (originally published in German as Akademie der Wissenschaften zu Berlin (1992) Umweltstandards, De Gruyter, Berlin)

Rascoff S, Revesz R (2002) The biases of risk tradeoff analysis: towards parity in environmental and health- and safety regulation. Univ Chicago L Rev 69: 1763–1836

Rat von Sachverständigen für Umweltfragen [SRU] (1996) Umweltgutachten 1996. Zur Umsetzung einer dauerhaft-umweltgerechten Entwicklung. Metzler-Poeschel, Stuttgart

Rat von Sachverständigen für Umweltfragen [SRU] (1999) Umwelt und Gesundheit. Risiken richtig einschätzen. Sondergutachten. Metzler-Poeschel, Stuttgart

Rat von Sachverständigen für Umweltfragen [SRU] (2002) Umweltgutachten 2002. Für eine neue Vorreiterrolle. Metzler-Poeschel, Stuttgart

Rehbinder E (1997a) Stoffrecht. In: Arbeitskreis für Umweltrecht (ed) Grundzüge des Umweltrechts. 2., völlig neu bearbeitete und wesentlich erweiterte Auflage, ergänzbare Ausgabe. Erich Schmidt, Berlin, Teil 13

Rehbinder E (1997b) Ziele, Grundsätze, Strategien und Instrumente. In: Arbeitskreis für Umweltrecht (ed) Grundzüge des Umweltrechts. 2., völlig neu bearbeitete und wesentlich erweiterte Auflage, ergänzbare Ausgabe. Erich Schmidt, Berlin, Teil 4

Rehbinder E (1999) Ein Modell für die Setzung von Umweltstandards. In: Czajka D, Hansmann K, Rebentisch M (eds) Immissionsschutzrecht in der Bewährung, 25 Jahre Bundes-Immissionsschutzgesetz. Festschrift für Gerhard Feldhaus. CF Müller, Heidelberg, pp141–158

Rehbinder E. (2003) Allgemeine Regelungen – Chemikalienrecht. In: Rengeling W (ed) Handbuch zum europäischen und deutschen Umweltrecht, 2. Aufl. Carl Heymanns, Köln, § 61

Rengeling W (2000) Bedeutung und Anwendbarkeit des Vorsorgeprinzips im europäischen Umweltrecht. Deutsches Verwaltungsblatt (DVBL) 115: 1473–1483

Renn O (2002) Vorsorge als Prinzip: besser in der Vorsicht irren als im Wagemut. Gaia 11: 44–45

Revesz R (1999) Environmental regulation, cost-benefit analysis and the discounting of human lives. Columbia L Rev 99: 941–1017

Risiko-Kommission (2002) Ad Hoc-Kommission „Neuordnung der Verfahren und Strukturen von Risikobewertung und Standardsetzung im gesundheitlichen Umweltschutz in der Bundesrepublik Deutschland." Erster Bericht über die Arbeit der Risiko-Kommission. Bundesministerium für Umwelt, Reaktorsicherheit und Naturschutz, Bonn

Rowe G (2001) Gerechtigkeit und Effizienz im Umweltrecht – Divergenz und Konvergenz. In: Gawel E (ed) Effizienz im Umweltrecht. Nomos, Baden-Baden, pp 303–337

Sachs M (2002) Artikel 20. In: Sachs M (ed) Grundgesetz, 3. Aufl. CH Beck, München

Salzwedel J (1987) Risiko im Umweltrecht – Zuständigkeit, Verfahren und Maßstäbe der Bewertung. Neue Zeitschrift für Verwaltungsrecht (NVwZ) 6: 276–279

Salzwedel J (1992) Rechtsgutachten. In: Länderausschuss für Immissiosschutz [LAI] und Ministerium für Umwelt, Raumordnung und Landwirtschaft des Landes Nordrhein-Westfalen (eds) Krebsrisiko durch Luftverunreinigung. Düsseldorf, Teil IV

Scheringer M (2000) Persistenz und Reichweite von Umweltchemikalien. Wiley-VCH, Weinheim

Scherzberg A (1993) Risiko als Rechtsproblem. Verwaltungsarchiv 84: 484–513

Schroeder W (2002) Die Sicherung eines hohen Schutzniveaus für Gesundheits-, Umwelt- und Verbraucherschutz im europäischen Binnenmarkt. Deutsches Verwaltungsblatt (DVBL) 117: 213–221

Schuldt N (1997) Rationale Umweltvorsorge. Ökonomische Implikationen vorsorgender Umweltpolitik. Economica, Bonn

Shere ME (1995) The myth of meaningful environmental risk assessment. Harvard Environ L Rev 19: 409–492

Sobotta C (2001) Transparenz in den Rechtssetzungsverfahren der Europäischen Union. Nomos, Baden-Baden

Steinberg R (1998) Der ökologische Verfassungsstaat. Suhrkamp, Frankfurt/Main

Suhr D (1990) Die Bedeutung von Kompensationen und Entscheidungsverknüpfungen. In: Hoffmann-Riem W, Schmidt-Aßmann E (eds) Konfliktbewältigung durch Verhandlungen. Nomos, Baden-Baden, pp113–138

Sunstein RC (1996a) Health-health tradeoffs. Univ Chicago L Rev 63: 1533

Sunstein RC (1996b) Congress, constitutional mandates and the cost-benefit state. Stanford L Rev 48: 247

Szczekalla P (2002) Die sogenannten grundrechtlichen Schutzpflichten im deutschen und europäischen Recht. Duncker & Humblot, Berlin

The Netherlands (2001) Strategy on management of substances (SOMS) – approved by the Cabinet on March 16, 2001. The Hague

Theuer A (1996) Risikobewertungsmodelle als Grundlage von Stoffverboten? Natur und Recht (NuR) 18: 120–130

Viehbrock J (1995) Öffentlichkeit im Verfahren der Chemikalienkontrolle am Beispiel PCP. Werner, Düsseldorf

Viscusi WK (1996) Regulating the regulators. Univ Chicago L Rev 63: 1423–1461

Von Holleben H, Schmidt G (2002) Beweislastumkehr im Chemikalienrecht. Neue Zeitschrift für Verwaltungsrecht (NVwZ) 21: 532–538

Von Holleben H, Schmidt G (2003) Shifting the burden of proof in chemicals legislation: the guiding principle of the reform debate under scrutiny. Eur Environ Law Rev 12: 19–30

Von Lersner H (1990) Verfahrensvorschläge für umweltrechtliche Grenzwerte. Natur und Recht (NuR) 12: 193–197

Voßkuhle A (1999) Das Kompensationsprinzip. Mohr-Siebeck, Tübingen

Wagner H (1980) Die Risiken von Wissenschaft und Technik als Rechtsproblem. Neue Juristische Wochenschrift (NJW) 33: 665–671

Wagner WE (1995) The science charade in toxic risk regulation. Columbia L Rev 95: 1613–1723

Wahl R, Appel I (1995) Prävention und Vorsorge. Von der Staatsaufgabe zur rechtlichen Ausgestaltung. In: Wahl R (ed) Prävention und Vorsorge. Economica, Bonn, pp1–216

Wiedemann P, Mertens J, Schütz H, Hennings W, Kallfass M (2001). Risikopotenziale elektromagnetischer Felder: Bewertungsansätze und Vorsorgeoptionen. Arbeiten zur Risiko-Kommunikation 81, Forschungszentrum Jülich

Wiegand G (1997) Die Schadstoffkontrolle von Lebensmitteln aus ökonomischer Sicht. Springer, Heidelberg

Wiener J, Rogers M (2002) Comparing precaution in the United States and Europe. J Risk Res 5: 317–349

Williamson GH, Hulpke H (2000) Das Vorsorgeprinzip. InternationalerVergleich, Möglichkeiten und Grenzen, Lösungsvorschläge. Umweltwissenschaften und Schadstoffregulierung (UWSF) – Z Umweltchem Ökotox 1(1,2): 27–39, 91–96

Winter G (1986) Gesetzliche Anforderungen für Grenzwerte für Luftimmissionen. In: Winter G (ed) Grenzwerte. Werner, Düsseldorf, pp 1–62

Winter G (1995) Maßstäbe der Chemikalienkontrolle im deutschen Recht und im Gemeinschaftsrecht. In: Winter G (ed) Risikoanalyse und Risikoabwehr im Chemikalienrecht. Werner, Düsseldorf, pp 1–62

Winter G (2000) Redesigning joint responsibility of industry and government. In: Winter G (ed) Risk assessment and risk management of toxic chemicals in the European Community. Nomos, Baden-Baden, pp 177–184

Winter G (2001) Über Nutzen und Kosten der Effizienzregel im öffentlichen Recht. In: Gawel E (ed) Effizienz im Umweltrecht. Nomos, Baden-Baden, pp 97–126

Winter G (2002) Neuere Entwicklungen des Umweltrechts der EU. Anwaltsblatt (AnwBl) 53: 75–86

Winter G (2003) Umweltrechtliche Prinzipien des Gemeinschaftsrechts. Zeitschrift für Umweltrecht (ZUR) 14: 137–145

Winter G, Ginzky H (1999) Nutzen und Kosten im deutschen und europäischen Chemikalienrecht. In: Winter G, Ginzky H, Hansjürgens B (eds) Die Abwägung von Risiken und Kosten in der europäischen Chemikalienregulierung. UBA Berichte 7/99. Erich Schmidt, Berlin, pp 1–119

Wissenschaftlicher Beirat der Bundesregierung Globale Umweltveränderungen [WBGU] (1999) Welt im Wandel: Strategien zur Bewältigung globaler Umweltrisiken. Springer, Berlin

Wulfhorst R (1994) Der Schutz "überdurchschnittlich empfindlicher" Rechtsgüter im Polizei- und Umweltrecht. Duncker & Humblot, Berlin

11 Summary

11.1 Introduction

The increases in the release of toxic agents from technical installations as well as from commercial and household uses of such agents have led to heavy pollution in some regions. In order to protect human health and the environment it is necessary to establish relevant environmental standards, which limit exposures to these toxic agents. The assessment and decision-making procedures have to be based on solid scientific data about the effects and mechanisms of these agents. For risk evaluation, the knowledge of the dose-response curve is an essential prerequisite. Dose-responses without a threshold dose are most critical in this connection. Such dose-responses are assumed for mutagenic and carcinogenic effects, which, therefore, dominate the discussion about environmental standards. In the important low-dose range, risk estimation can only be achieved by extrapolation from higher dose ranges with measurable effects. The extrapolation is accompanied with uncertainties which makes risk evaluation as well as risk communication frequently problematic. For the regulatory processes besides legal requirements and administrative procedures, ethical, social and economic aspects are important. In order to secure rational efficient and fair decisions a dialogue between disciplines, with the affected people and with the general public is necessary. The whole range of relevant aspects, starting with the ethical foundations, going to the scientific data, the contribution of theory of cognition, and concluding with the social and legal perspectives are dealt with in the report. It should give insights into the problems of standard setting to a broader readership of scientists, legislators, administrators and the interested public.

11.2 Ethical aspects of risk

The establishment of standards for tolerable or acceptable risks in connection with exposure to toxic agents might seem to be exclusively a matter of sound scientific methodology in estimating the probabilities for damages and reducing the uncertainties involved in extrapolating from the known to the unknown.

However, the setting of standards for acceptable risk is not just a matter of probabilities and degrees of uncertainty. Acceptability is a *normative* notion and involves a number of factors having to do with values and human autonomy that are discussed here. Whoever makes the decision, whether an individual, a group, an industrial company or a public institution or agency, there are a number of ethical norms and values that come into play.

This chapter starts with a decision theoretic model that aims at providing a way of bringing to bear on the decision all the various values and norms that should be taken into account in such decisions. The importance of considering alternatives is stressed and also the precautionary principle.

Thereafter some of the most important factors that enter into such decisions are discussed: time, life, health, environmental values, whether the persons affected are known or anonymous, freedom of choice, consent, distribution of benefits and burdens, compensation, background exposure, the value of actions, and finally, how all these various values can be compared and justified.

Time is a resource, it counts among the values that we have to take into account, and the time it takes to make the decision is sometimes crucial, in that spending much time on making a decision may eliminate some of the best alternatives.

The value of life comes in importantly in many decisions and has been the subject of much controversy. Some alternative approaches to the valuation of lives are discussed. The value of health raises somewhat similar problems. The crucial question is the same: how shall these values be weighted against other values?

The environment is gradually receiving more attention, but we are still far from recognising and taking into account the full range of values that are at stake when the environment is threatened by man-made activities.

A large group of important values is connected with our status as autonomous human beings, with freedom, responsibility and dignity. One important consideration in decisions about the acceptability of risks is whether people are free to choose whether they shall be exposed to the risk. Pollution and many kinds of low-dose risks are not chosen freely by those who are exposed to them, and the level of acceptability then has to be very low. Informed consent is normally required before anybody can be exposed to serious risk. How serious the risk shall be before consent is required and what kind of information one needs in order to qualify as informed are difficult questions that are discussed.

The distribution of burdens and benefits is highly important for the acceptability of risks and it is one of the factors that are most often neglected. Exposing somebody to a risk in order that somebody else shall benefit is almost never acceptable. This becomes particularly problematic when those who are exposed to the risk are already badly off, while those who benefit from the activity are among the better off already before the risky activity starts.

Compensation can sometimes help to overcome distributional problems. Yet, here too, there are difficulties that can be resolved. For example, should one be compensated for mere exposure to risk, or should one be compensated only if something actually goes wrong? Moreover, in the latter case, who bears the burden of proving whether the damage was caused by the risky activity? The person who was hurt or the person or institution responsible for the risky activity?

Some activities are claimed to be bad in themselves, regardless of consequences, for example harming others, even if one does it in order to pursue an important and good goal. Kant held that one should never treat any person merely as a means for something. What bearing does this have on the notion of acceptable risk?

In a concrete situation where we have to make a decision, all these values have to be weighed against one another. Is this possible and if so how? This difficult question is considered as well as a discussion of how judgements about values can be justified.

Finally, the chapter ends with a list of questions to ask in deliberations about the acceptability of risks which may be handled in concrete regulatory processes:

- What are the *alternatives*?
- *Who* is affected? Do we know *who* they are, or are they *anonymous, statistical* victims?
- How are they *experiencing* what happens? Can they *control* what happens to them? Have they given *informed consent*?
- *How are the benefits and burdens distributed? How are the worst-off affected? How are the silent victims heard? Are future generations taken into account?*
- *Are there better alternatives?*

11.3
Effects of ionising radiation in the low-dose range – radiobiological basis

For the evaluation of risk by exposures to toxic agents and for the connected judgement with respect to regulatory processes, the knowledge about the shape of dose-response curves is essential. In principle, two different classes of curves are considered: 1) Dose-response curves with a threshold and 2) dose-response curves without a threshold. For the consideration of risks after radiation exposures in the low-dose range those radiation effects which follow the first type of dose-response (acute effects, late deterministic tissue effects) are not relevant as the thresholds are in the range of several 100 mGy (mSv) of low-LET radiation, and radiation doses in the environment from natural as well as anthropogenic sources are usually in the range of 1 to 10 mSv or even lower. The second type of dose-response is proposed for hereditary effects, carcinogenesis and certain developmental effects (stochastic effects) and is generally used for regulatory processes after exposures to ionising radiation in the low-dose range.

The low-dose range can be defined on the basis of microdosimetric or biological terms. Microdosimetry describes the distribution of events of energy absorption after the passage of ionising particles through tissues and cells. In the dose range of 10 mSv and less (average tissue dose), the distribution of ionising events becomes very heterogeneous between cells in a tissue. With decreasing radiation doses less cells will be hit. This is especially the case with high-LET radiation (cf. α-particles, neutrons). Health effects can usually be observed only after radiation doses in the range of 100 mSv and higher. In special cases (cf. with radiosensitive individuals) radiation effects have been found after radiation doses in the lower range of 10 to 100 mSv. For example this is the case with thyroid cancer in children and with the induction of chromosomal aberrations after very careful studies with very qualified techniques.

For the development of stochastic effects, radiation damage in the DNA is of utmost importance. The various types of damage (polynucleotide breaks and base damage) have been chemically classified. Multiple damaged sites (clusters of damage) are very relevant for the development of biological effects. The repair of radiation damage of DNA can be very efficient, however, the repair of various types of DNA damage differs very much. The repair of DSBs is much more difficult and slower, -misrepair or even no repair may take place – than that of SSBs especially when the DSBs occur in multiple damaged sites (clusters). In this respect the LET of the radiation quality is important, as clusters of radiation damage are more frequent after exposures to high-LET radiation than to low-LET radiation (X-rays, photons, electrons).

Many radiobiological investigations have been performed with mammalian cells in vitro after irradiation with different radiation qualities and a range of dose rates. The induction of chromosomal aberrations, of cell transformations, of gene mutations and of other biological endpoints has been measured. The dose-response of such studies can best be described by dose-effect curves without a threshold. After exposures to high-LET radiation, linear dose-effect curves have been observed and with low-LET radiation linear as well as linear-quadratic dose-effect curves have been found.

Strong influences of dose rate have been reported for low-LET radiation but no influence or very little for high-LET. Thus, the temporal dose distribution may be important for risk estimates after exposures to low-LET radiation. Besides radiation quality, dose rate and DNA repair, a number of other parameters are known that modify the expression of DNA damage. These processes (adaptive response as well as apoptosis) can lead to a considerable reduction of radiation damage in the low-dose range. However, very little effects by these modifications are observed with high-LET radiation and even with low-LET radiation, well-defined conditions are necessary in order to achieve the modifications.

In recent years it has been observed that ionising radiation can increase the genomic instability of many cell generations after radiation exposures. This effect has mainly been investigated with chromosomal aberrations, cell death and developmental effects after prenatal exposures. This mechanism leads to an enhanced probability for the induction of later mutations by a factor of 10 to 10^4. As several mutations are necessary for the development of cancer and other late effects (multi-step processes) after irradiation, this mechanism may be important for radiation-induced carcinogenesis and other stochastic effects. Further, it has been observed that radiation responses (gene expression, chromosomal aberrations, cell death) can be increased in cells although they have not been hit by ionising particles (bystander effect). The significance of these bystander effects for the development of biological radiation effects has not been clarified up to now.

Individual radiosensitivity can vary considerably. Children are more radiosensitive than adults but also the genetic disposition is very important for the radiation response. Several genetic syndromes have been described of which the corresponding carriers show a marked increase in radiosensitivity. These effects are mainly studied with cells from such individuals *in vitro* and for the induction of secondary cancers in connection with cancer therapy by ionising radiation. A reduction or deficiency of certain pathways of DNA repair has frequently been found in such individuals. All these modifying phenomena and parameters show a high expression after exposure to low-LET radiation but little effect to high-LET radiation.

Animal experiments are very valuable for the evaluation of mechanisms and of the shape of dose-response curves. Therefore, extensive animal experiments have been performed by investigating the induction of cancers and hereditary effects mainly in mice and rats. As with humans it has been found that the radiosensitivity varies tremendously between different strains of mice and rats. In addition, the dose-response curves vary considerably but generally, they can best be described by curves without a threshold. After exposures to high-LET radiation, again a linear dose-response without threshold dose is usually observed and after low-LET radiation linear, linear-quadratic as well as quadratic dose-responses have been reported. With these latter radiation qualities, again the temporal dose distribution has a considerable influence.

For hereditary effects, which have been mainly performed with male mice, linear dose-effect curves without a threshold have been reported for low-LET as well as for high-LET radiation. Considerable dose rate effects have been observed for low-LET radiation but not for high-LET radiation. The radiation doses in these experiments have been in the range of 400 mSv to 8 Sv (local doses to the gonads). Radiation experiments with female mice have demonstrated an appreciable DNA repair capacity in oocytes. The types of mutations observed after radiation exposures are the same as the "spontaneous" mutations. This is in analogy to the radiation-induced cancers, chromosomal aberrations and other biological effects.

The experimental data with cells and with animals suggest that the dose-response curves for stochastic effects (cancer, hereditary effects, and some developmental effects) are best described by dose-effect curves without a threshold in most cases. This is especially the case with high-LET radiation but also holds generally for low-LET radiation. In the low-dose range, a linear dose-response seems to be a best estimate. The slope of the dose-response gets shallower with decreasing dose rates until a certain limit is reached in the case of low-LET radiation. All dose modifying parameters seem not to be universal. Limitations and exceptions occur with respect to exposure conditions, biological systems and endpoints.

As the observed effects do not differ from the corresponding background rates, biological effects cannot be measured in the lower dose range. An estimate of damage or risk can then only be obtained by extrapolation. This is usually done by the linear-no-threshold (LNT) model. The scientific data and the consideration of the precautionary principle support such a procedure although it should be taken into account that the LNT model has not beencannot be experimentally proven in a direct manner. For regulatory purposes, it should also be taken into account that all populations experience radiation exposures from natural sources. The average exposures from these sources are in the range of several mSv and the standard deviation of these exposures is less than 1 mSv. These values can be used as benchmarks for regulatory purposes.

11.4
Toxicology of chemical carcinogens

A chemical carcinogen may act either as the parent compound itself or as an ultimately active metabolite. Consequently, in performing low-dose extrapolations a complicated interplay of activating and inactivating enzyme systems must be considered in addition to the further processes starting for instance with the genotoxic action for a "direct" carcinogen. These and the following processes have a number of analogies as described in the foregoing section of ionising radiation. In addition, a complexity of mechanisms that may introduce "practical" or "perfect" thresholds must be recognised. Well investigated carcinogens are discussed in this volume to visualise the influence of factors modifying the dose-effect relationships, comprising examples of genotoxic carcinogens, with and without threshold mechanisms, and of non-genotoxic mechanisms of action.

Vinyl chloride is a well-established carcinogen. For experimental induction of liver tumours, there is experimental evidence that a linear low-dose extrapolation is in fact justified. The linear model with no threshold assumption appears also most appropriate for extrapolation of human cancer risk due to *diethylnitrosamine*. In addition, the example of the tobacco smoke-specific carcinogen *NNK* presents itself as that of a no-threshold carcinogen. Similar scenarios are observed with the genotoxins *2-acetaminofluorene* (AAF) and with *aflatoxin B1*. The example of *2-acetaminofluorene*, in addition, shows that low-dose-effect relationships do not only depend on the nature of the substance tested, but may also be tissue-specific. Thus, distinct differences have been shown for the carcinogenic effects of AAF in bladder and liver of mice for instance. The example of *4-aminobiphenyl* in rodents shows that occurrence of thresholds might be sex- and tissue-dependent.

The carcinogenicity of *inorganic arsenic* is not sufficiently well understood, although there are indications that indirect mechanisms of action are operative that could lead to non-linearity of the dose-response at low doses. In such unclear situations, for regulatory matters, the default assumption of a linear dose-response appears appropriate now. *Formaldehyde* represents a well-investigated example of a genotoxic chemical with a non-linear carcinogenic dose-response, implying a practical threshold. For *vinyl acetate* carcinogenesis, factors of cytotoxicity and regenerative cell proliferation are of importance, also arguing in favour of existing thresholds. Based on the nature of the interaction of topoisomerase with a toxicant (e. g. *phytooestrogens*), the existence of a "practical" threshold of a resulting genotoxic effect seems at least plausible. Interactions of chemicals (e. g. *lead, mercury*) with cytoskeletal proteins and, in consequence, the disturbance of chromosome segregation, should follow a conventional dose-response. Non-mutational types of chromosomal genotoxicity, based on chemical-protein interactions, should involve a threshold. There is almost general agreement that a distinction should be made between *genotoxic* and *non-genotoxic* chemicals when assessing cancer risk to humans.

For genotoxic carcinogens, the present case studies of chemicals demonstrate an array of different possibilities. Positive data of chromosomal effects only, e. g. aneugenicity or clastogenicity, in the absence of mutagenicity, may support the characterisation of a compound that produces carcinogenic effects only at high, toxic doses. In such cases, relevant arguments are in favour of the existence of "practical" thresholds. Also, genotoxicity may be relevant only under conditions of sustained local tissue damage and associated increased cell proliferation. In addition, in such cases practical thresholds can be derived.

Non-genotoxic carcinogens are characterised by a "conventional" dose-response, which allows derivation of a No-Observed-Adverse-Effect-Level (NOAEL). For instance, *peroxisome proliferators* exemplify compounds leading to experimental carcinogenicity based on a receptor-mediated effect, characterised by a threshold. A most prominent further example for a non-genotoxic, receptor-based carcinogen is *TCDD*. However, extremely large interspecies differences in TCDD-induced toxicity are known. The guinea pig is the most susceptible mammal known that is more than three orders of magnitude more sensitive than the hamster, which is the most resistant mammal. Humans appear to be less sensitive than most laboratory animals. Environmental exposure to background levels of TCDD is not likely to cause an increase in human cancer risk. Tumours arising mainly in consequence of *hormonal imbalance* generally imply a threshold mechanism; in such cases, no tumour formation should be expected below doses at the no-hormonal-effect-level.

There is now a general agreement that aspects of mechanisms involved should be much more considered within regulatory assessments of carcinogenic risks of chemicals.

11.5
Mathematical models of carcinogenesis

The process of carcinogenesis is too complex to be described in full detail by a mathematical model. Therefore, modelling carcinogenesis implies simplifications that try to identify the most important features. This procedure helps understanding the carcinogenic process by establishing hypotheses and testing resulting predictions in laboratory experiments or by analyses of epidemiological data.

Epidemiological data have limitations in describing the implications of age, dose and dose rate dependencies of excess risks for cancer after exposures to radiation or other carcinogens, especially at low doses. Mathematical models of carcinogenesis have the potential to contribute to a better understanding of these dependencies. Compared to conventional risk models used in epidemiology, as, e. g. the excess relative risk model, mathematical models of carcinogenesis have the advantage of a straightforward description of the hazard due to complex exposure scenarios. After identification of the action of a carcinogen on the parameters of the model, no additional parameters are necessary to calculate the effects of exposures that may change many times during the observation period. Also, mathematical models of carcinogenesis have in general a smaller number of free parameters.

Carcinogenesis is conventionally modelled by a stochastic process in which a susceptible normal cell undergoes several alterations to become a cancer cell. In most cases, such alterations are assumed to be genetic or epigenetic changes. The number of necessary genetic alterations appears to vary from cancer to cancer. In the case of retinoblastoma, two mutations that inactivate both copies of the *RB* tumour-suppressor gene suffice, for tumour development. In the case of colorectal cancer, alterations in three tumour-suppressor genes (*APC*, *p53* and *DCC*) and one proto-oncogene (*K-ras*) have been observed during progressive stages of tumour development. The presence of four or more mutations in the genesis of colorectal cancer suggests the existence of events that increase the rates of subsequent mutations, i. e. the induction of a genomic instability. Also, the increase of the division rate of intermediate cells in the carcinogenic process may lead to an increase of critical mutational events.

Phenotypically altered cells that are believed to represent intermediate stages of carcinogenesis may form clones as papillomas of the mouse skin, enzyme altered foci in the rodent liver, or adenomatous polyps in the human colon. Correspondingly, a large family of mathematical models of carcinogenesis contains a clonal expansion of the intermediate cells. Two-step models of carcinogenesis with a clonal expansion of intermediate cells are among the simplest models that can describe age and exposure dependencies in many observed data. This is possibly because the complex transition from a normal cell to a malignant cell can be approximated, at least in some cases, by two rate-limiting events. Independent of this, there is an increasing effort to develop and apply multi-step models that take cell proliferation kinetics into account.

The large potential of mathematical models of carcinogenesis in the application to epidemiological data lies in the assessment of risks in the low-dose range by taking epidemiological, radiobiological and molecular genetic data into account. Further, the models relate numbers and sizes of intermediate stages as adenomas or liver foci to cancer incidence or mortality rates in the same population.

Comparative studies with lung cancer data for rats that were exposed to radon progeny and with cancer incidence data for the atomic bomb survivors have shown that the statistical power of these data sets is not large enough to decide which is the most appropriate model. Simplicity, plausibility, and compatibility with radiobiology and molecular genetics are additional aspects to identify appropriate models.

Depending on the exposure scenario and on the exposure-rate dependence of the model parameters, mathematical models of carcinogenesis predict direct as well as inverse dose-rate effects, increases of excess risk with dose as well as decreases. At the same time, in most applications to epidemiological data, parameter estimates have a large uncertainty. Therefore, conclusions on age, time and exposure dependencies for single cancer types remain to be uncertain. The application of a variety of models (including phenomenological epidemiological models) offers, however, the possibility to identify common predications and their uncertainties including the model uncertainty.

Interactions of initiating and promoting carcinogens have been studied with the TSCE model. As to be expected, the degree of synergistic effects depends on the exposure scenarios, e. g. which carcinogen is applied first. However, according to the model, synergistic effects tend to become negligibly small if the doses of both carcinogens are small.

In the past one or two decades, several low-dose radiobiological effects have been discovered that have a non-linear dose dependence. These include adaptive response, low-dose hypersensitivity, bystander effects and genomic instability. Models of carcinogenesis offer the possibility to integrate such effects in the analysis of epidemiological data, which is not so easily done with phenomenological models. A first example of such a study is the integration of low-dose hypersensitivity concerning cell inactivation in the TSCE model and its application to lung cancer incidence among the atomic bomb survivors. The example demonstrates as a possible consequence a non-linear dose dependence of the cancer risk after low-dose exposures at older age. More work in this direction is to be expected in the future.

For a large class of models, the stochastic multi-stage models with clonal expansion, the exploration of its features is still in an early stage. A main weakness of the currently existing approaches in this field is a large number of possible processes and free parameters. The growing experimental experience on genetic and epigenetic pathways, however, may offer the possibility of a fruitful application of such models in the future. A first impressing step has been made with applying a four-step model to population data on colorectal cancer.

11.6
Epidemiological perspectives on low-dose exposure to human carcinogens

In the atomic bomb survivors and numerous studies of patients receiving radiotherapy, it has been convincingly shown that cancer incidence and mortality from cardio-vascular disorders are related to exposure to ionising radiation. We do not know if there are radiation doses below which there is no significant biological change, or below which the damage induced can be effectively dealt with by normal cellular processes. Analyses of Japanese data could not exclude a threshold of 60 mSv, i. e. below this dose limit no significant excess of cancer or leukaemia could be identified. The possibility that ecological studies of environmental exposure to ionising radiation can contribute to our knowledge of the effects of low-dose exposure is limited. The effects of natural background radiation are low and other risk factors will distort the results making data unreliable.

Age and gender are strong modifiers with respect to cancer causation by ionising radiation. The age effect has especially been observed for thyroid cancer with the exposed population in Hiroshima and Nagasaki as well as in Belarus and the Ukraine after the Chernobyl accident. This effect has also been found for breast cancer in females in Hiroshima and Nagasaki and with patients. Lung cancer has been extensively studied after exposures to radon and its radioactive daughter products. An increased cancer rate has been found in several studies with miners. Advances in molecular biology and genetics will hopefully increase the likelihood of finding the "true" effect of ionising radiation at low doses. Research will focus on understanding cellular processes responsible for recognising and repairing normal oxidative damage and radiation-induced damage.

Malignant melanoma is found to be related to excessive UV-light exposure. However, there are still questions to be answered, e. g. the difference in effect of accumulated UV-light exposure as in contrast to sunburns and the effect of accumulated exposure. The effect of UV-light immuno-suppression is also not understood.

Mobile phones were introduced in the mid 1980s and given the wide spread current use of mobile phones, any possible harmful effect is of considerable public health importance. In the largest study to date, no increased risk of CNS tumours was noticed. It was concluded that no evidence of an increased risk was found for a person using a mobile phone for 60 or more minutes per day or regularly for five or more years. The potential carcinogenic effects of extremely low frequency electromagnetic fields have been evaluated (brain cancer or leukaemia) but no increased risk has been observed.

Tobacco is probably the most studied human carcinogen and the largest preventable risk factor for morbidity and mortality in industrialised countries. Active smoking causes a number of health problems including cancer at several sites besides lung cancer. Elevated risks of lung cancer after passive smoking have been identified. In a large pooled analysis, an increased risk of lung cancer in non-smokers, living with a smoker, was seen. A significant dose-response relation was also seen and the risk increased with the number of cigarettes smoked per day and by duration of smoking.

The late health effects in humans exposed to pesticides have been extensively studied. DDT has been for many years the main focus. Today there is no firm evidence that pesticides cause cancer or any other adverse health effect.

Polychlorinated biphenyls accumulate in the adipose tissue and could theoretically influence the male reproduction system, and the risk of leukaemia and breast cancer. A definite confirmation of the toxicity is still lacking.

Polycyclic aromatic hydrocarbons and the most potent dioxin, TCDD, have been considered potent carcinogens. That is, a positive relationship has been observed between the exposure and cancer in studies in which chance, bias and confounding could be ruled out with reasonable confidence.

There is convincing evidence that asbestos causes lung cancer and mesothelioma and that the carcinogenic effect is increased by tobacco smoke. No study has identified other tumours to be related to asbestos exposure.

11.7
Carcinogenesis of ionising radiation and chemicals: similarities and differences

The development of cancer is a multi-step process, which apparently initiates with damaging events in the DNA. This is the case with ionising radiation and with genotoxic chemical carcinogens. While the energy absorption of ionising radiation leads to clustered DNA damage especially in the case of high-LET radiation, chemicals react in single events more equally distributed over the genome. With many carcinogenic chemicals it is necessary that they are activated by metabolic processes in order to interact with DNA. There also exist inactivating enzymes with other chemical carcinogens. This dependence of the carcinogenic activity of many genotoxic chemicals on metabolism complicates the determination of exposures to these chemicals in tissues and cells. The metabolism varies between species and makes extrapolations from animals to humans difficult. Further, metabolism differs between various organs and tissues and therefore the carcinogenic activity between organs is much more different with chemicals than with ionising radiation. Metabolic differences between individuals also occur which can have considerable significance for the individual susceptibility.

After these initial steps of carcinogenesis several modifying mechanisms like DNA repair, adaptive response, apoptosis, immunological activities and regulatory processes of the cell proliferation cycle exist which act principally in the same way after damage by ionising radiation as well as by chemicals. However, differences in details exist. In contrast, the further steps of cancer development, which are connected with the processes of promotion and progression are very similar or even identical in carcinogenesis induced by chemicals or ionising radiation. These events depend on further mutations but also very strongly on cell proliferation and its stimulation by endogenous or exogenous agents like hormones or other chemicals.

Genetic predisposition with mutations in DNA repair genes or tumour suppressor genes can increase the susceptibility for cancer proneness. Cancers which have been induced by exogenous factors can be distinguished from "spontaneous" cancers neither on clinical nor on cellular nor on molecular grounds. As the spontaneous cancer rate has a considerable spread, the exogenously induced cancers cannot be observed when the effect of the carcinogen becomes smaller than the spread of the spontaneous rate. The LNT model holds for the risk evaluation of ionising radiation in general, whereas with chemicals there are carcinogens for which the LNT concept is valid but for others it is not.

It is proposed that the chemical carcinogens are classified into genotoxic and non-genotoxic substances. Within the group of genotoxic chemicals a number of substances are found for which the scientific basis for the validity of the LNT concept is very high (vinyl chloride, dimethylnitrosamine, NNK from tobacco smoke). For other substances, the uncertainty of the dose-response in the low-dose range is very high. Differences between species and between tissues in the same species exist with respect to the shape of the dose-response. For these chemicals the LNT concept may be used under the provision of the precautionary principle (2-acetylaminofluorene, 4-aminobiphenyl). Further genotoxic chemicals are known where many scientific data point to a practical threshold. For these substances repetitive DNA damage in tissues is needed as well as further stimulation of cell proliferation before cancer development can be seen (formaldehyde, vinyl acetate). Also non-genotoxic carcinogens exist, which clearly show a threshold for carcinogenicity (TCDD, hormones).

For the carcinogenic effects of ionising radiation, risk factors have been derived for a number of human organs and tissues as well as for the total stochastic risk by defining the effective dose. For chemicals, risk factors have been estimated from epidemiological studies and animal experiments. Some of these risk values for ionising radiation (total stochastic risk) and some chemicals in drinking water are exemplified. The comparisons show that similar principles are involved in the development of cancer after exposures to ionising radiation and genotoxic chemicals. A unifying concept for risk evaluation is possible to a certain degree. However, the diversity even within the group of chemical carcinogens is so large that these differences have to be considered for regulatory purposes.

11.8
Hypothesis testing and the choice of dose-response

Scientists appeal to a whole suite of hypotheses, models, theories and laws when evaluating the degree to which low doses of radiation might be harmful to living things. Support for scientific claims may be drawn from many sources, including experimental studies, epidemiological data and mathematical modelling, and decisions need to be made concerning extrapolation, the existence or not of a threshold and the shape of the dose-response curve. The available information can appear contradictory (e. g. epidemiological studies may show a greater or smaller excess of cancer than predicted by the dose assessment; results from *in vitro* experiments may differ from those carried out *in vivo*), thus further judgements about the plausibility and robustness of the available scientific evidence will have to be made. Many disagreements between scientists reflect disputes as to the strength of the available evidence, but such disputes are often further aggravated by a lack of common ground and insufficient attention being paid to basic definitions and terminology.

It seems reasonable to describe the assumption of a linearity in the relationship between exposure and effect as a *model* for most low-dose radiation exposures, and to be clear that the actual shape of the dose-response curve will depend on a number of variables such as the effect in question, the species and the mechanism. This also applies to the various mathematical models used to describe the different stages in the development of cancer. On the other hand, the assumption that there is no threshold to the risk of cancer (or other stochastic effects) following exposure to radiation, may be best described as a *hypothesis*. Although it is one that is, statistically, difficult to falsify for cancer, it should not be seen as non-falsifiable. For some effects, the LNT hypothesis has been falsified (e. g. carcinogenic effects of promoting agents), but certainly not for all effects and all exposure situations. Hypotheses may also be used to explain other dose-response relationships and possible correlation between different effects and/or mechanisms. *Theories* refer to a collection of hypotheses and models, such as those invoked in the theory of cancer development or the various toxicological mechanisms.

There may be a number of different reasons behind a suggested *threshold*. An absolute (or perfect) threshold arises when there is a toxicant concentration or exposure below which no adverse effects will occur in the organism. A practical (or statistical) threshold has been used to describe situations where there is a concentration or level of exposure below which there is no *observable* level of damage above the expected "naturally occurring" or baseline incidence rate of effects. An apparent (or "protective") threshold can apply in cases where exposure to the toxin stimulates or enhances the occurrence of a protective or adaptive biological response that modifies or "cancels-out" another detrimental effect – typically one caused by exposure to endogenous and environmental toxins other than the toxin in question. Likewise, *extrapolation* is more complex than the mere question of how one extrapolates from "high dose" to "low dose". It includes judgements about correlations and connections between species (e. g. between animals and humans), effects (e. g. cell transformation to cancer development), mechanisms and models, as well as similar assumptions about the transition from environmental concentrations to dose calculation. Many of the apparent disagreements between scientists could be avoided with better clarification of the various hypotheses, models and assumptions.

Finally, scientists have different ways of judging *scientific plausiblity* and different disciplines appeal to different rules and have different standards, including (in toxicology) criteria concerning epidemiology, experimental evidence, biological plausibility, animal experiments and modelling, as well as more specific statistical "standards of proof" used in hypothesis testing. Thus, scientific plausibility is not a matter of appeal to one specific rule but rather set of different criteria wherein plausibility is a question of scale, depending largely on how many of those criteria are met. Depending on the type of judgements that will be made about exposures to toxins, and the possible consequences of those judgements (e. g. routine control as compared to decisions to build a new installation), decision-makers may use different stringencies in judgements of the scientific evidence. It follows that besides the importance of the scientific data there is certainly an influence from a diversity of values (politics, economics, aesthetics, etc.) and circumstances on the types of choices made in the protection of human health against exposure to harmful agents. Many of these decisions (such as what level of risk is acceptable) cannot be made on scientific grounds alone. Yet, it is vital that the best available science informs these decisions by providing relevant facts and information, and this includes transparency in the uncertainties and assumptions associated with that scientific knowledge. There can be valid and rational reasons for choosing different dose-response models in "science" and "policy making". Finally, the greater the uncertainty in prognosis of the consequences of exposure to radiation, the more important ethical values will become in the judgements about risk management.

11.9
Risk evaluation and communication

In risk analysis, several processes are necessary: the accurate determination of the potential harm – either quantitatively, i. e. statistical data, if possible, or based on probabilistic estimates. These insights form the basis for risk characterisation, which is bound to develop an understanding about how a particular risk can occur as objectively as possible and if and how it can be controlled. These insights are then incorporated into risk management and decision making.

Two basic types of risk have been taken into consideration: individual and societal. If, for a specific category of risks, all risk values for an individual would be considered, the result would be highly unrealistic. Therefore, potential risks have been identified for specified groups within the population or within geographical regions. A distinction has furthermore to be made between risks of mortality and morbidity and immediate and delayed effects. The knowledge of the causal agents for these effects is not always and for all areas well understood.

However, choices and decisions have to be made and in some countries, particularly in the EU, a precautionary principle has been developed and implemented, which stipulates that in case of doubt about effects the legislator should rather opt for the 'safe side' than wait for conclusive scientific evidence.

In case the particular risk has been characterised as unacceptable, i. e. too high, risk management is called for to reduce it to an acceptable level. This is predominantly an economic question and involves a reasonable setting of standards and priorities. An additional aspect for the decision-maker is the acceptance of the potential risk by society. Here, the aspect of risks versus benefits has to be taken into account, whereby the understanding prevails that only if the expected benefits are large enough that the option of abandoning the activity cannot be seriously considered, e. g. X-rays, the efforts to reduce the associated risks have to directed towards reducing the exposure to acceptable levels. In this context, it is sometimes helpful to resort to comparative aspects of interpreting risks, especially with regard to health impacts, since this is crucial for communication of decisions to the public.

Particularly communication of risk related issues is a difficult area in the political arena and should serve to achieve more democratic and participatory policy strategies. The components to be considered in a meaningful communication effort can be presented in form of the question: "Who says to whom what in which channel with what effects". The actual communicator needs to be credible for the audience and has to possess the skills necessary to relate the message in a way that is understandable and acceptable to the particular audience. Knowledge about the audience to be expected and adjustment of the information to their primary needs is an important aspect of successful communication. Therefore, consideration has to be given to the mechanisms of information processing of an individual and the various factors influencing intuitive risk perception. The message itself can be based on factual information, i. e. most aspects of the risk are known, or can consist of evaluative statements, which only attempt to put a particular risk into perspective with other, already familiar risks. Here, so called 'frames' play an important role, since they permit the individual to fit new information into an already existing structure of information. The problem here is that this individual structure can be very stable, i. e. hard to change. However, in addition it might contain false or misleading information. Thus, the overall effect of a communication and in particular about aspects of risk is difficult to measure.

11.10
Low doses in health-related environmental law

The process of risk assessment and risk management is only to a limited extent structured by environmental law. Protective obligations of the state and the European Union, the fundamental distinction between prevention (protection against clearly unacceptable risk) and precaution (reduction of undesirable risk), a common understanding of the notion of harm to health, and the obligation to consider individual risks at least in the form of protecting vulnerable groups set a framework. However, they leave a wide margin of appreciation of facts and discretion as to decisions.

In risk assessment, from a legal point of view, the definition of precaution and its delimitation from residual risk that has to be tolerated by all citizens poses difficult problems. The precautionary principle applies to situations of uncertainty. However, a classification of a situation of uncertainty as risk, which must be reduced, requires reasonable grounds for concern; a merely speculative suspicion is not sufficient, although the definition of what is speculative and what is science-based is somewhat fluid. Under this perspective, the "prudential" elements in scientific risk assessment, especially the use of "proxy" criteria to determine risk such as persistence and bio-accumulation, extrapolation factors and scientific risk models such as the no-threshold hypothesis for carcinogenic agents require legal scrutiny. With respect to these three problem areas, a differentiated assessment is appropriate. In particular, from a legal perspective, more normative harmonisation in the field of extrapolation appears desirable and should be envisaged where scientifically tenable. The no-threshold hypothesis is legitimate as long as there is no strong evidence that it is not valid for a particular agent.

Risk management is a complex normative process, which consists of the fundamental decision as to tolerability of risk and the decision on appropriate risk reduction measures. As regards health-related criteria, the frequently used concept of avoiding "significant risk" opens a wide margin of appreciation. In deciding on the tolerability of risk, to a limited extent, background risk and comparison with other types of risk for which a pertinent risk decision has already been taken may play a role. Moreover, at least in the field of merely precautionary regulation, balancing advantages and disadvantages is legitimate and often legally required. However, a formal cost/benefit analysis or any other formal balancing procedure is not prescribed, although in contrast to Germany, it plays an important role in the practice of the European Union. If applied with awareness of its methodological shortcomings and political risks, one could make wider use of cost/benefit analysis in risk management in order to focus attention to the costs of regulation, but also increase the degree of public acceptance of measures. The extent to which "soft" factors such as distributional justice and risk perception may play a role in the risk management process is subject to controversy; this depends to a large degree on the legal framework. While unequal risk imposed on people due to differences in susceptibility must be addressed by using the concept of vulnerable groups, it is more doubtful whether in the field of precautionary regulation there are clear-cut legal principles which require an equalisation of exposure to carcinogenic agents. The weight of human health and the impossibility of delimiting prevention and precaution in this field militate at least for a certain approximation of exposure. Subjective risk perception is not a factor the law expressly recognises as legitimate in risk management. However, to a limited extent, it may play a role, especially in setting priorities for action.

In Germany and the European Union, quantification concepts for determining tolerable risk of cancer, especially delimiting precaution and residual risk, are not widely used, although in practice a life-long individual risk of exposure to carcinogenic substances of 10^{-6} seems to be accepted. A quantification concept is in principle useful because it allows for a more transparent discussion of tolerability of a particular risk. As regards standards for carcinogenic substances, the emissions-oriented standard concept, especially minimisation, with respect to precaution preferred in Germany should at least be supplemented by an ambient quality standards concept – which is now underway due to action by the European Union.

The organisation of the process of risk assessment and risk management has recently been a subject of many debates. Apart from strengthening the participatory elements of the process both relating to the scientific discourse and interest and value presentation, there are a number of issues that have to be settled, especially the degree of centralisation and integration of the cognitive-predictive and normative phases and the composition of advisory bodies in risk assessment. With regard to all three questions, pragmatic solutions, which, however, recognise the specific role of science in the process, appear preferable.

12 Zusammenfassung

12.1 Einleitung

Mit der zunehmenden technischen Entwicklung unserer Gesellschaft ist auch die Freisetzung gesundheitsschädlicher Substanzen und Strahlung aus industriellen Einrichtungen und Privathaushalten stetig gewachsen. Dies hat in einigen Regionen der Welt zu einer starken Umweltbelastung geführt. Zum Schutz der menschlichen Gesundheit und der Umwelt ist es daher nötig, entsprechende Umweltstandards festzusetzen, die den Ausstoß schädlicher Substanzen begrenzen. Die Abschätzung der Risiken sowie die politischen Entscheidungsprozesse müssen sich auf solide wissenschaftliche Erkenntnisse über die Effekte dieser Substanzen stützen. Für den Prozess der Risikobewertung ist daher das Wissen über Dosis-Wirkungs-Beziehungen wichtig. Hierbei sind Dosis-Wirkungs-Beziehungen ohne eine Schwellendosis, die für mutagene und krebserzeugende Effekte angenommen werden, als die bedenklichsten einzustufen. Dosis-Wirkungs-Beziehungen ohne Schwellendosis beherrschen daher einen Großteil der Diskussion über Umweltstandards.

Bei der Festlegung von Umweltstandards ist man an dem niedrigen Dosisbereich interessiert, in dem keine oder nur sehr geringe Wirkungen vermutet werden. In diesem niedrigen Dosisbereich kann aber für Dosis-Wirkungs-Beziehungen ohne Schwellendosis keine eindeutige Aussage über das Risiko gemacht werden. Man ist hierbei auf eine Abschätzung der Wirkung durch Extrapolation von wesntlich höheren (in der Umwelt so gut wie nie auftretenden) Dosen angewiesen. Diese Extrapolation birgt viele Unsicherheiten in sich, welche die Risikobewertung und die Risikokommunikation oft problematisch machen. Für die Standardsetzung sind daher neben den rechtlichen Anforderungen und administrativen Verfahren auch ethische, soziale und wirtschaftliche Gesichtspunkte wichtig. Um vernünftige, leistungsfähige und adäquate Entscheidungen zu ermöglichen, ist ein Dialog zwischen den Disziplinen sowie mit den betroffenen Personenkreisen und der breiteren Öffentlichkeit nötig. Das gesamte Spektrum der beteiligten Betrachtungsweisen, beginnend mit den ethischen Grundlagen über die neuesten wissenschaftlichen Erkenntnisse bis hin zu erkenntnistheoretischen Betrachtungen, sozialen und rechtlichen Überlegungen, wird in diesem Bericht untersucht. Hierbei sollen einer breiteren Leserschaft, bestehend aus Wissenschaftlern, dem Gesetzgeber, Verwaltungsbeamten und der interessierten Öffentlichkeit, Einblicke in die Gesamtproblematik der Festlegung von Umweltstandards gegeben werden.

12.2
Ethische Überlegungen

Die Festlegung von Grenzwerten für zumutbare oder akzeptable Risiken aufgrund einer Exposition gegenüber giftigen Stoffen oder ionisierender Strahlung mag zunächst ausschließlich als Problem grundlegender wissenschaftlicher Methoden für die Abschätzung von Schadenswahrscheinlichkeiten und der Verminderung von Unsicherheiten, die mit den Extrapolationen von Bekanntem in Unbekanntes einher gehen, gelten.

Die Standardsetzung für akzeptable Risiken ist jedoch nicht nur eine Frage von Wahrscheinlichkeiten und Unsicherheitsgraden. Akzeptabilität ist ein *normativer* Begriff, der eine Anzahl von weiteren Faktoren umfasst, die Werte und menschliche Autonomie betreffen. Diese werden im Folgenden diskutiert. Wer immer die Entscheidung trifft, ob es sich um eine Einzelperson handelt, eine Gruppe, ein industrielles Unternehmen oder eine öffentliche Einrichtung, es gibt immer eine Anzahl von ethischen Normen und Werten, die beachtet werden müssen.

Die ethischen Betrachtungen beginnen mit einem entscheidungstheoretischen Modell, das einen Weg vorschlägt, welcher all die verschiedenen Werte und Normen, die für solche Entscheidungen wichtig sind, berücksichtigt. Hierbei wird die Bedeutung des Vorsorgeprinzips ebenso wie der Erwägung möglicher Alternativen betont.

Danach werden einige der wichtigsten Elemente für solche Entscheidungen besprochen: Zeit, Leben, Gesundheit, umweltrelevante Werte, der Bekanntheitsgrad, die Wahlfreiheit und die Zustimmung der betroffenen Personen, die Verteilung von Lasten und Nutzen, eventuelle Entschädigungsmaßnahmen, eine eventuell schon in der Umwelt existierende und daher zu ertragende Konzentration einer Substanz, der Wert der geplanten Aktivitäten und schließlich die Rechtfertigungs- und Vergleichsmöglichkeiten all dieser unterschiedlichen Werte.

Zeit ist ebenfalls ein Gut, das wir als Wert in Betracht ziehen müssen. Die Zeit, die bis zur Entscheidungsfindung benötigt wird, ist manchmal entscheidend, und zwar derart dass ein hoher Zeitaufwand für das Treffen einer Entscheidung einige der besten Alternativen eliminieren kann.

Der Wert des Lebens wird als ein sehr wichtiger Punkt in viele Entscheidungen einbezogen. Er ist daher das Thema vieler Kontroversen gewesen. Einige Alternativen für die Bewertung des Lebens werden besprochen. Der Wert der Gesundheit wirft ähnliche Probleme auf. Die entscheidende Frage ist hier dieselbe: wie sollen diese Werte gegenüber anderen Werte beurteilt werden?

Die Umwelt und ihr Schutz erhalten schrittweise mehr Aufmerksamkeit. Die Gesellschaft ist aber noch weit entfernt von der Anerkennung des vollen Wertespektrums, das auf dem Spiel steht, wenn die Umwelt durch menschliche Aktivitäten bedroht wird.

Eine große Gruppe wichtiger Werte ist mit dem Status der Menschen als autonome Wesen, mit Freiheit, Verantwortlichkeit und Würde verbunden. Die Freiheit der Betroffenen, zu wählen, ob sie einem Risiko ausgesetzt werden wollen, ist ein weiterer wichtiger Punkt in den Entscheidungen über die Akzeptabilität von Risiken. Unweltbelastungen und insbesondere Risiken durch niedrige Dosen werden nicht frei von denen gewählt, die diesen Risiken später ausgesetzt sind. Daher muss in diesen Fällen das Niveau der Akzeptabilität sehr niedrig gewählt sein. Zustimmung nach eingehender Information der Betroffenen ist normalerweise aus ethischer Sicht erforderlich, bevor jemand einem ernsthaften Risiko ausgesetzt werden kann. Wie „ernsthaft" ein Risiko sein muss, um eine Zustimmung erforderlich zu machen, und welche Art von Informationen nötig sind, um die Qualität „informiert" zu erreichen, sind schwierige Fragen, die erörtert werden.

Die Verteilung von Lasten und Nutzen ist sehr wichtig für die Akzeptabilität von Risiken, sie ist eines der Elemente, die sehr häufig vernachlässigt werden. Jemanden einem Risiko auszusetzen, so dass nur ein Anderer davon profitiert, ist aus ethischer Sicht fast nie akzeptabel. Das wird besonders problematisch, wenn die Personen, die dem Risiko ausgesetzt werden, bereits unter schlechten (gesundheitlichen, wirtschaftlichen, etc.) Bedingungen leben, während diejenigen, die von der Tätigkeit profitieren, bereits über bessere Bedingungen verfügen, bevor die risikoreiche Tätigkeit überhaupt beginnt.

Entschädigungen können manchmal helfen, solche Verteilungsprobleme zu überwinden. Jedoch gilt es auch hier, Schwierigkeiten zu überwinden. Es stellt sich z. B. die Frage, ob jemand schon entschädigt werden sollte, wenn er bloß einem Risiko ausgesetzt wird, oder nur, wenn er wirklich geschädigt wurde. Außerdem muss, wie im letzteren Falle, geklärt werden, wer – der Geschädigte oder der Verursacher – die Beweislast für die Schadensursache trägt.

Von einigen Aktivitäten wird behauptet, sie seien unabhängig von möglichen Folgen „an sich" schlecht, z. B. wenn Dritte geschädigt werden, selbst wenn dies für ein „wichtiges" und „gutes" Ziel geschieht. Denn laut Kant sollte niemand eine Person als bloßes Mittel zum Zweck behandeln. Welche Tragweite hat dieses für den Wert ‚akzeptables Risiko'?

In der konkreten Situation, in der eine Entscheidung zu treffen ist, müssen all diese Werte gegeneinander abgewogen werden. Der schwierigen Frage nach der Möglichkeit sowie nach der Art und Weise einer solchen Abwägung wird ebenso nachgegangen wie der Rechtfertigungsproblematik von Werturteilen.

Schließlich endet das Kapitel mit einer Reihe von Fragen, die bei Überlegungen zur Akzeptabilität von Risiken gestellt und in den konkreten reglementierenden Verfahren behandelt werden sollten: Was sind die *Alternativen*? Wer ist betroffen? Ist bekannt, *wer* die Opfer sind, oder sind *sie anonyme, statistische* Opfer? Wie *empfinden* die Betroffenen das, was geschieht? Können die Betroffenen *kontrollieren*, was mit ihnen geschieht? Haben sie aufgrund *eingehender Information* zugestimmt? Wie werden der *Nutzen und die Belastungen verteilt*? Wie sind die betroffen, die am *schlechtesten gestellt* sind? Wie werden die *„stummen Opfer"* berücksichtigt? Wird an *zukünftige Generationen* gedacht? Gibt *es bessere Alternativen*?

12.3
Wirkungen ionisierender Strahlung im niedrigen Dosisbereich – Strahlenbiologische Grundlagen

Für die Bewertung der Risiken durch Exposition mit toxischen Stoffen jeder Art und die damit verbundene Regulierung seitens der Behörden ist das Wissen über die Form der Dosis-Wirkungs-Kurven wesentlich. Prinzipiell lassen sich zwei unterschiedliche Kategorien von Kurven unterscheiden: 1) Dosis-Wirkungs-Kurven aus denen sich eine Schwellendosis ableiten lässt und 2) Dosis-Wirkungs-Kurven ohne Schwelle. Für Risikoerwägungen infolge von Strahlenexpositionen im niedrigen Dosisbereich sind jene Strahleneffekte, die dem ersten Typ der Dosis-Wirkungs-Kurve folgen (z. B. akute Strahlenschäden, späte deterministische Gewebeschäden) nicht relevant, da die Schwellendosis für locker ionisierende Strahlung im Bereich einiger 100 mGy (mSv) liegt. Strahlendosen aus natürlichen und künstlichen Quellen sind in der Natur jedoch in der Regel im Bereich von nur 1 bis 10 mSv zu finden. Der zweite Typ von Dosis-Wirkungs-Kurve wird für erbliche Effekte, Karzinogenese und bestimmte Effekte auf die Embryonalentwicklung (stochastische Effekte) angenommen. Sie wird im Allgemeinen für die reglementierenden Verfahren infolge einer Exposition mit niedrig dosierter ionisierender Strahlung angewandt.

Der niedrige Dosisbereich kann anhand von mikrodosimetrischen oder biologischen Überlegungen definiert werden. Mikrodosimetrie beschreibt die Verteilung von Energieabsorptionsereignissen nach der Durchquerung ionisierender Teilchen durch Gewebe und Zellen. Für Dosen (gemessen als durchschnittliche Gewebedosis) von 10 mSv und weniger wird die Verteilung der ionisierenden Ereignisse zwischen Zellen eines Gewebes sehr heterogen. Mit abnehmenden Strahlendosen werden weniger Zellen getroffen. Dies trifft besonders für dicht ionisierende Strahlung (z. B. α-Teilchen, Neutronen) zu. Signifikante Wirkungen auf die Gesundheit können normalerweise nur nach Strahlendosen von 100 mSv und mehr beobachtet werden. In speziellen Fällen (z. B. bei strahlenempfindlichen Personen) sind Wirkungen schon nach niedrigen Dosen im Bereich von 10 bis 100 mSv beobachtet worden. Dies ist z. B. der Fall für Schilddrüsenkrebs bei Kindern und für die Induktion von Chromosomenaberrationen, die in sehr sorgfältigen Studien mit sehr qualifizierten Techniken durchgeführt werden.

Für die Entwicklung stochastischer Effekte ist die strahlenbedingte Schädigung der DNA von äußerster Wichtigkeit. Die verschiedenen Arten der Schäden (Polynucleotidbrüche und Basenschädigung) sind chemisch klassifiziert worden. Bereiche der DNA mit mehrfachen Schädigungen („Schadenscluster") sind für die Entwicklung der biologischen Effekte gravierend. Die Reparatur der Strahlenschäden in der DNA kann sehr effizient sein. Die Reparaturfähigkeit unterscheidet sich jedoch erheblich in Abhängigkeit der verschiedenen Schadenstypen. So sind Doppelstrangbrüche (*double strand breaks* DSBs) viel schwieriger und langsamer zu reparieren, als Einzelstrangbrüche (*single strand breaks* SSBs) – es kann eine schlechte („misrepair") oder gar keine Reparatur stattfinden – insbesondere wenn DSBs Bereichen mit mehreren Schäden auftreten. In dieser Hinsicht ist der *l*ineare *E*nergie*t*ransfer (LET) der Strahlung wichtig, da Schadenscluster häufiger nach Exposition mit dicht ionisierender Strahlung auftreten als nach Exposition mit locker ionisierender Strahlung (Röntgenstrahlen, Photonen, Elektronen).

Viele strahlenbiologische Untersuchungen sind *in vitro* an Säugetierzellen durchgeführt worden, die mit unterschiedlichen Strahlenqualitäten und mit einem Bereich von Dosisleistungen bestrahlt worden sind. Die Induktion von Chromosomenaberrationen, von Zelltransformationen, von Genmutationen und anderen biologischen Endpunkten ist hierbei gemessen worden. Die Dosis-Wirkungs-Beziehungen in solchen Studien können dabei am besten ohne Schwellendosis beschrieben werden. Nach Exposition mit dicht ionisierender Strahlung sind lineare Dosis-

Wirkungs-Kurven beobachtet worden, mit locker ionisierender Strahlung sind außerdem auch linear-quadratische Kurven beobachtet worden.

Für locker ionisierende Strahlung hat die Dosisleistung einen starken Einfluss auf die Dosis-Wirkungs-Kurven ergeben, während dicht ionisierende Strahlung keinen oder nur einen schwacher Einfluss gezeigt hat. So kann für die Schätzung des Risikos bei einer Exposition mit locker ionisierender Strahlung die zeitliche Verteilung der Dosis wichtig sein. Neben der Strahlenqualität, der Dosisleistung und der DNA-Reparatur ist eine Anzahl weiterer Faktoren bekannt, welche die Genese von DNA-Schäden beeinflussen. Diese Prozesse (im Besonderen „*Adaptive Response*" und Apoptose) können zu einer beträchtlichen Minderung der Strahlungsschäden im niedrigen Dosisbereich führen. Jedoch werden bei dicht ionisierender Strahlung nur sehr geringe Wirkungen dieser Prozesse beobachtet und selbst bei locker ionisierender Strahlung sind sehr definierte Untersuchungsbedingungen nötig, um diese Verringerung zu erzielen.

Wie in den vergangenen Jahren beobachtet wurde, kann ionisierende Strahlung die genomische Instabilität selbst viele Zellgenerationen nach der eigentlichen Strahlenbelastung erhöhen. Dieser Effekt wurde hauptsächlich für Chromosomenaberrationen, Zelltod und Störungen der Embryonalentwicklung nach pränataler Exposition untersucht. Hierbei erhöht sich die Wahrscheinlichkeit für die Induktion späterer Mutationen um einen Faktor von 10 bis 10^4. Da nach einer Bestrahlung mehrere Mutationen für die Entwicklung von Krebs und anderen Späteffekten (Mehrstufenprozesse) erforderlich sind, kann der oben genannte Effekt für die strahleninduzierte Karzinogenese und andere stochastische Wirkungen wichtig sein. Ferner hat man beobachtet, dass Strahleneffekte (z. B. Genexpression, Chromosomenaberration, Zelltod) in Zellen verstärkt wurden, die überhaupt nicht von ionisierenden Teilchen getroffen worden waren (*bystander*-Effekt). Die Bedeutung dieser *bystander*-Effekte für die Entwicklung biologischer Strahlenwirkungen ist bis heute ungeklärt.

Die individuelle Strahlenempfindlichkeit kann beträchtlich schwanken. So sind Kinder empfindlicher als Erwachsene. Aber auch die genetische Veranlagung ist für die Reaktion auf eine Bestrahlung sehr wichtig. Verschiedene genetische Syndrome, deren Träger eine deutlich erhöhte Strahlenempfindlichkeit zeigen, sind beschrieben worden. Die Wirkungen dieser genetischen Veränderungen, wurden zum einen hauptsächlich *in vitro* an Zellen von Trägern dieser Veränderungen, zum anderen bei Strahlentherapiepatienten untersucht. Durch Beobachtungen von Zweittumoren infolge einer Strahlentherapie besteht jedoch kein Zweifel, dass eine erhöhte Strahlenempfindlichkeit auch hinsichtlich der Krebsinduktion existiert. In solchen Patienten wurden häufig Einschränkungen oder der komplette Ausfall bestimmter DNA-Reparaturwege gefunden. Alle diese verändernden Ereignisse und Faktoren treten häufig nach Bestrahlung mit locker ionisierender Strahlung auf, haben aber nur eine geringe Wirkung nach Einwirkung ionisierender Strahlung.

Tierexperimente sind für die Beurteilung der krebsbildenden Mechanismen und der Form der Dosis-Wirkungs-Kurven hervorragend geeignet. Folglich sind umfangreiche Tierexperimente durchgeführt worden, bei denen man die Induktion von Krebs und gentische Defekte hauptsächlich an Mäusen und Ratten erforschte. Wie beim Menschen fiel auch hier auf, dass die Strahlenempfindlichkeit extrem zwischen unterschiedlichen Mäuse- und Rattenstämmen schwankt. Die Dosis-Wirkungs-Beziehungen variieren stark in ihrer Form, aber im Allgemeinen können sie gut durch Kurven ohne eine Schwelle beschrieben werden. Nach Exposition mit dicht ionisierender Strahlung wird normalerweise eine lineare Dosis-Wirkungs-Beziehung beobachtet. Im Gegensatz dazu führt locker ionisierende Strahlung zu linearen sowie linear-quadratischen und sogar quadratischen Dosis-Wirkungs-Beziehungen. Für locker ionisierende Strahlung hat wiederum die zeitliche Dosisverteilung einen beträchtlichen Einfluss auf die Wirkung.

Für erbliche Effekte, die hauptsächlich an männlichen Mäusen untersucht wurden, sind Dosis-Wirkungs-Kurven ohne Schwelle für locker und dicht ionisierende Strahlung gefunden worden. Allerdings wurde ein starker Einfluss der Dosisleistung nur für locker, jedoch nicht für dicht ionisierende Strahlung beobachtet. Die Strahlendosen lagen in diesen Experimenten im Bereich von 400 mSv bis 8 Sv (lokale Dosen im Bereich der Gonaden). Strahlungsexperimente an weiblichen Mäusen haben eine beachtliche DNA-Reparaturtätigkeit in den Oozyten gezeigt. Die Mutationen, die nach Bestrahlung beobachtet werden, sind von der gleichen Art wie die „spontanen" Mutationen, die auch ohne Strahlenbelastung auftreten. Das ist analog zu den strahlungsinduzierten Krebsen, den Chromosomenaberrationen und zu anderen biologischen Effekten.

Die Ergebnisse aus *in vitro* Experimenten und Tierexperimenten veranlassen zu der Annahme, dass die Dosis-Wirkungs-Beziehungen für stochastische Effekte (Krebs, erbliche Effekte, sowie einige entwicklungsbiologische Effekte) in den meisten Fällen gut durch Dosis-Wirkungs-Kurven ohne eine Schwelle beschrieben werden können. Das ist besonders bei dicht ionisierender Strahlung der Fall, gilt aber im Allgemeinen auch für locker ionisierende Strahlung. Im niedrigen Dosisbereich scheint eine lineare Dosis-Wirkung die beste Annäherung zu sein. Die Steigung der Dosis-Wirkungs-Kurve wird mit abnehmender Dosisleistung flacher bis, im Fall vom locker ionisierender Strahlung, eine bestimmte Grenze erreicht wird. Keiner der Parameter, welche die Dosis-Wirkung modifizieren, scheint universell zu sein. Es treten Beschränkungen und Ausnahmen hinsichtlich der Bestrahlungsbedingungen, der betroffenen biologischen Systeme und der betrachteten Endpunkte auf.

Da die beobachteten Wirkungen sich nicht von der entsprechenden Hintergrundrate unterscheiden, können biologische Effekte im niedrigen Dosisbereich nicht gemessen werden. Eine Schadens- oder Risikoabschätzung kann daher nur durch Extrapolation erreicht werden. Hierzu wird normalerweise das *LNT*-Modell (*linear-no-threshold* oder lineares Modell ohne Schwelle) herangezogen. Die wissenschaftlichen Befunde untermauern diese Vorgehensweise. Allerdings muss in Betracht gezogen werden, dass das LNT-Modell experimentell nicht direkt bewiesen werden kann. Für die Regulierung sollte auch beachtet werden, dass jeder Mensch Strahlenbelastungen aus natürlichen Quellen ausgesetzt ist. Die durchschnittliche Exposition aus diesen Quellen liegt immerhin im Bereich einiger mSv mit einer Standardabweichung von weniger als 1 mSv. Diese Werte können als Vergleichswerte für regulative Zwecke verwendet werden.

12.4
Toxikologie chemischer Karzinogene

Ein chemisches Karzinogen kann entweder als die primäre Substanz selbst, oder als eigentlich aktives sekundäres Stoffwechselprodukt wirken. Daher muss bei Extrapolationen in den niedrigen Dosisbereich die komplexe Wechselwirkung von aktivierenden und inaktivierenden Enzymsystemen zusätzlich zu den weiteren Prozessen, die mit der gentoxischen Wirkung eines „direkten" Karzinogens beginnen, beachtet werden. Bei der gentoxischen Wirkung und den darauf folgenden Prozessen gibt es eine Reihe von Analogien mit den Vorgängen, die durch ionisierende Strahlung verursacht werden und denen der vorangegangene Abschnitt gewidmet ist. Bei chemischen Karzinogenen muss allerdings noch die Komplexität der Wirkungsmechanismen beachtet werden, die in der Folge zu „praktischen" oder „vollkommenen" Schwellen führen können. Daher werden in diesem Bericht einige gut erforschte Substanzen als Beispiele für gentoxische Karzinogene mit oder ohne Schwellenmechanismus und für nicht gentoxische Karzinogene vorgestellt, um den Einfluss der Faktoren sichtbar zu machen, welche die Dosis-Wirkungs-Beziehungen ändern.

Vinylchlorid gilt als Karzinogen. Für die experimentelle Induktion von Lebertumoren gibt es den Beweis, dass eine lineare Extrapolation in den niedrigen Dosisbereich tatsächlich gerechtfertigt ist. Das LNT-Modell erscheint auch für Extrapolationen der Krebsgefährdung durch Diethylnitrosamin beim Menschen am geeignetsten. Weiter erweist sich auch das beispielhaft gewählte tabakrauchspezifische Karzinogen 4-[*N*-methyl-*N*-nitrosoamino]-1-[3-pyridyl]-1-butanon (NNK*)* als ein Nicht-Schwellen-Karzinogen. Ähnliche Szenarien werden auch bei den gentoxischen 2-Acetaminofluoren (AAF) und Aflatoxin B_1 beobachtet. Das Beispiel von AAF zeigt außerdem, dass Dosis-Wirkungs-Beziehungen im niedrigen Dosisbereich nicht nur von der Natur der geprüften Substanz abhängen, sondern auch gewebsspezifisch sind. So wurden bei Mäusen eindeutige Unterschiede für die krebserzeugende Wirkung von AAF am Beispiel der Blase und der Leber gezeigt. Das Beispiel der Wirkung von 4-Aminobiphenyl bei Nagetieren zeigt, dass das Auftreten einer Schwelle geschlechts- und gewebsspezifisch sein kann.

Die krebserregende Wirkung von anorganischem Arsen ist nicht genügend erforscht, obgleich es Hinweise gibt, dass indirekte Wirkungsmechanismen wirksam sind, die zu einer Nicht-Linearität der Reaktion im niedrigen Dosisbereich führen könnte. In solch unklaren Situationen und mangels weiterer Erkenntnisse erscheint für regulative Belange die Annahme einer linearen Dosis-Wirkungs-Beziehung adäquat. Im Gegensatz dazu ist Formaldehyd ein gut erforschtes Beispiel einer gentoxischen Chemikalie mit einer nicht-linearen krebserzeugenden Wirkung, die außerdem auf die Existenz einer „praktischen" Schwelle hinweist. Für die krebserzeugende Wirkung von Vinylacetat sind Faktoren wie die Giftigkeit für die Zelle und die regenerative Zellvermehrung wichtig, die ebenfalls auf eine existierende Schwelle deuten. Aufgrund der Natur der Wechselwirkung von Topoisomerase mit anderen Giftstoffen (hier am Beispiel der Phyto-Östrogene), erscheint das Vorhandensein einer „praktischen" Schwelle für den resultierenden gentoxischen Effekt zumindest plausibel. Wechselwirkungen von Chemikalien (z. B. Blei, Quecksilber) mit Proteinen des Zytoskeletts und die daraus resultierende Störung der Chromosomentrennung sollten einer herkömmlichen Dosis-Wirkungs-Beziehung folgen. Die nicht-mutagenen Formen der chromosomalen Gentoxizität, basierend auf Wechselwirkungen zwischen Chemikalien und Proteinen, sollten eine Dosis-Wirkung mit einer Schwelle einschließen. Es entspricht einer fast allgemeinen Übereinstimmung, dass bei der Abschätzung des menschlichen

Krebsrisikos zwischen gentoxischen und nicht-gentoxischen Chemikalien unterschieden werden sollte.

Für gentoxische Karzinogene zeigen die vorgestellten Fallstudien eine Reihe unterschiedlicher Möglichkeiten. Positive Daten über chromosomale Veränderungen, z.B. numerischer oder klastogener Natur in Abwesenheit einer Mutagenität können die Charakterisierung einer Verbindung unterstützen, die kanzerogene Effekte nur bei hohen, toxischen Dosen hervorruft. In solchen Fällen deuten die relevanten Argumente auf die Existenz einer „praktischen" Schwelle hin. Auch mag Gentoxizität nur unter der Voraussetzung einer massiven lokalen Gewebeschädigung mit der dazugehörigen erhöhten Zellvermehrung relevant sein. Auch dann können „praktische" Schwellen abgeleitet werden.

Nicht-gentoxische Karzinogene zeichnen sich durch die „herkömmliche" Dosis-Wirkungs-Beziehung aus, welche die Ableitung eines Bereiches erlaubt, in dem keine nachteiligen Wirkungen (No-observed-adverse-effect-level NOAEL) mehr nachweisbar sind. Peroxisomenproliferatoren sind z. B. Substanzen, die unter experimentellen Bedingungen eine krebserzeugende Wirkung zeigen, die auf einem Rezeptor vermittelten und durch eine Schwelle gekennzeichneten Effekt basiert. Ein weiteres berühmtes Beispiel für ein nicht-gentoxisches, Rezeptor basiertes Karzinogen ist 2,3,7,8-tetrachlorodibenzo-p-dioxin (TCDD). Jedoch sind zwischen verschiedenen Tierarten extrem große Unterschiede bezüglich der durch TCDD ausgelösten Toxizität bekannt. So ist das Meerschweinchen bezüglich der Wirkung von TCDD das empfindlichste Laborsäugetier. Es ist mehr als drei Größenordnungen empfindlicher als der Hamster, der gegenüber TCDD das widerstandsfähigste Laborsäugetier ist. Menschen scheinen dagegen gegenüber TCDD weniger empfindlich als die meisten Labortiere zu sein. Eine Konzentration von TCDD, die auf dem Niveau der in der Umwelt ohnehin vorhandenen Hintergrundkonzentration liegt, verursacht höchstwahrscheinlich keine Zunahme des Krebsrisikos beim Menschen. Tumoren, die hauptsächlich infolge einer Störung des hormonellen Gleichgewichts entstehen, weisen in der Regel auf einen Schwellenmechanismus hin; in solchen Fällen sollte daher keine Tumorbildung unterhalb von Dosen erwartet werden, die keine hormonelle Wirkung mehr zeigen.

Es besteht eine allgemeine Übereinstimmung, dass in der Riskobewertung potentiell krebserregender Chemikalien die Aspekte der beteiligten Mechanismen viel mehr Beachtung finden sollten.

12.5
Mathematische Modelle der Krebsentstehung

Der Prozess der Krebsentstehung ist zu kompliziert, um im Detail durch ein mathematisches Modell beschrieben werden zu können. Folglich bedeutet das Modellieren des Prozesses der Krebsentstehung eine Vereinfachung, die versucht, die wichtigsten Eigenschaften zu identifizieren und zu integrieren. Dieses Verfahren hilft, den krebserregenden Prozess zu verstehen, indem es Hypothesen aufstellt und die daraus resultierenden Vorhersagen mit Hilfe von Laborexperimenten oder durch die Analyse epidemiologischer Daten prüft.

Epidemiologische Daten haben allerdings nur eine beschränkte Aussagekraft, wenn sie die Auswirkung des zusätzlichen Risikos (excess risk ER) einer Krebsentwicklung in Abhängigkeit von Alter, Dosis und Dosisleistung, besonders im niedrigen Dosisbereich, beschreiben. Mathematische Modelle der Karzinogenese können zu einem besseren Verständnis dieser Abhängigkeiten beitragen. Verglichen mit den herkömmlichen Risikomodellen, die in der Epidemiologie benutzt werden, wie z. B. das Modell für das zusätzliche relative Risiko (*excess relative risk model, ERR*), haben mathematische Modelle der Krebsentwicklung den Vorteil, dass sie das Risiko, das durch komplexe Expositions-Szenarien bedingt ist, direkt beschreiben können. Nachdem die Wirkung eines Karzinogens auf die Parameter des Modells festgestellt ist, sind keine zusätzlichen Parameter notwendig, um die Auswirkung der Exposition, die sich unzählige Male während der Beobachtungsperiode ändern kann, zu berechnen. Auch haben mathematische Modelle der Karzinogenese in der Regel eine kleinere Anzahl freier Parameter.

Karzinogenese wird gewöhnlich als stochastischer Prozess modelliert, in dem eine normale, aber empfindliche Zelle einige Veränderungen durchmacht, um zu einer Krebszelle zu werden. In den meisten Fällen wird vermutet, dass solche Änderungen genetischer oder epigenetischer Natur sind. Die Anzahl der notwendigen genetischen Veränderungen scheint von Krebs zu Krebs zu schwanken. Im Fall des Retinoblastoms genügen für die Tumorentwicklung zwei Veränderungen, die beide Kopien des *RB* Tumorsuppressorgens inaktivieren. Im Fall des kolorektalen Krebses wurden Änderungen in drei Tumoruntersuppressorgenen (*APC, p53* und *DCC*) und in einem Proto-Onkogen (*K-ras*) während der verschiedenen Etappen der Tumorentwicklung beobachtet. Die Existenz von vier oder mehr Veränderungen während der Entstehung des kolorektalen Krebses weist auf die Existenz von Ereignissen hin, welche die Rate der nachfolgenden Veränderungen, d. h. der Induktion einer genomischen Instabilität, erhöhen. Auch kann die Zunahme der Proliferationsrate der intermediären Zellen während der krebserregenden Prozesse zu einer Zunahme kritischer Mutationsereignisse führen.

Phenotypisch veränderte Zellen, für die vermutet wird, dass sie Zwischenstadien in der Karzinogenese sein könnten, bilden z. B. Klone Bindegewebsgeschwulste wie Papillome der Mäusehaut, wie Foci mit Enzymveränderungen in der Nagetierleber oder wie adenomatöse Polypen im menschlichen Dickdarm. Daher berücksichtigen viele mathematische Modelle der Karzinogenese die klonale Ausdehnung der intermediären Zellen. Zweistufige Modelle der Karzinogenese mit klonaler Ausdehnung der intermediären Zellen (TSCE) gehören zu den einfachsten Modellen, die Alters- und Bestrahlungsabhängigkeiten vieler Daten beschreiben können. Dies liegt vielleicht daran, dass der komplizierte Übergang von einer normalen Zelle zu einer bösartigen Zelle zumindest in einigen Fällen näherungsweise durch zwei, geschwindigkeitsbegrenzende, Ereignisse beschrieben werden kann. Unabhängig davon gibt es zunehmende Bemühungen, Mehrstufenmodelle, welche die Kinetik der Zellvermehrung mit einbeziehen, zu entwickeln und anzuwenden.

Das große Potential der mathematischen Modelle der Karzinogenese, wenn sie auf epidemiologische Daten angewendet werden sollen, liegt in der Einschätzung von Risiken im Niedrigdosisbereich, durch Einbeziehung epidemiologischer, strahlenbiologischer und molekulargenetischer Daten. Weiterhin setzen diese Modelle die Anzahl und Größen der Zwischenstadien wie z. B. von Adenomen oder Leberfoci mit der Erkrankungs- oder mit der Sterblichkeitsrate eines Krebses innerhalb derselben Bevölkerungsgruppe in Beziehung.

Vergleichende Studien von Daten zur Lungenkrebsentwicklung bei Ratten, die Radon und seinen Zerfallsprodukten ausgesetzt wurden, mit Daten zur Krebsentwicklung bei den Atombombenüberlebenden, haben gezeigt, dass die statistische Aussagekraft dieser Datensätze nicht groß genug ist, um zu entscheiden, welches das geeignetste Modell ist. Einfachheit, Plausibilität und Kompatibilität mit strahlenbiologischen und molekulargenetischen Erkenntnissen sind zusätzliche Punkte, um passende Modelle zu identifizieren.

Bedingt durch das Expositionsszenario und die Abhängigkeit von der Dosisleistung der Modellparameter sagen mathematische Modelle der Karzinogenese sowohl direkte als auch inverse Dosisleistungseffekte sowie sowohl Zu- als auch Abnahmen des zusätzlichen Risikos in Abhängigkeit von der Dosis voraus. Gleichzeitig bergen die Abschätzungen der Parameter bei den meisten Anwendungen der Modelle auf epidemiologische Daten eine große Unsicherheit. Dementsprechend bleiben Schlussfolgerungen auf Alter, Zeit- und Bestrahlungsabhängigkeiten für die einzelnen Krebsarten unsicher. Die Anwendung einer Vielzahl von Modellen (einschließlich phänomenologischer epidemiologischer Modelle) bietet jedoch die Möglichkeit, gemeinsame Aussagen und deren Unsicherheiten, inklusive der Unsicherheit des Modells, zu identifizieren.

Wechselwirkungen zwischen den initiierenden und promovierenden Karzinogenen sind mit dem TSCE Modell untersucht worden. Wie erwartet, hängt der Grad der synergistischen Effekte von den Expositionsszenarien ab, z. B. welches Karzinogen als erstes angewandt wird. Abhängig vom Modell neigen synergistische Effekte jedoch dazu, vernachlässigbar klein zu werden, wenn die Dosen beider Karzinogene niedrig sind.

In den vergangenen ein oder zwei Jahrzehnten sind einige strahlenbiologische Wirkungen im niedrigen Dosisbereich entdeckt worden, die eine nicht-lineare Dosisabhängigkeit haben. Diese schließen *adaptive response*, Hypersensitivität im niedrigen Dosisbereich, *bystander*-Effekte und genomische Instabilität ein. Modelle zur Karzinogenese bieten die Möglichkeit, solche Effekte in die Analyse epidemiologischer Daten zu integrieren. Dies ist mit phänomenologischen Modellen nicht leicht machbar. So wurde als ein erstes Beispiel die Hypersensitivität hinsichtlich der Zellinaktivierung in das TSCE Modell integriert und das Modell auf die Lungenkrebsinzidenz unter den Überlebenden der Atombombenabwürfe angewandt. Dabei zeigte sich, dass bei Opfern, die bei der Bestrahlung schon im fortgeschrittenen Alter waren, eine mögliche Konsequenz einer Bestrahlung im niedrigen Dosisbereich die nicht-lineare Dosisabhängigkeit des Krebsrisikos ist. Mehr Arbeiten in dieser Richtung werden für die Zukunft erwartet.

Für eine große Gruppe von Modellen, die so genannten stochastischen mehrstufigen Modelle mit klonaler Ausbreitung, ist die Erforschung ihrer Eigenschaften noch in einem frühen Stadium. Eine Hauptschwäche der zur Zeit bestehenden Annäherungen auf diesem Gebiet ist die Vielzahl der möglichen zu integrierenden Prozesse und freien Parameter. Die wachsende experimentelle Kenntnis der genetischen und epigenetischen Wege kann jedoch in Zukunft die Möglichkeit einer lohnenden Anwendung solcher Modelle bieten. Ein erster eindrucksvoller Schritt ist mit der Anwendung eines Vier-Schritt-Modells auf epidemiologische Daten zur kolorektalen Krebsentwicklung geleistet worden.

12.6
Epidemiologische Aspekte der Niedrigdosisexposition menschlicher Karzinogene

An den Überlebenden der Atombombenabwürfe und in den zahlreichen Studien an Patienten, die eine Strahlentherapie erhielten, ist auf überzeugende Weise gezeigt worden, dass die Wahrscheinlichkeit des Auftretens von Krebs und die Sterblichkeit durch kardiovaskuläre Erkrankungen mit der Exposition ionisierender Strahlung zusammenhängt. Es ist nicht bekannt, ob es Strahlendosen gibt, unterhalb derer es keine wesentliche biologische Veränderung mehr auftritt, oder ob der Schaden durch normale zelluläre Prozesse effizient repariert werden kann. Analysen der japanischen Daten konnten eine Schwelle von 60 mSv nicht ausschließen, d. h. unterhalb dieser Grenze konnte keine signifikante Erhöhung von Krebs- oder Leukämiefällen beobachtet werden. Die Möglichkeit, dass ökologische Studien über Expositionen der Umwelt durch ionisierende Strahlung zum Wissen über die Wirkung von niedrigen Strahlendosen beitragen können, ist begrenzt. Die Effekte der natürlichen Hintergrundstrahlung sind zu gering und andere Risikofaktoren verzerren die Resultate, so dass die Daten unzuverlässig werden.

Was die Krebserregung durch ionisierende Strahlung betrifft, sind Alter und Geschlecht stark beeinflussende Faktoren. Der Alterseffekt wurde besonders bei Schilddrüsenkrebs, sowohl in der betroffenen Bevölkerung von Hiroshima und Nagasaki, als auch in der von dem Tchernobyl-Unfall betroffenen Bevölkerung von Weißrussland und der Ukraine beobachtet. Aber auch für Brustkrebs bei Frauen aus Hiroshima und Nagasaki und bei Strahlentherapie-Patientinnen wurde dieser Effekt gefunden. Lungenkrebs wurde als Folge der Exposition mit Radon und seinen radioaktiven Zerfallsprodukten eingehend studiert. In mehreren Studien an mit Radon exponierten Bergleuten wurde eine erhöhte Krebsrate gefunden. Fortschritte in der Molekularbiologie und in der Genetik erhöhen erfolgversprechend die Wahrscheinlichkeit des Auffindens des „wirklichen" Effektes ionisierender Strahlung nach niedrigen Dosen. Die Forschung konzentriert sich auf das Verstehen der zellulären Prozesse, die für das Erkennen und die Reparatur der normalen oxidativen und der strahlungsinduzierten Schäden verantwortlich sind.

Bösartiger Hautkrebs wird mit übermäßiger UV-Bestrahlung in Zusammenhang gebracht. Jedoch gibt es immer noch offene Fragen wie z. B. den Unterschied in der Wirkung von akkumulierter UV-Bestrahlung und Sonnenbränden. Der Effekt der UV-induzierten Schwächung des Immunsystems wird auch noch nicht verstanden.

Mobiltelefone wurden Mitte der achtziger Jahre eingeführt. Angesichts der derzeitigen weiten Verbreitung von Mobiltelefonen ist jede Möglichkeit einer schädlichen Wirkung von beträchtlicher Bedeutung für das öffentliche Gesundheitswesen. Die bisher größten Studie konnte kein erhöhtes Risiko für Tumore des zentralen Nervensystems nachweisen. Daraus wurde schließlich gefolgert, dass es keinen Beweis für ein erhöhtes Risiko gibt, selbst für Personen, die Mobiltelefone für mindestens 60 Minuten pro Tag oder regelmäßig für mindestens fünf Jahre benutzten. Die Möglichkeit krebserregender Effekte (Gehirntumor oder Leukämie) durch extrem niederfrequente elektromagnetische Felder wurde ebenfalls bewertet, die Untersuchungen gaben aber keinen Hinweis auf ein erhöhtes Risiko.

Tabak ist vermutlich das meist studierte Karzinogen für den Menschen und in industrialisierten Ländern der größte vermeidbare Risikofaktor sowohl für die Beeinträchtigung der Gesundheit als auch für die Mortalität. Das aktive Rauchen verursacht eine Reihe von Gesundheitsproblemen einschließlich diverser Krebse in verschiedenen Organen, nicht nur Lungenkrebs. Ein erhöhtes Risiko für Lungenkrebs wurde auch für Passivraucher festgestellt. In einer großen kombinierten Untersuchung hatten Nichtraucher, die mit einem Raucher zusammen lebten, ein erhöhtes Lungenkrebsrisiko. Eine signifikante Dosis-Wirkungs-Beziehung wurde gefunden und das Risiko stieg mit der Anzahl der pro Tag gerauchten Zigaretten und der Dauer des Rauchens an.

Die gesundheitlichen Spätfolgen von Schädlingsbekämpfungsmitteln sind ausgiebig studiert worden. Dichlordiphenyltrichlorethan (DDT) ist hierbei viele Jahre im Zentrum des Interesses gestanden. Heute gibt es aber immer noch keinen festen Beweis dafür, dass Schädlingsbekämpfungsmittel Krebs verursachen oder sonstige nachteilige Einflüsse auf die menschliche Gesundheit haben.

Polychlorierte Biphenyle sammeln sich im Fettgewebe an. Sie können theoretisch die männliche Fortpflanzung beeinflussen und das Risiko von Leukämie und Brustkrebs erhöhen. Eine definitive Bestätigung ihrer Toxizität fehlt aber noch.

Polyzyklische aromatische Kohlenwasserstoffe und das sehr wirksame Dioxin, TCDD, gelten als starke Karzinogene. In Studien, in denen Zufall, äußere Einflüsse und Wahrscheinlichkeit, Verzerrungen der Ergebnisse und Vermischungen verschiedener Faktoren mit angemessener Sicherheit ausgeschlossen werden konnten, zeigte sich eine positive Korrelation zwischen der Exposition und dem Krebs.

Überzeugende Beweise wurden dafür geliefert, dass Asbest Lungenkrebs und Mesotheliom verursacht und dass der krebserregende Effekt durch Tabakrauch noch verstärkt wird. Keine Studie konnte jedoch bisher andere Tumoren mit der Asbestexposition in Verbindung bringen.

12.7
Karzinogenese durch ionisierende Strahlung und Chemikalien. Gemeinsamkeiten und Unterschiede

Krebsentwicklung ist ein Mehrstufenprozess. Diese Entwicklung beginnt offenbar mit Schadensereignissen in der DNA. Das ist sowohl bei ionisierenden Strahlen als auch bei gentoxischen chemischen Karzinogenen der Fall. Während die Energieabsorption, besonders im Fall von dicht ionisierender Strahlung, zu g DNA-Schäden in Clustern führt, verursachen Chemikalien eher Einzelschäden, die gleichmäßiger über das Genom verteilt sind. Viele krebserregende Chemikalien müssen erst durch metabolische Prozesse aktiviert werden, um auf die DNA einwirken zu können. Bei einigen chemischen Karzinogenen handelt es sich aber auch um inaktivierende Enzyme, die z. B. Reparaturmechanismen blockieren. Die metabolische Abhängigkeit der Aktivität vieler gentoxischer Chemikalien erschwert daher die Bestimmung von Expositionen, die durch diese Stoffe in den Geweben und in den Zellen verursacht werden. Der Metabolismus ist von Tierart zu Tierart unterschiedlich. Das erschwert Extrapolationen vom Tierexperiment auf den Menschen. Außerdem unterscheidet sich die metabolische Aktivität auch innerhalb eines Individuums zwischen verschiedenen Organen und Geweben. Dies hat zur Folge, dass die krebserregenden Wirkungen von Chemikalien zwischen Organen viel mehr schwanken, als das für ionisierende Strahlung der Fall ist. Und schließlich haben metabolische Unterschiede zwischen verschiedenen Individuen ebenfalls eine erhebliche Bedeutung für die individuelle Empfindlichkeit.

Auf die eben genannten Anfangsschritte der Karzinogenese folgen verschiedene modifizierende Mechanismen wie z. B. DNA-Reparatur, Anpassungsreaktionen (*Adaptive Response*), Apoptose, Reaktionen des Immunsystems und Mechanismen, die in die regelnden Prozesse des Zellproliferationszyklus eingreifen. Die Funktionsweise dieser Mechanismen ist prinzipiell die gleiche, ob die Schäden nun durch Chemikalien oder ionisierende Strahlung verursacht werden. Jedoch bestehen in Details auch Unterschiede. Demgegenüber sind die weiteren Schritte der Krebsentwicklung, die mit den Prozessen der Promotion und der Progression zusammenhängen, bei ionisierender Strahlung und Chemikalien sehr ähnlich oder gar identisch. Diese Ereignisse hängen nämlich sehr stark sowohl von weiteren Mutationen als auch von der Zellvermehrung und ihrer Stimulation durch endogene oder exogene Substanzen, wie z. B. Hormone oder andere Chemikalien, ab.

Genetische Prädisposition durch Mutationen der DNA-Reparaturgene oder der Tumorsuppressorgene kann die Anfälligkeit für die Krebsentstehung erhöhen. Krebse, die durch exogene Faktoren verursacht werden, können von „spontanen" Krebsen weder auf klinischer, zellulärer noch auf molekularer Grundlage unterschieden werden. Da die spontane Krebsentwicklung eine große Schwankungsbreite zeigt, können exogen verursachte Krebse nicht als solche erkannt werden, wenn die Wirkung des exogenen Karzinogens kleiner als die Schwankungsbreite der spontanen Krebsentwicklung ist, und somit im Rauschen verschwindet. Für die Risikobewertung ionisierender Strahlung gilt generell das LNT-Konzept; bei Chemikalien gibt es jedoch auch Karzinogene, für die das LNT-Konzept nicht gültig ist, und für die verschiedene Typen von Schwellen identifiziert werden können.

Es wird daher vorgeschlagen, dass chemische Karzinogene in gentoxische und nicht-gentoxische Substanzen unterschieden werden. Innerhalb der Gruppe der gentoxischen Chemikalien gibt es eine Anzahl von Substanzen, für die aufgrund wissenschaftlicher Erkenntnisse die Gültigkeit des LNT Konzeptes sehr wahrscheinlich ist (Vinylchlorid, Dimethylnitrosamin, NNK im Tabakrauch). Für andere Substanzen ist die Unsicherheit der Dosis-Wirkungs-Beziehung im niedrigen Dosisbereich sehr groß. Es bestehen dort Unterschiede in der Form der Dosis-Wirkungs-Beziehung zwischen verschiedenen Tierarten und zwischen verschiedenen Geweben innerhalb der gleichen Art. Für diese Chemikalien (AAF, 4-aminobiphenyl) sollte das LNT-Konzept unter der Beachtung des Vorsorgeprinzips verwendet werden. Untersuchungen an anderen gentoxischen Substanzen (Formaldehyd, Vinylazetat) geben einen begründeten wissenschaftlichen Hinweis auf eine praktische Schwelle. Bei diesen Substanzen ist eine sich wiederholende DNA-Schädigung in den Geweben sowie eine Stimulierung der Zellvermehrung erforderlich, bevor eine Krebsentwicklung beobachtet werden kann. Außerdem existieren auch nicht-gentoxische Karzinogene (TCDD, Hormone), die eine deutliche Schwelle für Karzinogenizität zeigen.

Für die krebserregenden Effekte von ionisierender Strahlung sind Risikofaktoren für eine Anzahl von Organen und Geweben sowie für das gesamte stochastische

Risiko abgeleitet worden, indem man die „effektive Dosis" definiert hat. Für Chemikalien sind Risikofaktoren anhand epidemiologischer Studien und Tierexperimenten abgeleitet worden. Einige Risikowerte werden für ausgewählte Chemikalienkonzentrationen im Trinkwasser und für das gesamte stochastische Risiko durch ionisierende Strahlung beispielhaft illustriert.

Die Vergleiche zeigen, dass ähnliche Prinzipien bei der Krebsentwicklung nach Exposition mit ionisierender Strahlung und mit gentoxischen Chemikalien gelten. Ein vereinheitlichendes Konzept für die Risikobewertung ist daher bis zu einem gewissen Grad möglich. Jedoch sind die Unterschiede bei chemischen Karzinogenen zum Teil sogar innerhalb einer Chemikaliengruppe so groß, dass sie bei der Festlegung von Grenzwerten beachtet werden müssen.

12.8
Über das Testen von Hypothesen und die Wahl von Dosis-Wirkungs-Modellen

Zur Bewertung der Gefährdung von Lebewesen durch niedrige Strahlendosen ziehen Wissenschaftler ein ganzes Arsenal von Hypothesen, Modellen, Theorien und Gesetzen heran. Unterstützung für die wissenschaftlichen Forderungen wird aus einer Vielzahl von Quellen gezogen, einschließlich experimenteller Studien, epidemiologischer Daten und mathematischer Modelle. Es müssen Entscheidungen getroffen werden hinsichtlich der Extrapolation, der Existenz oder des Fehlens von Schwellendosen und der Form der Dosis-Wirkungs-Kurve. Die vorhandenen Daten können unvereinbar sein (z. B. können epidemiologische Studien auf eine Erhöhung an Krebsfällen hinweisen, die über oder unter der vorausgesagten Höhe der Dosiswirkung liegt, Resultate aus Experimenten *in vitro* können anders als *in vivo* ausfallen). Daher müssen weitere Urteile über die Plausibilität und die Haltbarkeit der vorhandenen wissenschaftlichen Evidenz mit in die Bewertung einfließen. Viele Widersprüche zwischen Wissenschaftlern sind Ausdruck der Debatten über die Aussagekraft der vorhandenen Daten. Häufig werden solche Debatten aber noch durch einen Mangel an gemeinsamer Grundlagen und an unzulänglicher Aufmerksamkeit gegenüber den grundlegenden Definitionen und der Terminologie verschlimmert.

Es scheint vernünftig, die Annahme einer Linearität in der Beziehung zwischen Exposition und Effekt als ‚Modell' für die meisten Strahlenexpositionen im niedrigen Dosisbereich zu beschreiben. Jedoch sollte man dabei stets beachten, dass die wirkliche Form der Kurve von einer Vielzahl von Variablen wie z. B. dem in Frage stehenden Effekt, der untersuchten Tierart und dem Mechanismus abhängt. Dieses trifft auch auf die verschiedenen mathematischen Modelle zu, die benutzt werden, um die unterschiedlichen Stadien in der Entwicklung des Krebses zu beschreiben. Andererseits kann die Annahme, dass es keine Schwelle für das Risiko einer Krebsentwicklung (oder anderer stochastischer Effekte) nach Bestrahlung gibt, am besten als ‚Hypothese' bezeichnet werden. Obwohl sie eine ist, die aus statistischer Sicht für Karzinogenität schwer zu falsifizieren ist, sollte sie nicht als falsifizierbar gelten. Für einige (z. B. krebserregende Effekte von Promotoren), aber zweifellos nicht für alle Effekte und alle Expositionssituationen ist die LNT-Hypothese falsifiziert worden. Hypothesen können auch zur Erklärung anderer Dosis-Wirkungs-Beziehungen und möglicher Wechselwirkungen zwischen verschiedenen Effekten und/oder Mechanismen genutzt werden. ‚Theorien' beziehen sich auf eine Sammlung von Hypothesen und Modellen, wie sie in der Theorie der Krebsentwicklung oder der Theorie über verschiedene toxikologische Mechanismen zu Hilfe gerufen werden.

Es kann eine Reihe von unterschiedlichen Ursachen für eine vermutete *Schwelle* geben. Eine absolute (oder vollkommene) Schwelle entsteht, wenn es eine Konzentration oder Dosis gibt, unterhalb derer keine schädlichen Wirkungen im Organismus auftreten. Eine praktische (oder statistische) Schwelle wird zur Beschreibung von Situationen herangezogen, in denen es eine Konzentration oder eine Expositionshöhe gibt, unterhalb derer es kein *wahrnehmbares* Anzeichen einer Schädigung gibt, das über der zu erwartenden „natürlich vorkommenden" Häufigkeit von Effekten liegt. Die Bestimmung einer offensichtlichen (oder „schützenden") Schwelle kann auf Fälle zutreffen, in denen durch die Exposition mit dem toxischen Agens das Auftreten von Schutz- oder Anpassungsmechanismen, die einen anderen schädlichen Effekt beeinflussen oder verhindern, angeregt oder erhöht wird. Dieser andere schädliche Effekt wird üblicherweise durch endogene Substanzen oder andere als die zu untersuchenden Umweltschadstoffe verursacht. Ebenso umfasst das Problem der ‚Extrapolation' weit mehr als nur die Frage, wie man die Wirkung vom hohen in den niedrigen Dosisbereich extrapoliert. Es fließen hierbei auch Urteile über Wechselwirkungen und Beziehungen zwischen Arten (z. B. Extrapolation vom Tier zum Menschen), Effekten (z. B. von der Zelle zum Krebs), Mechanismen und Modellen sowie Vermutungen über die Umrechnung von Umweltkonzentrationen in Dosiswerte mit ein. Viele der vordergründigen Widersprüche zwischen Wissenschaftlern könnten durch eine bessere Klärung der verschiedenen benutzten Hypothesen, Modelle und Annahmen vermieden werden.

Schließlich gehen Wissenschaftler unterschiedliche Wege bei der Beurteilung der *wissenschaftlichne Plausibilität*. Unterschiedliche Disziplinen berufen sich auf unterschiedliche Regeln mit unterschiedlichen Standards zur Prüfung einer Hypothese. Das schließt (z. B. in der Toxikologie) sowohl Kriterien hinsichtlich der experimentellen Beweisfähigkeit epidemiologischer Daten, der biologischen Plausibilität von Tierexperimenten und des Modellierens als auch spezifischere statistische „Beweisregeln" ein. So ist wissenschaftliche Glaubwürdigkeit nicht eine Frage der Anwendung einer spezifischen Regel, sondern vielmehr eine Reihe unterschiedlicher Kriterien, in denen Plausibilität eine Frage des Maßstabs ist, der in großem Maße davon abhängig ist, wie viele jener Kriterien berücksichtigt werden. Abhängig von der Art der Beurteilung einer eventuellen Belastung mit toxischen Agenzien und den möglichen Folgen dieser Beurteilung (z. B. Routinekontrollen im Unterschied zu Entscheidungen über den Bau einer neuen Anlage) können Entscheidungsträger unterschiedliche Strenge in der Beurteilung der wissenschaftlichen Beweise walten lassen. Daraus folgt, dass neben dem Wert wissenschaftlicher Erkenntnisse zweifellos auch ein Einfluss verschiedener anderer Werte (z. B. aus der Politik, der Volkswirtschaft, der Ästhetik, usw.) und Umstände auf die Art der Entscheidungen zum Schutz gegen schädliche Expositionen besteht. Viele dieser Entscheidungen (wie z. B. die Bestimmung des akzeptablen Risikoniveaus) können nicht aufgrund wissenschaftlicher Erkenntnis allein getroffen werden. Dennoch ist es unerlässlich, dass die bestmögliche Wissenschaft diese Entscheidungen beratend unterstützt, indem sie relevante Fakten und Informationen zur Verfügung stellt. Dies schließt die Transparenz über bestehende Unsicherheiten und Annahmen, die mit diesem Wissen einher gehen, mit ein. Es kann daher triftige und vernünftige Gründe dafür geben, dass „Wissenschaft" und „Politik" unterschiedliche Dosis-Wirkungs-Modelle wählen. Und schließlich, je größer die Ungewissheit in der Prognose der möglichen expositionsbedingten Folgen, desto wichtiger werden ethische Werte bei Entscheidungen über Maßnahmen des Risikomanagements.

12.9
Risikobewertung und -kommunikation

In der Risikoanalyse müssen mehrere Verfahren angewandt werden. Zunächst geht es um die sorgfältige Bestimmung des möglichen Schadens – entweder quantitativ, d. h. wenn möglich, mit statistischen Daten, oder auf Grund von Wahrscheinlichkeitsschätzungen. Diese Erkenntnisse bilden die Grundlage für die Risikobeschreibung. Diese muss ein möglichst objektives Verständnis für die Ursachen und die Kontrollmöglichkeiten eines bestimmten Risikos entwickeln. Diese Erkenntnisse werden dann in das Risikomanagement und in die Entscheidungsprozesse integriert.

Zwei grundlegende Risikoarten werden betrachtet: individuelle und gesellschaftliche Risiken. Wenn für eine bestimmte Risikokategorie alle Risikowerte auf der Grundlagen individueller Risiken betrachtet würden, wäre das Ergebnis in hohem Maße unrealistisch. Folglich werden potentielle Risiken für genau beschriebene Gruppen innerhalb der Bevölkerung oder innerhalb einer geographischen Region beschrieben. Weiter muss zwischen dem Risiko zu sterben oder dem, eine gesundheitliche Beeinträchtigung zu erfahren, ebenso unterschieden werden wie zwischen akuten und verzögert auftretenden Wirkungen. Das Wissen über die eigentlichen Ursachen der betreffenden Wirkungen ist nicht immer, und für alle Bereiche, in ausreichendem Maße vorhanden.

Jedoch muss auch bei unvollständigem Wissen ausgewählt und entschieden werden. Daher ist in einigen Ländern, und besonders in der Europäischen Union, das Vorsorgeprinzip entwickelt und eingeführt worden, das, falls Zweifel über mögliche schädliche Effekte niederen Ausmaßes bestehen, dem Gesetzgeber nahelegt, sich eher für die „sichere Seite" zu entscheiden, als auf einen abschließenden wissenschaftlichen Beweis zu warten.

Falls ein bestimmtes Risiko als inakzeptabel, d. h. zu groß, bestimmt worden ist, wird vom Risikomanagement verlangt, dieses Risiko auf ein akzeptables Niveau zu reduzieren. Das ist hauptsächlich eine ökonomische Frage und schließt die Festsetzung von angemessenen Standards und Prioritäten mit ein. Ein zusätzliches Argument für die Entscheidungsträger ist die Akzeptanz eines möglichen Risikos durch die Gesellschaft. Hier muss das Risiko gegen den Nutzen abgewogen werden. Dabei herrscht das Verständnis vor, dass nur dann, wenn der erwartete Nutzen besonders groß ist, nicht ernsthaft erwogen werden kann, eine risikobehafteten Tätigkeit zu unterbinden. So können z. B. die Bemühungen, die mit Röntgenstrahlen verbundenen Risiken zu verringern, nur in Richtung einer Verringerung der Exposition auf ein annehmbares Niveau gehen. In diesem Zusammenhang ist es manchmal nützlich, auf Risikovergleiche zurückzugreifen, besonders hinsichtlich der Auswirkungen auf die Gesundheit, da dies für die Kommunikation von Entscheidungen in der Öffentlichkeit entscheidend ist.

Die Kommunikation von risikobezogenen Themen ist ein schwieriger Bereich in der politischen Arena und sollte der Verwirklichung demokratischer und partizipativer Politikstrategien dienen. Die verschiedenen Bestandteile einer sinnvollen Risikokommunikation, können in Form der Frage dargestellt werden: Wer sagt was, in welchem Kanal mit welcher Wirkung? Der eigentlich Mitteilende muss für das Publikum glaubwürdig sein. Er muss die nötigen Fähigkeiten besitzen, die Aufklärung in einer Weise zu betreiben, die für ein genau festgelegtes Publikum verständlich und akzeptabel ist. Das Wissen um das zu erwartende Publikum und die entsprechende Anpassung der Erklärung an dessen Grundbedürfnisse ist ein wichtiger Bestandteil einer erfolgreichen Risikokommunikation. Folglich müssen der Mechanismus der individuellen Informationsverarbeitung und die verschiedenen Faktoren, welche die intuitive Risikowahrnehmung beeinflussen, beachtet werden. Die Nachricht selbst kann sich entweder auf konkrete Informationen stützen (d. h. die meisten Aspekte des Risikos sind bekannt) oder sie kann aus bewertenden Aussagen bestehen, die nur versuchen, ein bestimmtes Risiko ins Verhältnis zu anderen, bereits vertrauten, Risiken zu setzen. Hier spielen so genannte „Rahmen" eine wichtige Rolle, die einer Person ermöglichen, neue Informationen in eine bereits vorhandene Informationsstruktur einzufügen. Das Problem hierbei ist, dass diese individuelle Struktur sehr starr, d. h. schwer zu ändern, sein kann. Zusätzlich könnte sie falsche oder irreführende Informationen enthalten. So ist die Gesamtwirkung einer Kommunikation, insbesondere über Risiken und deren Bewertung, schwer zu ermessen.

12.10
Niedrige Dosen im gesundheitsbezogenen Umweltrecht

Der Prozess der Risikobewertung und des Risikomanagements wird nur in einem begrenzten Umfang durch die Rechtsordnung strukturiert. Schutzverpflichtungen des Staates und der Europäischen Union, die grundlegende Unterscheidung zwischen ‚Gefahrenabwehr' (Schutz gegen ein offenbar nicht akzeptables Risiko) und ‚Vorsorge' (Verminderung eines nicht wünschenswerten Risikos), das allgemeine Verständnis des Begriffes ‚Gesundheitsschaden' und die Orientierung an der Gefährdung des Individuums, zumindest in Form des Schutzes vulnerabler Gruppen, bilden hierbei den Rahmen. Jedoch lassen sie viel Spielraum für die Beurteilung von Tatsachen und für die daraus resultierenden Entscheidungen.

Im Risikomanagement wirft aus juristischer Sicht die Definition von ‚Vorsorge' und ihre Abgrenzung zum ‚Restrisiko', das von allen Bürgern toleriert werden muss, schwierige Probleme auf. Das Vorsorgeprinzip bezieht sich auf die Sachlage der ‚Unsicherheit'. Jedoch erfordert die Einstufung eines ‚Zustands der Unsicherheit' als ‚Risiko', das verringert werden muss, einen angemessenen Verdachtsmoment. Ein bloß spekulativer Verdacht ist unzureichend, obgleich die Definition von dem, was ‚spekulativ' und was ‚wissenschaftlich begründet' sein soll, recht fließend ist. Daher erfordern diejenigen Elemente der wissenschaftlichen Risikobewertung eine genaue juristische Prüfung, die einen umsichtigen Umgang mit Unsicherheit ermöglichen, d. h. insbesondere die Verwendung von „Proxy"-Kriterien zur Riskobestimmung (z. B. Persistenz und Bioakkumulation), von Extrapolationsfaktoren und von wissenschaftlichen Modellen (z. B. die LNT-Hypothese für krebserregende Stoffe). Bezüglich aller drei Problembereiche ist eine differenzierte Bewertung angebracht. Insbesondere aus juristischer Sicht erscheint eine normative Harmonisierung im Bereich der Extrapolation wünschenswert und sollte immer dort, wo sie wissenschaftlich haltbar ist, in Betracht gezogen werden. Die LNT-Hypothese ist legitim, solange es keinen starken Beweis dafür gibt, dass sie für ein bestimmtes Mittel ungültig ist.

Risikomanagement ist ein komplexer normativer Prozess, bei dem sowohl eine grundsätzliche Entscheidung hinsichtlich der Zumutbarkeit von Risiken als auch hinsichtlich angemessener Maßnahmen zur Risikoverringerung gefordert ist.

Was gesundheitsbezogene Kriterien betrifft, öffnet das häufig verwendete Konzept der Vermeidung eines „signifikanten Risikos" einen großen Ermessensspielraum. Bei der Entscheidung über die Zumutbarkeit eines Risikos, können Hintergrundrisiken und Vergleiche mit anderen Arten von Risiko, für die eine angemessene Entscheidung über deren Verringerung bereits getroffen worden ist, eine Rolle spielen. Außerdem ist, zumindest auf dem Gebiet bloßer Vorsorge, die Abwägung von Vor- und Nachteilen legitim und sogar rechtlich gefordert. Jedoch ist weder eine formale Kosten-Nutzen-Analyse noch irgendein anderes formales vergleichendes Verfahren vorgeschrieben, obgleich diese, im Gegensatz zu Deutschland, in der Praxis der Europäischen Union eine wichtige Rolle spielen. Wenn sie im Bewusstsein ihrer methodologischen Fehler und möglichen politischen Gefahren angewandt würde, könnte man im Risikomanagement mehr Gebrauch von der Kosten-Nutzen-Analyse machen, um eine größere Aufmerksamkeit auf die Kosten einer Regulierung zu lenken. Gleichzeitig könnte auch der Grad der allgemeinen Akzeptanz von Maßnahmen innerhalb der Bevölkerung erhöht werden.

Der Umfang, in dem „weiche" Faktoren wie z. B. Verteilungsgerechtigkeit- und Risikowahrnehmung eine Rolle im Prozess des Risikomanagement spielen können, ist Gegenstand von Kontroversen; dies hängt in hohem Maße vom rechtlichen Rahmen ab. Während ungleiche Gefahren durch krebserregende Stoffe, welchen Menschen auf Grund ihrer unterschiedlichen Empfuindlichkeit ausgesetzt sind, berücksichtigt werden müssen, indem man das Konzept der „vulnerablen Gruppen" verwendet, ist es zweifelhafter, ob es auf dem Gebiet der Vorsorge strikte rechtliche Regeln gibt, die eine Angleichung solcher Risiken erfordern. Der Rang der menschlichen Gesundheit und die Unmöglichkeit einer wirklichen Abgrenzung von ,Gefahrenabwehr' und ,Vorsorge' auf diesem Gebiet sprechen zumindest für eine gewisse Angleichung der Expositionen. Subjektive Risikowahrnehmung ist kein Faktor, den das Gesetz im Risikomanagement ausdrücklich als legitim anerkennt. Jedoch kann sie in begrenztem Umfang eine Rolle spielen, besonders, wenn sie Prioritäten für Maßnahmen festgelegt werden.

In Deutschland und in der Europäischen Union sind Quantifizierungskonzepte für die Bestimmung des zumutbaren Krebsrisikos, besonders zur Abgrenzung von Vorsorge und Restrisiko, nicht sehr verbreitet, obgleich in der Praxis eine lebenslanges Individualrisiko für eine Belastung mit krebserregenden Substanzen von 10^{-6} akzeptiert worden scheint. Ein Quantifizierungskonzept ist prinzipiell nützlich, weil es eine transparentere Diskussion über die Zumutbarkeit eines bestimmten

Risikos erlaubt. Das in Deutschland für die Begrenzung von Karzinogenen mit Blick auf den Vorsorgegedanken bevorzugte emissionsorientierte Konzept und besonders die Minimierung sollte zumindest durch ein Konzept von Umweltqualitätsstandards ergänzt werden – wie es jetzt zur Umsetzung von Richtlinien der Europäischen Union erforderlich wird.

Die Organisation des Verfahrens der Risikobewertung und des Risikomanagements ist in jüngster Zeit ein Thema vieler Diskussionen gewesen. Abgesehen von der Stärkung der partizipativen Elemente des Verfahrens in Bezug auf die wissenschaftliche Begründung und die Darlegung von Interessen und Werten gibt es eine Anzahl von Punkten, über die noch Einigkeit erzielt werden muss. Hier ist besonders der Grad der Zentralisierung und der Integration der kognitiv-vorhersagenden und normativen Phasen und die Struktur der Beratungsgremien in der Risikobewertung zu nennen. Hinsichtlich aller drei Fragen scheinen pragmatische Lösungen, die jedoch die spezifische Rolle der Wissenschaft im Prozess anerkennen müssen, vorzugswürdig.

13 Recommendations for Environmental Risks to Human Health: Evaluation and Regulatory Processes

The discussions of the Working Group on "Environmental Standards – Risks and Regulatory Processes in the Low-Dose Range" showed that the interdisciplinary dialogue and interaction is of utmost importance for dealing with and setting environmental standards. In the modern, industrialised world with high human population densities, it is necessary to achieve a balance between the development and the operation of efficient technologies on the one hand and the maintenance of a healthy and sustainable environment for today's life and for future generations on the other hand. Therefore regulations; including the setting of environmental standards; are very important.

These challenges cannot be achieved without rational decision processes and should be based on:

1. solid scientific data about the effects of toxic agents and their mechanisms,
2. discussion of ethical and social implications including the uncertainties of the scientific evaluations, as well as a communication of risks and benefits under conditions of transparency,
3. procedures that promote rationality, provide legitimacy and ensure consistency with legal, especially constitutional, mandates and constraints.

In consideration of these fundamental requirements, and acknowledging the actual state of knowledge, the Working Group has come to the following recommendations:

> 1. Environmental standard setting has to be based on reliable, sound scientific data.

Critical assessments of the dose-responses have to be based on solid epidemiological and experimental investigations. In this connection, the knowledge about the shape of the dose-response curves in the low-dose range is essential. Two principal dose-responses are distinguished: A dose-response with a threshold dose and a dose-response without a threshold dose.

> 2. The distinction of harmful agents into two groups, those that act *with a threshold* and those that act *without a threshold* should be maintained.

The question of a threshold is crucial for risk evaluation especially in the low-dose range and has therefore to be investigated intensively. These issues differ from agent to agent with respect to the biological effects and specialities of the various agents must be considered thoroughly.

> 3. When a *threshold dose* can be verified on the basis of scientific data, regulatory standards should be set *below* the threshold level. Safety factors should be used in these cases to acknowledge uncertainties in the evaluation procedures and individual differences in susceptibility

Such dose-responses are usually observed for acute effects and non-malignant tissue damage, originating from many damaged cells.

In contrast, for mutagens and carcinogens with genotoxic mechanisms, a dose-response without a threshold appears very likely, which makes those agents more difficult to regulate through standard setting. The complex processes of carcinogenesis are far from being fully understood.

Therefore, setting standards for carcinogens cannot be based on a solid scientific basis alone. Conventions have to be made that contain as much actual scientific knowledge as possible.

> 4. For mutagens and most genotoxic carcinogens, e. g. ionising radiation, where there is *no evidence for a threshold dose*, the LNT (*lin*ear-*n*o-*t*hreshold) model[1] remains the most appropriate convention for risk evaluation by extrapolation into the low-dose range and for the control of carcinogens. Its simplicity for risk management is advantageous. Its application is therefore recommended for risk estimates of genotoxic agents.

This recommendation is not based on the assumption that the dose-response curve is linear in the whole range from low to high doses but it can be taken as a linear curve in the important low-dose range. Indeed, many radiobiological effects have been found in the past decade that do not exhibit linearity in the higher dose ranges. These include expression of different genes at lower than at medium/higher doses, DNA-damage-repair capacity, adaptive responses, induction of genomic instability, bystander effects and low-dose hypersensitivity. No clear conclusion could be achieved up to now, how these effects influence the shape of the dose-response curve.

The possible alternatives for dose-responses which are discussed in the scientific community and for which scientific data are available should be considered within risk evaluation processes. For standard setting, however, such models should be used that have a universal and broad validity and that include groups with different susceptibilities with respect to gender, age groups and genetic dispositions, as well as various tissues. This general validity is probably best covered for genotoxic agents by the LNT model in order to achieve a profound and safe protection of the environment and human health.

---

[1] Due to the inherent complexity regarding mechanisms and postulated biological endpoints, evaluations of dose-response relationships will need to include a number of assumptions, auxiliary hypotheses and simplifications. Thus the linear extrapolation of dose-response curves at low doses is better described as a LNT *model* rather than the LNT *hypothesis*.

However, there are a number of different mechanisms that may lead to a threshold in dose-response curves, and, even for chemical carcinogens, at least three different types of thresholds may be identified. When discussing thresholds it is important to be clear whether the hypothesis or model postulated refers to an *absolute* (no adverse effect below a certain exposure level), *practical* (no *observable* effect below a level) or *apparent* threshold (due to a protective or adaptive response initiated by the toxin). This leads the Working Group to suggest a classification of three groups of carcinogens based on the above-described distinction of thresholds and on the mechanisms of the chemicals which have been discussed in Chapter 7:

---

5. <u>*Group 1*</u>
Genotoxic carcinogens without a threshold, where a LNT model should be applied (see also chapter 4.).

<u>*Group 2*</u>
Genotoxic substances, where *sufficient scientific data,* concerning the underlying mechanisms, suggest existence of *a practical threshold.*

<u>*Group 3*</u>
Non-genotoxic carcinogens: These compounds display a clear threshold of carcinogenicity, and health-based exposure limits may be established by starting from "no observed adverse effect level" NOAEL and the introduction of a safety/uncertainty factor, if this is applicable.

The dose-responses after low exposures can usually only be obtained by extrapolations from observed effects in higher dose ranges. This proves true for all mutagens and carcinogens. Since the extrapolation procedures carry inherent uncertainties but at the same time they are unavoidable, a better understanding of the complex mechanisms in multistep processes is very important for an improved prediction of health effects in the low-dose range. Knowledge of these mechanisms gives guidance for the extrapolation.

---

6. Further research of the mechanisms involved in the multistep process of carcinogenesis is needed to elucidate the different mechanisms of the various agents to reduce the uncertainties.

The question of scientific plausibility regarding the toxicity of carcinogens is not something that can be settled by appeal to a single factor. Decisions need to be based on a multi-criteria evaluation, including consideration of the available knowledge on epidemiology, animal experiments, cellular and molecular studies, biological mechanisms, and models. Safety or extrapolation factors used for setting standards should reflect the level of uncertainty in the various criteria.

7. A harmonisation of the principles of standard setting within the fields of ionising radiation and genotoxic chemicals is recommended as the development of mutations and cancers is similar after exposures to ionising radiation and genotoxic chemicals, except for the primary events. Exposures from natural sources can be used as "benchmarks" for standard setting. Differences in susceptibility of groups and individuals must be considered.

Except for the primary events, the development of mutations and cancers is similar after exposures to ionising radiation and genotoxic chemicals. These similarities of mechanisms invite to harmonise the principles of standard setting within these two fields. Harmonisation of crucial criteria for risk assessment such as considerations for selecting extrapolation factors, use of risk models, treatment of carcinogenic substances with a (practical) threshold, and consideration of susceptibility in the absence of data is desirable. This concerns both, the national and the European levels. By contrast, harmonisation of "technical" criteria such as conversion factors should be avoided, as there is no scientific basis for standardisation.

8. *Ambient quality standards* should not only be set for threshold substances (*Groups 2 and 3*) *but also* for substances that are not associated with a threshold (*Group 1*), such as most carcinogenic (genotoxic) substances. However, ambient quality standards for no-threshold substances should be *supplemented by minimisation requirements* in order to further reduce a risk that cannot be adequately ascertained. In case of carcinogenic substances that are associated with an *absolute* or *practical* threshold (*Groups 2 and 3*) this additional safety valve is not necessary.

Ambient quality standards for no-threshold substances should be based on a *common measure* of quantified *tolerable risk*. At least gross divergences in tolerable risk figures for different carcinogens require special justification.

Exposures from natural sources and their possible risks should be considered for comparisons with exposures from man-made sources and for standard setting. This is especially possible in the case of ionising radiation but also of some toxic chemicals.

9. Differences of susceptibility must be attributed more weight both in risk assessment and risk management.

However, the delimitation and reasonable size of a susceptible group is not at all clear. It is certainly not advisable and feasible to take into account very small susceptible groups leading in its extreme to an individual. The use of standard groups of susceptibility (vulnerable groups) is legitimate, yet the risk assessor and regulator should also aim to include diffuse groups (e. g. persons with immune defects) where this is practicable and not inconsistent with effective protection of the vast majority of persons at risk. One possible response to unknown or diffuse variations in susceptibility is the use of an additional safety factor.

> 10. A weighing of societal costs against benefits is necessary when decisions are made concerning the acceptance and acceptability of risks. Not only monetary values but also other values, e. g. of ethical nature, have to be taken into account. Distribution of risks and benefits should be considered without neglecting practicality.

The consideration of societal costs associated with risk regulation and comparison with the expected benefits including, to the extent practicable, side effects and alternative actions to reduce risks constitutes a legitimate component of precautionary risk management. However, formal cost/benefit analysis should only be used with great caution as monetarisation may distort the whole balancing process. It is crucial that not just monetary values are being taken into account in such analyses, but the whole range of values that are relevant to the decision.

The distribution of costs and benefits is highly important for what risks can be considered ethically acceptable. It is very problematic to impose risks on people who are not getting their share in the benefits that accrue from the imposition of risks. From an ethical point of view one might therefore want differentiated levels of acceptable risk, one for people who themselves benefit from the risk and one for enterprises that primarily benefit others. Distributional concerns are particularly important when those who benefit are already well off, while those who suffer the risk are badly off, for example live in areas that are exposed to pollution or other negative effects of the activity. It is a challenge for legislators to develop rules that reflect such distributional concerns. However, such rules must consider practicality and generic benefits from which people adversely affected in this living or working environment may profit.

> 11. Acceptability is a normative notion and has a high value. In order to achieve acceptability, an effective risk communication together with a high degree of transparency regarding the preparation of decisions on risk assessment and risk management is necessary.

For the whole procedure of setting environmental standards it is highly important that both, the composition of the relevant bodies, and the procedures they follow in their deliberations and decisions are transparent. It must be as clear as possible how they reach their decision, so that the general public can be confident that their interests are well taken care of. In the process of risk assessment a scientific dialogue, in the process of risk management an open presentation of interests and values must be organised in order to achieve more rationality, legitimacy and acceptance of the relevant decisions.

Firstly, transparency is imperative in the scientific discourse, when the risk is assessed, characterised, and evaluated. Consequently, the scientific community should be able to control and repeat the relevant studies in order to avoid debates about the results. Secondly, when the results of the study are used by the risk manager and/or decision maker, transparency is needed for involving the public in the discussion and for communicating to the public on what basis the decisions were made and the standards were set.

Since the ultimate goal of risk communication is acceptance of the decision by the public at large, careful planning is necessary. This involves: preparation of the information about the scientific findings and the whole decision making process with its background in an understandable way, selection of speakers and audiences, providing information about the beneficial as well as deleterious aspects and constructive relations to the media. The mechanisms to employ could be: public hearings, press releases, scientific journals, modern information technology, statements of trustworthy officials, i. e. scientists, politicians, medical doctors, teachers, etc. One of the problems encountered in the past is that there can be opinions prevailing within a smaller group of the public on the topic in question which are not based on scientific insights but rather on extreme emotional interpretations. Even though such a minority can create an opposition, the decision-makers would be well advised to give such irrational movements of very small and unrepresentative groups or individuals only limited attention as this will increase the opponents' credibility for the public.

---

12. In the risk evaluation and standard setting process the rights and potential concerns also of those who cannot be directly heard should be considered.

"Silent victims", people who are handicapped or have little access to media, and no influence in elections, must never be forgotten when policies are made. One should see to it that silent victims are represented by people who know their experiences and concerns.

Future generations have no possibility to influence our discussions and decisions. Yet we owe them a decent share of the world's resources and should not turn our pollution and environmental problems over to them. When decisions are made, we should therefore pay close attention to environmental concerns, especially irreversible processes. Two aspects have especially to be considered: the long-lasting persistence of toxic agents and hereditary effects of the agents. Such concerns should be adequately reflected in legislation.

---

13. The bodies that give advice or make decisions concerning setting of environmental standards should include representatives from all competent disciplines. The interests of affected groups should be represented as well. It has to be made sure that the experts are knowledgeable and experienced on scientific grounds. They have to be up to date with the latest findings.

Scientists including ecologists, ethicists, legal specialists, other experts such as economists and representatives of affected groups or affected individuals should participate in the decision making process. Scientists have to be familiar with the scientific background for the specific risk evaluation. They have to be aware of the feasible alternatives and the probabilities of the various consequences. The expertise of ecologists has to be included. Where there is disagreement between competent and well-informed scientists, all main views should be represented. Ethicists should be familiar with the latest research on what different factors are ethically relevant in connection with the issue at hand and how they should be weighted against one another. Legal specialists must be familiar with legislation and jurisprudence in this area and should also have knowledge of the likely social impacts of alternative kinds of regulation. Other experts will often be needed, for example economists, who can help to clarify the often complicated economic effects of the various alternatives and who can also suggest and evaluate various economic incentives and compensation schemes that might be considered. Representatives of the groups that might be affected have to be involved in the process of decision making and should get a clear understanding of the situation. One should find ways to ensure that "silent victims" are represented, such as under-privileged and handicapped, who often are forgotten when policies are made. It is also highly important that the interests of *future generations*, which cannot attract attention, be adequately represented. Elected constitutional bodies should make the ultimate decisions.

One of the main difficulties is certainly located in balancing the manifold aspects, which have been discussed in the foregoing text. It should be kept in mind that the scientific knowledge about the effects and the dose-responses is the most import basis for setting environmental standards for toxic agents. However, it must also be clear that all toxic exposures are only permissible and justified if they are associated with a benefit to the society and if the risks are tolerable. The decisions about standards have to be made on the basis of constitutional mechanisms, which include societal acceptance.

14 Glossary

α	Conditional probability of → type 1 error. H_0 is only rejected if $\alpha \le 0{,}05$. → Significance level
Absolute risk	The difference in the rate of occurrence of a particular health effect in an exposed population and an equivalent population with no radiation exposure (rate difference or absolute risk).
Absorbed dose (D)	Radiobiology: Fundamental dosimetric quantity, defined as: $D = \frac{d\bar{\varepsilon}}{dm}$, where $d\bar{\varepsilon}$ is the mean energy imparted by ionising radiation to matter in a volume element and dm is the mass of matter in the volume element. Absorbed dose is defined at a point; for the average dose in a tissue or organ, → organ dose. Unit: Gray (Gy)=J kg^{-1} (formerly, rad was used). → dose
Acceptability	Normative notion, to set obligatory criteria for the evaluation of tolerability of conditions or actions; an option is acceptable if it satisfies the corresponding criteria of acceptance. → also acceptable risk, acceptance
Acceptable Daily Intake (ADI)	Estimate of the amount of a substance in food that can be ingested daily over a lifetime by humans without appreciable health risk. ADIs are derived from laboratory toxicity data, and from human experiences of such chemicals when this is available, and incorporate a → safety factor.
Acceptable risk	Probability of suffering disease or injury that ought to be consented by an individual, group, or society. The acceptability of risk depends on scientific data, social, economic, and political factors, and on the perceived benefits arising from a chemical or a radioactive process. → risk concepts, acceptability, accepted risk
Acceptance	Actual assent to a condition or an action (empirical notion); an option is accepted if it is acted in accordance with it.

Accepted risk	Probability of suffering disease or injury, which is actually approved by an individual. → acceptable risk, acceptance
Acoustic neuroma	A tumour that originates in the acoustic nerve cells.
Acute	1. Short-term, in relation to exposure or effect. In experimental toxicology, "acute" refers to studies of two weeks or less in duration (often less than 24 h).
	2. In clinical medicine, sudden and severe, having a rapid onset. A sudden onset of symptoms or disease.
	Usually contrasted with → chronic and → sub-chronic
Acute lymphoblastic leukemia (ALL)	A form of leukemia characterised by an increase in the number of B-lymphocytes in the blood and bone marrow.
Acute myeloid leukemia (AML)	A type of leukemia that affects the neutrophils (a type of white blood cell).
Adapting dose	Dose that produces an → adaptive response
Adaptive response	Responses to exposures to some toxic agents in the low-dose range that increases the resistance against a second exposure of a toxic agent.
Adenoma	Usually → benign → tumour in glandular tissue.
Adenomatosis polyposis coli	A non-cancerous tumour with a cauliflower like growth that develops in the mucous membrane lining of the colon.
Adverse health effect	Abnormal, undesirable, or harmful effect to an organism, indicated by some result such as mortality, altered food consumption, altered body and organ weights, altered enzyme levels, or visible pathological change. An effect may be classed as adverse if it causes functional or anatomical damage, causes irreversible change in the homeostasis of the organism, or increases the susceptibility of the organism to other chemical or biological stress. A non-adverse effect will usually be reversed when exposure ceases.
Adversity	Property of an → adverse health effect
Agent	Toxicology, Radiobiology, Epidemiology: A force or substance that produces an effect or change to

the condition of a subject (human being, animal, plant, nature, cultural good etc.).

Philosophy, Sociology: Somebody who does something or causes something to happen. In the frame of this book the philosophical agent will be changed to "actor".

ALARA principle, "As low as reasonably achievable"	Principle to reduce a potential harm to the lowest level reasonably possible. In Anglo-Saxon literature sometimes opposed to → ALARP principle → Minimisation/Optimisation
ALARP principle, "as low as reasonably practicable"	Principle mostly found in Anglo-Saxon literature to be opposed to the → ALARA principle. The attribute "practicable" seems to have a looser meaning than "achievable". → Minimisation/ Optimisation
Alkylating Agent	Any substance that introduces an alkyl → radical into a compound in place of a hydrogen atom.
Allele	Used for two or more alternative forms of a gene resulting in different gene products and thus different phenotypes.
Alpha (α)-radiation	A positively charged particle ejected spontaneously from the nuclei of some radioactive elements. It is identical to a helium nucleus, but of nuclear origin. It comprises two neutrons and two protons. → radiation
Angiogenesis	Blood vessel formation. Tumour angiogenesis is the growth of blood vessels from surrounding tissue to a solid tumour. This is caused by the release of chemicals (cytokines) by the tumour.
Angiosarcoma	A type of cancer that begins in the lining of blood vessels.
Annual dose	The dose due to external exposure in a year together with the → committed dose from intakes of toxic substances or radionuclides in that year. The → individual dose, unless otherwise stated. This is not, in general, the same as the dose actually delivered during the year in question, which could include doses from toxic substances or radionuclides remaining in the body from intakes in previous years, and could exclude doses delivered in future years from intakes during the year in question. → dose

Annual risk	The probability that a specified health effect will occur at some time in the future in an individual as a result of radiation exposure received or committed in a given year, taking account of the probability of exposure occurring in that year.
	This is not the probability of the health effect occurring in the year in question; it is the lifetime risk resulting from the annual dose for that year.
Antagonism	Combined effect of two or more factors, which is smaller than the solitary effect of any one of those factors.
Apoptosis	A normal series of enzymatic events in a cell induced by signal transduction starting at the cell membrane that lead to its death (and its successive replacement).
Asbestosis	→ Pleural mesenthelioma
Asymptotic value	An asymptotic function gradually approaches a constant value (the Asymptote); after some time period, the function is arbitrarily close to the constant.
Ataxia telangiectasia	A rare, inherited genetic disorder that results in damage to the nerves. It is associated with an increased risk of developing cancer. The individuals have a high radiosensitivity with arepair deficiency of DNA damage.
Auger electrons	→ secondary electrons
Availability heuristic	The ease with which probabilities of occurrences are remembered biases risk perception.
β	Conditional probability of → type 2 error.
Background dose	The → dose or → dose rate (or an observed measure related to the dose or dose rate), attributable to all sources other than the one(s) specified. Background is used generally, in any situation in which a particular source (or group of sources) is under consideration, to refer to the effects of other sources. → background incidence, → natural background
Background incidence	Rate of occurrence of a damage that is attributable to all sources other than the one under consideration. → background

Balancing	Comparison and weighing of alternatives.
Baseline	A baseline would be an incidence of effects that are associated to a specific risk factor without a significant contribution from that risk factor.
Becquerel (Bq)	Name for the SI unit of radioactivity, equal to one radioactive decay per second. Supersedes the curie (Ci): 1 Bq $\approx$ 27 pCi (2.7 10^{-11} Ci).
Belief system	Set of opinions about an object, a person, or a situation.
Benchmark dose BMD_x	Dose associated with a designated level or percent (x) of response adjusted for background.
Benign	Any growth which does not invade surrounding tissue. $\rightarrow$ Malignancy, tumour
Beta (β)-radiation	Particulate $\rightarrow$ Radiation of $\rightarrow$ electrons and positrons emitted by radioactive materials.
Bioaccumulation	Process by which some endogenous or exogenous substances, present in small quantities, increase in concentration in an organ, an organism, a food chain, or an ecosystem.
Bioactivation	Any metabolic conversion of an agent to a more toxic derivative.
Bioavailability	The degree to which a substance is available to be absorbed and metabolised by an exposed living organism.
Burden of proof	Obligation or burden imposed on an authority or a person to substantiate and prove facts relevant for a right, empowerment or claim, e. g. $\rightarrow$ causation.
Bystander effect	A range of biological phenomena whereby mammalian cells irradiated in culture produce damage-response signals, which are communicated to their non-irradiated neighbours.
Carcinogenesis	Carcinogenesis is the process, which leads to development of cancer. Carcinogenesis may be a matter of induction by chemical, physical, or biological agents of $\rightarrow$ neoplasms
Carcinoma	A general term applied to a $\rightarrow$ malignant epithelial $\rightarrow$ tumour.
Cardiomyocytes	Heart muscle cells

Case-control study	A study which starts with the identification of persons with the disease (or other effect) of interest, and a suitable control (comparison, reference) group of persons without the disease. The relationship of an attribute to the disease is examined by comparing the diseased and non-diseased with regard to how frequently the attribute is present or, if quantitative, the levels of the attribute, in the two groups. Compared to a cohort study, a case-control study allows for better control of → confounding, however the selection bias is increased.
Causality	Relationship between cause and effect.
Causation	The bringing about of one event by another. Law: → causality, → burden of proof
Cell proliferation	Rapid and repeated production of cells by a rapid succession of cell divisions.
Challenging Dose (CD)	A second radiation dose which is applied several hours after an → adapting dose which has increased the cellular resistance.
Chromatin	The composite of DNA and proteins that comprise chromosomes.
Chromosomal aberration	Any abnormality of chromosome number or structure.
Chronic	Normally used to refer to continued exposures occurring over an extended period of time, or a significant fraction of the test species' or of the group of individuals', or of the population's lifetime as a result of long lived agents in the environment. The adjective 'chronic' relates only to the <u>duration</u> of exposure, and does not imply anything about the <u>magnitude</u> of the → doses involved. Also applied to the consequences of this exposure: effects, toxicity and risks. ↔ subchronic, acute
Clastogen	Agent causing chromosome breakage and/or consequent gain, loss or rearrangement of pieces of chromosomes.
Clonal expansion	Proliferation of cells in order to form a → clone.
Clone	Population of genetically identical cells having a common ancestor. → clonal expansion

Clonogenicity	Refers to the ability of the → stem cells including malignant cells to develop colonies.
Cohort	Component of the population born during a particular period and identified by period of birth so that its characteristics (such as causes of death and numbers still living) can be ascertained as it enters successive time and age periods. The term "cohort" has been broadened to describe any designated group of persons followed or traced over a period of time, as in the term cohort study (prospective study).
Cohort study	Comparison of exposed and non-exposed cohorts or of a cohort with a wide range of exposure, using the lowest exposed group for comparison. Observation of the population for a sufficient number of → person-years to generate reliable incidence or mortality rates in the population subsets.
Cokayne-syndrome	Rare genetic disease with increased radiosensitivity.
Collective dose	The total radiation → dose incurred by a population. This is the sum of all of the individual doses to members of the population. Unless otherwise specified, the relevant dose is normally effective dose. Unit: man sievert (man Sv).
	Sometimes it is also falsely expressed as the product of the number of individuals exposed to a source and their average radiation dose. Indeed, in order to calculate the average dose it would normally be necessary to calculate the collective dose (by summing the individual doses and integrating over time if necessary) and divide it by the number of individuals. → Individual dose.
Committed dose	The lifetime → dose expected to result from an intake of radioactivity with a long physical and biological half life.
Compensation	Physiology: adaptation of an organism to changing conditions of the environment (especially chemical) is accompanied by the emergence of stresses in biochemical systems, which exceed the limits of normal (homeostatic) mechanisms.
	Law: Physical or monetary restitution of harm by the person or institution responsible for it.

Computational model	A calculation tool that implements a → mathematical model.
Conceptual model	A set of qualitative assumptions used to describe a system (or part thereof). These assumptions would normally cover, as a minimum, the geometry and dimensionality of the system, initial and boundary conditions, time dependence, and the nature of the relevant physical, chemical and biological processes and phenomena.
Confidence interval/level	A confidence interval is the estimated range, being calculated from a given set of sample data, that is likely to include an unknown population parameter. → $1 - \alpha$. H_0 is only rejected if $1 - \alpha \geq 0.95$ α is also called the confidence level.
Conformational change	A change in the physical position of atoms in a molecule without changing the makeup of the molecule, but the structure of the molecule and thereby the biological activity is altered.
Confounding	A → risk factor, associated with the one under study, that is differently distributed among exposed and non-exposed
Conjugate	Derivative of a substance formed by its combination with compounds such as acetic acid, glucuronic acid, glutathione, glycine, sulphuric acid etc.
Conjugation	A process where two cells come into contact and exchange genetic material.
Consensus conference	A deliberative procedure wherein participants come together to identify areas of consensus and dissent, and to identify reasons for disagreement.
Consent	Acceptance of a conduct by, or, at very minimum, noninterference with the conduct of other individuals, the public or authorities. Consent can be seen as a normative transaction, often considered as a self-assumed obligation. Consent must be given in the absence of coercion and mostly under knowledge condition by a person that is mentally and legally competent.
Consequentialism	Moral view or theory which bases its evaluations of acts solely on consequences. In contemporary philosophical usage, it refers mostly to the view that the rightness of an act depends on whether

the consequences of it are at least as good as (or better than) those of any alternative.

Conservative
Approach/Assumption: Tendency to interpret an analysis with emphasis on protection of human health.

Control, personal
A person's ability to decide about or at least influence what is happening.

Cost–benefit analysis
A systematic evaluation of the positive effects (benefits) and negative effects (costs) of undertaking an action. A decision-aiding technique commonly used in the optimisation of protection and safety.

Countermeasure
An action aimed at alleviating the consequences of an accident.

Critical group
A group of members of the public which is reasonably homogeneous with respect to its exposure to a given source and exposure pathway and is typical of individuals receiving the highest effective dose or equivalent dose (as applicable) by the given exposure pathway from the given source. A group can be homogeneous with respect to dose but not risk, and, more importantly, vice versa. A commonly adopted solution is to define a critical group – often a hypothetical critical group – that is reasonably homogeneous with respect to risk, and is typical of those people who might be subject to the highest risk. → vulnerable group

Crosslink
A constitutional unit connecting two parts of a macromolecule that were earlier separate molecules.

Cumulative exposure
Combined exposure to a group of chemicals that share a common mechanism by all relevant pathways and routes of exposure.

Curie (Ci)
Unit of radioactivity, equal to 3.7×10^{10} Bq (exactly). Superseded by the Becquerel (Bq).

Cytosol
Cell liquid of the cytoplasm.

Cytotoxicity
Toxic to cells.

Danger
Legal term: a condition, situation or an event which can cause damage to human beings, to the environment or other goods with sufficient probability and that must be prevented unconditionally.

DBA/2 mice	Strain of mice with a special mutation.
De minimis	Completely *de minimis non curat lex* "the law does not care about small matters". It can be used to describe a component part of a wider problem, where it is in itself insignificant or immaterial to the problem as a whole, and will have no legal relevance or bearing on the result.
Degree of evidence	Degree of probability or plausibility to which a fact relevant for a right, empowerment or claim is established.
Delphi procedure	A method for the systematic solicitation and collation of informed judgments on a particular topic. A set of carefully designed sequential questionnaires interspersed with summarised information and opinions feedback derived from earlier responses will be sent to the relevant group.

Possible objectives:

1. To determine or develop a range of possible alternatives.
2. To explore or expose underlying assumptions or information leading to differing judgments.
3. To seek out information which may generate a consensus of judgment on the part of the respondent group.
4. To correlate informed judgments on a topic spanning a wide range of disciplines.
5. To educate the respondent group as to the diverse and interrelated aspects of the topic.

Densely ionising radiation	Radiation that has a considerably higher → biological effectiveness (i. e. produces more complex DNA lesions) per deposited energy → α-particles, neutrons.
Deterministic Effect	An → effect for which generally a threshold level of dose exists above which the number of effects and the severity of the effect increase with the dose. The level of the threshold dose is characteristic of the particular health effect but may also depend, to a limited extent, on the exposed individual.

The term → *non-stochastic effect* is used in some older publications, but is now superseded.

Detriment	Estimated measure of the expected harm or loss associated with an adverse event, usually in a manner chosen to facilitate meaningful addition

over different events. It is generally the integrated product of arbitrary values of risk and hazard which is often expressed in terms such as costs in US dollars, loss in expected years of life or loss in productivity. It is needed for numerical exercises such as cost-benefit analysis.

Direct radiation damage

The energy of radiation is deposited directly in the target molecule (DNA) resulting in excitation or ionisation. This starts a chain reaction with chemical changes in the target molecules which may or may not be fatal to the cell (e. g. may be repaired).

Distribution

Allocation or spread of resources, goods, benefits, harms, budgets etc. amongst a population or between different sorts of options.

DNA adduct

A chemical mutagen binds to a specific nucleotide and induces a form of DNA damage. This damage can lead to mutations.

DNA replication

The process of copying the DNA.

DNA transcription

The synthesis of an RNA copy from a sequence of DNA (a gene); the first step in gene expression.

Dose

Radiobiology: A measure of the energy deposited/ absorbed by radiation (degree of → exposure) in a target. Measured as absorbed energy (J) kg^{-1} body mass.

Pharmacology and toxicology: administered (ingested, absorbed, inhaled) amount of a substance (degree of → exposure), measured for instance in mg kg^{-1} body mass.

For definitions of the most important such measures, see dose quantities → Absorbed dose, → Effective dose, → Equivalent Dose, → Organ dose and dose concepts: → Annual dose, → Collective dose, →Committed dose, → Effective dose, → Individual dose, →Lifetime dose, → Residual dose.

Dose-effect curve

Graph drawn to show the relationship between the dose of an → agent and the magnitude of the graded effect that it produces.

Dose fractionation

A dose is given in several portions.

Dose rate

→ Dose per unit time i. e. Gy min^{-1} or mSv yr^{-1}.

Dose rate effectiveness factor (*DREF*)	The ratio between the risk per unit effective dose for high dose rates and that for low dose rates. Superseded by dose and dose rate effectiveness factor (*DDREF*).
Dose-response curve	Graph to show the relation between the dose of an → agent and the degree of response it produces, as measured by the percentage of the exposed population showing a defined, often quantal, effect. If the effect determined is death, such a curve may be used to estimate an → LD_{50} value.
Dysplasia	Cells that have developed abnormally. Their size, shape and organisation may be abnormal.
ED_{50} value	Dose that causes 50 % of effect (frequency), this parameter is often used to characterise → dose-effect curves.
Effect	→ acute, → chronic, → deterministic, → direct, → indirect, → non-stochastic, → stochastic, → subchronic, → *hereditary,* → *somatic.*
Effective dose (*E*)	Radiobiology: Summation of the tissue equivalent doses, each multiplied by the appropriate tissue. weighting factor: $E = \sum_{T} w_T \cdot H_T$ where H_T is the equivalent dose in a tissue T and w_T is the tissue weighting factor for tissue T. Unit: Sievert (Sv) = J kg^{-1}, or rem (older unit) = 0.01 Sv. Effective dose is a measure of dose designed to reflect the amount of radiation → detriment likely to result from the dose. Values of effective dose from any type(s) of radiation and mode(s) of exposure can be compared directly. → dose
Emission standard	This regulatory value is a quantitative limit on the emission or discharge of a potentially toxic substance from a source expressed in concentration or volume. The simplest form for regulatory purposes is a uniform emission standard (UES) where the same limit is placed on all emissions of a particular contaminant. → Limit value
Endocrine Disruptor	An exogenous substance that causes adverse health effects in an intact organism or its offspring, subsequent to changes in endocrine function.

Endpoint	Used, somewhat loosely, to describe a range of different results or consequences. For example, the term 'biological endpoint' is used to describe a health effect (or a probability of that health effect) that could result from exposure.The researcher decides what is the endpoint under study.
Environmental standards	Legal norms, administrative rules or private provisions which transpose broad statutory terms of environmental and health laws (such as → "danger", → "precaution" or → "minimisation") into operational, especially quantified provisions (often the maximum concentration of a potentially harmful substance which can be allowed in an environmental compartment, i.e air). They represent quantitative specifications of qualitative objectives that shall be achieved in the environment. → Emission standard; → environmental quality standard
Epigenetic mechanism	Alterations in the action of genes are called epigenetic mechanisms. Epigenetic transformation refers to those processes, which cause normal cells to become tumour cells without any mutations having occurred.
Epistemic objectivity	An unbiased knowledge, i. e. knowledge where one's attitude would not be changed if one had fuller knowledge of the event or thing in question.
Equivalent dose, H_T:	A measure of the dose to a tissue or organ designed to reflect the amount of harm caused. Taking into account the → absorbed dose and the → radiation weighting factor (w_R) or the → relative biological effectiveness (*RBE*) of the organ ($H_T = D_T \times w_R$). Values of equivalent dose to a specified tissue from any type(s) of radiation can therefore be compared directly.
	Unit: Sievert (Sv) = J kg^{-1}, or rem = 0.01 Sv.
Ethical objectivity	An unbiased evaluation of what is considered to be moral.
Etiology	Cause or origin of disease.
Evaluative information	↔ factual information.
Evidence	Reason for believing that a fact is established.
Excess absolute risk (*EAR*)	The difference between the incidence of a specified → stochastic effect observed in an exposed

group to that in an unexposed control group. → risk→ Excess risk, Absolute risk

Excess relative risk (*ERR*) The ratio of the → excess risk of a specified stochastic effect to the probability of the same effect in the unexposed population, i. e. the relative risk minus one.

Exposure The act or condition of being subject to radiation or other agents. → acute, → chronic, → external, → internal, → medical, → natural, → public, → subchronic

External exposure → Exposure due to a source outside the body.

F344 rats Strain of rats with special mutations.

Factual information Information about possible consequences and the associated probabilities. ↔ evaluative information.

Fanconi-anaemia Rare genetic disease with increased radiosensitivity.

Fetotoxicity Toxic to the fetus.

Fibroblast A connective tissue cell that makes and secretes collagen proteins.

Fibrosarcoma A type of soft tissue → sarcoma that begins in fibrous tissue, which holds bones, muscles, and other organs in place.

Fischer rats Strain of rats with special mutations.

Fluence A measure of the strength of a radiation field. Commonly used without qualification to mean particle fluence (density of particles in a radiation field).

Foci The medical term applied to a small group of cells occurring in an organ and distinguishable, either in appearance or histochemically, from the surrounding tissue.

Frames Perspectives adopted and predisposing the person to focus on certain aspects of the situation in controversies.

Framing The way of conceptualisation, defining or structuring of a problem.

G^+ cell	Initiated cells marked by selective staining of single rat hepatocytes for the presence placental glutathione S-transferase.
G_1-Phase	The period during → interphase (when the cell is not reproducing) in the cell cycle between → mitosis and the → S-phase (when DNA is replicated). Also known as the "decision" period of the cell, because the cell "decides" to divide when it enters the S-phase. The G stands for gap.
G_2-Phase	The period during interphase in the cell cycle that is between the → S-phase (when DNA is replicated) and mitosis (when the nucleus, then cell, divides). At this time, the cell checks the accuracy of DNA replication and prepares for → mitosis. The G stands for gap.
Gamma (γ)-radiation	Short-wavelength electromagnetic radiation of nuclear origin released in connection with radioactive decays.
Gaussian distribution	The "normal distribution", an example of which is the symmetrical bell shape.
Genetic predisposition	An inherited increase in the risk of developing a disease due to the presence of one or more gene alterations. → Genetic susceptibility
Genomic instability	Abnormally high rates (possibly accelerating rates) of genetic change occurring serially and spontaneously in cell-populations, as they descend from the same ancestral cell. By contrast, normal cells maintain genomic stability by operation of elaborate systems which ensure accurate duplication and distribution of DNA to progeny-cells and which prevent duplication of genetically abnormal cells.
Genotoxic	Adjective applied to any substance that is able to cause harmful changes to DNA.
Geographical correlation (or ecological) study	An epidemiological study design where the unit of analysis is a group, often defined geographically.
Glioma	A group of cancers of the brain and spinal cord, which arise from the glial, cells. These cells surround neurones and provide support and electrical insulation between neurones.
Gray (Gy)	Unit of the absorbed dose, equal to 1 J kg^{-1}.
Guide limit	→ recommended limit

H$_0$	Hypothesis testing: Null hypothesis, to be accepted if it cannot be rejected with a degree of certainty $< \alpha$ (which is the probability of rejecting a true null hypothesis).
Han/Wistar rats	Strain of rats with special mutations.
Harmonisation policy	Adoption of common rules, standards or methodologies by the organs of the European Union.
Hazard	General term for anything which has the ability to cause injury or for the potential to cause injury. The hazard associated with a potentially toxic substance is a function of its toxicity and the potential for exposure to the substance.
Hepatoblastoma	A type of liver tumour found in infants and children.
Hepatocellular Carcinoma	A cancerous (malignant) tumour of the liver.
Hepatocytes	Liver cells.
Hereditary effect	Health → effect that occurs in a descendant of the exposed person. The less precise term 'genetic effect' is also used, but hereditary effect is preferred. Hereditary effects are normally → Stochastic effects.
Heuristic (or peripheral) information processing	Superficial, brief consideration of information. ↔ Systematic (or central) information processing.
High-LET	Radiation with high linear energy transfer normally assumed to comprise protons, neutrons and α-particles (or other particles of similar or greater mass). → Linear Energy Transfer
Histiogenic origin	Germ cell layer of the embryo from which a given adult tissue develops.
Hodgkin's disease	A cancer that affects the lymph nodes.
Holistic risk judgement	Integration of evaluation into an overall risk judgement, instead of separate evaluation of aspects.
Homologous chromosome	A pair of chromosomes containing the same linear gene sequences, each derived from one parent.
Hormesis	Supposed stimulatory effect of small doses of an agent that is harmful in larger doses. The risk of harm would be negative below some threshold value of dose or dose rate; i.e. low doses and dose rates would protect individuals against stochastic effects and/or other types of harm.

Hyperplasia	Excessive cell or tissue growth in a particular area.
Hyperradiosensitivity (HRS)	Exaggeration of effects due to ionising radiation. State in which an individual reacts to lower doses of radiation or, at a given dose, with a more severe effect than "normal" individuals.
Hyperthyroidism	Too much thyroid hormone. Symptoms include weight loss, chest pain, cramps, diarrhoea, and nervousness. Also called overactive thyroid.
Hypothetico-deductive method	Proposition of hypotheses whose logical consequences are deduced, followed by experimental testing of those consequences against experience.
Ignorance	Lack of knowledge of a possible severe, but not ascertainable future harm.
Immune response	The immune response is the general reaction of the body to substances that are foreign or treated as foreign. It may take various forms e. g. antibody production, cell-mediated immunity, immunological tolerance, or hypersensitivity (allergy).
In utero	In the womb.
In vitro	Isolated from the living organism and artificially maintained, as in a test tube. ↔ *in vivo*.
In vivo	Occurring within the living organism. ↔ *in vitro*
Incidence rate	The rate at which new cases of disease arise in a population. The incidence rate is the number of new cases of disease during a period of time when the change from non-disease to disease is being measured, divided by the person-time-at-risk throughout the observation period. The denominator for incidence rate (person-time) changes as persons originally at risk develop the disease during the observation period and are removed from the denominator.
Indirect radiation damage	Involves the production of a free → radical in water, which is the most common effect, since the cell consists mostly of water.
Individual dose	The dose incurred by an individual. For contrast with → collective dose. → dose
Individual risk	The probability that an individual person will experience an adverse effect risk. Estimation of

	probabilities about consequences to representative individuals.
Individualistic model of fundamental rights	Focus on the individual rather than the community in protecting fundamental rights.
Information bias	Bias due to non-identified individuals that have been lost to follow-up, i. e. via death or migration. These individuals will continue to contribute person years at risk without being at risk of developing a disease.
Initiating event/Initiation	Primary lesion of the DNA. Accumulation of genetic damages leading to partial abrogation of growth control.
Initiator	Any agent, which starts the process of tumour formation, usually by action on the genetic material, is called an initiator.
Inter alia	Latin term which means "amongst other things".
Intermediate cells	Cells at all steps after the initiation that have the ability to further proliferate.
Internal exposure	→ Exposure due to a source within the body.
Interphase	The period in the cell cycle when the cell is not reproducing (undergoing → mitosis). During this time, the cell is growing rapidly, increasing its size, making new molecules, and all other activities that are not associated with cell division. The period is subdivided into the → G_1 phase, the → S-phase, and the → G_2-phase.
Intuitive risk perception	Does not rely on technical or quantitative aspects of risk, but on personal memory or experience and is based on subjective criteria.
Isoform	A protein having the same function and similar or identical sequence.
Isozyme	An enzyme performing the same function as another enzyme but having a different set of amino acids. The two enzymes may function at different speeds.
Justification	The process of determining whether a practice is, overall, beneficial, i. e. whether the benefits to individuals and to society from introducing or continuing the practice outweigh the harm resulting from the practice.

Kinesin	One of the types of motor proteins that use the energy of ATP hydrolysis to move along microtubules.
knockout	Informal term for the generation of a mutant organism in which the function of a particular gene has been completely eliminated.
Kolmogorov forward differential equation	Mathematical equation also called Fokker-Planck equation.
Kuhnian social construction	The proposition by the philosopher of science, Thomas Kuhn, that science and the conclusions of scientific research are not objective, as scientists often assume, but subjective social constructions.
Kupffer cells	Specialised macrophage in the liver.
Latency	Time interval between the occurrence and the effects of a damage i. e. between initiation and the development of cancer.
LD_{50} value	Statistically or experimentally derived single dose of an agent that can be expected to cause death in 50 % of a given population of organisms under a defined set of experimental conditions → ED_{50}, → dose-response curve.
Life span study (LSS)	The follow-up of around 120,000 survivors of the atomic bombings of Hiroshima and Nagasaki.
Lifetime dose	The total dose received by an individual during his/her lifetime. In practice, often approximated as the sum of the → annual doses incurred. Because annual doses include → committed doses, some parts of some of the annual doses may not actually be delivered within the lifetime of the individual, and therefore this may overestimate the true lifetime dose. For prospective assessments of lifetime dose, a lifetime is normally interpreted as 70 years.
Lifetime risk	The probability that a specified health effect will occur at some time in the future in an individual as a result of radiation exposure.
Li-Fraumeni Syndrome	An inherited (genetic) disorder that is associated with a predisposition for developing certain types of cancer. A high proportion of families with Li-Fraumeni syndrome carry a germline $p53$ mutation in one allele. This condition occurs in chil-

dren and young adults. They develop clusters of particular cancers, including breast cancer, acute leukemia, and sarcomas of the arms and legs.

Ligand

A molecule or ion that can bind another molecule.

Limit value

→ Environmental standard

Linear dose-response

Relationship between dose and biological response that is a straight line. In other words, the rate of change (slope) in the response is the same at any dose. A linear dose-response is written mathematically as follows: if Y represents the expected, or average response and D represents dose, then $Y = aD$ where a is the slope, also called the linear coefficient.

Linear energy transfer (LET)

A measure of how, as a function of distance, energy is transferred from radiation to the exposed matter. Defined generally as:

$L_\Delta = \left(\dfrac{dE}{d\ell}\right)_\Delta$, where dE is the energy lost in tra-traversing distance $d\ell$ and Δ is an upper bound on the energy transferred in any single collision.

A high value of LET indicates that energy is deposited within a small distance. → low-LET, → high-LET.

Linear-quadratic dose-response

Curved relationship between dose and biological response. This implies that the rate of change in response is different at different doses. The response may change slowly at low doses, for example, but rapidly at high doses. A linear-quadratic dose-response is written mathematically as follows: if Y represents the expected, or average response and D represents dose, then $Y = aD + bD^2$ where a is the linear coefficient (or slope) and b is the quadratic coefficient (or curvature). → Linear dose-response

Linear regression analysis

Way of measuring the relationship between two or more data sets. Linear Regression attempts to explain a relationship using a straight-line fit to the data and then extending that line to predict future values.

Linear–no-threshold (LNT) hypothesis/model

The hypothesis that the risk of → stochastic effects is directly proportional to the dose especially in the low-dose range of → dose and → dose rate (below those at which → deterministic effects occur), hence that they show a → linear dose-response and that any non-zero dose implies a non-zero risk of stochastic effects. It is not proven – indeed it is probably not provable – for low doses and dose rates, but it is generally considered the most radiobiologically defensible assumption on which to base safety standards.

Locus	The position of a gene on a chromosome.
Low-LET	Radiation with low linear energy transfer, normally assumed to comprise photons (including X-rays and γ-radiation), electrons, positrons and muons. → Linear Energy Transfer
Lung fibrosis	Relative increase in formation of interstitial fibrous tissue in the lung.
Lymphocytes	Type of white blood cell, which makes substances to fight germs, foreign tissues or cancer cells.
Lymphoma	The type of cancer that begins in lymph nodes.
Malignancy	Cancerous growth, a mass of cells showing uncontrolled growth, a tendency to invade and damage surrounding tissues and an ability to extend daughter growths to sites remote from the primary growth.
Malignant conversion	Acquisition of further cellular changes leading to malignant cells → Transformation
Malignoma	→ Malignant tumour.
Markovian process	Mathematical tool.
Mathematical model	A set of mathematical equations designed to represent a → conceptual model.
Maximum Tolerated Dose (MTD)	Highest amount of a substance that, when introduced into the body, does not kill test animals (denoted by LD_0).
Mechanism	Chain of effects or parts of the chain.
Mediation	In mediation, an experienced neutral meets with the parties to help them resolve their differences. The mediator encourages such resolution by helping the parties open communications, helping them to focus on their real interests and attempting to find a pathway towards a resolution satisfactory to both sides. It is a non-binding process;

neither party must accept any recommendation the mediator might make for settlement.

Medical exposure

→ Exposure incurred by patients as part of their own medical or dental diagnosis (diagnostic exposure) or treatment (therapeutic exposure); by persons, other than those occupationally exposed, knowingly exposed while voluntarily helping in the support and comfort of patients; and by volunteers in a programme of biomedical research involving their exposure.

Melanoma

The most dangerous type of skin cancer. Appearing first as a pigmented mole, a melanoma can spread locally and to distant organs rapidly.

Meningioma

Usually a non-cancerous (benign) slow growing brain tumour. Meningiomas occur in the membranes that surround the brain or the spinal cord.

Mesothelioma

A rare cancer in which malignant cells are found in the sac lining of the chest (pleura) or abdomen (peritoneum). Most cases of malignant mesothelioma are attributable to asbestos exposure.

Metabolic activation

Transformation of relatively inert chemicals to biologically reactive metabolites.

Metabolism

Sum of the physical and chemical changes that take place in living organisms. These changes include both synthesis (anabolism) and breakdown (catabolism) of body constituents. In a narrower sense, the physical and chemical changes that take place in a given chemical substance within an organism. It includes the uptake and distribution within the body of chemical compounds, the changes (biotransformations) undergone by such substances, and the elimination of the compounds and their metabolites.

Metaplasia

Abnormal transformation of an adult fully differentiated tissue of one kind into a differentiated tissue of another kind.

Methylation

Biochemical reaction resulting in the addition of a methyl group ($-CH_3$) to another molecule.

Microsome

Artefactual spherical particle, not present in the living cell, derived from pieces of the endoplasmic reticulum present in homogenates of tissues or cells.

Minimisation	→ optimisation/ALARA
Mitosis	The natural process by which cells of the body grow and divide including equal distribution of chromosomes into the daughter cells.
Model	An analytical representation or quantification of a real system and the ways in which phenomena occur within that system, used to predict or assess the behaviour of the real system under specified (often hypothetical) conditions. → computational model, → conceptual model, → mathematical model
Monte-Carlo Calculation	Computer models based on "chance".
Morbidity	Any departure, subjective or objective, from a state of physiological or psychological well being: in this sense, "sickness", "illness", and "morbid condition" are similarly defined and synonymous.
Multiplicative risk	The joint effect of two or more agents is the product of their effects. For example, if one factor multiplies risk by a and a second factor by b, the combined risk of the two factors is ab.
Mutagen	Agent causing → mutations.
Mutation	Any heritable change in genetic material. This may be a chemical transformation of an individual gene (a gene or point → mutation), which alters its function. On the other hand, this change may involve a rearrangement, or a gain or loss of part of a chromosome, which may be microscopically visible. This is designated a chromosomal → mutation. Most mutations are harmful.
Myeloid leukemia	A type of leukemia that affects the neutrophils (a type of white blood cell).
Naive theories	Theories based on the understanding and experience of the world around lay people. Theory based on several elements of information where wrong information can lead to a biased opinion.
Naked DNA	DNA without a telomere (a special tip or end of a chromosome).
Natural background	Doses, dose rates, activity concentrations of frequency of an agent in nature, as distinct from presence resulting from inputs from human

activities. The contamination of the natural environment by some man-made compounds may be so widespread that it is practically impossible to get access to biota with a truly natural level; only "normal" levels can be measured, those which are usually prevalent in places where there is no obvious local contamination. → background

Natural exposure	→ Exposure due to natural sources.
Necrosis	Mass death of areas of tissue surrounded by otherwise healthy tissue.
Neoplasia	New and abnormal formation of tissue as a tumour or growth by cell proliferation that is faster than normal and continues after the initial stimulus that initiated the proliferation has ceased.
Neoplasm	Any new formation of tissue associated with disease such as a → tumour.
Neurofibromatosis	Inherited disorder characterised by changes in the nerves, muscles, bones and skin. Also associated with an increased risk for certain types of cancer.
No observed (adverse) effect level (NO(A)EL)	This is the greatest concentration or quantity of an agent found by experiment or observation, that causes no detectable → (adverse) alteration of morphology, functional capacity, growth, development, or life span of the target organism.
	The maximum dose or ambient concentration which an organism can tolerate over a specified period of time without showing any detectable → (adverse) effect and above which (adverse) effects are apparent.
No observed effect concentration (NOEC)	→ NOAEL
Non-genotoxic carcinogen	Carcinogen that acts on an epigenetic level (e. g. promotors).
Non-Hodgkin's lymphoma	A group of related cancers that arise from the lymphatic system. Each type is characterised by the appearance, structure and genetic features of the lymphocyte seen in a lymph node biopsy.
Non-stochastic Effect	→ Deterministic effect

Nucleosome	The basic structural unit of → chromatin in which DNA is wrapped around a core of histone molecules.
Nucleotide	A subunit of DNA or RNA consisting of a nitrogenous base (purine in adenine and guanine, pyrimidine in thymine, or cytosine for DNA and uracil cytosine for RNA), a phosphate molecule, and a sugar molecule (deoxyribose in DNA and ribose in RNA). Thousands of nucleotides are linked to form a DNA or RNA molecule.
Occupational exposure	All → exposure of workers incurred in the course of their work, with the exception of excluded exposures and exposures from exempt practices or exempt sources.
Occupational Exposure Limits (OEL)	Regulatory level of exposure to substances, intensities of radiation etc. or other conditions, specified appropriately in relevant government legislation or related codes of practice.
Odds ratio	The ratio of two odds. It is used frequently in case control studies where it is the ratio of the odds in favour of getting the disease, if exposed, to the odds in favour of getting the disease if not exposed.
Oestrogen	A female hormone mostly produced by the ovaries. It influences such female sexual characteristics as breast development, and it is necessary for reproduction.
Oncogene	Gene that can cause → neoplastic transformation of a cell; oncogenes are slightly changed equivalents of normal genes known as → proto-oncogenes. Many oncogenes are involved, directly or indirectly, in controlling the rate of cell growth.
Optimisation	→ minimisation/ALARA
Organ dose	The mean →absorbed dose D_T in a specified tissue or organ T of the human body, given by: $D_T = \frac{1}{m_T} \int_{m_T} D\,dm$ where m_T is the mass of the tissue or organ and D is the absorbed dose in the mass element dm. → dose

Papillomas — Non-cancerous tumours of the voice box (larynx).

Pathogenesis — The development of morbid conditions or of disease.

Perceived risk — Subjective interpretation of risk based on personal experience or information.

Percentile — One of 99 actual or notional values of a variable dividing its distribution into 100 groups with equal frequencies; the 90th percentile is the value of a variable such that 90 % of the relevant population is below that value.

Perfect threshold — → real threshold

Persistence — Property of a substance to stay in the environment without being degraded for a long period of time.

Person-time — Person-time is an estimate of the actual time-at-risk in years, months, or days that all persons contributed to a study.

Pharmacodynamics — Way in which drugs exert their effects on living organisms. A pharmacodynamic study aims to define the fundamental physicochemical processes, which lead to the biological effect observed → toxicodynamics.

Pharmacokinetics — Process of the uptake of drugs by the body, the transformation they undergo, the distribution of the drugs and their metabolites in the tissues, and the elimination of the drugs and their metabolites from the body. Both total amounts and tissue and organ concentrations are considered. → toxicokinetics

Phase I reactions — This group of reactions comprises every possible stage in the enzymatic modification of an agent by oxidation, reduction, hydrolysis, hydroxylation, dehydrochlorination and related reactions catalysed by enzymes of the → cytosol, of the endoplasmic reticulum (→ microsomal enzymes) or of other cell organelles.

Phase II reaction — Binding of a substance or its metabolites from a phase 1 reaction, with endogenous molecules (conjugation, i. e. combination with glucuronic acid, glutathione, sulphate, acetate, glycine etc.). This creates more water-soluble derivatives that may be excreted in the urine or bile.

Physiologically-Based Pharma- — A type of pharmacokinetic model that integrates

cokinetic (PBPK) Model	physiological parameters (e. g. blood flow, tissue perfusion rates, ventilation), to allow better extrapolations to other species.
Plausible	Apparently reasonable, valid, truthful, trustworthy or believable.
Plausible reasons for concern	More than a merely hypothetical risk consideration. Sufficient evidence based on available scientific data is necessary. Termed also: reasonable grounds for concern, scientifically based suspicion, credible scientific warning, scientifically serious assumptions.
Point mutation	A change in a single base pair of DNA.
Poisson distribution	A probability distribution that characterises discrete events occurring independently of one another in time.
Polycthemia vera	A rare condition characterised by overproduction of red blood cells in the bone marrow. It is considered to be a precursor of acute myelogenous leukemia.
Polymorphism	Difference in DNA sequence among individuals. Applied to many situations ranging from genetic traits or disorders in a population to the variation in the sequence of DNA or proteins.
Popperian falsification	Variant of the → hypothetico-deductive method suggested by Karl Popper. Scientific hypotheses have to posses the ability to be falsified.
Practical threshold	A threshold type dose-response that might be at least → plausible.
Precaution	Measures taken to implement the → precautionary principle.
Precautionary principle	Principle of environmental law which mandates to take regulatory measures in a situation of uncertainty about future harm, i. e. when there is some slight reason to believe that future harm may occur (see Sections 2.2.3, 9.2.4 and 10.3).
Preneoplastic	Before the formation of a tumour.
Primary cancer	The first tumour in order or in time of development.
Primary electrons	Electrons emitted by the nucleus of an atom. → β-radiation ↔ secondary electrons

Principle of Equality	Constitutional right to the extent that all persons in a like situation shall be treated equally.
Principle of proportionality	Constitutional principle whereby official action that intervenes into a fundamental right must be suited to reach its objectives, a less burdensome alternative does not exist and there must be an adequate relation between the weight of the objective pursued and the regulatory burden imposed on the holder of the fundamental right to safety.
Probabilistic risk assessment (PRA)	Provides an average estimate of how many undesirable events may be expected. Technique to predict the probability of undesirable events.
Procarcinogen	Substance that has to be metabolised before it can induce malignant tumours.
Progeny	Cellular biology: result of cell division. → radon progeny,
Promoter	Agent that increases tumour production by another substance when applied to susceptible organisms after their exposure to the first substance.
Promotion	→ Clonal growth of the number of → initiated or → intermediate cells.
Protraction effect	The effect due to a spreading out of a radiation dose over time by continuous delivery at a lower dose rate.
Proto-oncogene	Normal gene → Oncogene
Proxy variable	A variable used when the variable under study is not available, difficult to study or gives insufficient information.
Public exposure	→ Exposure incurred by members of the public from radiation sources, excluding any occupational or medical exposure and the normal local natural background radiation but including exposure from authorised sources and practices and from intervention situations.
Quiescent cells	Still or inactive cells.
Rad	Former unit of absorbed dose, equal to 0.01 Gy. Superseded by → Gray. Abbreviation of *radiation absorbed dose*.

Radiation	The emission and propagation of energy through space or through a material medium in the form of waves (e. g. the emission and propagation of electromagnetic waves or of sound and elastic waves), usually referring to electromagnetic radiation. Such radiation is commonly classified according to frequency, as microwaves, infrared, visible (light), ultraviolet, → X-rays and → γ-rays and, by extension, corpuscular emission, such as → α– and → β–radiation, neutrons, or rays of mixed or unknown type, such as cosmic radiation.
	In the context of this book mostly applied to ionising radiation i. e. radiation capable of producing ion pairs in biological material(s).
Radical	A highly reactive molecule or atom with an unpaired electron.
Radical scavengers	Molecules that bind to → radicals, thus eliminating them from the metabolism.
Radiotherapy	X-ray treatment that damages or kills cancer cells.
Radon progeny	Short-lived radioactive decay products. → Progeny
Real threshold	Threshold which is well established and supported by studies of underlying mechanism(s).
Recessive mutations	A mutation that is only expressed/visible when two copies of that particular gene are present – one from the mother and one from the father.
Recommended limit	This regulatory value is the maximum concentration of a potentially toxic substance that is believed to be safe and therefore is recommended. Such limits often have no legal backing.
Relative biological effectiveness (*RBE*)	Analysis of the *RBE*, is a useful way to compare and contrast the results observed in animal studies. The *RBE* for a given test radiation, is calculated as the dose of a reference radiation, usually X-rays, required to produce the same biological effect as was seen with a test dose, of another radiation:

$$RBE = \frac{Dose\ from\ reference\ radiation}{Dose\ from\ test\ radiation,\ DT}.$$

I. e.: ratio of the dose of 200 mGy X-rays required to produce a given effect to the dose required of any radiation (i. e. 20 mGy of neutrons) to produce the same effect:

$RBE = \frac{200}{20} = 10$.

→ Quality Factors (ICRP 1977), → Radiation Weighting Factors (ICRP 1991).

Relative risk

The ratio between the incidence of a specified stochastic effect observed in an exposed group and that in an unexposed control group.

Residual dose

In a chronic exposure situation, the dose expected to be incurred in the future after countermeasures have been taken → dose.

Residual risk

Remaining risk after risk reduction measures and positive acceptance decision. According to the legislation , it has to be borne by the citizens as a general burden of human civilisation in a variety of situations.

Retinoblastoma

Hereditary → malignant → tumour of the retina that develops during childhood, is derived from retinal germ cells, and is associated with a chromosomal abnormality.

Reversible

Invertible effect restricted in time, i. e. reactions of transmitter substances and hormones initiated through weak binding forces.

Rhabdomyosarcoma

A malignant tumour arising from the skeletal muscle cells and occuring mostly in the head and neck areas.

Risk

Depending on the context, the term risk may be used to represent a quantitative or a qualitative concept. This term must not be confused with the term → "hazard". It is most correctly applied to the predicted or actual frequency of occurrence of harmful or injurious consequences associated with actual or potential exposures to a chemical or other hazard. It relates to quantities such as the probability of harm and the magnitude and character of such consequences. The health effect(s) in question must be stated – e. g. risk of fatal cancer, risk of serious hereditary effects, or overall radiation detriment – as there is no generally accepted 'default'.

Risk Analysis

The whole process of → risk assessment through → risk evaluation to → risk management (including a preliminary procedure to limit the problem

	and setting the frame) and to determine measures of control.
Risk Assessment	A set of mathematical and scientific techniques for identifying, estimating, the probability of a threat. → "risk evaluation".
Risk Evaluation	Risk evaluation involves the establishment of a qualitative or quantitative relationship between → risks and benefits, involving the complex process of determining the significance of the identified → hazards and estimated risks to those organisms or people concerned with or affected by them. This process is sometimes not distinguished from → risk assessment for its part of scientific evaluation and from → risk management for its part of societal and political evaluation.
Risk factor	Anything that increases a person's chance of developing a disease. Some examples of risk factors for cancer include a family history of cancer, use of tobacco products, and certain foods, being exposed to radiation or other cancer-causing agents, and certain genetic changes.
Risk Management	Risk management is the decision-making process involving considerations of political, social, economic and engineering factors so as to identify and compare technological and/or regulatory options and to select the optimal response for safety from that → hazard.
Risk Perception	Process of subjective reception, transformation and evaluation of risk related information that stem from proper experience, direct observation, transmitted messages and direct communication between each other.
Safety	Safety is the practical certainty that injury will not result from exposure to a → hazard under defined conditions: in other words, the high probability that injury will not result.
Safety factor	Factor that is introduced to reduce a primarily derived → limit value. It is meant to acknowledge different types of uncertainty (i. e. extrapolation from animal to humans), variation of sensitivity between animals and humans and within a human

	population, analytical inaccuracies, or mixed exposures. → uncertainty factor.
Sarcoma	A malignant tumour of muscles or connective tissue such as bone and cartilage.
Schistosoma	A parasite.
Secondary cancer	Cancer that has spread from the organ in which it first appeared to another organ. → Primary tumour.
Secondary electrons	Electrons expulsed from the electron sheath of an atom by the action of primary radiation ↔ primary electrons, ↔ β–radiation. Selection bias Bias that occurs if the underlying cause of exposure influences the studied effect i. e. in populations medically exposed to ionising radiation. Sensitisation
	Exposure to an → agent (allergen) which provokes a response in the immune system such that disease symptoms will ensue on subsequent encounters with the same substance.
Separation model	Model of separation of the phases of → risk assessment and → risk management.
Sievert (Sv)	Unit of the → effective dose.
Significance	Hypothesis testing: A statistical analysis which reveals an effect unlikely to occur by chance alone. → significance level. Biology: Any effect observed in a study or experiment carries with it some degree of uncertainty, or imprecision, because of randomness and variability in most biological phenomena. Statistical techniques evaluate an observed effect in view of its precision to determine with what probability it might have arisen by chance (the level of significance). A biologically significant effect has a noteworthy impact on health or survival. Values with a low probability of occurring by chance are called "statistically significant" and are thought to represent a real effect.
Significance level	The level of significance → (α) refers to the degree to which the result could be explained by chance. At the 0.01 (1 %) level, the result could have occurred by chance only 1 time in 100.
Simple risk	→ Three-tier-concept: A risk below the → threshold of danger in a situation of → uncertainty that

	is undesirable and must be further reduced; → precautionary principle.
Sister chromatid exchange	Process producing reciprocal exchange of DNA between the two DNA molecules of a replicating chromosome.
Slope factor	Value, in inverse concentration or dose units, derived from the slope of a dose-response curve; in practice, limited to carcinogenic effects with the curve assumed to be linear at low concentrations or doses. The product of the slope factor and the exposure is taken to reflect the probability of producing the related effect.
Societal risk	Term sometimes used to express the combination of → risks to a number of people made up from long term environmental and short term i. e. accident risks. Estimation of the probability distribution for the magnitudes of adverse health and environmental effects for a defined area.
Solid tumour	Term used to distinguish between a localised mass of tissue and → leukemia. Types of solid tumours are → carcinomas, → sarcomas and → lymphomas.
Somatic effect	Health → effect that occurs in the exposed person. This includes effects occurring after birth that are attributable to exposure in utero. → Deterministic effects are normally somatic effects; stochastic effects may be somatic effects or → hereditary effects.
Sparsely ionising radiation	X- and γ-radiation induce mostly less damaging effect than → densely ionising radiation.
Speciation	The ability of a chemical to exist in several forms (e. g. charge, valence state, and complexation).
S-phase	The period during → interphase in the cell cycle when DNA is replicated. The S stands for synthesis.
Spontaneous damaging event	Damaging event occurring without the performance of an agent.
Squamous Cell	Flat thin cells that cover many of the outer and inner surfaces of the body, including the outer layer of skin and the lining of the throat.
Standard of proof	Science: Degree of scientific proof of a hypothesis H that habitually requires a → significance

level of → $\alpha \leq 0{,}05$. → Determines how much uncertainty is tolerated.

Law: The level to which something must be proven in court. In criminal law, the prosecution case must be proven beyond reasonable doubt. In civil law, the plaintiff's case must be proven on the balance of probabilities.

Standardise mortality/morbidity ratio (*SMR*) and Standardised incidence ratio (*SIR*)incidences)	Ratio of the number of (effects: death, morbidity, observed in the study group or population to the number of effects expected if the study population had the same specific rates as the standard population.
Stem cell	An unspecialised cell that gives rise to differentiated cells.
Stochastic effect	An → effect, the probability of occurrence of which is greater for a higher radiation dose and the severity of which (if it occurs) is independent of dose. Stochastic effects may be → somatic effects or → hereditary effects, and generally occur without a threshold level of dose.
Subchronic (sometimes called subacute) toxicity test	Animal experiment serving to study the effects produced by the test material when administered in repeated doses (or continually in food, drinking water, air) over a period of up to about 90 days.
Subchronic exposure	→ Exposure that is too protracted to be described as acute exposure, but does not persist for many years (in tests about 10 % of the lifetime of the test organism) → Subchronic toxicity test.
Sublinear hypothesis	The risk of → stochastic effects at low doses and/or dose rates is less than that implied by the → LNT hypothesis ↔ superlinear hypothesis.
Sufficient cause	A cause that nothing further is needed to bring an event to happen.
Superlinear (or supralinear) hypothesis	The risk of → stochastic effects at low doses and/ or dose rates is greater than that implied by the → LNT hypothesis ↔ sublinear hypothesis.
Synergism	Pharmacological or toxicological interaction in which the combined biological effect of two or more substances is greater than expected on the basis of the simple summation of the toxicity of each of the individual substances.

Systematic (or central) information processing	Thoughtful and careful consideration of information ↔ Heuristic (or peripheral) information processing.
Systemic effect	Consequence that is of either a generalised nature or that occurs at a site distant from the point of entry of a substance: a systemic effect requires absorption and distribution of the substance in the body.
Technische Richtlinien Konzentration (TRK)	Guideline values of concentrations dependent on the technical feasibility of reduction → minimisation concept.
Teratogenesis	Generation of non-heritable birth defects.
Three-tier concept of risk	In German Environmental Law: distinction between → danger that must be avoided, simple risk that must be reduced and → residual risk that must be tolerated.
Threshold	The highest dose of an agent that can be continuously administered to test organisms without producing biologically significant harmful effects. The risk of harm would be zero below some threshold value of dose or dose rate. Any larger dose would exceed the threshold and be expected to produce harmful biological effects in a certain proportion of the overall population. Because the individual susceptibility of test animals or humans varies across a range of threshold doses, the threshold is not an absolute quantity that can be measured directly. Instead, it must be estimated indirectly from experimental dose-response studies that determine the → NOAEL in several dose groups of test animals.
	In the frame of this report three types of threshold are distinguished, regarding the certainty of the above relationship:
	→ Apparent, → Real and → Practical.
Time-weighted average exposure/concentration	Concentration in the exposure medium at each measured time interval multiplied by that time interval and divided by the total time of observation: for occupational exposure a working shift of eight hours is commonly used as the averaging time.

Tolerance	Ability to experience exposure to potentially harmful amounts of an agent without showing an → adverse effect.
	Tolerance with respect to risk is basically the degree of uncertainty to suffer from an activity.
Toxicodynamics	→ Pharmacodynamics with respect to all → agents.
Toxicokinetics	→ Pharmacokinetics with respect to all → agents.
Transduction	Gene transfer from one cell to another.
Transformation	→ Malignant conversion.
Translocation	Transfer of a segment of a chromosome to a non-→ homologous chromosome.
Tumour (neoplasm)	Any growth of tissue forming an abnormal mass. Cells of a benign tumour will not spread and will not cause cancer. Cells of a malignant tumour can spread through the body and cause cancer. → Neoplasm
Tumour growth	Growth of → malignant cells to a detectable → tumour.
Tumourigenesis	Formation of → tumours.
Two-tier concept	Distinction between unacceptable or undesirable risk (→ danger, simple risk) that must be avoided or reduced and → residual risk that must be tolerated.
Type 1 Error	Probability of wrongly rejecting the → null hypothesis H_0, that is the probability of false positive.
Type 2 Error	Probability of wrongly *not* rejecting the → null hypothesis H_0, that is the probability of false negative.
Ubiquity	Omnipresence of a substance in the environment.
Uncertainty factor	Used in either of two ways depending upon the context (i) Mathematical expression of uncertainty applied to data that are used to protect populations from hazards which cannot be assessed with high precision. (ii) → safety factor.
Unit risk factor	The upper-bound → excess → lifetime cancer risk, due to continuous exposure to a chemical at

a concentration of 1 μg m^{-3} in air, or to 1 μg l^{-3} in water. This factor is generally extrapolated from a dose-effect curve. For example, if the *Unit Risk Factor* = 2 × 10^{-6} μg^{-1} m^{-3}, then a person exposed daily for a lifetime to 1 μg of the chemical in 1 m^3 of air would have an increased risk of cancer equal to 2 in a million.

Utilitarianism	Prominent, controversial theory about the fundamental basis of morality. It holds that:

(i) the right action is the one that brings most *utility,*

(ii) everybody counts *equally,*

(iii) utility is the only value, hence *distribution* does not matter.

Two main versions of utilitarianism are:

(a) *Sum* utilitarianism: it isthe total sum of utility that matters.

(b) *Average* utilitarianism: it is the average utility per person that matters.

One also distinguishes:

(1) *Act* utilitarianism: it si the consequence of each individual act that matters

(2) *Rule* utilitarianism: hat mattrs is not the individual act in isolation, but the effect that an individual act will have on people's tendency to usually, or as a *rule,* act in this way.

Values	A basic notion in ethics, which is interdefinable with other basic terms in ethics: an event, a thing or a person said to have a (positive) value if it is something *good,* something *worth striving for.* A *right* action sometimes defined as an action that brings about more (positive) value than any alternative action.
Victims, statistical	Victims of premature deaths calculated to occur because of a particular exposure to a hazard, although the people killed are not identifiable, because their deaths are often not immediate and they are usually not under medical examination. In theory, everybody could be the victim.

Vulnerable group	A group of people with common characteristics, which is most in need of protection because it is most susceptible to a given agent, i. e. infants, children, elderly people, pregnant women, groups with special genetic defects, ... → critical group.
Wild type genotype	In this context, applied to laboratory animals, that have not been genetically modified and should be similar to those animals found in the wild. ↔ knockout
Working level (WL)	Unit of dose used in uranium mining. A unit of measure for documenting exposure to radon or thoron progeny in air. $1\ WL = 2.1 \times 10^5\ J\ m^{-3}$.
Working Level Month (WLM)	Measure used to determine cumulative exposure to radon or thoron progeny. 1 WLM = to an average of 1 WL for 170 h.
X-rays	Penetrating electromagnetic radiation whose wavelengths are very much shorter than those of visible light. They are produced when an inner orbital electron falls from a high energy level to a low energy level.
Xenobiotic	Strictly, any substance interacting with an organism that is not a natural component of that organism.
Xeroderma pigmentosum	A group of rare inherited skin disorders characterised by a heightened reaction to sunlight (photosensitivity) with skin blistering occurring after exposure to the sun.

15 Abbreviations

α	In statistics: degree of certainty, in modelling: symmetrical cell division rate
	In dose-response curves a constant of the term for linearity of dose
β	In modelling: death or differentiation rate
	In dose-response curves a constant of the quadratic dose term
λ	Mutation rate
κ	Number of cellular changes
$\delta(a)$	Dirac δ-Distribution
$\gamma(d)$	Initiating mutation rate
2-AAF	2-Acetylaminofluorene
4-ABP	4-Aminobiphenyl
a	In some formulas: Age
A	Alternative
ACGIH	American Conference of Industrial Hygienists
AD	Adapting Dose
ADI	Acceptable Daily Intake
AFB1	Aflatoxin B_1
AFB_1-FAPY	AFB_1 formamidopyrimidine
AFB_1-N^7-Gua	8,9-Dihydro-8-(N^7guanyl)-9-hydroxyaflatoxin B_1
ALARA	As low as reasonably achievable
ALARP	As low as reasonably practicable
AML	Acute myeloid leukemia
a_r, a_s	Coefficients of linear dose term
ARNT	Aryl hydrocarbon nuclear translocator
AT	Ataxia telangiectasia
ATP	Adenosine triphosphate

AUC	Area under the curve of blood (tissue) concentrations vs. Time
BEIR	(Committee on) Biological Effects of Ionising Radiation
BMJFG	Former "Bundesministerium für Jugend, Familie und Gesundheit" of the Fed. Rep. of Germany
BNFL	British Nuclear Fuels plc.
Bq	Becquerel
C	Consequence
CCME	Canadian Council of Ministers of the Environment
CD	Challenging Dose
CEA	Commissariat à l'Energie Atomique
CHO	Chinese Hamster Ovary
CI	Confidence Interval
Ci	Curie
CNS	Central nervous system
CoA	Coenzyme A
CTPV	Coal tar Pitch
CYP	Cytochrome P450
D	Energy Dose, absorbed energy of ionising radiation per mass
d	Average exposure rate
$DDREF$	Dose and dose rate effectiveness factor
DDT	Dichloradiphenyltrichloroethane
DEN	Diethylnitrosamine ($-N$-Nitroso-diethylamine)
DFG	Deutsche Forschungsgemeinschaft
dG-C8-A7	N-(Deoxyguanosine-8yl)-2-aminofluorene
DNA	Desoxyribonucleic acid
DPX	DNA-protein crosslinks
$DREF$	Dose rate effectiveness factor
DSB	Double strand break in DNA
Dt	Threshold dose

D_{TR}	Absorbed dose delivered by radiation type R averaged over a tissue or organ T
E	Effective dose
EAR	Excess absolute risk
ED_{50}	Dose that causes an effect in 50 % of the cases
EMF	Electromagnetic fields
EPA	Environmental Protection Agency (USA)
ERR	Excess relative risk
FAP	Familial adenomatous polyposis
GDF	Growth and differentiation factor
GST	Glutathione S-transferase
Gy	Gray, unit of the energy dose (1 Joule absorbed per kg tissue)
H	Modelling: Hazard, Scientific hypothesis testing: hypothesis, Radiation protection: equivalent Dose (energy dose $D \times$ radiation weighting factor w_R)
H_0	Hypothesis testing: Null hypothesis, Modelling: Baseline Hazard
HCB	Hexachlorbenzene
HCH	Hexachlorhexane
HRS	Hyperradiosensitivity
H_T	Equivalent dose in a tissue
IAEA	International Atomic Energy Agency
IARC	International Agency for Research on Cancer
ICRP	International Commission on Radiological Protection
IPPC	Integrated Pollution Prevention and Control
IQ	2-Amino-3-methylimidazo[4,5-f]quinoline
IRR	Increased radio resistance
JBRC	Japanese Bioassay Research Center

LAI	Länderausschuss für Immissionsschutz
LD_{50}	Dose that causes an effect in 50 % of the cases
LET	Linear energy transfer
LNT	Linear-no-threshold model for dose-response
LOAEL	Lowest observable adverse effect level
LOCEL	Lowest observable cancer effect level
LSS	Life span study on survivors of atomic bombs in Hiroshima an Nagasaki
MAK	Maximale Arbeitsplatzkonzentration
MeIQx	2amino-3,8-dimethylimidazo[4,5-*f*]quinoxaline
MGMT	O_6-Methylguanine-DNA-methyltransferase
MTD	Maximum Tolerated Dose
MTOC	Microtubule-organising centre
NIH	US National Institute of Health
NNK	4-[*N*-Methyl-*N*-nitrosoamino]-1-[3-pyridyl]-1butanone
NNM	*N*-Nitrosomorpholine
NOAEL	No observed adverse effect level
NOEC	No observed effect concentration
n_s	Number of susceptible cells
O^4-Et-dT	O^4-Ethylthymidine
OECD	Organisation for Economic Co-operation and Development
OEL	Occupational Exposure Limits
p	In some formulas: Probability
PAH	Polycyclic aromatic hydrocarbons
PAPS	3´-phosphoadenosine-5´-phosphosulfate
PBPK	Physiologically-based pharmacokinetic
PCB	Polychlorinated biphenyl
PEC	Predicted environmental concentration
PhIP	2-Amino-1-methyl-6-phenylimidazo[4,5-*b*]pyridine

PNEC	Predicted no-effect concentration
PNNL	Pacific Northwest National Laboratory
PP	Precautionary principle
PPI	Paternal Pre-conceptual irradiation
PRA	Probabilistic risk assessment
PVC	Polyvinyl chloride
Q	Radiation quality
R	Radiation type
R_0	Baseline lifetime hazard
Rad	Older unit of absorbed dose (100 rad = 1 Gy)
RBE	Relative biological effectiveness of radiation
RIVM	Rijksintituut voo Volksgezondheit en Milieu (The Netherlands)
RNA	Ribonucleic acid
S(a)	Survival function
SCE	Sister chromatid exchange
SEER	US Surveillance Epidemiology and End Results
SIR	Standardised incidence ratio
SMR	Standardised mortality/morbidity ratio
SNP	Single nucleotide polymorphism
SSB	Single strand break in DNA
Sv	Sievert ($J\ kg^{-1}$), unit of equivalent dose
T	Tissue or organ average
<u>t</u>	In some formulas: Time
TCDD	2,3,7,8-Tetrachlorodibenzo-*p*-dioxin
TD_{50}	Target dose that causes an effect in a target organ in 50 % of the cases
TPA	Phorbol ester
TRK	Technische Richtlinien Konzentration
TSCE	Two-step clonal expansion

TSH	Thyroid stimulating hormone
TSN	Tobacco-specific nitrosamine
UDS	Unscheduled DNA synthesis
UNSCEAR	United Nations Scientific Committee on the Effects of Atomic Radiation
UV	Ultra-violet
v	Value
VC	Vinyl chloride
WHO	World Health Organisation
WL	Working level
WLM	Working Level Month
w_R	radiation weighting factor
WTO	World Trade Organisation

16   List of Authors

Bolt, Hermann M., Professor Dr. med., Dr. rer. nat., studied medicine at the University of Cologne (1962–1967) where he gained his MD in 1968. He studied in biochemistry at the University of Tuebingen (1968–1971) and gained his PhD in 1973. From 1971 to 1979 he was scientific assistant and lecturer there and achieved his habilitation in Biochemical Pharmacology in 1974. From 1979 to 1982 he was Head of Section Toxicology at the Institute of Pharmacology, University of Mainz. Then he became a Director at the Institute of Occupartional Health, University of Dortmund, and Chairman of the Department of Toxicology and Occupational Medicine. With the reorganisation of the institute in 1997 he became its Director. He was offered also a number of merits and awards, particularly the German Federal Order of Merit (2000) and the Merit Award of EUROTOX (Federation of European Societies of Toxicology, 2001) to name a few. Professor Bolt is member of a number of commissions and committees, i. e. the German MAK-Commission or the Committee of Dangerous Substances of the Federal Ministry of Labour and Social Affairs, the Scientific Committee on Occupational Exposure Limits (SCOEL) and of the EU Specialised Experts Group in the Field of Carcinogenicity, Mutagenicity, and Teratogenicity (until 1998). He is also engaged in a number of scientific societies like FEST/EUROTOX since 1988, whose President he was from 1996 to 1998 or the German Society of Pharmacology and Toxicology and German Society of Occupational Medicine from 1985 to 1991. As member of numerous editorial boards (amongst others Journal of Biochemical Toxicology, International Archives of Occupational and Environmental Health) and Editor-in-Chief of the Archives in Toxicology he ensures the scientific quality of publications. Address: Universität Dortmund, Institut füre Arbeitsphysiologie, Ardeystr. 67, D-44139 Dortmund.

Føllesdal, Dagfinn, Professor, Ph.D., studied science and mathematics in Oslo and Göttingen and worked in ionospheric physics for two years before he went to Harvard in 1957 to study philosophy with Quine. After his Ph.D. at Harvard in 1961 he taught there until he returned to his native Norway Since 1967 he has been professor in Oslo and since 1968 also in Stanford where he became C.I. Lewis Professor of Philosophy in 1976. He is best known for his contributions to the so-called "New Theory of Reference" and to phenomenology. His reflections on these topics together with his works on theory of action, ethics, and philosophy of science have been published in more than 20 books and 150 articles. He has been an editor of *The Journal of Symbolic Logic,* 1970–82 and is serving on the editorial board of numerous journals, including the Europäische Akademie's new journal *Poiesis & Praxis.* He has received several international research prizes and a honorary doctorate and is a member of academies in many countries, including USA, England

and Germany. He was President of the Norwegian Academy of Science in 1993, 1995 and 1997. Address: Dept. of Philosophy, Stanford University, Stanford CA 94305, USA.

Hall, Per, Ph.D., Associate Professor, he took his Ph.D. in Cancer Epidemiology at the Karolinska Institutet, Stockholm, in 1991. He worked as a medical oncologist between 1984 and 1998 at the Department of Oncology, Karolinska Hospital. Between 1996 and 1998 Dr. Hall acted as Head of the Cancer Epidemiological Unit, Dept. of Oncology/Radiumhemmet, Karolinska Hospital and has since 1998 been a Senior Researcher and Deputy Chairman at the Dept. of Medical Epidemiology and Biostatistics, Karolinska Institutet. He has co-ordinated a number of projects funded by the European Commission, National Institutes of Health, US Army, and the Swedish Cancer Society. Dr. Hall also has several ongoing projects with the same funding agencies (EC: 2000–2003 "Late health effects among individuals exposed to ionising radiation in the southern Urals", 2002–2004 "Expanding cause of death registers for chronically radiation exposed populations in the Russian Federation: Coverage, quality and analysis", NIH: 2003–2007 "Molecular epidemiology of secondary lung cancer"; NIH: 2003–2005 "Molecular epidemiology of radiation induced gastro-intestinal cancer"; US Army: 2003–2004 "Breast cancer – a disease of a susceptible subgroup of women?"; Swedish Cancer Society: 2003–2005 "Risk of cardio-vascular disorders in women with breast cancer"). Dr. Hall is a current member of the following committees: the Medical advisory board at the National Board of Health and Welfare and the Swedish Radiation Protection Institute (since 1992), the Cancer Society, Stockholm (since 1996). Dr. Hall is also, since 1997, on the Editorial Advisory Board for Radiation and Environmental Biophysics. Address: Dept. of Medical Epidemiology and Biostatistics, Karolinska Institutet, Box 281, SE-17177 Stockholm.

Hengstler, Jan G., Dr. med., born in 1965, studied medicine at the University of Mainz (1984–1990) where he gained his MD in 1991. From 1990 to 2002 he was scientific assistant at the Institute of Toxicology, University of Mainz (head: Franz Oesch). Since 2003 he is Professor for Molecular and Forensic Toxicology and Head of the Center for Toxicology, University of Leipzig. His research interests include molecular mechanisms of carcinogenesis, conditional expression of oncogenes and their influence on tumour development as well as drug metabolism by primary hepatocytes. Jan Hengstler is coordinator of the BMBF research network "Systems Biology of the Hepatocyte" and is editor of Experimental and Clinical Sciences (EXCLI J). Address: Center for Toxicology, Institute of Legal Medicine and Rudolf Boehm Institute of Pharmacology and Toxicology, Härtelstraße 16–18, D-04107 Leipzig.

Jacob, Peter, Dr. rer. nat., gained a Ph.D. in physics at the Technical University of Munich. He co-ordinated a number of projects funded by the European Commission (currently: EPR dose reconstruction, 2000–2003) and has been project leader of a number of projects of the Federal Office of Radiation Protection (currently 3): 1. "Implications of the radiobiological effects adaptive response, bystander effect and low-dose hypersensitivity for the cancer risk at low doses", 2002–2004, 2.

"Thyroid doses and cancer risk of Belorussian and Ukrainian children and adolescents exposed due to the Chernobyl accident", 2000–2004, 3. "Range of applicability of epidemiological studies with aggregated data", 2001–2004). In addition, he is Member of the German Radiation Protection Commission and vice-chairman of its committee on radioecology. Dr. Jacob was Chairman of two report committees of the International Commission on Radiation Units and Measurements: 1. "Gamma-ray spectrometry in the environment", ICRU Report 53, 1992–1995), and 2. "Retrospective Assessment of Exposures to Ionising Radiation", ICRU Report 68, 1999–2002). He is also Chairman of the expert's group "Long-Term Countermeasure Strategies in Rural Areas Affected by the Chernobyl Accident" in the frame of the IAEA Technical Cooperation Programme. Currently, he is Deputy Director of the Institute of Radiation Protection and Head of the Section of Risk Analysis, at the GSF National Research Center for Environment and Health, D-85764 Neuherberg.

Oughton, Deborah H., Professor/Dr., of the Department of Plant and Environmental Sciences, Agricultural University of Norway and the University of Oslo's Ethics Programme gained a PhD degree in radiochemistry at the University of Manchester in 1989, and is presently Professor in Environmental Chemistry at the Agricultural University of Norway. She has participated in and co-ordinated a number of projects funded by the European Commission covering radiation protection, ecotoxicology and the environmental chemistry of both radionuclides and other pollutants. She was a member of the IAEA group on Ethical Considerations on Protection of the Environment from the Effects of Ionisng Radiation (2000–2002), and co-chaired the Oslo Consensus Conference on Protection of the Environment from Ionising Radiation (2001). She is a Board member of the International Union of Radioecology and associate editor of the Journal Ethics in Science and Environmental Policy (ESEP). Address: Dept. of Chemistry and Biotechnology, Agricultural University of Norway, N-1432 Aas.

Prieß, Kathrin, Dr. rer. nat., studied Biology and Applied Oceanography in Berlin, Barcelona and Perpignan, where she graduated in 1992 (M ès Sc.). She gained a PhD in Marine Environmental Sciences in 1997, under the auspices of both the "Université de la Méditerranée" in Marseilles and the University Christian-Albrecht of Kiel. For her work on growth variations of massive reef corals she was awarded the Scientific Award of the Doctoral School "Environmental Sciences – Earth System". She did her post-doctoral research on coral reef ecology at the Centre d'Océanologie de Marseille (1997–1998) and at the Interuniversity Institute for Marine Sciences of Eilat/Tel Aviv University (1999). In 1998 she was a Ramón y Cajal Scholar and EPTA co-ordinator in the Office for Scientific and Technology Options Assessment (STOA, DG4) at the European Parliament in Luxembourg. She is member of the scientific staff of the Europäische Akademie Bad-Neuenahr-Ahrweiler GmbH since November 1999. She has co-ordinated two project-groups "Biodiversity – scientific foundations and social Relevance" and "Environmental standards low-dose effects and their risk evaluation" and where she has organised conferences on the above topics as well as on "The uniqueness of humankind". In addition she has been also in charge of European Research Policy issues. Address: Europäische Akademie GmbH, Wilhelmstr. 56, D-53474 Bad Neuenahr-Ahrweiler.

Rehbinder, Eckard, Dr. jur., is Professor of Business Regulation, Environmental, and Comparative Law at the Faculty of Law of the Johann Wolfgang Goethe University Frankfurt/Main. He is the Director of the Research Centre for Environmental Law at the Faculty of Law. Professor Rehbinder gained a Ph.D. degree in Law at the Johann Wolfgang Goethe University Frankfurt/Main in 1965. Between 1987 and 2000 he was member, between 1996 and 2000 the chairman of the German Council of Environmental Advisors. Between 1996 and 1999 he participated in the Academy Project "Environmental Standards" and between 2000 and 2003 he coordinated the EU mobility project "Enforcing Environmental Policy". He was awarded the Elizabeth Haub Prize for Environmental Law in 1978. Address: Johann Wolfgang Goethe University, Senckenberganlage 31, D-60054 Frankfurt/Main.

Streffer, Christian, Professor Dr. rer. nat. Dr. med. h.c., studied Chemistry and Biochemistry at the Universities of Bonn, Tübingen, Munich, Hamburg and Freiburg; Ph.D. in Biochemistry January 1963. Postdoctoral fellowship at the Department of Biochemistry, University of Oxford, 1971 Professor (C3) for Radiobiology University of Freiburg i.Br. From 1974 until 1999 he was full Professor for Medical Radiobiology at the University of Essen, 1988–1992 Rektor of the University, 1999 Emeritus. Guest Professorships: 1985 University of Rochester, N.Y. USA, 2000 University of Kyoto, Japan. Honorary Member of several Scientific Societies, 1995 Honorary Doctor of University of Kyoto. Professor Streffer is on the board of the Institute for Science and Ethics of the Univrsities of Bonn and Essen, Member of the International Commission on Radiological Protection (ICRP). His main research interests are the radiation risk especially during the prenatal development of mammals; combined effects of radiation and chemical substances; experimental radiotherapy of tumours, especially with predictive tests for human tumours. Address: Auf dem Sutan 12, 45239 Essen, Germany.

Swaton, Elisabeth, Dr. phil. received her Ph.D. in Psychology and Anthropology from the University in Vienna. She worked for 2 years at IIASA (International Institute for Applied Systems Analysis) in the Energy Project, particularly in the area of 'Risk Perception'. Since 1977 she worked with the IAEA (International Atomic Energy Agency) in the Safety Assessment Section of the Division of Nuclear Safety where, for many years, she was responsible for the programme on Risk Perception and Evaluation. This activity carried out and monitored surveys on attitudes of the public towards various energy sources in several countries (e. g. Austria, Germany, Philippines, USA, Brazil, Japan) with emphasis on risks and benefits. She was involved in activities dealing with man-machine interface and human factors. She is also Member of the Society for Risk Analysis and presented papers at various conferences world wide on Risk Perception, Human Factors, Public Attitudes towards Energy Systems, etc. Dr. Swaton is retired since May 2003. Address: Reisner Str. 33, A-1030 Wien, Austria.

In der Reihe *Wissenschaftsethik und Technikfolgenbeurteilung* sind bisher erschienen: